Frontispiece. *Archaeopteryx* emblem. What more appropriate symbol of this International *Archaeopteryx* Conference than this amusing drawing by Professor Dr. H. K. ERBEN of the Paleontological Institute of the University of Bonn? It appears to reflect the divided opinions and controversial issues that were debated during the Conference. This drawing appeared in "Miscellany Geological" by CRAIG and JONES, published in 1983. It is reproduced here with the permission of the artist and the publisher, Orbital Press of Oxford, England.

THE BEGINNINGS OF BIRDS

Proceedings of the International
Archaeopteryx Conference
Eichstätt 1984

M. K. Hecht, J. H. Ostrom,
G. Viohl, and P. Wellnhofer, Editors

Eichstätt 1985

Published by
Freunde des Jura-Museums Eichstätt, Willibaldsburg,
D - 8078 Eichstätt

Printed in Germany by Brönner & Daentler KG, D - 8078 Eichstätt
Jacket: RENATE KLEIN-RÖDDER

© Freunde des Jura-Museums Eichstätt 1985
ISBN 3-9801178-0-4

Preface

It probably is the first time that a scientific conference was devoted to a single fossil species: the focal point, *Archaeopteryx lithographica*, the geologically oldest-known bird. The "International *Archaeopteryx* Conference" was initiated by one of us (J. H. O.) a few years ago in response to increased interest in this famed fossil and its possible significance. The resulting Conference was organized by others of us (G. V. & P. W.) and the staff of the Jura Museum of Eichstätt, Federal Republic of Germany. It was held September 11 to 15, 1984 in Eichstätt. The location was ideal, situated in the center of the quarry district of the Solnhofen Limestones (Upper Jurassic) where the discovery sites of all five specimens of *Archaeopteryx* are located. This made it possible for two field excursions during the Conference to visit these discovery sites, a unique and valuable experience for all who attended. Also, Eichstätt's Jura Museum is the home of one of the most complete (and most recently recognized) specimens of *Archaeopteryx* – a specimen not previously seen by most of the Conferees.

The Conference was timely because of recent increased interest in these specimens. For the first seventy-five years after the original discovery, the London (and then the Berlin) specimens generated much interest and controversy. That was followed by nearly half a century of relatively little interest in *Archaeopteryx* – famed though it was. Perhaps the scientific world was satisfied that all questions had been answered? But the Phoenix rose again in a flurry of papers during the last fifteen years. These presented a variety of new and some of the old ideas on the "Urvogel", and on the questions of bird origins and how bird flight began. What better circumstances to invite these authors and other interested scientists together to express their views and debate each other? The original plan called for the attendance of five key participants – the five presently known specimens of *Archaeopteryx* (for critical collektive examination) – but only the Teyler (Haarlem, Netherlands) and the Eichstätt specimens could attend. Also present were the solitary feather imprint attributed to *Archaeopteryx*, and the unique specimen of the possibly related dinosaur – *Compsognathus* – both from the Bavarian Paleontological State Collections in Munich.

The Conference was attended by sixty-six paleontologists, zoologists, ornithologists, physiologists, embryologists, geologists and physicists from thirteen countries. This diverse audience provided an ideal expert "sounding board" for the more than thirty speakers to test their views. Unfortunately, the animated discussions that followed each paper could not be included in this volume, although that was our original intention. Their inclusion would have caused an unacceptable delay in the publication of these Proceedings.**

There was an attempt at the last session of the Conference to arrive at a consensus, at least on some of the key issues. While there was general agreement on some items, no one expected unanimity on all points. Most participants agreed that *Archaeopteryx*, and possibly later birds, probably evolved from an archosaurian reptile, but there was no consensus as to which archosaur. It was widely accepted that *Archaeopteryx* was a true bird, but some maintained that it was a coelurosaurian dinosaur. The origin of avian flight was not resolved, but the arboreal theory seems to have gained ground. But despite the apparent disagreements on some details expressed in the various studies published here, the Conferees did agree unanimously to the declaration that: Organic evolution is a fundamental process of biology and we recognize the importance of the *Archaeopteryx* contribution to that problem. The Conferees also signed a declaration addressed to the Bavarian Government requesting the extension of present State laws to cover important fossil finds, such as *Archaeopteryx*, in order to preserve them and guarantee that such specimens in private collections will be assured unrestricted access for scientific research.

Max K. Hecht, John H. Ostrom, Günter Viohl & Peter Wellnhofer

** Tape recordings of these discussions are held in the archives of the Jura Museum and the Peabody Museum of Yale University.

Acknowledgements

The International *Archaeopteryx* Conference could not have been held without the aid of several institutions and individuals. For the financial and organizational support we owe a special debt of gratitude to Landrat (County-Supervisor) KONRAD REGLER who also made the conference room available; the Mayor of the City of Eichstätt, Oberbürgermeister LUDWIG KÄRTNER; the Director General of the Bavarian State Natural History Collections, Professor Dr. WOLFGANG ENGELHARDT; the Director of the Episcopal Seminary, Dr. LUDWIG MÖDL; and the President of the association "Freunde des Jura-Museums Eichstätt", ALFRED FICHTL. The Director of The Teylers Museum in Haarlem (Netherlands), Dr. E. EBBINGE, was so kind as to make the Teyler specimen of *Archaeopteryx* available to the Conference, and Dr. JOHN DE VOS brought it to Eichstätt. Dr. ALAN CHARIG and Dr. CYRIL WALKER from the British Museum (Natural History) presented a detailed and magnificent videofilm of the London specimen. The firm GRUNDIG made a videounit available which was installed by Mr. WERNER KNÖR, Eichstätt. The latter also loaned us his tape-recorder for recording all talks of the Conference. Several volunteers and the staff of the Jura-Museum helped in organizing the Conference. The association "Freunde des Jura-Museums Eichstätt" made the publication of this volume possible. We would like to thank all mentioned persons and institutions.

The Editors

Fig. 1
Photograph of the Humboldt Museum specimen of the solitary feather imprint reported by HERMANN VON MEYER in 1861. Scale units = 1 mm.

John H. Ostrom

Introduction to *Archaeopteryx*

The conferees at the Eichstätt International *Archaeopteryx* Conference required no introduction to *Archaeopteryx*. All those in attendance have studied one or more of the five known skeletal specimens or have deliberated on some aspect of this most ancient bird and its biology, or evolutionary importance. Indeed, some participants have studied in minute detail all of the known specimens and published lengthy analyses on *Archaeopteryx* and its meaning.

Since not all readers of these Proceedings will have had such first-hand acquaintance with the specimens on which this Conference focused, it seems appropriate to provide an introduction to *Archaeopteryx*. It must be noted that this introduction was not presented at the Conference. It is published here as an abbreviated and modified version of several earlier papers by this author (OSTROM, 1974, 1975 a, 1975 b) to acquaint readers with the reasons behind this Conference and all the interest of the past century – and most importantly, the evolutionary significance of the specimens of *Archaeopteryx*.

The story of *Archaeopteryx* begins with the first discovery of a fossil print or image of a solitary feather in the quarry of the community of Solnhofen in the summer of 1860 (see Fig. 1.) This feather was first

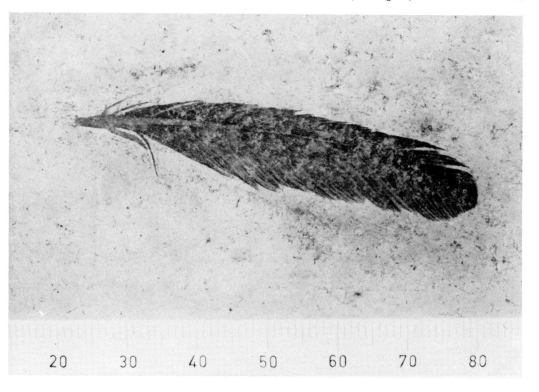

Fig. 2
Main slab of the London specimen of *Archaeopteryx lithographica* reported by HERMANN VON MEYER in 1861. Scale units = 5 cm. (Phot. P. WELLNHOFER)

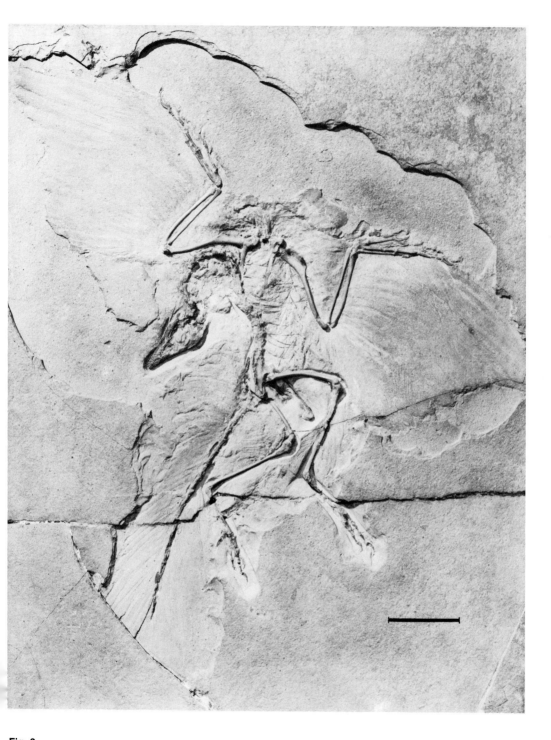

Fig. 3
Main slab of the Berlin specimen of *Archaeopteryx* (first described as *Archaeornis*) discovered in 1877, and first announced by E. HÄBERLEIN that same year. Scale units = 5 cm. (Phot. P. WELLNHOFER)

Fig. 4
Main slab of the "Maxberg" specimen of *Archaeopteryx* described by FLORIAN HELLER in 1959. Scale units = 1 mm.

announced in Neues Jahrbuch für Mineralogie, Geologie und Paläontologie by paleontologist HERMANN VON MEYER of Frankfurt (1861 a). The main slab (shown in Fig. 1.) is now in the Humboldt Museum für Naturkunde in East Berlin. The counterpart slab is in the Bayer. Staatssammlung für Paläontologie und historische Geologie in Munich.

One month later, VON MEYER (1861 b) published in the same journal the discovery of the first skeletal remains with clear impressions of feathers and referred to it as *Archaeopteryx lithographica*, in reference to the lithographic limestone in which it was preserved. It too was found in a quarry in the Langenaltheim region of Bavaria. Two years later, this specimen, together with many other Solnhofen fossils, was sold for 700 British Pounds by its owner Dr. KARL HÄBERLEIN of the town of Pappenheim to the British Museum (Natural History) in London, where it now resides. Quite naturally, this skeleton, with its distinct feather impressions (Fig. 2.), created much interest and was the subject of study by two of England's greatest 19th century scientists, Sir RICHARD OWEN and THOMAS HENRY HUXLEY. DEBEER described it in detail in 1954.

A second skeletal specimen (HÄBERLEIN, 1877) of *Archaeopteryx* was found in 1877 (Fig. 3) in a quarry on the Blumenberg just outside of the city of Eichstätt, approximately 30 km east of the Langenaltheim

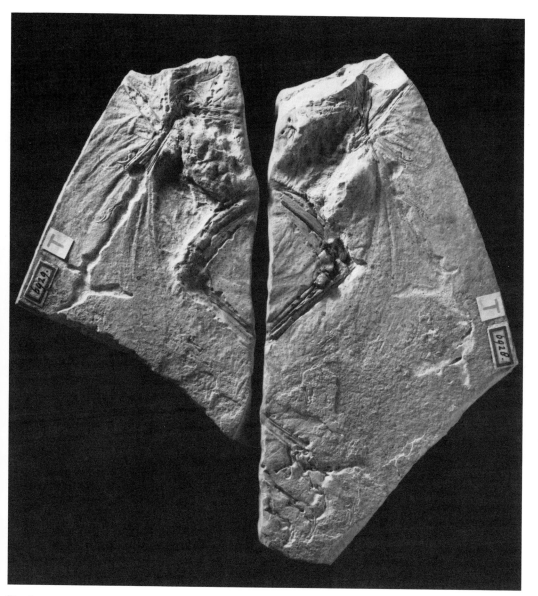

Fig. 5
The Teyler or Haarlem specimen of *Archaeopteryx* reported by the author in 1970.

localities. It was soon purchased by ERNST HÄBERLEIN, the son of Dr. HÄBERLEIN, for 140 Deutsch Marks. To prevent its loss to other nations, it was purchased by Werner Siemens a short time later for 20 000 Marks and presented to the Humboldt Museum of Berlin's Humboldt University. That is its official residence today. Without question this is the most famous and widely published specimen – fossil or modern – of any kind of creature, and has been illustrated repeatedly as a perfect intermediate or transitional animal kind between reptiles and birds, having anatomical features characteristic of both.

In 1955, a third fragmentary skeleton was found in another Langenaltheim quarry very close to the site of the London specimen. In this find the bones are in disarray and the feather impressions are not as clear as those in the London and Berlin specimens. Consequently, it was not immediately recognized as *Archaeopteryx*. Following its study by HELLER (1959), it was for many years displayed in the Maxberg Museum of the Solenhofer Aktienverein near the town of Solnhofen. At the present time it is in the private possession of its owner EDUARD OPITSCH of Pappenheim and is not available to the scientific community. (see Fig. 4.)

The fourth specimen of *Archaeopteryx* was discovered on display in the Teyler Museum of Haarlem, Netherlands in 1970 (OSTROM, 1970, 1972). It was displayed as a distinct pterosaur species, *Pterodactylus crassipes*. Because it is very fragmentary (parts of one hand, some arm and leg bones and the feet) plus the fact that the feather impressions are very faint, it had been mis-identified for more than a century. Originally found in 1855, it is believed to have come from a large quarry at Jachenhausen near the city of Kelheim, well to the east of the other sites. It was described by VON MEYER in 1859 and subsequently purchased by the Teyler Museum where it is currently on display. (see Fig. 5.)

The last known specimen discovered also was misidentified. This is the famed Eichstätt specimen which was found near Workerszell just north of Eichstätt. Initially thought to be a juvenile specimen of the theropod dinosaur *Compsognathus*, it was finally correctly identified in 1973 by F. X. MAYR the initiator of the Jura Museum in Eichstätt. That is where this specimen now resides in a place of honor. Although extremely well preserved, it was not correctly identified at first because the feather impressions are simply not recognizable. Moreover, there is no sign of a furcula, despite the high quality of preservation of the skeletal remains. Thus, there was no reason at the time of its discovery in 1950 to suspect that it was avian. A detailed study of this latest specimen of *Archaeopteryx* was published by WELLNHOFER in 1974. (see Fig. 6.)

These six specimens were the focal points of the Eichstätt Conference in September 1984, but to a large extent, that Conference dealt with questions and issues that had been debated, at least in part, over the years since the first recognized find in 1861. Students of *Archaeopteryx* during the past century have pondered such questions as: What was the evolutionary origin of birds? What was the ancestry of *Archaeopteryx*? How did bird flight evolve? Could *Archaeopteryx* fly? and how? and how well? As the geologically oldest known specimens of what are recognized as primitive but true birds, it is natural that these specimens of *Archaeopteryx* have been the focal points in addressing these and similar questions about the ancient history of birds. What follows is a brief summary of the history of thoughts about these famous specimens and their significance pertaining to some of the above questions.

With his announcement of the discovery of the London specimen, VON MEYER (1861 b) clearly recognized this as the remains of a bird. But that same year, ANDREAS WAGNER (1861) of Munich, who never saw that specimen, concluded that it was not a bird, but merely a feathered reptile and proposed the name *Griphosaurus* for these remains. VON MEYER (1862) seems to have wavered a bit with his comment that "the fossil feather of Solenhofen therefore, even if agreeing perfectly with those of our [modern] birds need not necessarily be derived from a bird". But OWEN (1862, 1863) and HUXLEY (1868a, 1868b, 1870) firmly believed in an avian identification.

It must be remembered that just two years before the discovery of the London specimen, a momentous event took place; the publication of DARWIN's "On the Origin of Species by Means of Natural Selection" in 1859. In fact, that event prompted WAGNER (1861) to write:

> "In conclusion, I must add a few words to ward off Darwinian misinterpretation of our new Saurian. At first glance of the *Griphosaurus* we might certainly form a notion that we had before us an intermediate creature, engaged in the transition from the Saurian to the bird. DARWIN and his adherents will probably employ the new discovery as an exceedingly welcome occurrence for the justification of their strange views upon the transformation of animals. But in this they will be wrong." (Translated by W. S. DALLAS, 1862.)

Fig. 6
The Eichstätt specimen of *Archaeopteryx* reported by F. X. MAYR in 1973 and described by P. WELLNHOFER in 1974.

HUXLEY, who was more concerned with defending DARWIN's theory on the origin of species than with the origins of birds, cited the similarities between the first specimen of *Archaeopteryx* and the small carnivorous dinosaur *Compsognathus* (which probably was collected from the same Solnhofen Limestone quarry at Jachenhausen that produced the Teyler specimen of *Archaeopteryx*). He recognized that as contemporaneous species, one could not be descended from the other, but he reasoned that as a reptile-like bird and a bird-like reptile, *Archaeopteryx* and *Compsognathus* closed the gap between the reptilian and avian classes.

During the ensuing decades various theories were proposed about the origin of *Archaeopteryx* and birds. VOGT (1879), WIEDERSHEIM (1884, 1885) and PETRONIEVICS (1921, 1927, 1950) derived *Archaeopteryx* from a lizard ancestry. OWEN (1875), SEELEY (1881) and WIEDERSHEIM (1883, 1886) proposed a pterosaur-like ancestor for *Archaeopteryx*. Following the implications of HUXLEY, an unspecified dinosaurian relationship was advocated by MARSH (1877), GEGENBAUR (1878), WILLISTON (1879), T. J. PARKER (1882) and BAUR (1883, 1884, 1885, 1886). In more recent years, the dinosaur link was advocated by BOAS (1930), LOWE (1935, 1944) and HOLMGREN (1955). Others, such as SEELEY (1881), DOLLO (1882, 1883), DAMES (1884), W. K. PARKER (1887) and FURBRINGER (1888) opposed a dinosaurian connection arguing that any anatomical similarities between *Archaeopteryx* (or birds in general) were adaptive only and thus had no evolutionary significance. MUDGE (1879) pointed to the large number of non-avian features present in dinosaurian kinds and made the (then) very astute observation that "The dinosaurs vary so much from each other that it is difficult to give a single trait that runs through the whole. But no single genus or set of genera, have many features in common with birds or a single persistent typical element or structure which is found in both." (MUDGE, 1879, p. 225)

In the century that has elapsed since MUDGE's assessment, many discoveries have been made and impressive new evidence within one group of dinosaurs, collectively referred to as coelurosaurs, casts considerable doubt on MUDGE's proclamation. As new dinosaur kinds came to light, it became apparent that the "Dinosauria" were much more diverse and less closely related than OWEN and HUXLEY had originally supposed. But FURBRINGER (1888), in his classic monograph on birds concluded that direct descent of birds from any known type of dinosaur was not possible; birds were monophyletic; resemblances between dinosaurs and birds are all "convergent analogies" and "parallels"; and the stem of birds lies in a common sauropsid ancestor lying between the Dinosauria, the Crocodilia and the Lacertilia.

FÜRBRINGER's concept was the beginning of a compromise explanation – the common ancestor hypothesis – subsequently advocated by OSBORN (1900), BROOM (1908, 1913) and culminated by HEILMANN's (1926) compelling book "The Origin of Birds". At the time of FÜRBRINGER's monograph, the ancestry of the dinosaurs was unknown. (It still is unknown, although a general consensus places a common – or dual – ancestry for the saurischian and ornithischian dinosaur orders within the Order Thecodontia.) But with that novel idea, FÜRBRINGER planted seeds that led OSBORN (1900) to postulate a hypothetical ancestor among primitive rhynchocephalian reptiles of Permian age as the source of all archosaurs including birds, and who argued that the avian line most probably separated off the "bipedal" dinosaurian stem. BROOM (1908) then wrote that "probably both dinosaurs and birds were derived" from Triassic bipedal "Rhynchocephaloid reptiles".

With this turn of the century recognition of progressive thecodontians (i. e. *Ornithosuchus*), BROOM (1913) was the first to claim:

> "There cannot, I think, be the slightest doubt that the Pseudosuchia have close affinities with the Dinosaurs, or at least with the Theropoda [bipedal carnivores]. . . . In fact, there seems to me little doubt that the ancestral Dinosaur was a Pseudosuchian. . . . There is still another group to which some Pseudosuchian has probably been ancestral, namely the Birds. For a time one or other of the Dinosaurs was regarded as near the avian ancestor . . . Seven years ago . . . I argued that the bird had come from a group immediately ancestral to the Theropodous Dinosaurs. The Pseudosuchia, now that it is better known, proves to be such a group as is required. In those points where we find the Dinosaur too specialized we see the Pseudosuchia still primitive enough. (BROOM, 1913, pp. 630–631.)

BROOM's statement set the stage for GERHARD HEILMANN's classic treatise on "The Origin of Birds", which firmly emplaced the hypothesis of a common ancestor of dinosaurs and birds. The impact of HEILMANN's book cannot be exaggerated. On the question of bird origins, its impact has been second only to the original discovery of *Archaeopteryx*. After reviewing the evidence of fossil and modern birds and reptiles, HEILMANN concluded that *Archaeopteryx* most closely resembled coelurosaurian (small theropod) dinosaurs, but because the coelurosaurs apparently lacked clavicles (the supposed future furcula of birds), they could not possibly be the ancestors of *Archaeopteryx* or modern birds. He suggested that bird ancestry lay in a group of reptiles closely akin to the coelurosaurs, which he stated were possibly the pseudosuchians.

With few exceptions, HEILMANN's pseudosuchian ancestry of birds was widely accepted until the early 1970's. GEORGE GAYLORD SIMPSON (1946) appeared to have stilled the controversy with his footnote:
"Almost all the special resemblances of some saurischians to birds, so long noted and so much stressed in the literature, are demonstrably parallelisms and convergences. These cursorial forms developed strikingly bird-like characters here and there in the skeleton and in one genus or another. They never showed a general approach to avian structure (as do *Archaeopteryx* and *Archaeornis* [= the Berlin specimen] . . . It is not a matter for argument but a simple fact of observation (if one accepts the published data of Heilmann and other authorities not questioned by Lowe) that *Archaeopteryx* is intermediate between the pseudosuchian *Euparkeria* and *Columba* (pigeon) in every one of these basic characters." (SIMPSON, 1946, pp. 94–95).

But in the early 1970's the issue was re-activated. The renewed debate pits three camps against each other: those who argue for a crocodilian – avian relationship, first proposed by WALKER (1972) and adopted by MARTIN ET AL (1980); those who favor a theropod origin of birds, advocated by this author (OSTROM 1973, 1975a, 1975b, 1976) and those who argue for a pseudosuchian origin of birds, advocated by TARSITANO and HECHT (1980) and subsequent papers (1982, 1983).

More than a century after the recovery of the first two specimens of *Archaeopteryx* the debate rages on about the origin of that "Urvögel" and its significance for the question of bird origins. Those remarkable specimens have also been invoked in the debate on the evolutionary origin of avian flight, an issue that was first addressed by WILLISTON (1879), who favored a cursorial origin of avian flight, and MARSH (1880), who hypothesized an arboreal origin of flight. From this vantage point a century later, it is difficult to decipher which of these models was favored then, but two notable papers by NOPCSA (1907, 1923) seem to have "grounded" the running theory of bird flight origins. NOPCSA suggested that proto-wings with somewhat enlarged scales or proto-feathers provided enlarged thrusting surfaces as the fore limbs beat ("oared") against the air – in other words, they served as propellers adding acceleration and velocity to the cursorial hind limbs. Theoretically, this led to sufficient velocity for "pro-avis" to become airborne. It is not at all surprising that this idea received very little support, and that the gravity-powered birth of avian flight from arboreal roosts achieved almost universal acceptance. BOCK (1965) presented a most logical model by which avian flight might have evolved by way of a series of steps – each of which was fully adapted: 1) an ancestral ground-dwelling quadrupedal reptile; 2) a bipedal ground-dweller; 3) a bipedal tree-climber; 4) a bipedal tree-dweller leaping between branches; 5) leaping between trees and parachuting between perches or to the ground; 6) active flapping flight.

This model has been widely accepted, but the cursorial theory of avian flight origins has made a come back (OSTROM, 1974, 1979, In press) and this issue had become a matter of intriguing debate prior to and during the 1984 *Archaeopteryx* Conference in Eichstätt. A key question at that Conference was whether the specimens of *Archaeopteryx* do in fact provide any clues about the earliest adaptations toward avian powered flight. The final answer is not yet in, but those specimens still remain as the only real evidence that is close in time to the beginnings of bird flight. They cannot be ignored. Recent studies by CAPLE, BALDA und WILLIS (1983) strengthen that view.

From this historical perspective, the reader can appreciate the intensity of interest by protagonists on all sides of these intriguing questions during the International *Archaeopteryx* Conference. From August 1861, when the cientific community first became aware of the existence of birds more than 140 million

Fig. 7
The well-known reconstruction of *Archaeopteryx* executed by RUDOLF FREUND for an article by KENNETH PARKES (1966), "Speculations on the origin of feathers". The artist deliberately incorporated several of the more controversial issues about *Archaeopteryx*, namely: could it perch in trees and use a grasping hallux? how many primary and secondary feathers did it possess? could it fly? and what was its diet?

years ago, until September, 1984, the remarkable specimens of *Archaeopteryx* have been a source of wonder, intensive study and stimulating scientific debate. No doubt these will continue.

References Cited

BAUR, G. (1883): Der Tarsus der Vögel und Dinosaurier. – Morphol. Jahrb. **8**: 417–456; Leipzig.
BAUR, G. (1884): Dinosaurier und Vögel. – Morphol. Jahrb. **10**: 446–454; Leipzig.
BAUR, G. (1885): Bemerkungen über das Becken der Vögel und Dinosaurier. – Morphol. Jahrb. **10**: 613–616; Leipzig.
BAUR, G. (1886): Zur Vögel – Dinosaurier-Frage. – Zool. Anz. **8**: 441–443; Leipzig.
BOAS, J. E. V. (1930): Über das Verhältnis der Dinosaurier zu den Vögeln. – Morphol. Jahrb. **64**: 223–247; Leipzig.

Bock, W. J. (1965): The role of adaptive mechanisms in the origin of higher levels of organization. – Syst. Zool. **14:** 272–287; Lawrence, Kansas.

Broom, R. (1908): On the early development of the appendicular skeleton of the ostrich with remarks on the origin of birds. – Trans. So. Africa Phil. Soc. **16:** 355–368. Cape Town.

Broom, R. (1913): On the South-African pseudosuchian *Euparkeria* and allied genera. – Proc. Zool. Soc. London **1913:** 619–633; London.

Caple, G., Balda, R. P. & Willis, W. R. (1983): The physics of leaping animals and the evolution of pre-flight. – Amer. Nat. **121:** 455–476; Chicago.

Dallas, W. S. (1862): On a new fossil reptile supposed to be furnished with feathers. – Ann. Mag. Nat. Hist. (3) 9: 261–267. London. (Translation of J. A. Wagner, 1861 listed here.)

Dames, W. (1884): Ueber *Archaeopteryx*. – Palaeontol. Abh. Bd. **2,** 3: 119–198; Berlin.

Darwin, C. (1859): On the Origin of Species by Means of Natural Selection. – John Murray, London, 1–479.

de Beer, G. R. (1954): *Archaeopteryx lithographica*. – Brit. Mus. (Nat. Hist.), London, 1–68.

Dollo, L. (1882): Première note sur les dinosauriens de Bernissart. – Bull. Mus. Roy. Hist. Natur. Belgique **I:** 161–180; Brussells.

Dollo, L. (1883): Note sur la présence chez les oiseaux du "troisième trochanter" des dinosauriens et sur la fonction de celui-ci. – Bull. Mus. Roy. Hist. Natur. Belgique **II:** 13–20; Brussells.

Fürbringer, M. (1888): Untersuchungen zur Morphologie un Systematik der Vögel. – Holkema, Amsterdam, 1–1751.

Gegenbaur, C. (1878): Grundriß der vergleichenden Anatomie. – W. Engelmann, Leipzig. 1–655.

Häberlein, E. (1877): Neue Funde von *Archaeopteryx*. – Leopoldina **13:** 80; Halle.

Hecht, M. K. & Tarsitano, S. (1982): The paleobiology and phylogenetic position of *Archaeopteryx*. – Geobios Spec. Mem. **6:** 141–149; Lyon.

Hecht, M. K. & Tarsitano, S. (1983): *Archaeopteryx* and its paleoecology. – Acta Palaeo. Polon. **28:** 133–136; Warsaw.

Heilmann, G. (1926): The Origin of Birds. – Appleton, New York, 1–210.

Heller, F. (1959): Ein dritter *Archaeopteryx* Fund aus den Solnhofener Plattenkalken von Langenaltheim/Mfr. – Erlanger Geol. Abh. **31:** 1–25; Erlangen.

Holmgren, N. (1955): Studies on the phylogeny of birds. – Acta Zool. **36:** 243–328; Stockholm.

Huxley, T. H. (1868 a): Remarks upon *Archaeopteryx lithographica*. – Proc. Roy. Soc. London **16:** 243–248; London.

Huxley, T. H. (1868 b): On the animals which are most nearly intermediate between the birds and reptiles. – Ann. Mag. Natur. Hist. **4:** 2: 66–75; London.

Huxley, T. H. (1870): Further evidence of the affinity between the dinosaurian reptiles and birds. – Quart. Jour. Geol. Soc. London, **26:** 12–31; London.

Lowe, P. R. (1935): On the relationships of the Struthiones to the dinosaurs and to the rest of the avian class, with special reference to the position of *Archaeopteryx*. – Ibis **13:** 398–432; London.

Lowe, P. R. (1944): An analysis of the characters of *Archaeopteryx* and *Archaeornis*. Were they reptiles or birds? – Ibis **86:** 517–543; London.

Marsh, O. C. (1877): Introduction and succession of vertebrate life in America. – Proc. Amer. Assoc. Advan. Sci. 1877: 211–258; Washington, D. C.

Marsh, O. C. (1880): Odontornithes: a monograph on the extinct toothed birds of North America. – Prof. Paper Engineer. Dept. U. S. Army **18:** 1–201; Washington, D. C.

Martin, L. D., Stewart, J. D. & Whetstone, K. N. (1980): The origin of birds: structure of the tarsus and teeth. – Auk **97:** 86–93; Lawrence, Kansas.

Mayr, F. X. (1973): Ein neuer *Archaeopteryx* Fund. – Paläontol. Zeit. **47:** 17–24; Stuttgart.

Meyer, H. von (1860): Zur Fauna der Vorwelt. Reptilien aus dem lithographischen Schiefer des Jura in Deutschland und Frankreich. – 64–66; H. Keller, Frankfurt am Main.

Meyer, H. von (1861 a): Vogel-Federn und *Palpipes priscus* von Solnhofen. – Neues Jahrb. Mineral. Geol. Palaeontol. 1861: 561. Stuttgart.

Meyer, H. von (1861 b): *Archaeopteryx lithographica* (Vogel-Feder) und *Pterodactylus* von Solnhofen. – Neues Jahrb. Mineral. Geol. Palaeontol. 1861: 678–679; Stuttgart.

Meyer, H. von (1862): *Archaeopteryx lithographica* aus dem lithographischen Schiefer von Solnhofen. – Palaeontographica **10:** 53–56; Stuttgart.

Mudge, B. F. (1879): Are birds derived from dinosaurs? – Kansas City Rev. Sci. **3:** 224–226; Kansas City.

Nopcsa, F. (1907): Ideas on the origin of flight. – Proc. Zool. Soc. London. 1907: 223–236; London.

Nopcsa, F. (1923): On the origin of flight in birds. – Proc. Zool. Soc. London. 1923: 463–477; London.

Osborn, H. F. (1900): Reconsideration of the evidence for a common dinosaur-avian stem in the Permian. – Amer. Natur. **34:** 777–799; Chicago.

OSTROM, J. H. (1970): *Archaeopteryx*: Notice of a "new" specimen. – Science **170**: 537–538; Washington, D. C.
OSTROM, J. H. (1972): Description of the *Archaeopteryx* specimen in the Teyler Museum, Haarlem. – Proc. Kon. Neder. Akad. Wetensch. **B75**: 289–305; Amsterdam.
OSTROM, J. H. (1973): The ancestry of birds. – Nature **242**: 136; London.
OSTROM, J. H. (1974): *Archaeopteryx* and the origin of flight. – Quart. Rev. Biol. **49**: 27–47; Stony Brook, N. Y.
OSTROM, J. H. (1975a): On the origin of *Archaeopteryx* and the ancestry of birds. – Proc. Centre Nat. Rech Sci. **218**: 519–532; Paris.
OSTROM, J. H. (1975b): The origin of birds. – In: F. A. DONATH (Ed.) Ann. Rev. Earth Planet. Sci. **3**: 55–77; Palo Alto, Calif.
OSTROM, J. H. (1976): *Archaeopteryx* and the origin of birds. – Biol. Jour. Linnean Soc. **8**: 91–182; London.
OSTROM, J. H. (1979): Bird flight: How did it begin? – Amer. Sci. **67**: 46–56; New Haven, Conn.
OSTROM, J. H. (1985, In press): The cursorial origin of avian flight. – Proc. Calif. Acad. Sci. (In press); San Francisco.
OWEN, R. (1862): On the fossil remains of a long-tailed Bird (*Archaeopteryx macrurus* Ow.) from the lithographic slate of Solnhofen. – Proc. Roy. Soc. London **12**: 272–273; London.
OWEN, R. (1863): On the *Archaeopteryx* of VON MEYER, with a description of the fossil remains of a long-tailed species from the lithographic stone of Solnhofen. – Philos. Trans. **153**: 33–47; London.
OWEN, R. (1875): Monograph of the fossil reptiles of the Liassic formations II. Pterosauria. – Palaeontol. Soc. Monogr. 41–81; London.
PARKER, T. J. (1882): On the skeleton of *Notornis mantelli*. – Trans. Proc. N. Z. Inst. **14**: 245–258; Wellington.
PARKER, W. K. (1887): On the morphology of birds. – Proc. Roy. Soc. London **42**: 52–58; London.
PARKES, K. C. (1966): Speculations on the origin of feathers. – The Living Bird **5**: 77–86; Pittsburgh, Pa.
PETRONIEVICS, W. K. (1921): Über das Becken, den Schultergürtel und einige andere Teile der Londoner *Archaeopteryx*.-Genf. Buchhandl. George, Leipzig. 1–31.
PETRONIEVICS, W. K. (1927): Nouvelles recherches sur l'osteologie des Archaeornithes. – Ann. Paleontol. **16**: 39–55; Paris.
PETRONIEVICS, W. K. (1950):Les deux oiseaux fossiles les plus anciens (*Archaeopteryx* et *Archaeornis*). – Ann. Geol. Pen Balkan **18**: 89–127; Palermo.
SEELEY, H. G. (1881): Prof. CARL VOGT on the *Archaeopteryx*. – Geol. Mag. (2) **8**: 300–309; London.
SIMPSON, G. G. (1946): Fossil Penguins. – Bull Amer. Mus. Nat. Hist. **87**: 1–95; New York.
TARSITANO, S. & HECHT, M. K. (1980): A reconsideration of the reptilian relationships of *Archaeopteryx*. – Zool. Jour. Linnean Soc. **69**: 149–182; London.
VOGT, C. (1879): *Archaeopteryx*, ein Zwischenglied zwischen den Vögeln und Reptilien. Naturforscher **42**: 401–402; Berlin.
WAGNER, J. A. (1861): Über ein neues, angeblich mit Vogelfedern versehenes Reptil aus dem Solnhofener lithographischen Schiefer. – Sitzungsber. Bayer. Akad. Wiss. **2**: 146–154; Munich.
WALKER, A. D. (1972): New light on the origin of birds and crocodiles. – Nature **237**: 257–263; London.
WELLNHOFER, P. (1974): Das Fünfte Skelettexemplar von *Archaeopteryx* – Palaeontographica A **147**: 169–216; Stuttgart.
WIEDERSHEIM, R. E. E. (1883): Lehrbuch der vergleichenden Anatomie der Wirbelthiere aus Grundlage der Entwicklungsgeschichte. – Fischer, Jena, 1–905.
WIEDERSHEIM, R. E. E. (1884): Die Stammesentwicklung der Vögel. Biol. Zentralbl. **3**: 654–695; Leipzig.
WIEDERSHEIM, R. E. E. (1885): Über die Vorfahren der heutigen Vögel. – Humboldt **4**: 212–224; Stuttgart.
WIEDERSHEIM, R. E. E. (1886): Lehrbuch der vergleichenden Anatomie der Wirbelthiere auf Grundlage der Entwicklungsgeschichte. – Fischer, Jena, 1–890.
WILLISTON, S. W. (1879): "Are birds derived from dinosaurs?" – Kansas City Rev. Sci. **3**: 457–460; Kansas City.

Author's address: Prof. Dr. JOHN H. OSTROM, Peabody Museum of Natural History, Yale University, 170 Whitney Avenue, P.O. Box 6666, New Haven, Conn. 06511, U.S.A.

Alan J. Charig

Analysis of the Several Problems Associated with *Archaeopteryx*

Abstract

The many *Archaeopteryx* problems have often been approached illogically or unscientifically, and they have sometimes been confused with each other. It might therefore be helpful to analyse them, separate them – interconnected though they be – and classify them in logical order:

1. Problems concerning the material itself. This covers its collection in the field, its development in the laboratory, and the taxonomic identity of the several specimens.

2. Interpretation of the structure. This is effected by reference to other animals, extant and fossil. It includes the identification of doubtful elements and the postulation of homology (really the same thing!) through supposed transformation series in stratigraphical sequence and through the ontogeny of living forms; it includes also the inference of function and (rarely) physiology.

Especially important is the locomotor system, obviously directed towards some sort of flight. Was *Archaeopteryx* a glider or a flapper? How well could it fly? Other topics of interest are the feathers and the structure of the wings, the inferred flight musculature, aerodynamics, and bioenergetics. Most important of all is a comparison of the hypothesized evolutionary pathways to flight, i. e. the several variations on the "cursorial" and "arboreal" scenarios.

3. Phylogeny and classification. This is necessarily preceded by two related discussions. One of those is on the demarcation of zoological taxa (which, in palaeontology, is preferably effected by what I shall call "evolutionary delimitation", as opposed to diagnostic delimitation and to typification); the other is on the demarcation of the particular taxon Aves (effected primarily on the presence of feathers and secondarily on the furcula). Consideration of the phylogeny itself, based mainly on character distribution analyses but also on any other available evidence, includes both internal relationships (i. e. of *Archaeopteryx* to other major groups of birds) and external relationships (i. e. of the birds as a whole and of *Archaeopteryx* in particular to certain groups of archosaurs – chiefly Thecodontia, Crocodylia, and Theropoda – and also to a few non – archosaurs, including mammals). A Simpsonian approach to classification is recommended.

4. Other topics. These might include stratigraphy/chronology (though those are not really in doubt) and, more fruitfully, subjects related to the environment: sedimentology, palaeoclimatology, taphonomy, the sympatric flora and fauna, palaeoecology, and the probable habits of *Archaeopteryx*.

Few, if any, of the many problems associated with *Archaeopteryx* are likely to be resolved at this Conference. But we may edge a little nearer towards an eventual solution of some of them.

Zusammenfassung

Die zahlreichen durch *Archaeopteryx* aufgeworfenen Probleme wurden öfters ohne Logik oder unwissentschaftlich behandelt und nicht selten miteinander verwechselt. Es wäre deshalb von Nutzen, wenn man sie, wie immer verkettet sie auch untereinander sein mögen, getrennt analysieren und logisch anordnen würde:

1. Materialbedingte Probleme. Es werden Fragen des Aufsammelns am Fundort, der Präparation im Laboratorium und der taxonomischen Identität der verschiedenen Belegstücke behandelt.

2. Die Deutung der Struktur. Sie wird erreicht durch Bezugnahme auf andere, rezente oder ausgestorbene, Tiere. Das schließt die Bestimmung zweifelhafter Elemente ein und das Postulieren von Homologien (wirklich derselbe Prozeß!) auf der Basis von Transformationsreihen in stratigraphischer Abfolge und der Ontogenie lebender Formen; auch mutmaßliche Funktionen (inklusive, aber selten, Physiologie) sind in Betracht zu ziehen.

Von besonderer Bedeutung ist das zweifellos auf eine Art des Fliegens hingerichtete Fortbewegungssystem. War *Archaeopteryx* ein Gleitflieger oder ein Flatterflieger? Wie gut konnte er fliegen? Andere interessante Themen sind Federn und Flügelbau, die mutmaßliche Flugmuskulatur, Flugdynamik und bioenergetische Faktoren. Von besonderer Wichtigkeit ist ein Vergleich der Entwicklungstheorien des Fluges – der vielen Variationen der "Läufertheorie" und "Baumspringertheorie".

3. Phylogenie und Klassifikation. Dem gehen notwendigerweise zwei miteinander verwandte Themen voraus, nämlich die Festlegung zoologischer Taxa (in der Paläontologie vorzugsweise erreicht durch das, was ich, im Gegensatz zur diagnostischen Sippenumgrenzung und zur Typifikation, als "evolutionäre Umgrenzung" bezeichnen werde) und speziell die Festlegung des Taxons Aves (erstens begründet durch das Vorhandensein von Federn und zweitens des Gabelbeins). Betrachtungen über die Phylogenie selbst, in erster Linie auf Analysen der Merkmalsverteilung, aber auch auf anderen Indizien, beruhend, umfassen die interne Verwandschaft (d. i. von *Archaeopteryx* mit anderen wichtigen Vogelgruppen) und die externe Verwandschaft (d. i. der Vögel als Ganzes und von *Archaeopteryx* im besonderen mit gewissen Archosauriergruppen – hauptsächlich Thecodontiern, Krokodilen und Theropoden – und auch mit einigen Nicht-Archosauriern, einschließlich Säugetieren). Die Simpsonsche Klassifikationsmethode wird empfohlen.

4. Andere Themen. Es könnten behandelt werden Fragen der Stratigraphie/Chronologie (hier scheint allerdings fast alles problemlos zu sein) und, nutzbringender, die Fragenkomplexe zur Umwelt des *Archaeopteryx*: Sedimentologie, Paläoklimatologie, Taphonomie, die vorherrschende Flora und Fauna der Periode, Paläoökologie und die möglichen Verhaltensweisen des Urvogels.

Im Hinblick auf die vielen mit *Archaeopteryx* verknüpften Probleme kann diese Konferenz nur ein Anfang sein. Keines der Probleme wird wohl gelöst werden; aber vielleicht kommen wir eventuellen Lösungen etwas näher.

Introduction

There are many problems associated with *Archaeopteryx*, the earliest fossil bird; most of them will be addressed by this Conference. In the past, however, those problems have often been approached in a manner that is not entirely logical or scientific, and it is also true that they have sometimes been confused with each other; this applies particularly within the fields of phylogeny and classification. Thus, for example, "The origin of *Archaeopteryx*" is not necessarily the same question as "The origin of birds", and both those questions are certainly very different from "The origin of avian flight". Because of this it may be helpful, before we get down to the actual substance of our meeting, to try to analyse the situation: to unravel the tangled skein, to separate the various problems and to classify them under distinct heads – even though, to some extent, they remain interconnected. It also seems important (at least, it seems important to me) to present the questions in a logical order.

One begins with the material itself and the anatomical facts to be ascertained therefrom. One then proceeds to the interpretation of those details of anatomy. Which structures might reasonably be regarded as homologous with certain structures in other animals, both fossil and extant (as evinced by transformation series in stratigraphical sequence and by the ontogeny of living forms)? What functions might the various structures serve (as suggested by comparisons with living animals), and what might they tell us about the animal's physiology (again on a basis of comparison with the animals of today)? Of especial interest in the fields of function and physiology is the animal's locomotor system, which obviously enabled flight of some sort or another (gliding or flapping flight); and to study the aerial capabilities of *Archaeopteryx* requires a sound knowledge of both aerodynamics and bioenergetics. This leads on to another major category of problem: those concerning phylogeny and classification. Incidentally, we can take nothing for granted; consider the views of GARDINER (1982) on "Haemothermia".

The distinction between the 'phylogenetic' and 'taxonomic' problems on the one hand and, on the other, the 'functional' and 'ecological' problems is perfectly clear to everyone. The difference between the 'phylogenetic' and the 'taxonomic' problems themselves is not quite so pronounced, while the 'functional' and 'ecological' problems show some tendency to grade into each other. What is much less clear is the divisibility of the 'phylogenetic' problems into two. For example, OSTROM's very important paper of 1976 was called *"Archaeopteryx and the origin of birds"*; but it is not about the origin of birds, it is about the origin of *Archaeopteryx*, which, conceivably, might have nothing whatever to do with the origin of birds. Thus it would appear that we are faced with two distinct phylogenetic problems: (a) what were the ancestors of *Archaeopteryx*, and (b) was *Archaeopteryx* ancestral, or near-ancestral, to the rest of the birds as we know them? Indeed, it is not even as simple as that. The whole subject requires systematic analysis.

Problems (1): Concerning the material itself

One question that has been asked repeatedly in recent years concerns reports of the alleged discovery of a sixth specimen of *Archaeopteryx* (HOWGATE 1982, BENTON 1983). These have now been discounted (HOWGATE 1983).

More pertinent would be purely technical problems relating to the collection of the specimens in the field and their development in the laboratory – the sort of questions that might best be addressed to PETER WHYBROW, who has recently carried out such brilliant preparation upon the cranium of the holotype (WHYBROW 1982, WHETSTONE 1983). This category would also include the problem of the taxonomic identity of the several specimens, i. e. whether the other specimens known from the Solnhofen Limestone of Bavaria are conspecific, or at least congeneric, with the holotype. In other words, are we all dealing with the same thing? Most people are aware that two of the other specimens (of which there are no fewer than five in all, including the original feather, i. e. six including the holotype) have been described under different names: the Teyler fossil originally as *Pterodactylus* (!) *crassipes* VON MEYER 1857, and the Berlin specimen in 1897 as *Archaeopteryx siemensi* DAMES. (SEELEY had earlier expressed the opinion, in 1881 and 1882, that the London and Berlin specimens might be specifically if not generically distinct.) *A. siemensi* was later segregated still farther, into the separate genus *Archaeornis*, by PETRONIEVICS (1917). (Incidentally, it should be noted here that *crassipes* is the oldest specific name associated with any of these specimens, and that OSTROM, bearing that in mind, successfully applied to the International Commission on Zoological Nomenclature in 1972 for the suppression of *crassipes* and the retention of the younger name *lithographica*.) It has even been suggested (PETRONIEVICS 1921) that *Archaeopteryx* and *Archaeornis* belonged to entirely different groups of birds, *Archaeopteryx* being related to the ratites and *Archaeornis* to the carinates. DE BEER, on the other hand, opted cautiously (1954) for specific identity. Despite the recognition or discovery of three more specimens since DE BEER's day, almost everyone has agreed until now (at least in print) that we are dealing with only one species. U n t i l n o w; for one worker has the present intention of proposing a new genus and species for the Eichstätt specimen.

Problems (2): Interpretation of the structure

These problems are purely anatomical. They include the identification of doubtful elements and the postulation of homology, which are essentially the same thing. Suppose that we identify a particular element in *Archaeopteryx* with an element in the skeleton of another animal; this we may do on the grounds of their structure and spatial relationships, which may be better understood in the case of extant animals through a study of their ontogeny. Then, if we are evolutionists, we are implying that the structures are homologous and that a phylogenetic relationship exists between *Archaeopteryx* and the other animal. The more restricted the distribution of that particular structure, the closer the relationship. A good example of this is the old problem concerning the identification of the three digits in the manus of theropods, *Archaeopteryx* and Recent birds respectively (the last as studied through embryology): are they nos. 1, 2 and 3, or nos. 2, 3 and 4? Our phylogenetic reconstruction will depend upon the answers to those three questions, on whether they are the same or different. The danger of circular argument should be noted here; we may identify structures and postulate homology on the grounds of phylogenetic relationships that are themselves based upon the homologies in question.

Anatomical questions, however, may relate not only to homology and thus to phylogeny but also to function. That means, almost invariably in the case of *Archaeopteryx*, the subject of locomotion; and that I prefer to deal with under a separate sub-heading. Meanwhile it should be noted that functional considerations may depend not only upon the presence of structures and their precise form but also upon their absence (see below).

Anatomy has even been used as a guide to physiology. Thus, for example, the supposition that *Archaeopteryx* was an endotherm has been based not only on the presence of feathers but also on its relative brain size, as evinced by the capacity of its endocranial cavity.

One other question of an anatomical nature concerns the dimensions and mass of the animal in life, the answer to which is not necessarily such a simple and straightforward matter.

Problems (2a): Locomotion, in particular flight

If *Archaeopteryx* is to be regarded as a bird, then it is the earliest bird known; in which case its flying capabilities and the evolutionary pathways by which it acquired them are matters of intense interest. Needless to say, they are also matters of intense controversy.

Perhaps the first question we should ask ourselves is whether *Archaeopteryx* could fly at all. Few people seem inclined to doubt that proposition if we interpret the word "fly" in its broadest sense; but there is certainly a great deal of disagreement as to whether it merely glided, or was capable of "powered" flight, and, if the latter, whether it was restricted in its aerial activities to nothing more than a few fluttering leaps, or could fly properly (albeit weakly), or fly really well. Clues to these problems have been found not only in the shape of the fore-limbs (one hesitates to call them wings for fear of being accused of begging the question) but also in the bony structure of the pectoral girdle and associated elements, which might reasonably be expected to give some indication of the extrinsic musculature of the fore-limb. Thus it has been claimed (OSTROM 1974, 1976a) that *Archaeopteryx* could scarcely fly at all, the reasons (*inter alia*) being (a) that it lacked a keel to the sternum for the attachment of powerful pectoralis muscles and (b) that it lacked an acrocoracoid process on the coracoid to function as a pulley for the tendon of the supracoracoideus muscle, which in Recent birds elevates the wing. OLSON & FEDUCCIA (1979) have refuted both those points. The belief that *Archaeopteryx* was a perfectly adequate powered flier was shared by FEDUCCIA & TORDOFF (1979), who based their conclusion upon the twin observations that in modern birds there seems to be a correlation between the degree of feather asymmetry and flying ability (i. e. flightless birds have symmetrical feathers) and that the feathers of *Archaeopteryx* are asymmetrical. Other workers, meanwhile, have concentrated their attentions upon questions of aerodynamics and flight energetics.

What is the evidence for and against the various hypothesised evolutionary pathways to flight? Here the controversies are even greater and depend upon a number of factors that are themselves uncertain, apart from the actual flying abilities of *Archaeopteryx* itself. Was *Archaeopteryx* derived from quadrupedal or bipedal ancestors? Were they arboreal or cursorial in habit? Was *Archaeopteryx* itself arboreal or cursorial? Here again anatomy and function have been interpreted differently by different authors. Thus OSTROM, in several publications, has stated his belief that the claws on the manus and the pes of our animal were not suited to arboreal habits. F. X. MAYR, on the other hand (1973), opined that the form of the claws on both extremities was of a kind never possessed by running birds and that, on the contrary, the claws were characteristic of tree-dwelling animals (tree-creepers, nuthatches, woodpeckers, squirrels and pine-martens).

The old theories of the evolution of avian flight, the classical "cursorial" theory originating with WILLISTON (1879) and the classical "arboreal" theory originating with MARSH (1880), have meanwhile been elaborated upon and modified to produce several distinct variations. OSTROM (1974) suggested that the large feathers on the fore-limbs of *Archaeopteryx* transformed those limbs into natural insect nets and enhanced the prey-catching function of the hands; fluttering leaps on to larger prey eventually evolved into proper flight. HARRISON (1976) envisaged *Archaeopteryx* as a creature that ran bipedally along branches to escape pursuing predators, launched itself clumsily into the air and fluttered erratically down into the undergrowth – the very unpredictability of this type of behaviour proving advantageous. (He compared it to a modern coucal.) CAPLE, BALDA & WILLIS (1983) suggested that the original function of the fore-limbs was to stabilise the animal during prey-catching lunges with the head and jaws (not with the hands) and that downward-and-forward strokes of the fore-limb made with an aerodynamically suitable aerofoil would produce vortices and consequent lift. Their ideas were explained in more popular form by PADIAN (1982), referring to their then unpublished work.

Problems (3): Phylogeny and classification

Let us assume that all other presently known birds, fossil and Recent, form a clade. We shall assume also that that clade arose from some common ancestor within the archosaurs, the first "true bird" perhaps, and that *Archaeopteryx* arose likewise, though not necessarily from the same ancestor. (Obviously we could define a clade that included not only the other birds but *Archaeopteryx* too; such a clade, however, might have to include other archosaurs that we should not regard as birds.) We cannot continue our analysis of the *Archaeopteryx* problems until we have agreed on a definition of what constitutes a bird.

But first we must divert ourselves from our analysis to consider the demarcation of zoological taxa. There are three approaches to this subject:

1. The first is by typification, the designation of a type. The Rules of Zoological Nomenclature require that this be done for all taxa up to the level of the family; for the purpose of demarcation, however, the practical value of types is largely restricted to the level of the species, and even there it is concentrated on fossil rather than extant forms. At the generic and familial levels the value of types is almost exclusively for purposes of nomenclature, and above the level of the family types do not exist.

2. The second approach is by diagnostic delimitation, the listing of characters typical of the taxon in question; there are generally several such characters, sometimes dozens. This is the approach employed by almost all neontologists and most palaeontologists, no matter which systematic creed they favour; they differ only in the rigidity of their requirements as to whether all the members of a taxon should possess all its diagnostic characters. In general, however, it may be said that each individual character is not necessarily exclusive to the taxon, but its particular combination of characters must be exclusive to it. In a supraspecific taxon not every constituent member need possess every character listed as diagnostic, but all characters so listed should be present in most of the members and all members should possess most of the characters so listed. SNEATH (1962) coined the term "polythetic" for taxa defined in this manner; he wrote that "a group may be perfectly natural, and yet have not a single character common to all the species that compose it."

The diagnostic method appears also in a different guise when people define a category simply by listing its constituent members, e.g. "I know what birds are, they include sparrows, eagles, parrots, ostriches . . ." The speaker is listing those elements that he subconsciously recognizes as sharing a number of unique characters (or at least a unique combination of characters) with each other; he subconsciously uses those characters (or the combination of characters) to diagnose the category.

3. The third approach is by evolutionary delimitation, which is based on the hypothesized phylogenetic tree (we cannot recognise real phylogenetic trees). A taxon is delimited from the taxon of equal rank from which it is presumed to have evolved by the appearance of one single new character, an "evolutionary novelty"; and other taxa of equal rank which are presumed to have evolved subsequently from the taxon in question are each likewise delimited by the appearance of their respective "evolutionary novelties". Thus the class Reptilia is delimited from the preceding class Amphibia by the appearance of an amnion, just as the following class Aves is delimited from the Reptilia by the appearance of feathers and, in another evolutionary direction, the following class Mammalia is delimited in similar fashion by the appearance of a functional jaw articulation between the dentary and the squamosal. (Not all authorities on the earliest mammals agree as to which "evolutionary novelty" makes the best marker for the reptile-mammal boundary.)

When we are dealing with fossils of known stratigraphical position (and hence, at least approximately, chronological age) the method of delimitation is logically preferable. After all, it would be absurd to believe that all the several character-states listed as diagnostic of a newly evolved taxon arose at precisely the same time, we must suppose that they arose at several different stratigraphical/ chronological levels; and that would mean that our attempts to draw a taxonomic boundary between adjacent taxa (provided we knew what those levels were) would be confused by the simultaneous employment of several different criteria. This creates practical difficulties too, because fossils will be found — if they have not been found already — that show only s o m e of the characters in the state diagnostic of the new taxon, others being in the contrasting state typical of its predecessor. Should they be assigned to the emerging taxon or be left still outside it? On the other hand, the use of delimitation, which is tantamount to the employment of a single diagnostic character, is also fraught with difficulties; new fossils will be found that are not sufficiently complete to show the delimiting character, or, if they do show it well enough, its state may be ambiguously intermediate. Feathers are a case in point in so far as the actual feathers are never preserved fossil and even their impressions are extremely rare.

The definition and composition of the Aves

Let us define a class Aves as the clade that is demarcated from its antecedents by the appearance of the evolutionary novelty "feathers". It is unlikely, by the nature of things, that *Archaeopteryx* was the earliest bird of all, which means that the clade would have originated before *Archaeopteryx*. The earliest members of this clade, indeed, including the very earliest of all, may be forms that we know already, but in which we do not yet know of the presence of feathers (a possible example is *Syntarsus*); on the other hand, they may be forms that we do not know at all. Either way the clade will contain a subordinate clade originating with *Archaeopteryx*; that clade may include later birds, known or unknown to us, or, since it is perfectly possible that no later bird is descended directly from *Archaeopteryx*, it may not. If the second possibility is the correct one then it is also possible (but by no means inevitable) that all later birds known be united into a single clade that excludes *Archaeopteryx*, i. e. that all the others shared a common ancestor that was n o t an ancestor of *Archaeopteryx*.

I have stated above that in palaeontology the demarcation of higher taxa is preferably effected by means of e v o l u t i o n a r y d e l i m i t a t i o n, by defining the origin of a taxon on the first appearance of a chosen evolutionary novelty, what I am proposing to call a h o l o d i a g n o s t i c (CHARIG, in prep.). It cannot be denied that the possession of feathers is generally accepted as the holodiagnostic character of the class Aves. Strictly speaking the possession of feathers is a p a n d i a g n o s t i c, i. e. it may be apparent in fossil birds (notably in *Archaeopteryx* itself!) as well as in Recent forms and it should therefore suffice as a primary indicator; on the other hand, feathers are so rarely preserved that it seems prudent to choose another pandiagnostic, a n o n - c o n t r o v e r t e d paradiagnostic, for use as a l e c t o d i a g n o s t i c in that vast majority of instances where there is a *prima facie* case for "birdiness" but where there is no indication whatever of either the presence of feathers or their absence. The obvious choice for such a lectodiagnostic falls upon the possession of an ossified furcula, the merrythought or wishbone, a median structure that appears to be derived by the fusion of the paired clavicles. In our consideration of the limits of the class Aves we need not concern ourselves (as some workers have done) with the entirely hypothesized presence of feathers or "thermoplumes" — whatever the latter may be! — in certain dinosaurs, e. g. in *Syntarsus* (by BAKKER, 1975) and in *Massospondylus* (by COOPER, 1981); to do so would be to indulge at best in a circular argument and at worst in pure fantasy. More relevant would be the claims of BARSBOLD (1983) that some theropod dinosaurs possessed conjoined clavicles, closely resembling the furcula of *Archaeopteryx*, but those I propose to ignore (without denigrating the work of BARSBOLD in any way) until they have been substantiated more fully and the homology more firmly established.

Other workers, at times, have proposed the incorporation of additional groups into the birds on the grounds of supposed common ancestry, thus abandoning the traditional concept of the class. THULBORN (1975) provided an outstanding example of this when he suggested incorporating the theropod dinosaurs into the birds; the class Aves would then be divided into two subclasses, the Archaeornithes with the orders Theropoda and Archaeopterygiformes (the latter including only the one genus *Archaeopteryx*) and the Neornithes (including all the other birds). One objection to this arrangement (CHARIG 1976) was that it implied general acceptance of the theropod origin of birds; that theory d o e s have many supporters, but it is still far from being universally accepted. A further point, which I failed to make at that time, is that THULBORN's subdivision of his "Aves" implied a closer relationship between *Archaeopteryx* and the theropods than between *Archaeopteryx* and other birds; and that contention, whatever THULBORN himself might believe, would be positively accepted at present by only a minority. By contrast, BAKKER & GALTON (1974) intended no change in the composition of the taxon Aves as such, they wished only to demote its rank in the hierarchy from class to subclass and to place that subclass within their proposed new class Dinosauria.

I still maintain, however, that scientists, no matter how well-intentioned, cannot alter the fundamental meanings of popular concepts like "bird" or "dinosaur"; the man-in-the-street will ignore the new classification, whatever the arguments for and against. To quote my own words (CHARIG 1976: 101), "just as the layman will refuse to accept BAKKER & GALTON's suggestion that a sparrow is a dinosaur, so will he balk at THULBORN's idea of *Tyrannosaurus rex* as a bird." I shall now go still further than that: the layman won't even know of the proposed changes!

Having agreed that the possession of feathers should be the holodiagnostic for birds, we are bound to agree also that *Archaeopteryx lithographica* – which, in the first instance, means the holotype (i. e. the so-called London specimen) and in the second instance all the referred material – is an undoubted bird *sensu nostro*.

We may now begin to attempt our phylogeny reconstruction, using the only method available to us – character distribution ("cladistic") analysis, which relies upon the subjective evaluation of alleged similiarities; we shall, however, view them in the light of other helpful information (such as stratigraphical position – and hence chronological age – and the hypothesized evolution of function). Each stage in the proceedings is subjective; what, for example, constitutes identity or similarity? If it is deemed to be present, is it due to a close phylogenetic relationship (i. e. a true synapomorphy), or are we dealing with a "primitive" character common to a much wider group (a symplesiomorphy), or is it due to some other cause such as parallelism or convergence (homoplasy), or is it due merely to coincidence? (See CHARIG 1982.) In any case, the pattern that gives maximal congruence (i. e. the most parsimonious, the one with the fewest discordant character-states) need not reflect the correct phylogenetic situation.

There are two types of phylogenetic problem to be solved, internal and external to the clade Aves.

Internal relationships

Here we have the problem of constructing a tree that shows the phylogenetic relationships between *Archaeopteryx* and other major groups of birds. The latter would comprise the Enantiornithes ("opposite birds") from the Cretaceous of South America, newly described by C. A. WALKER (1983); the Odontornithes, the Cretaceous toothed birds; and the Neornithes or modern birds, subdivided into ratites and carinates.

External relationships

Here we may consider birds as a whole, including *Archaeopteryx*, but we must also consider *Archaeopteryx* in particular; the problem is to relate birds to other, non-avian groups. The matter is rendered more difficult by the fact that many of the characters found in *Archaeopteryx* are not present in later birds. That could be due to *Archaeopteryx*'s having been on a sterile side-line off the main lineage leading to modern birds, so that it evolved its own autapomorphic characters; alternatively it could be that those characters were common to all early birds and were subsequently lost in the later forms known to us. (It is also true, of course, that many of the characters present in modern birds are absent in *Archaeopteryx*.)

All the chief candidates for the closest relationship to *Archaeopteryx* and the birds are groups classified as Archosauria; within the last fifteen years or so the theccodontians, crocodilians, theropods, and ornithischians have each had their supporters. The thecodontians remain one of the two most popular candidates (e. g. with TARSITANO & HECHT 1980, and with HECHT & TARSITANO 1982); which particular thecodontians should be linked to the birds in this fashion remains a mystery. (It should not be forgotten that the enigmatic little *Megalancosaurus* from the Upper Triassic of Italy, described as a thecodontian by CALZAVARA, MUSCIO & WILD in 1981, has also been considered in this connection.) Perhaps even more popular as putative bird ancestors are the theropod dinosaurs, put forward for this rôle by OSTROM (1973 and 1976b); within the theropods both the coelurosaurs and the deinonychosaurs are strongly fancied. The crocodilians, for long supported staunchly by A. D. WALKER (1972), have lost favour with him but have meanwhile gained favour with WHETSTONE & MARTIN (1979). As recently as 1970 the ornithischians too were suggested for the part, but the author in question (GALTON) later recanted; that theory might now be forgotten. Among the non-archosaurs proposed as ancestors or close relatives of the birds were a strange little reptile from the Middle Triassic of Spain (*Cosesaurus aviceps* P. ELLENBERGER & DE VILLALTA 1974) and, believe it or not, the mammals (GARDINER 1982); neither of those propositions, however, merits serious consideration.

A solution to the puzzle of the origin of avian flight might well help solve this puzzle of the origin of birds. Also related to both those problems is the question of the structural and functional origin of feathers (PARKES 1966, BOCK 1969). The question of function is especially important: what were feathers originally for – thermal insulation, flight, protection against abrasion, trapping insects, or some sort of display?

Problems (4): Other topics

As far as *Archaeopteryx* is concerned, matters of stratigraphy and chronology seem to raise no problems. Subjects related to the environment (sedimentology, palaeoclimatology, taphonomy, the whole rich flora and fauna of the Solnhofen Limestone, palaeoecology and the probable habits of *Archaeopteryx*) are still in an early stage of study and we may yet hope to learn much more about those aspects of this early bird.

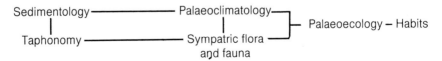

Conclusion

I have sometimes been accused, with reference to other controversies entirely unconnected with *Archaeopteryx*, of being a moral coward, afraid of coming down on one side or the other, of "sitting on the fence". But, if one is trying to be objective, to follow the basic principles of logic and science, what else can one do? The very existence of a controversy implies that some of the alleged evidence points in one direction and some in another; we can have opinions as to the relative weights that should be attributed to the respective items of evidence, but, unless we can discredit old items of evidence or produce altogether new ones, we can do no more than express those opinions (with or without certain reservations). If the evidence presently available were already capable of resolving the controversy, then *ipso facto* the controversy would not exist. That is why I do not expect many – to be frank, I do not expect any – of these *Archaeopteryx* controversies to be resolved at the present Conference. We shall all learn more about those fields in which we ourselves have not specialized; we shall understand more, and shall have a better perspective of the subject as a whole. Perhaps, in relation to some of the controversies, we may even edge a little nearer towards their eventual solution.

References Cited

BAKKER, R. T. (1975): Dinosaur renaissance. – Scient. Am., **232**, 4: 58–78; New York.
BAKKER, R. T. & GALTON, P. M. (1974): Dinosaur monophyly and a new class of vertebrates. – Nature, **238**: 81–85; London.
BARSBOLD, R. (1983): O "ptichnikh" chertakh v stroenii khishchnykh dinozavrov. – In Iskopaemye reptilii Mongolii: 96–103. Moscow (Izdatel'stvo "Nauka").
BEER, G. R. DE (1954): *Archaeopteryx lithographica*, a study based upon the British Museum specimen. – 68 pp. London (British Museum [Natural History]).
BENTON, M. J. (1983): No consensus on *Archaeopteryx*. – Nature, **305**: 99–100; London.
BOCK, W. J. (1969): The origin and radiation of birds. – Ann. N. Y. Acad. Sci., **167**: 147–155; New York.
CALZAVARA, M., MUSCIO, G. & WILD, R. (1981): *Megalancosaurus preonensis*, a reptile from the Norian of Friuli, Italy. – Gortania. Atti del Museo Friulano di Storia Naturale, **2**: 49–64; Undine.
CAPLE, G., BALDA, R. P. & WILLIS, W. R. (1983): The physics of leaping animals and the evolution of preflight. – Am. Nat., **121**, 4: 455–476; Chicago.
CHARIG, A. J. (1976): "Dinosaur monophyly and a new class of vertebrates": a critical review. – In Morphology and biology of reptiles [eds. BELLAIRS, A. d'A. & COX, C. B.], Linn. Soc. Symp., **3**: 65–104. London (Academic Press).
CHARIG, A. J. (1982): Systematics in biology: a fundamental comparison of some major schools of thought. – In Problems of phylogenetic reconstruction [eds. JOYSEY, K. A. & FRIDAY, A. E.] Syst. Ass. spec. Vol. **21**: 363–440. London (Academic Press).
CHARIG, A. J. (in preparation): The demarcation of zoological taxa.
COOPER, M. J. (1981): The prosauropod dinosaur *Massospondylus carinatus* OWEN from Zimbabwe: its biology, mode of life and phylogenetic significance. – Occ. Pap. natn. Mus. Rhod., **6**: 689–840; Bulawayo.
DAMES, W. (1897): Ueber Brustbein, Schulter- und Beckengürtel der *Archaeopteryx*. – Sber. preuss. Akad. Wiss., **1897**, 2: 818–834; Berlin.
ELLENBERGER, P. & DE VILLALTA, J. F. (1974): Sur la présence d'un ancêtre probable des oiseaux dans le Muschelkalk supérieur de Catalogne (Espagne). Note préliminaire. – Acta geol. hispan., **9**, 5: 162–166; Barcelona.
FEDUCCIA, A. & TORDOFF, H. B. (1979): Feathers of *Archaeopteryx*: asymmetric vanes indicate aerodynamic function. – Science, **203**: 1021–1022; New York.
GALTON, P. M. (1970): Ornithischian dinosaurs and the origin of birds. – Evolution, **24**, 2: 448–462; Lawrence, Kansas.
GARDINER, B. (1982): Tetrapod classification. – Zool. J. Linn. Soc., **74**, 3: 207–232; London.
HARRISON, C. J. O. (1976): Feathering and flight evolution in *Archaeopteryx*. – Nature, **263**: 762–763; London.
HECHT, M. K. & TARSITANO, S. (1982): The paleobiology and phylogenetic position of *Archaeopteryx*. – Geobios, Mém. spéc., **6**: 141–149; Lyons.
HOWGATE, M. (1982): News and views. – Nature, **300**: 469; London.
HOWGATE, M. (1983): *Archaeopteryx* – no new finds after all. – Nature, **306**: 644–645; London.
MARSH, O. C. (1880): Odontornithes: a monograph on the extinct toothed birds of North America. – Prof. Pap. Engineer. Dept. U. S. Army, **18**: 1–201; Washington.
MAYR, F. X. (1973): Ein neuer *Archaeopteryx*-Fund. – Paläont. Z., **47**, 1/2: 17–24; Berlin, &c.
MEYER, H. VON (1857): Beiträge zur näheren Kenntnis fossiler Reptilien. – Neues Jb. Miner. Geol. Paläont., **1857**: 437; Stuttgart.
OLSON, S. L. & FEDUCCIA, A. (1979): Flight capability and the pectoral girdle of *Archaeopteryx*. – Nature, **278**: 247–248; London.
OSTROM, J. H. (1972): *Pterodactylus crassipes* MEYER 1857 (Aves): Proposed suppression under the plenary powers. Z. N. (S.) 1977. – Bull. zool. Nom., **29**: 30–31; London.
OSTROM, J. H. (1973): The ancestry of birds. – Nature, **242**: 136; London.
OSTROM, J. H. (1974): *Archaeopteryx* and the origin of flight. – Q. Rev. Biol., **49**: 27–47; Baltimore.
OSTROM, J. H. (1976a): Some hypothetical anatomical stages in the evolution of avian flight. – Smithson. Contr. Paleobiol., **27**: 1–21; Washington.
OSTROM, J. H. (1976b): *Archaeopteryx* and the origin of birds. – Biol. J. Linn. Soc., **8**, 2: 91–182; London.
PADIAN, K. (1982): Running, leaping, lifting off. – The Sciences, May-June **1982**: 10–15; New York.
PARKES, K. C. (1966): Speculations on the origin of feathers. – Living Bird, **5**: 77–86; Ithaca, N. Y.
PETRONIEVICS, B. (1921): Über das Becken, den Schultergürtel und einige andere Teile der Londoner *Archaeopteryx*. – 31 pp. Geneva.
PETRONIEVICS, B. & WOODWARD, A. S. (1917): On the pectoral and pelvic arches of the British Museum specimen of *Archaeopteryx*. – Proc. zool. Soc. Lond., **1917**: 1–6.
SEELEY, H. G. (1881): On some differences between the London and Berlin specimens referred to *Archaeopteryx*. – Geol. Mag., **2**, 8: 454–455; London.

SEELEY, H. G. (1882): On a restoration of a skeleton of *Archaeopteryx*, with some remarks on the differences between the Berlin and London specimens. – Rep. Brit. Ass., **1881:** 618; London.
SNEATH, P. H. A. (1962): The construction of taxonomic groups. – In Microbial classification, Symp. Soc. gen. Microbiol., **12:** 289–332. Cambridge (Cambridge University Press).
TARSITANO, S. & HECHT, M. K. (1980): A reconsideration of the reptilian relationships of *Archaeopteryx*. – Zool. J. Linn. Soc., **69,** 2: 149–182; London.
THULBORN, R. A. (1975): Dinosaur polyphyly and the classification of archosaurs and birds. – Aust. J. Zool., **23,** 2: 249–270; Melbourne.
WALKER, A. D. (1972): New light on the origin of birds and crocodiles. – Nature, **237:** 257–263; London.
WALKER, C. A. (1983): New subclass of birds from the Cretaceous of South America. – Nature, **292:** 51–53; London.
WHETSTONE, K. N. (1983): Braincase of Mesozoic birds: I. New preparation of the "London" *Archaeopteryx*. – J. vert. Paleont., **2,** 4: 439–452; Norman, Oklahoma.
WHETSTONE, K. N. & MARTIN, L. D. (1979): New look at the origin of birds and crocodiles. – Nature, **279:** 234–236; London.
WHYBROW, P. J. (1982): Preparation of the cranium of the holotype of *Archaeopteryx lithographica* from the collections of the British Museum (Natural History). – Neues Jb. Geol. Paläont. Mh., **1982,** H. 3: 184–192; Stuttgart.
WILLISTON, S. W. (1879): "Are birds derived from dinosaurs?" – Kansas City Rev. Sci., **3:** 457–460.

Author's address: Dr. ALAN J. CHARIG, Department of Palaeontology, British Museum (Natural History), Cromwell Road, London SW7 5BD, England.

Günter Viohl

Geology of the Solnhofen Lithographic Limestone and the Habitat of *Archaeopteryx*

Abstract

Stratigraphy, paleogeography, and paleoecology of the Solnhofen Lithographic Limestones are surveyed. Some evidences, such as clay minerals, landflora, and fauna, suggest a semiarid, monsoonal climate with a change of dry and rainy seasons. The habitat of *Archaeopteryx* was bushland of conifers and Bennettitales alternating with areas of sparse vegetation. Because of the lack of forests trunk-climbing as the prevailing mode of life of *Archaeopteryx* can be ruled out. The protobirds possibly climbed about on bushes in search of insects using the claws of their forelimbs for clinging to branches. *Archaeopteryx* was undoubtedly capable of powered flight. The complete preservation especially of the Berlin and Eichstätt specimen suggests a rapid burial and precludes long transport. Flying across the sea the animals were probably caught by a storm and drowned.

Zusammenfassung

Es wird ein kurzer Überblick über Stratigraphie, Paläogeographie und Palökologie der Solnhofener Plattenkalke gegeben. Verschiedene Befunde wie Tonmineralien, Landflora und Fauna deuten auf ein semiarides Monsunklima mit einem Wechsel von Trocken- und Regenzeiten hin. Der Biotop von *Archaeopteryx* war Buschland aus Koniferen und Bennettiteen, das mit Flächen spärlicher Vegetation wechselte. Infolge des Fehlens von Wäldern scheidet Stammklettern als vorherrschende Lebensweise für *Archaeopteryx* aus. Auf der Suche nach Insekten dürften die Urvögel auf Büschen herumgeklettert sein, wobei sie die Krallen der Vorderextremitäten zum Festhaken an den Zweigen benutzten. *Archaeopteryx* war zweifellos bereits ein aktiver Flieger. Die vollständige Erhaltung besonders des Berliner und Eichstätter Exemplars läßt auf eine rasche Einbettung schließen. Längerer Transport scheidet somit aus. Die Tiere wurden wahrscheinlich während eines Fluges über das Meer von einem Sturm überrascht und ertranken.

Archaeopteryx is the most famous fossil of the Solnhofen Lithographic Limestones. Its full understanding requires the knowledge of the geologic-paleogeographic setting and the depositional environment of the enclosing sediment. Any biological theory making assertions about the habitat of this animal, its mode of life, and the selectional constraints, being effective, should at least not contradict the geological-paleontological data obtained from the Solnhofen Limestones. In the following it is therefore attempted to reconstruct the environment of *Archaeopteryx* proceeding from these data.

Stratigraphic position and facies of the Solnhofen Lithographic Limestones

Stratigraphically the Solnhofen Lithographic Limestones (Malm Zeta 2) belong to the lower part of the Lower Tithonian, to the zone of *Hybonoticeras hybonotum* comprising at most a half ammonite zone. If one estimates the duration of an Upper Jurassic ammonite zone with BARTHEL (1964, 1978) to one million years, the whole series of the Solnhofen Limestones represents at best 500,000 years, probably much less.

The occurrence of the Solnhofen Limestones is restricted to the Southern Franconian Alb (Fig. 1). Their depositional area has an extent of about 70 x 30 km. They were deposited in depressions between algal-sponge reefs, so-called "Wannen" (Fig. 2). Therefore their total thickness varies considerably, ranging from 0 to 95 m. Growth of the reefs started during the Oxfordian at the margins of submarine highs and extended widely during the Middle Kimmeridgian. With the shallowing of the sea in the Upper

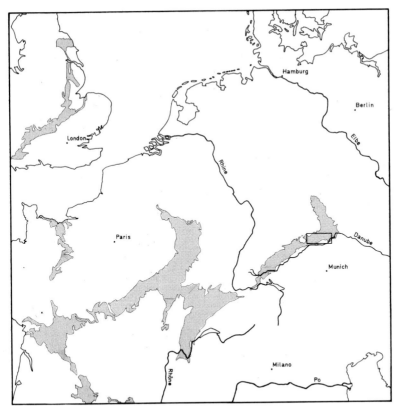

Fig. 1
Distribution of the Jurassic in Central Europe. The rectangle marks the sedimentation area of the Solnhofen Lithographic Limestones.

Kimmeridgian the development of algal-sponge reefs became regressive. Finally, in the Lower Tithonian, the reefs died, partially covered by the Solnhofen Lithographic Limestones. To some extent algal-sponge reefs were replaced by coral reefs. This took place in the eastern part of the Southern Franconian Alb already in the uppermost Middle Kimmeridgian, further west only in the Lower Tithonian. The typical Solnhofen Lithographic Limestones consist of even-layered pure limestone slabs ("Flinze") with an internal microbedding and irregularly intercalated calcareous fine-layered marls ("Fäulen") (Fig. 3 + 4). There are remarkable alterations of facies in the Lower Tithonian flaggy limestones subsumed under the name "Solnhofen Lithographic Limestones" (for an overlook see V. FREYBERG 1968 and MEYER & SCHMIDT-KALER 1983), so that individual layers cannot be correlated from basin to basin or even within the same basin (FESEFELDT 1962). Only within the Eichstätt deposition area such a correlation is possible to some degree (V. EDLINGER 1964). In the Eichstätt and Solnhofen area the Flinz layers are uniform micritic (concerning their sedimentology see DE BUISONJÉ, this volume), whereas in the eastern basins they are coarser containing much reef detritus from the adjacent coral reefs.

In the region of Eichstätt and Solnhofen the Solnhofen Limestones can be divided into two sequences with a slumping unit ("Krumme Lage") on the top of each: the "Untere Schiefer" with "Trennende Krumme Lage" and the "Obere Schiefer" with "Hangende Krumme Lage". As these slumping units can be traced over a distance of 20 km in the same stratigraphic level, they cannot be due to local events. Probably they were caused by earthquakes. At present only the "Obere Schiefer" are quarried, since the "Untere Schiefer" are not suitable for technical purposes because of their higher content of marl. Whereas in the lower sequence traces of life can be found, in the upper sequence autochthonous macrobenthos is completely lacking, thus indicating that the paleoenvironment became hostile.

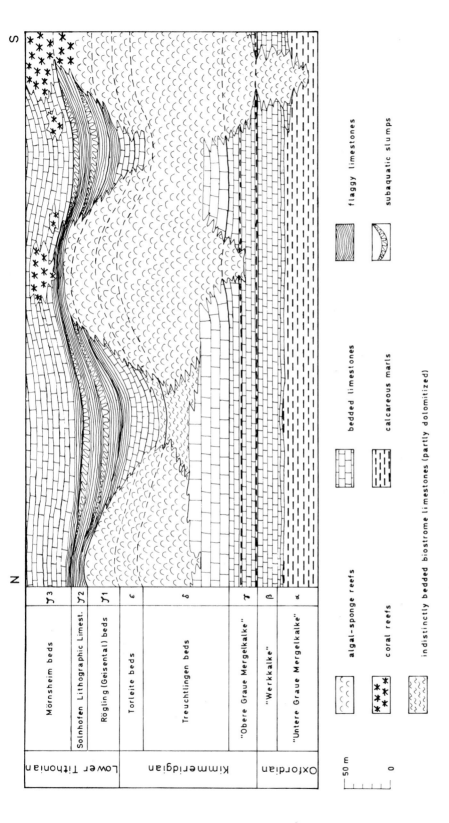

Fig. 2
Stratigraphic scheme of the Upper Jurassic in the Southern Franconian Alb.

Fig. 3
Solnhofen Lithographic Limestones from the center of the Solnhofen basin at the Maxberg quarry. Irregular alternation of limestone layers and calcareous marls.

Fig. 4
Solnhofen Lithographic Limestones from Blumenberg (Eichstätt sedimentation area).

Paleogeography

During the Lower Tithonian, Central Europe was occupied by some larger islands with arms of the sea between them (Fig. 6). South of Munich the European shelf sloped towards the deep sea of the Tethys. The depositional area of the Solnhofen Lithographic Limestones was located on the southern part of this shelf southeast of the Central German Swell and west of the Bohemian Island. The more detailed paleogeographic reconstruction of this area (Fig. 7) is based on papers of V. FREYBERG (1968), BARTHEL (1970, 1978), MEYER (1977, 1981), SCHAIRER & BARTHEL (1981) as well as my own considerations (VIOHL 1983).

The deposition area of the Solnhofen Limestones was bordered in the east, partly in the south, and possibly in the west (area now occupied by the Ries meteorite crater) by coral reefs. Hence the area of the flaggy limestones had a backreef position (V. FREYBERG 1978, BARTHEL 1970). It can be regarded as a "lagoon" but in a very broad sense, because there existed connections with the Tethys in the south, and possibly with the sea in Northern Germany and Poland (ZEISS 1968).

Fig. 5
Quarry area "Langenaltheimer Haardt". Here the London and Maxberg specimens of *Archaeopteryx* were discovered.

The eastern coral reefs along with some small islands were situated on a southeast directed submarine high, the Parsberg-Landshut Swell. In its prolongation the Central German Swell and perhaps some offshore islands emerged. Even within the deposition area of the Solnhofen Limestones, isolated algal-sponge-reefs might have stood above the sea level as islands.

The main connections between the "lagoon" and the Tethys probably existed via a zone of deeper water southwest of the Parsberg-Landshut Swell and parallel to it, and across a barrier of living algal-sponge reefs between the depositional area of the Solnhofen Limestones and a deeper bay ("Zementmergel-Meer") in the southwest. This reef barrier was part of a platform of shallow water (mean water depth probably 40–50 m) extending from the present river Danube about 100 km southwards (MEYER in SCHMIDT-KALER 1979 and MEYER 1981).

Paleoclimate

In the Lower Tithonian the sedimentation area of the Solnhofen Limestones lay near the northwestern rim of the Tethys between 25° and 30° northern latitude (Fig. 8). That is within the girdle of the subtropical high and northeastern trade winds. In summer the large landmasses of Asia and North America probably caused a northward shift of the intertropical convergence zone and therefore monsoon winds directed from the Tethys to both continents. Monsoon wind blowing from SE could have affected the "lagoon" of Solnhofen and the Central German Island bringing them periodical rainfalls. But as higher mountain ranges were lacking, the precipitation was presumably low.

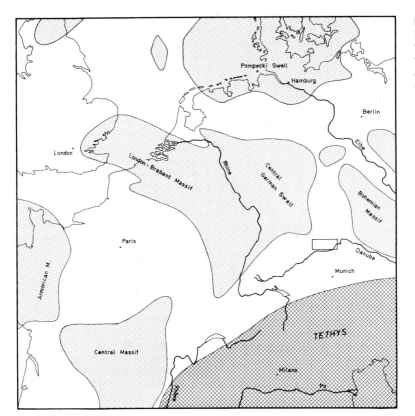

Fig. 6
Paleogeography of Central Europe during the Lower Tithonian after ENAY et al. (1980), MEYER (1981), and ZIEGLER (1982).

This consideration, though highly speculative due to the many factors controlling the climate, suggests a warm semiarid climate with long dry and short rainy seasons. This is corroborated by the following evidence found in the Solnhofen Limestones:

1. The great accumulation of carbonate as well as the occurrence of thermophile organisms (corals, pterosaurs, large insects) indicate high temperatures.

For the Middle Kimmeridgian of the Eichstätt region ENGST (1961) has calculated a mean water temperature of 26,4° C from the $16_0/18_0$ ratio. The value for the Lower Tithonian may not have been very different.

2. Clay mineral investigations carried out by HÜCKEL (1974), BARTHEL (1976), BAUSCH (1980), and SALGER (not yet published) showed varying, but generally low contents of kaolinite, and relative high percentages of montmorillonite. Although the distribution of clay minerals is controlled by different circumstances, such as the exposed rocks in the areas of supply, the differential settling tendencies, and the sedimentary environment, that result can be best interpreted by a dry, but not desert climate. Motmorillonite requires for its formation a neutral to alkaline environment, kaolinite an acid one.

3. Despite the vicinity of land suggested by the fossils (terrestrial plants and animals), the amount of terrigenous sediment material was very little, thus indicating the lack of persistent rivers. Hence a low precipatation rate can be inferred.

4. Previous authors (MUTSCHLER 1923, HIRMER 1924, MEYER 1974) have already pointed out that the landplants preserved in the Solnhofen Lithographic Limestones display xeromorphic characteristics. Leaves were covered by thick cuticles, and in the conifers *Brachyphyllum* and *Palaeocyparis*, they were reduced to scales to restrict evaporation. Even the fern-shaped leaves of *Cycadopteris* (belonging to the

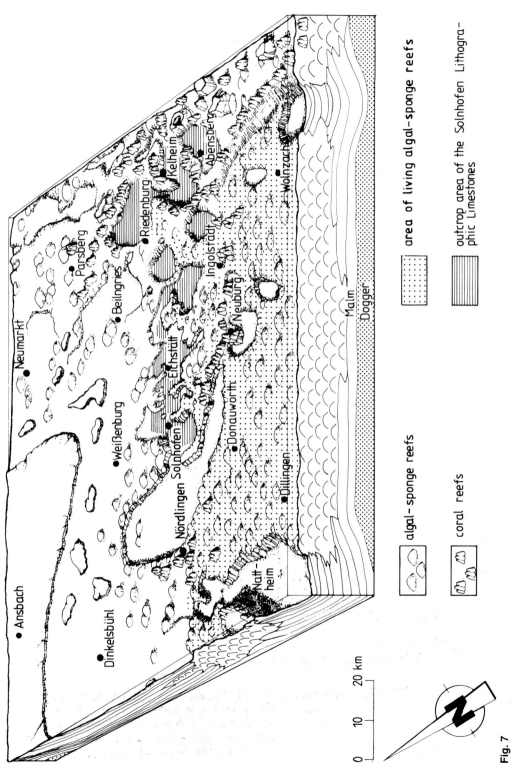

Fig. 7
Paleogeographic reconstruction of the sedimentation area of the Solnhofen Limestones.

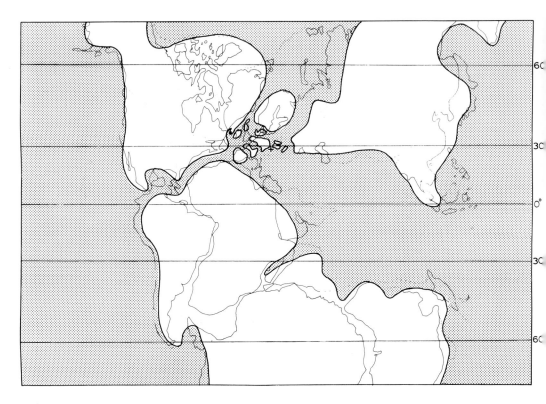

Fig. 8
Approximate distribution of land and sea (stippled) in the Lower Tithonian (after HALLAM 1975 with little changes). The sedimentation area of the Solnhofen Limestones is indicated by a cross.

Pteridospermae), show on the top side a leathery cuticular coating, whereas the stomata lie in sunken fields on the underside. Real ferns have never been found. *Furcifolium longifolium* described by KRÄUSEL (1943) as a ginkgophyte exhibits thorns at the base of the clustered bifurcate leaves which can also be regarded as an xeromorphic feature.

JUNG (1974) has shown that the conifers *Brachyphyllum* and *Palaeocyparis*, by far the most frequent species of all landplants, were stem succulents. They had only a woody central cylinder, whereas the most part of the stems and branches were occupied by water-storing tissue.

After being embedded in the lime mud of the "lagoon" the soft tissue decayed slowly and the resulting cavity was filled with calcite. Fig. 9 shows a crotch of a branch consisting mainly of such a calcite filling. At two ends of it, traces of the jutting-out woody central cylinders are detectable. The plants had the same structure as living cacti have (Fig. 10).

From the little portion of woody tissue JUNG argued rightly that for static reasons these conifers could not have been tall trees, as was assumed previously. They must have been shrubs not exceeding 3 m. (Fig. 11).

But, evidently influenced by WALTHER (1904), JUNG considers these plants as halophytes growing on the higher part of the shallows in the manner of a mangrove. This conclusion is, however, not well-founded. It is true, halophytes are mostly succulent, but most succulents are not halophytes. BARALE (1981) also rejects the idea of an Upper Jurassic mangrove, though he admits that some plants might have been halophytes. The stem succulence of *Brachyphyllum* and *Palaeocyparis* can be regarded as an adaptation to aridity.

Fig. 9
Crotch of a succulent conifer with traces of the woody central cylinder at two ends (explanation see text). Jura-Museum Eichstätt (No. 151.68).

Fig. 10
A cactus (*Echinocereus* sp.) gnawed by an animal displays the woody central cylinder. Organ Pipe National Monument, Arizona.

Evidence of trees are missing in the Solnhofen Lithographic Limestones. Larger pieces of drift wood originating from logs are not known. Only remains of branches with a maximum thickness of 3 to 4 cm have been found (see also JUNG 1974).

5. In contrast with the landplants, there seem to be insects normally dependent on fresh water: the Odonata, Ephemeroptera, and Trichoptera. Where did they develop? Actually larvae of dragonflies have been observed in brackish coastal waters and even in salt lakes (KLOTS, KLOTS & FORSTER 1965). Therefore it might be that the numerous dragonflies of the Solnhofen Limestones have developed in the lagoonal waters near the shore, though we have no evidence of their larvae. But that seems improbable for the Ephemeroptera and Trichoptera which always need fresh water biotopes today. These may have existed under a relatively dry climate, at least temporarily, supplied by springs and periodical precipitations.

Compared with the Early and Middle Jurassic, in the Late Jurassic, a change from a humid to a much dryer climate took place. That can be deduced, from among other things, the decrease of kaolinite and an increase of montmorillonite (SALGER 1959, STARKE 1976). Still, during the Dogger in Central Europe, fine-grained sandstones, and high iron concentrations, testify to an intense weathering on the land areas which required a warm, humid climate. According to VAKHRAMEEV (1962), with the beginning of the Late

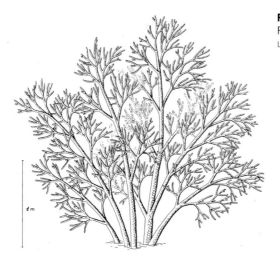

Fig. 11
Reconstruction of *Brachyphyllum*. Drawing by H. Thiele. From Jung (1974).

Jurassic epoch, the climate became more arid over a considerable part of the Indo-European region, resulting in the cessation of coalbearing bed formation, in the appearance of red beds, and locally of gypsum-bearing deposits practically devoid of plant remains. A dry belt extended from France to Kazakhstan, as is suggested by the fossil flora (compare also Barale 1981). Only in Scotland the occurence of ferns indicates a more humid climate.

Paleoenvironment and origin of the Solnhofen Lithographic Limestones

Conclusions on the paleoenvironment of the Solnhofen Lithographic Limestones can be drawn from the following observations:

1. In the Solnhofen Limestones, indigenous macrobenthos is almost lacking. Nearly all macrobenthic forms were washed in. Of them, only the crustacean *Mecochirus longimanatus*, and the horseshoe crab *Mesolimulus walchi* – evidently more resistant species – were still alive when reaching the bottom, for they are occasionally found at the end of their trails.

2. Autochthonous microbenthos (foraminifers and ostracods) has been described from the marly "Fäulen", respectively the bedding planes by Groiss (1967) and Gocht (1973). The foraminifers show, after Groiss (1967), in the compound of species, and adaptation to a special environment. The number of species and individuals decreases remarkably from east to west (oral communication of Prof. Groiss). In addition to the microfauna, some small, mostly juvenile gastropods of the species *Spinigera spinosa* have been found on the bedding-planes of some Flinz layers in the Eichstätt area. These snails bore two long spines on each whorl, perpendicular to the longitudinal axis, obviously an adaptation to the soft substratum. They can therefore be regarded as autochthonous.

3. A rich nannoplankton is concentrated on the bedding-planes (Keupp 1977 a + b). The coccoliths were arranged in accumulations and to a large extent deposited by the way of fecal pellets.

4. Also, abundant crinoids of the genus *Saccocoma*, coprolites (chiefly of fishes) and many small juvenile ammonites can be found on bedding-planes, indicating a temporary planktonic and nektonic life. These ammonites could not have undergone a longer transport, as their aptychi are still present.

5. With few exceptions all macrofossils lie on the bedding-planes.

6. Measurements of fossil remains at the digging site of the Jura-Museum near Schönau (Eichstätt area) did not yield a preferential current direction. Aptychi and fragments of ammonite shells all lie with the convex side down. This indicates a very calm bottom environment which can similarly be inferred from settling marks next to the fossil (Mayr 1967) and the absence of drag marks.

The aforesaid data can be interpreted on the basis of the paleogeographic and climatic conditions. Due to high evaporation rates under the hot climate and restricted water exchange by the backreef position and the strong relief of the seabottom, the water in the lagoonal basins was hypersaline. Density stratification developed which caused stagnant conditions in the bottom zone (lack of current indicators). For most of the time, the basins offered a hostile environment to all organisms. But as the coprolites, the nannoplankton, and especially the microbenthos show, the conditions changed from time to time. Onshore summer monsoons blowing from the SE with heavy rainfalls added great quantities of normal sea water and fresh water into the "lagoon". KEUPP (1977 a + b) has pointed out what mechanism could have then led to nearly normal marine conditions in the basins. Initially only the superficial water layers were influenced, while the deeper hypersaline water remained intact due to density stratification. The superficial water body of minor density acted as a burning-lense and heated the lower hypersaline water body of high density. At great differences of temperatures a turn over took place, and for a short time nearly normal marine conditions were established. Nektonic animals (e. g. fishes, cephalopods) immigrated and microfauna could settle on the botton. The decrease of microfauna from east to west show that the water exchange became less intense in this direction. That can easily be understood, if monsoonal winds blowing from the SE were held to be responsible for it.

According to the model of KEUPP such turn over events were represented by the bedding-planes (including internal microbedding). They did not occur every year. The duration of nearly normal marine conditions was too short for the immigration of major benthic animals. Then evaporation created again a hypersaline environment. This might also explain the extremely high percentage of juvenile individuals among the ammonites, the above mentioned gastropods, and even the swimming crinoids *Saccocoma* which had insufficient time to grow up.

During the summer monsoons, probably most organisms from the living reefs, the open sea, and the land habitats, were driven into the basins too, and therefore can be found on the bedding-planes. Insects and pterosaurs were caught in flight by monsoonal storms and drowned. Clay minerals enriched on bedding-planes were washed in from the Central German Island also during the rainy seasons.

Concerning the origin of the Solnhofen Lithographic Limestones there still are different opinions among modern authors. BARTHEL who has previously developed the stagnant basin model (BARTHEL 1964, 1970) advocated an allochthonous origin of the "Flinze" by suspension fallout after hurricanes (BARTHEL 1972, 1978), while regarding the "Fäulen" as the normal background sediment. DE BUISONJÉ (1972 and this volume) considers recurrent coccolithophorid blooms responsible for the formation of the limestone layers. According to KEUPP (1977 a + b) there is no essential difference between "Flinze" und "Fäulen" except the denser sequence of bedding-planes in the "Fäulen" and thereby a higher clay content. He thinks the whole sediment to be a formation of coccoid bluegreen algae (= cyanobacteria).

This is not the place to evaluate these theories (for discussion see VIOHL 1983). All can explain some facts, but are obviously inconsistent with others. To develop a new, more satisfactory model, which should be more differentiated, further sedimentological data are necessary.

The habitat of *Archaeopteryx*

The habitats of *Archaeopteryx* lay on the island formed by the Central German Swell and the London-Brabant-Massif, on the Bohemian Island, and perhaps some smaller offshore islands. The just emerged coastal territories must have been occupied by wide flat plains cut by dry drainage channels. Only on those high areas which were exposed to weathering and erosion since the Early and Middle Jurassic (London-Brabant Massif, Rhenish Massif, central parts of the Bohemian Island), inselbergs might have developed under the warm and humid climate of that time. Along the shoreline of the slowly emerging land in the NW, and the offshore islands, marine abrasion probably formed cliffs, though we have no direct evidence.

A clue to such cliffs could be the pebbles attached to seaweeds which were described by MAYR (1953). Previous authors (e. G. ABEL 1922, BARTHEL 1978) assumed mostly a flat coast, but without substantiation. Flat coasts generally evolve, if submergence and accumulation prevail. Here, however, upheaval of land occurred, and because of the lack of persistent water courses, no accumulation of terrestrial sediments took place. This suggests the development of cliffs along the shore line.

As stated above, the Solnhofen Lithographic Limestones do not provide any evidence for trees. If there had been any trees whatever, they can be expected to have thrived only in such places where water was more abundant, e. g. around springs or along periodic water courses. Major continuous forests were surely missing. The main vegetation was probably a bushland of conifers and Bennettitales alternating with open plains, with only sparse vegetation.

An incomplete soil covering is indicated by wind-transported quartz grains (BARTHEL 1976).

Such an environment would be consistent with a ground-dwelling existence of *Archaeopteryx* as suggested by the cursorial theory for the evolution of birds' flight (OSTROM 1974 and 1979; PETERS & GUTMANN 1976; PETERS 1984), or with a hopping *Archaeopteryx* (RIETSCHEL, this volume). But it does not fit an arboreal mode of life. Here a big problem arises. YALDEN (this volume) has clearly demonstrated that the claws of the manus of *Archaeopteryx* had a climbing function. He regards this animal as a trunk climber. Trunkclimbing, however, makes sense only in larger forests, not in bushland possibly with scattered trees. As the arguments of YALDEN are convincing, another kind of climbing must be taken into consideration.

Probably *Archaeopteryx* fed on insects and other small animals. More insects could be found on plants, especially shrubs, than on the bare ground. While foraging there *Archaeopteryx* had to climb about on the branches and to leap or fly from one bush to another. In this activity it might have used the claws of its forelimb for clinging to branches and its apposable hallux for perching. It is also possible that *Archaeopteryx* fed on small animals living at the shore (worms, crayfish, fishes etc.) and climbed on cliffs where it might have had its nesting-places.

Archaeopteryx was undoubtedly capable of powered flight, as the presence of asymmetrical flight feathers and the furcula show, but the lack of the supracoracoideus muscle suggests that it could not take off from level ground (FEDUCCIA 1980). Its flight ability is also indicated by the complete preservation especially of the Berlin and Eichstätt specimens. This precludes long transport and required rapid burial. It can be assumed that the birds flew across the sea, perhaps from one island to another, when they were caught in a storm or monsoonal shower and drowned. The lungs filled with water, and the plumage was soaked. Only in this case could the carcasses sink down quickly, as has already been stated by RIETSCHEL (1976). Otherwise they would have floated for a prolonged time (30–40 days), becoming largely decomposed (SCHÄFER 1962, 1976).

From the paleoecological data the issue cannot be decided, whether a gliding stage was a prerequisite of the powered flight of birds (U. M. NORBERG and RAYNER, this volume) or not (BALDA, Caple & Willis, this volume). In either case evolution of flapping flight could have occurred in a semiarid environment, as sketched above, prevailing on the Central and Western European islands, during the entire Late Jurassic (over a period of nearly 20 million years). In such an environment, gliding from trees must, of course, be precluded, but powered flight can have been developed from jumps down slopes, as suggested by PETERS (1984, and this volume). In a xerothermic habitat, with little shade, the hypothesis of REGAL (1975, and this volume) on the evolution of feathers as heat-shields, also becomes plausible. If, however, the flight of birds originated by gliding from trees, it could have evolved only during the Early or Middle Jurassic or in other more humid regions at the time of the Late Jurassic (e. g. Northern Europe).

References Cited

ABEL, O. (1922): Lebensbilder aus der Tierwelt der Vorzeit. – 643 pp., 507 fig.; Jena (Gustav Fischer).
BARALE, G. (1981): La Paléoflore Jurassique du Jura français: étude systématique, aspects stratigraphiques et paléoécologique. – Docum. Lab. Géol. Lyon, **81**, 467 p., 59 fig., 12 tabl., 66 pl.; Lyon.
BARTHEL, K. W. (1964): Zur Entstehung der Solnhofener Plattenkalke (unteres Untertithon). – Mitt. Bayer. Staatsslg. Paläont. hist. Geol, **4:** 37–69, pls. 8–11, 1 fig.; München.
BARTHEL, K. W. (1970): On the deposition of the Solnhofen lithographic limestone (Lower Tithonian, Bavaria, Germany). – N. Jb. Geol. Paläont., Abh., **135**, 1: 1–18, pls. 1–4, 2 figs., 1 tab.; Stuttgart.
BARTHEL, K. W. (1972): The genesis of the Solnhofen lithographic limestone (Low. Tithonian): further data and comments. – N. Jb. Geol. Paläont., Mh., **1972**, 3: 133–145, 4 figs.; Stuttgart.
BARTHEL, K. W. (1976): Coccolithen, Flugstaub und Gehalt an organischen Substanzen in Oberjura-Plattenkalken Bayerns und SE-Frankreichs. – Eclogae geol. Helv., **69**, 3: 627–639, 4 pls., 2 figs.; Basel.
BARTHEL, K. W. (1978): Solnhofen – Ein Blick in die Erdgeschichte. – 393 pp., 16 colourpls., 64 pls., 50 figs.; Thun (Ott Verlag).

Bausch, W. M. (1980): Tonmineralprovinzen in Malmkalken. – Erlanger Forsch. B. Naturwiss. Med. **8,** 78 pp., 33 figs.; Erlangen.
Buisonjé, P. H. de (1972): Recurrent red tides, a possible origin of the Solnhofen Limestone (I/II). – Proc. k. nederl. Akad. Wetensch. (B), **75,** 2: 152–177, 2 figs.; Amsterdam.
Edlinger, G. v. (1964): Faziesverhältnisse und Tektonik der Malmtafel nördlich Eichstätt/Mfr. – Erlanger geol. Abh., **56,** 72 pp., 2 pls., 37 figs.; Erlangen.
Enay, R., Mangold, C., Cariou, E., Contini, D., Debrand–Passard, S., Donze, P., Gabilly, J. Lefavrais–Raymond, A., Mouterde, R. & Thierry, J. (1980): Synthèse paléogéographique du Jurassique Français. – Docum. Lab. Géol. Lyon, H. S. **5, 1980,** 210 pp., 42 pls., 1 fig.; Lyon.
Engst, H. (1961): Über die Isotopenhäufigkeit des Sauerstoffes und Meerestemperatur im Süddeutschen Malm-Delta. – 184 pp., 41 figs., Diss.; Frankfurt a. M.
Feduccia, A. (1980): The age of birds. – 196 pp.; Cambridge/Mass. and London (Harward University Press).
Fesefeld, K. (1962): Schichtenfolge und Lagerung des oberen Weißjura zwischen Solnhofen und der Donau (Südliche Frankenalb). – Erlanger geol. Abh., **46,** 80 pp., 2 pls., 30 figs., 2 tables; Erlangen.
Freyberg, B. v. (1968): Übersicht über den Malm der Altmühl-Alb. – Erlanger geol. Abh., **70,** 40 pp., 4 pls., 5 figs.; Erlangen.
Gocht, H. (1973): Einbettungslage und Erhaltung von Ostracoden-Gehäusen im Solnhofener Plattenkalk (Unter Tithon, SW-Deutschland). – N. Jb. Geol. Paläont., Mh., **1974,** 4: 189–206, 28 figs.; Stuttgart.
Groiss, J. Th. (1967): Mikropaläontologische Untersuchung der Solnhofener Schichten im Gebiet um Eichstätt (Südliche Frankenalb). – Erlanger geol. Abh., **66:** 75–93, 1 pl., 2 figs.; Erlangen.
Hallam, A. (1975): Jurassic environments. – 269 pp.; Cambridge (University Press).
Hirmer, M. (1924): Zur Kenntnis von *Cycadopteris* Zigno. – Palaeontographica, **66**: 127–162, pl. IX-XII, 27 figs.; Stuttgart.
Hückel, U. (1974): Vergleich des Mineralbestandes der Plattenkalke Solnhofens und des Libanon mit anderen Kalken. – N. Jb. Geol. Paläont., Abh., **145,** 2: 153–182, 17 figs., 3 tables; Stuttgart.
Jung, W. (1974): Die Konifere *Brachyphyllum nepos* Saporta aus den Solnhofener Plattenkalken (unteres Untertithon), ein Halophyt. – Mitt. Bayer. Staatsslg. Paläont. hist. Geol., **14:** 49–58, pl. 3–4, 3 figs.; München.
Keupp, H. (1977 a): Ultrafazies und Genese der Solnhofener Plattenkalke (Oberer Malm, Südliche Frankenalb). – Abh. Naturhist. Ges. Nürnberg, **37,** 128 pp., 30 pls., 19 figs.; Nürnberg.
Keupp, H. (1977 b): Der Solnhofener Plattenkalk – ein Blaugrünalgen Laminit. – Paläont. Z., **51,** 1/2: 102–116, pl. 8, 4 figs.; Stuttgart.
Klots, A. B., Klots, E. B. & Forster, W. (1965): Insekten. – Knaurs Tierreich in Farben, 350 pp., 294 figs.; München und Zürich (Droemersche Verlagsanstalt).
Kräusel, R. (1943): *Furcifolium longifolium* (Seward) n. comb., eine Ginkgophyte aus dem Solnhofener Jura. – Senckenbergiana, **26,** 5: 426–433, 6 figs.; Frankfurt a. M.
Mayr, F. X. (1953): Durch Tange verfrachtete Gerölle bei Solnhofen und anderwärts. – Geol. Bl. NO-Bayern, **3,** 4: 113–121, pl. 8; Erlangen.
Mayr, F. X. (1967): Paläobiologie und Stratinomie der Plattenkalke der Altmühlalb. – Erlanger geol. Abh., **67,** 40 pp., 16 pls., 8 figs.; Erlangen.
Meyer, R. K. F. (1974): Landpflanzen aus den Plattenkalken von Kelheim (Malm). – Geol. Bl. NO-Bayern, **24,** 3: 200–210, 9 figs.; Erlangen.
Meyer, R. K. F. (1977): Stratigraphie und Fazies des Frankendolomits und der Massenkalke (Malm). – Erlanger geol. Abh., **104,** 40 pp., 5 pls., 10 figs.; Erlangen.
Meyer, R. K. F. (1981): Malm (Weißer oder Oberer Jura). In: Erläuterungen zur Geologischen Karte von Bayern 1:500 000. – 168 pp., 6 pls., 29 figs., 21 tables; München (Bayer. Geol. Landesamt).
Meyer, R. & Schmidt-Kaler, H. (1983): Erdgeschichte sichtbar gemacht. Ein geologischer Führer durch die Altmühlalb. – 260 pp., 260 figs., 2 pls.; München (Bayer. Geol. Landesamt).
Mutschler, O. (1927): Die Gymnospermen des Weissen Jura von Nusplingen. – Jber. Mitt. oberrh. geol. Ver., **16:** 25–50, pl. I/1, 26 figs.; Stuttgart.
Ostrom, J. H. (1974): *Archaeopteryx* and the origin of flight. – Quart. Rev. Biol., **49:** 27–47, 10 figs.
Ostrom, J. H. (1979): Bird flight: How did it begin? – Am. Scient., **67,** 1: 46–56, 10 figs.
Peters, D. S. (1984): Konstruktionsmorphologische Gesichtspunkte zur Entstehung der Vögel. – Natur u. Museum, **114,** 7: 199–210, 8 figs.; Frankfurt a. M.
Peters, D. S. & Gutmann, W. F. (1976): Die Stellung des "Urvogels" *Archaeopteryx* im Ableitungsmodell der Vögel. – Natur u. Museum **106,** 9: 265–275, 2 figs.; Frankfurt a. M.
Regal, P. J. (1975): The evolutionary origin of feathers. – Quart. Rev. Biol., 50: 35–66; Baltimore.
Rietschel, S. (1976): *Archaeopteryx* – Tod und Einbettung. – Natur u. Museum, **106,** 9: 280–286, 9 figs.; Frankfurt a. M.

SALGER, M. (1959): Der Mineralbestand von Tonen des fränkischen Keuper und Jura. – Geologica Bavarica, **39:** 69–95, 5 figs.; München.
SCHÄFER, W. (1962): Aktuo-Paläontologie nach Studien in der Nordsee. – 666 pp., 36 pls., 277 figs.; Frankfurt a. M. (Waldemar Kramer).
SCHÄFER, W. (1976): Aktuopaläontologische Beobachtungen. 10. Zur Fossilisation von Vögeln. – Natur u. Museum, **106,** 9: 276–279, 7 figs.; Frankfurt a. M.
SCHAIRER G. & BARTHEL, K. W.: Die Cephalopoden des Korallenkalks aus dem Oberen Jura von Laisacker bei Neuburg a. d. Donau. V. *Torquatisphinctes, Subplanites, Katroliceras, Subdichotomoceras, Lithacoceras* (Ammonoidea, Perisphinctidae). – Mitt. Bayer. Staatsslg. Paläont. hist. Geol., **21:** 3–21, pl. 1–5, 4 figs.; München.
SCHMIDT-KALER, H. (1979): Geologische Karte des Naturparks Altmühltal, Südliche Frankenalb 1:100 000. – München (Bayer. Geol. Landesamt).
STARKE, R. (1976): Verteilung, Bildung und Umwandlung der Tonminerale in Sedimentgesteinen. – Schriftenr. geol. Wiss., **5:** 213–222; Berlin.
VAKHRAMEEV, V. A. (1962): Jurassic floras of the Indo-European and Siberian botanical-geographical regions. – Colloque du Jurassic, Luxembourg 1962: 411–421; Luxembourg.
VIOHL, G. (1983): Forschungsprojekt "Solnhofener Plattenkalke". – Archaeopteryx, **1983:** 3–23, 12 figs.; Eichstätt.
WALTHER, J. (1904): Die Fauna der Solnhofener Plattenkalke. Bionomisch betrachtet. – Festschrift zum siebzigsten Geburtstage von ERNST HAECKEL: 135–214, pl. VIII, 21 figs,; Jena (Gustav Fischer).
ZEISS, A. (1968): Untersuchungen zur Paläontologie der Cephalopoden des Unter-Tithon der Südlichen Frankenalb. – Abh. Bayer. Akad. Wiss., Mathe.-naturwiss. Kl., N. F. **132:** 190 pp., 27 pls., 17 figs., 6 tables; München.
ZIEGLER, C. A. (1982): Geological Atlas of Western and Central Europe. – 130 pp., 38 encl.; Shell.

Author's address: Dr. GÜNTER VIOHL, Jura-Museum, Willibaldsburg, 8078 Eichstätt, Federal Republic of Germany.

Paul H. de Buisonjé

Climatological Conditions During Deposition of the Solnhofen Limestones

Abstract

The Solnhofen Limestones are marine backreef sediments, deposited in a longitudinal basin, landlocked along its northwestern side and semi-separated from the open Tethys by discontinuous coral reefs along its eastern and southern side. Types of sediments, their origin and diagenesis are briefly discussed. Remarks are given on the taphocoenose and biostratinomy of macrofossils.

In the vertebrate taphocoenose three different types of preservation are discerned. The distribution of these types over the Solnhofen depositional area, together with peculiarities in sedimentology, taphocoenose and biostratinomy are explained as results of seasonal, midsummer coccolithophorid blooms.

Such yearly blooms, in their turn, have been largely dependent on certain paleogeographical and paleoclimatological conditions, causing upwelling of deeper water, bringing inorganic nutrients to the surface and producing a high fertility especially in the eastern parts of the Solnhofen backreef basin. A seawater circulation model is given for the Solnhofen basin.

Palaeoclimatological conditions making possible the coccolithophorid blooms, were essentially the same as those inferred from the paleoecology of fossils. They indicate an arid, warm tropical paleoclimate in the vicinity of the Solnhofen basin, with the possibility that less arid, semi-arid conditions prevailed more landward from the arid coastal belt along the northwestern side of the Solnhofen basin.

Zusammenfassung

Die Solnhofener Plattenkalke wurden als marine "Backreef"-Sedimente in einem langgestreckten Becken abgelagert, das im Nordwesten von Land begrenzt und entlang seines Ost- und Südrandes von der offenen Tethys durch einen unterbrochenen Korallenriff-Gürtel unvollständig abgetrennt war. Sedimenttypen, ihre Entstehung und Diagenese werden kurz diskutiert. Außerdem werden die Taphozönose und Biostratinomie von Makrofossilien erörtert.

In der Wirbeltiertaphozönose lassen sich drei verschiedene Erhaltungstypen unterscheiden. Die Verteilung der Typen über den Solnhofener Ablagerungsraum sowie Eigentümlichkeiten in Sedimentologie, Taphozönose und Biostratinomie werden als Resultate hochsommerlicher Coccolithophoriden-Seeblüte erklärt.

Solche jährlichen Seeblüten waren ihrerseits weitgehend abhängig von gewissen paläogeographischen und paläoklimatischen Bedingungen. Diese verursachten ein Aufsteigen von Tiefenwasser, welches anorganische Nährstoffe zur Oberfläche brachte und eine hohe organische Produktivität besonders im östlichen Teil des "Backreef"-Beckens bewirkte. Es wird ein Modell der Seewasser-Zirkulation für das Solnhofener Becken vorgestellt.

Die paläoklimatischen Bedingungen, welche Coccolithophoriden-Seeblüten ermöglichten, waren dieselben wie die, welche aus der Paläokologie der Fossilien erschlossen werden können. Sie zeigen ein arides, warmes tropisches Klima in der unmittelbaren Umgebung des Solnhofener Beckens an. Jedoch besteht die Möglichkeit, daß weniger trockene, semiaride Bedingungen landeinwärts des ariden Küstengürtels im Nordwesten des Solnhofener Beckens vorherrschten.

Introduction

The Solnhofen Limestones occur in a longitudinal, east-northeast to west-southwest belt, 80 km long and 30 km wide, roughly between Kelheim-Riedenburg at the eastern side and Solnhofen-Langenaltheim at the western side (Fig. 1).

Considered as a whole, the Solnhofen Limestone depositional area was a marine backreef basin of a roughly east-northeast to west-southwest, longitudinal shape. This backreef basin was subdivided into smaller, secondary basins. The basin was landlocked along its northwestern side by the Mid-European Landmass. (see also VIOHL/1985, this volume).

During most of its history, the Solnhofen basin was in free connection with the open Tethys, especially near the southwestern limit of the basin.

Sedimentology: types of sediments, their origin and diagenesis.

Sediments from the Upper Solnhofen Limestone, Malm zeta 2b, can be roughly divided into two different types: marly limestones with 10 to 20 % clay minerals and about 3 % quartz, the "Fäulen", and almost pure limestones with about 3 % clay minerals and about 0.4 % quartz, the "Flinze".
The marly limestones have a grain size optimum between 1 and 3 µm and have a high porosity, due to the presence of hollow spheres with an inner diameter of 5 to 15 µm, considered as fossil coccoid Cyanophycaea by KEUPP (1977). The marly limestones are finely laminated, each lamina between 0.05 and 0.20 mm thick. They show only initial stages in cementation and contain excellently preserved coccoliths on their bedding planes (KEUPP, 1977; BARTHEL, 1976).

The pure limestone beds with a thickness of individual beds between about 1 mm and several centimetres, are separated from each other by such marly lamina as described above. The pure limestones also have a grain size optimum between 1 and 3 µm, sometimes with a slight secondary optimum between 3.5 and 5 µm (KEUPP, 1977). An almost identical grain size curve for the Upper Solnhofen Limestone was given by FLÜGEL & FRANZ (1967) and from the number of coccoliths observed in electron microscope photographs), they calculated the presence of 500,000 coccoliths per cubic millimetre. The same authors noted that coccoliths in the limestones often show as relicts, difficult to identify as broken up coccoliths. Other authors too, mention that coccoliths in the pure limestones are often disaggregated and show effects of recrystallization (BARTHEL, 1976; KEUPP, 1977).

Our own scanning electron microscope photographs also reveal that within the pure limestones the coccoliths are virtually always present, but scarce. And in most cases they show signs of disaggregation into single micelles, overgrowth and recrystallization (Fig. 2). On the other hand, the recrystallization and cementation of the micritic limestones has not reached the final stages of the diagenetic model given by SCHLANGER & DOUGLAS (1974).

The pure limestones still have a certain permeability, due to the almost complete absence of clay minerals, together with the incomplete cementation. It is exactly this permeability which makes the pure limestones so useful for lithographic purposes, and on the other hand, such limestones can not be used as outdoor paving tiles: penetrating water and frost will split such tiles open!

Some remarks must be made here concerning the differences in sedimentation rates for the marly limestones and the pure limestones. The marly limestones were deposited at a much slower sedimentation rate than the pure limestones (BANZ, 1969; BARTHEL, 1970, 1972, 1976, 1978; DE BUISONJÉ, 1972; VAN STRAATEN, 1971; GOLDRING & SEILACHER, 1971). Although on different grounds, these authors consider the marly limestones as the "normal" sediments, deposited rather slowly, and consider the pure limestones as "abnormal" sediments, possibly deposited within a time span of less than one year, for instance within a few days, weeks or months. The abnormal character of the pure limestone beds, the "Flinze", shows most clearly in two sedimentological features: the effects of syneresis and the extremely high initial compaction factor.

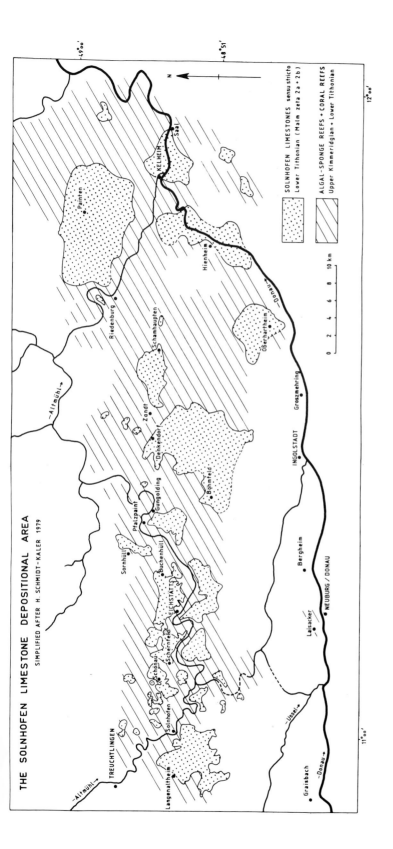

Fig. 1
The Solnhofen Limestone Depositional Area.
Individual basins with typical Solnhofen lithofacies on top of a pre-existing relief of dead algal-sponge reefs. Living coral reefs along the eastern side of the depositional area and also in the south, roughly following the present occurred course of the Donau. The depositional area was landlocked along the northwestern side.

Fig. 2
Scanning electron microscope photographs from the pure limestone beds.
Left: Langenaltheim   Right: Blumenberg near Eichstätt
In both photos the coccoliths show signs of disaggregation, dissolution, recrystallization, and overgrowth with newly formed calcite. Coccoliths are often the only recognizable microfossils in the pure limestones.

Syneresis is caused by expulsion of a solvent or a fluid from a gel or from a suspension (JANICKE, 1969). It shows at the upper bedding planes of the limestone beds as irregularly anastomozing, non-directionally arranged, sharply protruding rims. These rims encircle more flat, slightly sunken areas.

Such syneresis features are identical with the perhaps more familiar feature of a wrinkled film that developed on top of a quantity of paint left in a tin that was insufficiently closed after having been used.

Syneresis features are frequently found in the top surfaces of the individual limestone beds and indicate that each limestone bed originally was deposited rather suddenly as a homogeneous suspension or ooze, still containing a high amount of interstitial fluids, these fluids being expelled soon after the deposition of the suspension or ooze.

Another character, most typical for the Solnhofen Limestone, is the strong compaction which took place in two different phases. In normal sediments the compaction is a gradual process that proceeds with depth of burial. But for the Solnhofen Limestone it can be demonstrated that considerable compaction already took place immediately after deposition of one single limestone bed, before any overburden of later beds could exert compressional forces.

This initial compaction, directly related with the syneresis features at the upper bedding planes, must be sharply distinguished from the later compaction, the latter being mainly responsible for the well-known socles, pedestal phenomena under collapsible fossils.

Initial compaction is clearly demonstrated where fossils, such as ammonites or fish, are covered up by two or more successive but thin limestone layers. Macrofossils occur almost exclusively in the lower bedding planes, at the underside of the limestone beds. When for instance an ammonite came to rest in a horizontal position on the bottom, it was lying on a completely flat surface. Shortly afterwards, the ammonite was covered up by a micrite ooze. This ooze settled from a suspension, but essentially in a vertical direction: the lower umbilicus of the ammonite was not filled and even underneath most of the ammonite no sediment was deposited.

Fig. 3
Tharsis dubius (Blv.) Blumenberg near Eichstätt.
In this 3.5 mm thick limestone slab, the fossil fish is preserved slightly sunken into the lower bedding plane, not visible here. In the upper bedding plane, given in this photograph, the fish underneath only shows its contours. Note the zigzag fractures along the backbone and the star-shaped openings in the abdominal region, both filled with drusy calcite. The adipocire-pseudomorphose of this fisch collapsed only under considerable sediment cover.

Sln. 335, Coll. Jura Museum

In most cases the settling suspension or ooze completely covered up the ammonite. The ammonite was "snowed in" and did not show at the top surface of the ooze. But when the initial compaction started, the situation changed. Because less sediment was present above the ammonite and the ammonite was not pressed down into the already stiff and flat layer underneath, selective compaction resulted in faint ammonite outlines at the surface of the compacting ooze. When later, after an interval during which a thin lamina of marly composition was deposited, a second lime-ooze was deposited, the ammonite showed as a protruding boss at the top surface of the already compacted first ooze. This boss has a shape that vaguely follows the outlines of the ammonite underneath.

This second ooze in its turn possessed a thickness that was slightly less where the boss in the underlying bed occurred. In the same way, especially when relatively thin limestone beds occur on top of each other, the lower ones show a decrease in thickness above the ammonite underneath.

Much later in the compactional history, when several more layers were deposited above the ammonite and the beds surrounding the ammonite were of about the same consistency, the limestone beds below the ammonite were pressed upwards, the beds above the ammonite downwards. Together with collapse of the ammonite, often in a two-phase collapse, the now completely or nearly completely flattened ammonite came to rest upon a socle in the underlying bed and slightly sunken in the lower bedding plane of the overlying bed (SEILACHER et al., 1976).

Body chambers of ammonites in most cases never became filled with sediment. They possess only a small wedge of sediment in the aperture of the body chamber. Most likely this is a result of the fact that the body chamber still contained soft parts even up to the moment when the body chamber collapsed. Now all collapse features are non-geopetal, adjacent bedding planes above and below collapsible bodies are deflected towards the completely flattened fossil. This indicates that the collapse must have taken place under considerable sediment cover.

More fragile ammonite types show a two-phase collapse (SEILACHER et al., 1976). Here the body chamber collapsed first, mostly with deflection of the beds above and below, indicating that the beds still possessed a certain degree of plasticity. The phragmocone could withstand the compression longer and often became partly filled with drusy calcite. When at last the phragmocone collapsed too, the sediment was already cemented, was brittle and reacted by fracturing along microfaults.

In the case of fossil fish and also in completely flattened Macrura and Teuthoidea, the final collapse must have taken place at about the same level of burial, where in more fragile ammonites such as *Neochetoceras*, the body chamber collapsed. For in all such cases the collapse features are connected with deflection of the surrounding beds, rather than with fracturing in the shape of calderas with microfaulting.

That even fossil fish sometimes kept their three-dimensional shape and volume when already relatively deeply buried and collapsed only under considerable sediment cover, could be verified in a specimen of *Tharsis dubius* (BLV.) from Blumenberg (Fig. 3). Here a relatively thin, 3.5 mm thick limestone bed covers with a thickness of only 1.5 mm the completely flattened fish and shows fracturing in a zigzag line, roughly following the vertebral column. This fracture line and also some star-shaped openings above the abdominal region of the fossil fish, are filled with drusy calcite. This means that the fractures were kept open during compaction and final collapse, indicating that the overlying limestone bed was already rigid and reacted with fracturing.

That even fish under considerable overburden of several limestone beds still possessed their original three-dimensional shape is considered to be a result of a combination of factors. Relatively low temperatures of the stagnant bottom water was one of these factors. Another factor was that dead fish, lying on the bottom, shortly after death were sealed off by a swiftly settling and compacting micritic suspension. As a result anaerobic bacteria could transform fleshy parts of carcasses into relatively rigid and chemically stable adipocire. Macrofossils, such as fish, teuthoids, and crustaceans, where found with a socle underneath, all had passed through an intermediate stage of adipocire-pseudomorphose before they finally collapsed.

From fossils flattened by collapse and lying on a socle or pedestal, the minimum amount of total compaction can be calculated. For instance in the above mentioned case of *Tharsis dubius* (BLV.), the 2 millimeters difference between the thickness directly above the fossil and at some distance from the fish, must represent the original width of the fish, estimated at 30 mm. This gives a compaction factor of 15.

If an ammonite is cut vertically (Fig. 4), the limestone bed underneath the fossil is pressed upwards only, but keeps its thickness both outside the ammonite as well as below the fossil. But the limestone bed above the completely flattened ammonite shows strong variations. Some distance away from the ammonite, the bed has a constant thickness. But above the ammonite the thickness varies. Here the thickness is at a maximum in the umbilical region, but much less where the limestone covers the

Fig. 4
Neochetoceras steraspis (Oppel). Schernfeld near Eichstätt.
This completely flattened ammonite rests slightly sunken in the lower bedding plane of overlying bed **a** and rests on a socle in bed **b**. This underlying bed **b** is of constant thickness. The difference in thickness of bed **a** above the

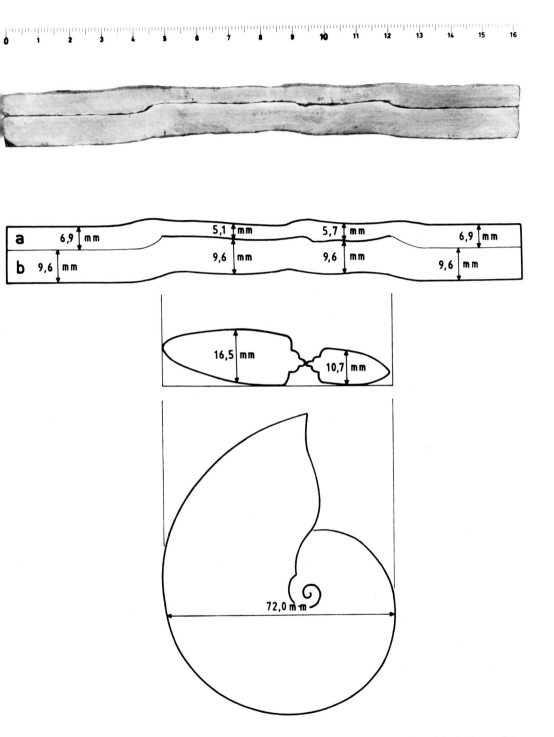

ammonite and outside the ammonite, multiplied with the compaction factor, represents the original volume of the ammonite. The 1.8 mm difference in sediment thickness left of the umbilicus and outside the ammonite, represents originally 16.5 mm. This gives a compaction factor of about 9.

Sln. 019, Coll. Jura Museum

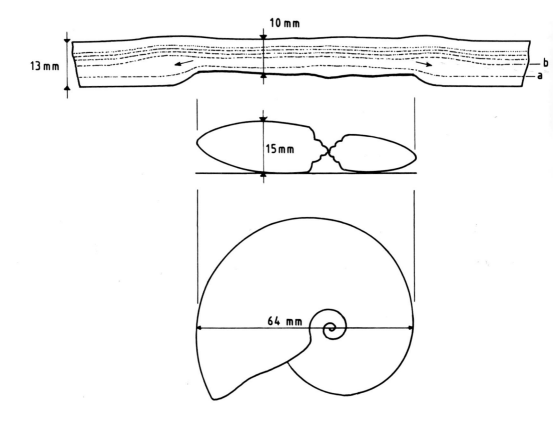

Fig. 5
Neochetoceras steraspis (Oppel). Wintershof near Eichstätt.
Here only the overlying limestone bed is given. This limestone bed shows several crypto-laminae. A thickened rim, a sulcus, was formed around the ammonite venter, due to lateral flow of sediment originally deposited on top of the ammonite. This lateral flow occurred when the upper parts of the micritic ooze already had lost much of their interstitial fluid content, but the lower parts, between lamina a and b were still highly mobile.   Sln. 138, Coll. Jura Museum.

originally swollen chamber cross sections. Using the dimensions and cross sections of non-compressed, normally preserved ammonites of the same species (BARTHEL & SCHAIRER, 1977), a simple calculation gives a compressional factor of 9. This means that the limestone bed, when just deposited, had a thickness of at least 9 times the present thickness of the bed as measured outside the fossil.

Such calculated factors always give a minimum value for the compaction. For in vertical cross sections of horizontally embedded and flattened ammonites, it can be demonstrated that some of the sediment, during the initial phase of the compaction, was pressed from its position on top of the ammonite to a position just outside the ammonite proper. Here the limestone shows a slightly thickened rim, the marginal sulcus (SEILACHER et al., 1976). This sulcus completely encircles the ammonite and is only lacking near the apertural plug where some sediment has been pressed into the distal part, the aperture of the body chamber.

This marginal sulcus, lacking near the aperture, was formed during the initial compaction of the still mobile ooze simultaneously with the plug in the body chamber. The formation of this sulcus is demonstrated in an ammonite from Wintershof (Fig. 5).

Here the limestone bed above a specimen of *Neochetoceras steraspis* (OPPEL) shows repeated crypto-lamination. Especially in the lower parts of this single limestone bed the crypto-lamination – irregularities of regional extent within one single lime-mud or ooze deposition –, shows bulging features just outside the ammonite venter. This indicates that during the initial compaction, the top surface of the compacting ooze already achieved a certain degree of thixotropy, whereas the lower parts of the ooze still were mobile. Here, between lamina a and b in Fig. 5, some of the sediment originally on top of the ammonite has been pressed sideways, down and lateral to the ammonite venter. Again we can draw a parallel with the insufficiently closed tin with some paint left in it: below the wrinkled film on top, the paint can still be used, is still in its liquid state!

That some sediment, originally deposited on top of the ammonite, has been pressed sideways – and this is nearly always the case –, makes the calculations of the total compaction factor always result in a minimum value. When in the above described specimen of *Neochetoceras* the lowest lamina (a) is used as an indicator of an originally flat top surface, a compaction factor of 15 is calculated. Using lamina b, just above the level with lateral bulging effects, a compaction factor of only 6.5 is found. And using the whole bed above the ammonite the compaction factor is even only 5.

The compaction factors mentioned above give the combined total compaction, the swift initial compaction together with the later more gradual compaction as a result of considerable overburden. It seems justified to suppose that most of the calculated compaction took place during the initial phase, immediately after deposition of one single micritic ooze. This is corroborated by the fact that syneresis features, results of expulsion of fluids from the ooze, already were present before the next suspension or ooze was deposited (DE BUISONJÉ, 1972). This initial loss of fluids in micritic oozes, close to the surface of the sediment, is a well known fact (GINSBURG, 1957).

From the compactional history of collapsible bodies in the Solnhofen Limestone follows that recrystallization and cementation took place rather late: reaction of the sediment with microfaulting around phragmocones indicates a certain degree of cementation. From its deposition as a suspension or ooze with a high content of interstitial fluids up to the time when cementation started, the limestones possessed a permeability allowing free movement of interstitial fluids. As there was also ample time to destroy the original components of the ooze, it is astonishing that the limestones still show some of their original coccolith content. Probably this is also due to the fact that coccoliths consist of biogenetically crystallized low-magnesium calcite, a stable form of calcite.

The permeability in the marly limestones has been strongly reduced, due to the relatively high content of clay minerals. As a result the marly limestones not only show much less recrystallization and cementation, but also a much better state of preservation of the nannofossils. Coccoid blue green algae formed the framework of the laminae in the marly limestones and are the only benthos that could live in the hypersaline bottom water (KEUPP, 1977).

Calcareous nannoplankton is concentrated on the bedding planes of the marly limestones and consists mainly of well-preserved coccoliths (BARTHEL, 1976, 1978; KEUPP, 1975, 1976, 1977).
The marly limestones, the „Fäulen", are the normal sediments. Their lamination, with thicknesses between 0.05 and 0.20 mm, reflects normal, yearly fluctuating sedimentation of nannoplankton, added to the benthonic blue green algae.

Yearly sedimentation rates as mentioned here for the Solnhofen marly limestones, are of the same magnitude as those for recent coccoliths. Calculated for a seawater depth of 100 metres, the normal yearly coccolith production adds between 0.057 and 0.228 mm of sediment (BERNARD & LECAL-SCHLAUDER, 1953). The same authors mention that coccoliths, in areas where clay minerals are deposited, are the main contributors to the sedimentation. They constitute between 20 % and 80 % of recent muds between 50 and 500 metres depth in the Mediterranean. Another important observation is, that recent coccoliths, even before they reach the bottom, may disaggregate into calcite rhombs or grains. On the bottom too, coccoliths are broken up into single micelles due to bacterial action.

Although the Solnhofen marly limestones, compared with the pure limestones, have a much higher percentage of insoluble residue, the ratios Illite: Montmorrilonite: Kaolinite are essentially the same in marly limestones and in the pure limestones. Even the quartz percentages, all composed of eolian transported grains (BARTHEL, 1976), vary with the insoluble residue (HÜCKEL, 1974).

From the relatively low percentage of land-derived insoluble residue in the pure limestones, the conclusion can be drawn that these limestones have been deposited in a shorter time span than marly limestones of the same thickness. But with this statement an explanatory difficulty arises. For on the one hand there are several observations, indicating that each limestone bed has been deposited in a short time interval, on the other hand pure limestone beds are much thicker than the yearly laminae of the marly limestones. As a result, one single limestone bed of 2 cm thickness, would contain about 40 times more land-derived clay and quartz than one single marly lamina of 0.1 mm thickness.

An explanation is found when we take into consideration that the total Solnhofen depositional area was subdivided into several smaller, secondary basins separated from each other by dead algal-sponge reefs. The original suspensions from which the limestone beds condensed, were capable of flow, to settle sideways away from the dead algal-sponge reefs and mounds. As a result single limestone beds within one of the smaller basins all wedge out in the direction of the dead algal-sponge reefs. But when we pass such reefs or mounds and enter the next basin, the same limestone beds can be traced back again.

This means that each limestone bed resulted from an event of regional character, occurring somewhere at a much shallower depth, and of an extent much larger than the size of the smaller, individual basins of the bottom relief. Other authors also mentioned that during the deposition of the pure limestone beds they originally were suspensions that flowed down by their own weight, as "turbidity currents of mostly rather low density and small velocity" (VAN STRAATEN, 1971), or with "non-directional turbulence that might have developed in the final phase of a density current" (SEILACHER et al., 1976).

It is this slow basin-ward flow of the original suspensions that was responsible for the tumbling into a horizontal position of macrofossils which stood vertically on the bottom before the micritic suspensions came down.

Due to this syn-sedimentary flow of micritic suspensions in the direction of basinal centres, the present limestone beds show exaggerated thicknesses. Again the limestone beds are the abnormal sediments!

Although recognizable coccoliths within the pure limestones are scarce, they show over and again under the electron microscope, but always with signs of disaggregation, recrystallization, overgrowth and cementation.

When the numbers and sizes of the single micelles of the two most common coccolithophorids, *Ellipsagelosphaera* and *Cyclagelosphaera*, are put into a theoretical grain size curve, this curve corresponds exactly with the grain size curve for the Solnhofen Limestone (KEUPP, 1977).

Disaggregation of coccoliths most probably took place during the flow of the original suspensions in the direction of the basinal centres, or during the settling immediately afterwards, the initial phase in the compaction. During the compaction, from the swift initial compaction up to the final cementation, the pure limestones were in the realm of the so-called "high diagenetic potential" (SCHLANGER & DOUGLAS, 1974) These authors define the diagenetic potential of a sediment as the length of the diagenetic pathway the sediment has left to traverse before it becomes a crystalline aggregate.

Because the Upper Solnhofen Limestones never became deeply buried in later Cretaceous or Tertiary times, the Solnhofen Limestones only came in the Shallow-burial Realm V of SCHLANGER & DOUGLAS. In this realm, oozes establish firm grain contacts, dissolution of fossils starts, together with initiation of overgrowths. Compared with the pure limestones of Solnhofen, the marly limestones possessed a much lower diagenetic potential. As a result the pure limestones, when compared with the marly limestones, show stronger effects of dissolution, stronger recrystallization, overgrowth and cementation.

Owing to these differences in diagenetic potential, coccoliths are much better preserved in the marly limestones but largely dissolved and overgrown in the pure limestones.

It is not necessary to accept the explanation given by KEUPP (1977) for the origin of the limestones. KEUPP is of the opinion that, although the relatively large spheres of coccoid blue green algae are virtually absent in the pure limestones, these limestones nevertheless originated from disaggregated blue green algae, such as found in the marly limestones.

Taking into account the peculiar depositional and compactional history of the pure limestones, their grain size curve exactly matching the theoretical curve for disaggregated coccoliths and moreover coccoliths far outnumbering all other recognizable microfossils, I consider the pure limestones as the result of large scaled phytoplankton blooms in which coccolithophorids played a rôle.

Coccolithophorid blooms are not only known from recent observations, but also in their fossil state. Recent coccolithophorid blooms are recorded for the North Sea (BRAARUD et al., 1953) and for the Atlantic near Senegal (BERNARD, 1949). From the Kimmeridgian of Northwestern Europe several coccolith-rich bands, used as correlation levels in oil drillings and mostly composed of one species only, are mentioned as fossil blooms (GALLOIS, 1976).

Because recognizable coccoliths in the pure limestones constitute only a few promille of the volume, one could suppose that coccolithophorids were the victims of blooms, caused by phytoplankton that was not capable of fossilization. But accepting coccolithophorids as victims only, the grain size curve for the limestones, matching the theoretical curve for disaggregated coccoliths, becomes inexplicable. Because coccoliths within the limestones always show signs of disaggregation, overgrowth and recrystallization, only one conclusion is possible: the pure limestone beds of the Solnhofen Limestone represent short-lived outbursts of exceptionally high organic productivity, here as coccolithophorid blooms, in surface waters of a large backreef basin, semi-separated from open sea by discontinuous coral reef barriers.

Paleontology: remarks on taphocoenose and biostratinomy of macrofossils

It is not necessary to discuss here in detail the taxa represented in the taphocoenose. Excellent reviews are given by several authors (VON GÜMBEL, 1891; WALTHER, 1904; MALZ, 1976; KUHN, 1977; BARTHEL, 1978) and innumerable, often more specific short notes and monographs were published. Only some general remarks will be given here.

1. Both in the marly limestones as well as in the pure limestones any evidence of life activities of epi- or endobenthos are absent. Lamination of the sediments is never disturbed by bioturbation. For the extremely seldomly found endobiont *Eunicites* it can be shown that they were already dead when they were transported into the sedimentation area.

Epibionts are slightly less scarce as fossils. But even for types like *Geocoma, Comatula* and *Ophiopsammus* it is clear that they did not live where they are found now. Life activities are absent and from the position of their arms, mostly lying all in the same direction, it follows that they were transported when already dead.

Conclusion: the bottom of the Solnhofen Limestone depositional area was constantly an azoic realm for epi- and endobenthos. Only blue green algae and certain bacteria could flourish here, at least during the sedimentation of the marly limestones.

2. Macrofossils almost exclusively belong to free swimming or drifting marine types. They are mostly found at the lower bedding planes of the pure limestones. Death in macro-organisms, their arrival at the bottom and the onset of the micritic sedimentation of the pure limestone beds are intimately connected with each other. As will be shown later this is a causal relationship.

Preservational type A

In the pure limestones the majority of the fossils are found completely flattened and lying on top of a socle in the underlying bed, slightly sunken into the lower bedding plane of the overlying limestone bed. Such fossils are always preserved with all their skeletal elements, for instance scales of fish, lying in natural position. Signs of any prolonged post-mortem transport are absent.

Mass mortalities, for instance with tens or even hundreds of small fish like *Leptolepides* on a bedding plane of one square metre, are without exception of this preservational type.

That fish of this preservational type have been killed rather suddenly is obvious in specimens of *Caturus* which often contain in their intestinal region the undigested smaller fish they had swallowed (MAYR, 1967; VIOHL, 1983).

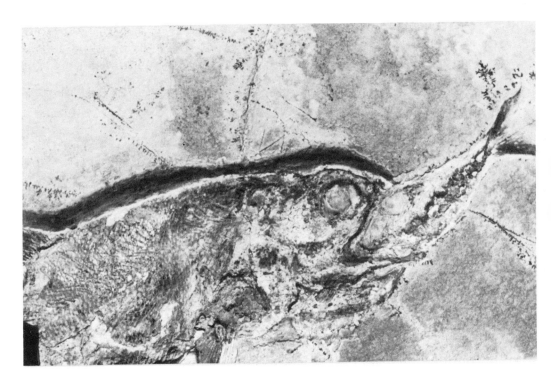

Fig. 6
Caturus sp. with its prey halfway in its mouth. Museum Bergér.
Here an already deadly poisoned fish-of-prey still managed to seize another, smaller fish and then sank with its victim to the bottom.
Photo with permission of the director, Mr. F. Bergér.

Instructive cases of sudden death, most probably caused by poisoning, are the exceptional findings of *Caturus* with its prey halfway in its mouth (Fig. 6).

Especially in the central and western parts of the Solnhofen Limestone area, between Pfalzpaint-Eichstätt-Solnhofen, mass mortalities are recorded in *Saccocoma, Leptolepides* and *Rhizostomites*. Animals with low swimming speeds and animals that merely drifted were preferentially killed by blooms. This means that for instance, fish of small sizes, whether mature or immature larger species, are over-represented in the taphocoenose.

Fossil fish are mostly found in stable position, compressed laterally. But not infrequently they have a dorsally bent shape with the caudal fin rotated or broken free from the rest of the body (Fig. 7). Such dorsally bent fish, shortly after death, possessed some buoyancy in the abdominal cavity. They stood in reversed position with the belly directed upwards, "hanging" dorsally bent and with only the dorsal tip of the caudal fin and the head touching the bottom. At the onset of the sedimentation of the micritic suspension they came to rest in a horizontal position, now resting in stable position on their side, but without any further transport.

From the moment they were killed up to the moment they came to lie in a horizontal position, such fish were surrounded by water poor in oxygen or oxygen-free. For otherwise an increasing gas pressure in the abdominal cavity would have caused dead fish to rise to the surface.

A more or less identical biostratinomy is shown when ammonites are found associated with an impression of the ventral side of the body chamber for when teuthoids are preserved with impressions of the tentacle ring nearby.

In both cases the fossil itself is lying horizontally embedded and close to the impressions made. Here again carcasses were kept in an upright position during a certain period, due to buoyancy. Only later, when this buoyancy gradually decreased, or when oozes started to settle from the suspensions and gently pushed the carcasses into a horizontal position, they came to rest in the bedding plane. I prefer the last mentioned explanation because the imprints made by the dead animals standing upright, mostly do not show any asymmetry in the direction of the fossil.

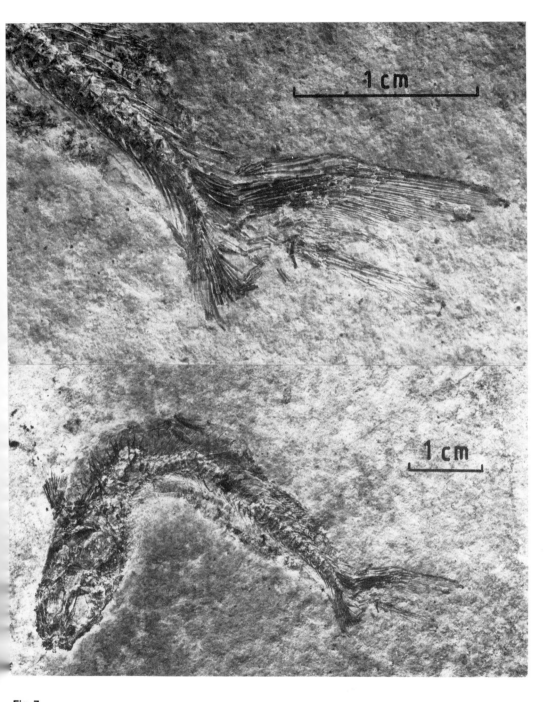

Fig. 7
Thrissops sp. Wintershof near Eichstätt
Due to buoyancy in the abdominal cavity, this dead fish stood vertically, with its ventral side upwards, dorsally bent on the bottom. Only the dorsal tip of the caudal fin and the tip of the head touched the bottom.
When shortly afterwards a micritic suspension started to settle, the fish was gently pushed into a horizontal position.
Sln. 054, Coll. Jura Museum.

A remark must be made here concerning the famous trace fossils, made by *Mecochirus* and *Mesolimulus*, often with the body fossil at the end of the track. An explanation which considers the body fossils of *Mesolimulus* as exuviae and the tracks as normal life activities during molting (ZEISS, 1975), is inadmissable because all other life activities, such as foraging trails of *Mesolimulus* are never found.

Because all indications of other benthonic life are absent, the explanation of BARTHEL (1970, 1974) that *Mesolimulus* specimens are true body fossils of animals, trapped in a toxic enviroment, and stopped dead at the end of the tracks, is highly acceptable. BARTHEL moreover mentions that the tracks, often more or less in circles, indicate that disorientation of the dying specimens of *Mesolimulus* took place. Again death of macrofossils, caused by poisoning, seems a plausible explanation.

Now living recent limulids possess a high resistance to fluctuations in salinity, oxygen content, and temperature. From the undertracks mentioned by GOLDRING & SEILACHER (1971) it became evident that *Mesolimulus*, in several cases found its death when the settling of the micritic ooze already had started. Often some sediment has to be scraped away from the lower bedding plane of a limestone slab to reveal the fossil.

This gives us some information concerning the time that elapsed between death in fossil fish and their being covered up by micritic ooze afterwards. For fish, resting dead on the bottom, were always covered up only afterwards. But the extremely resistant *Mesolimulus* could survive somewhat longer and became embedded in the settling ooze itself. This time elapse, probably some days, could also afford for the time necessary to achieve the dorsally bent shape in dead fish, resting free on the bottom just previous to the settling of the slowly sinking micritic ooze.

It is plausible that animals with low swimming speeds, such as the jellyfish *Rhizostomites*, or animals that drifted like *Saccocoma*, were especially liable to be killed when blooms occurred.

In the upper sections of the Malm zeta 2 b, between Eichstätt and Schönau, several limestone beds show lower bedding planes, crowded with the remains of *Saccocoma*. Often such mass mortalities are composed of juvenile specimens only. Not improbably this feature is connected with a one-year life cycle in *Saccocoma*. Seasonal, midsummer blooms would have killed only immature specimens.

Rhizostomites is found not infrequently in the Eichstätt-Schernfeld area. Here they occur in the lower bedding planes of the limestone beds, but never with the socle formation underneath. Clearly the animal tissues of these jellyfish could not be transformed into adipocire!

In the Pfalzpaint-Gungolding area some limestone beds contained innumerable excellently preserved specimens of *Rhizostomites*. Here they were found not at the lower bedding planes, but within the limestone beds, lying in the crypto-lamination of such beds. Now the crypto-laminations are the results of short-lived interruptions or stops in one single seasonal bloom. It is evident that jellyfish, picked up within one single ooze, settling from suspension, possessed much better preservational possibilities than dead jellyfish resting at least for some days free on the bottom. The specimens from Pfalzpaint-Gungolding often show dehydration wrinkles, indicating that during the settling of the ooze, less concentrated fluids were extracted from the animal tissue (BARTHEL, 1970). Such dehydration wrinkles are not the results of compaction, but are due to osmosis, probably caused by a relatively high salinity of the settling ooze. In any case the excellent preservation of jellyfish, especially when embedded within one single limestone bed, indicates a rapid deposition of the original suspension or ooze.

Preservational type B

Fossils with this type of preservation are found, just as fossils with type A, in the lower bedding planes of the limestone beds. But here they are found with their skeletal elements lying scattered around. In the case of fossil fish for instance, all the scales, even the tiniest ones are present, but lying disarticulated from each other. In fossils with a preservation in type B, a prolonged period of decay took away all the soft tissues prior to sedimentation of a micritic layer. For a prolonged period they were lying free at the sediment surface. Most probably they were covered by sediment one or more years after death. Fossils with this type B preservation are moreover never associated with the formation of socles underneath.

Scavengers were absent in the time interval and bacteria had ample time to remove animal tissues, except the bony parts. Strong currents evidently were absent and only extremely weak currents were capable of scattering the skeletal elements such as scales.

To this type of preservation also belong the rather exceptional cases of so-called "life activities immediately prior to death" (excluding the tracks of *Mecochirus* and *Mesolimulus* with body fossil at the end). Such "life activities" are mentioned in older literature, for instance in the case of a fossil fish which was seen as "stranded" on a mud flat, but prior to death still "flapped around with its tail" (MAYR, 1964). In reality such "life activities" must be explained as imprints made by the body of dead fish, slowly moving around the point where they stuck in the viscous layer underneath (BARTHEL, 1966). In such cases the fossil itself is always preserved in type B.

Preservation in type B is considered to be in a causal relation with the rather exceptional cases where individual limestone beds wedge out between limestone layers above and below.

Preservational type C

It is only in fossils of this preservational type that we are dealing with incomplete skeletons. For instance in fossil fish of this type, only the bones of the head, the pectoral fins, the backbone with the ribs and the tail are preserved. But all other elements, such as scales, dorsal fin and pelvic girdle with the fins are missing.

Such fossils are considered to represent carcasses that drifted for some time near the surface. Only when in a state of advanced disarticulation and decay, did they come to rest on the bottom of the basins. From that moment on, in the absence of scavengers and strong currents, such incomplete carcasses escaped further disarticulation. It is obvious that fossils of type C preservation are never associated with socles underneath.

Fossil fish with this incomplete preservation are not infrequently found in the Solnhofen-Langenaltheim basin. In more easterly basins, such as near Eichstätt, preservation of type C is much less common.

This unequal distribution of preservational types over the total Solnhofen Limestone area, with prevalence of type A in the easterly and central basins and prevalence of type C in the western parts, indicates that near-surface currents in the backreef basin were directed from east to west. This net effect of wind-driven surface currents from east to west is even visible in our admittedly small series of *Archaeopteryx* specimens (BARTHEL, 1970; RIETSCHEL, 1976), More completely preserved specimens were found between Kelheim and Eichstätt, whereas incomplete skeletons of *Archaeopteryx* are all from the Solnhofen-Langenaltheim area.

Other animals that certainly drifted for some days after death were the Pterodactyloidea. WELLNHOFER (1970) is of the opinion that they drifted into the sedimentation area as mummified carcasses which got their awkward backward twist of the neck when they dried, lying dead on a shore line.

In my opinion the same effects may be the result when pterodactyls, killed directly by a bloom or otherwise killed by eating poisoned fish or poisoned invertebrates, drifted for some days before they sank to the bottom. Drifting at the surface, some buoyancy in the chest region kept them afloat, the head and neck hanging dorsally bent. Essentially in the same position they came to rest on the bottom when they finally sank.

An identical drifting position is also mentioned for the dead *Archaeopteryx* specimens (RIETSCHEL, 1976).

In Pterodactyloidea too, preservational type C (corresponding with types 2 a, 2 b and 3 of WELLNHOFER, 1970) again prevails in the Solnhofen-Langenaltheim area, whereas type A prevails in more easterly areas. Again a westerly drift of carcasses floating at the surface is expressed in the distribution of preservational types (WELLNHOFER, 1970 and WELLNHOFER in BARTHEL, 1970, p. 11).

3. Macrofossils of terrestrial habitat, especially *Archaeopteryx*, but also the Pterodactyloidea and Rhamphorhynchoidea, rendered the Solnhofen Limestones their worldwide fame. They need not be discussed here again.

The Solnhofen Limestones are also renowned for the excellently preserved terrestrial insects, of which about 180 species are known.

Fossil insects formerly were seen as evidence supporting the lagoonal theory, with temporary emergence of mud flats (MAYR, 1967). Nowadays the fossil insects are considered to have settled on the seabottom after having floated for some days at the surface (BARTHEL, 1964, 1970).

Most insects, just like macrofossils from marine habitat, came to rest on the bottom and were shortly afterwards covered by micritic ooze. They often display in an excellent state of preservation, type A, and are found again at the lower bedding plane of overlying limestone beds.

But not infrequently they are only vaguely visible at this lower bedding plane because they are lying a few millimetres above this plane, just inside the pure limestone. In such cases some sediment has to be scraped away to exibit the insect in all its details. Here again we have an indication, not only that blooms were disastrous also for animals of terrestrial habitat, but also that it took some days before micritic suspensions settled near the bottom of the basins.

In such cases, dead insects floated for some time at the surface and their sinking speed was slow. As a result they did not reach the bottom before the ooze settled, but during the settling of the micritic suspension.

Fossil insects are especially found in the basins near Eichstätt. Taking into account their drifting time necessary to let them sink, it seems that they had their habitat in more easterly parts of the Solnhofen depositional area.

Fossil insects near Eichstätt are mostly preserved in type A. In the Solnhofen-Langenaltheim area they are less frequently found and it can not be established whether preservational type C, which is rather seldomly seen, prevails in more westerly parts of the Solnhofen Limestone area.

On the other hand, the presence of well preserved fossil insects indicates that scavengers were absent, at least during blooms, from the surface of the seawater down to the bottom of the basins. For in most cases they reached the bottom intact.

Seawater blooms and the paleogeographical and paleoclimatological setting of the Solnhofen Limestone depositional area.

Recent seawater blooms occur only in certain regions of the oceans where upwelling of seawater rich in inorganic nutrients occurs, in combination with a relatively high temperature, at least during part of the year. The driving force of upwelling is a wind of constant direction. Such winds are the trade winds. Upwelling usually occurs along the west coasts of the continents, but similar phenomena, the so-called wind-driven estuarine-like circulation, occur also in semi-enclosed marine basins which have their long axes more or less parallel to the direction of the wind, and the opening downwind (BRONGERSMA-SANDERS, 1965, 1966, 1969; BRONGERSMA-SANDERS & GROEN, 1970).

An example of a basin with such an estuarine-like circulation is the 70 km long Gulf of Cariaco on the north coast of Venezuela. Flow away of the water from the inner part of the basin results in a shortage of water at the inner end of the basin. In order to replenish this shortage, deeper, subsurface water flows in. This leads to a high accumulation of inorganic nutrients and a high productivity (BRONGERSMA-SANDERS, 1965, 1966, 1969). In this Gulf of Cariaco, mortality occurs every summer and small fish die as a result of the blooms (BRONGERSMA-SANDERS, 1957).

In the Solnhofen Limestone backreef basin, the situation has been almost identical with the circumstances in the Gulf of Cariaco. Here too, a longitudinal basin existed, landlocked along the northwestern side and semi-separated from the open ocean by a coral reef barrier in the east and a discontinuous barrier along the south side.

From the distribution of preservational types, an easterly paleo-wind is deduced. An easterly wind direction is also concluded from northeast to southwest trending spurs and grooves in transitional stages between the Upper-Kimmeridgian algal-sponge mounds and the Lower-Tithonian coral reefs in the surroundings of Saal near Kelheim (BARTHEL, 1977). In a quarry north of Painten, also in the eastern part of the Solnhofen area, an independent proof of near-surface paleo-currents from east to west was ascertained. Here north-south ripple marks and roll-marks of ammonites, transported from east to west, both indicating near surface water movements from east to west (BARTHEL, 1964). Identical current directions from east to west were also deduced from asymmetrical ripple marks in a quarry near Gungolding (JANICKE, 1969).

In the backreef basin, west of the coral reef barrier near Kelheim-Riedenburg, ripple marks and roll-marks are much less common. Here, in the deeper parts of the backreef basin, well below wave base, the more typical Solnhofen lithofacies could develop. The deepest part of the backreef basin was situated near the western or southwestern limit of the longitudinal basin, near the places of Solnhofen and Langenaltheim. In this setting the basin was in open connection with the sea to the south and it was here, that the Solnhofen Limestones developed their maximum thickness of about 90 metres (Upper Solnhofen Limestone 60 meters).

Fig. 8 ▶
All year long, bottom waters in the Solnhofen backreef basin were largely stagnant and were lethal for macro-organisms. Under normal conditions, with a countercurrent bringing nutrients to the surface, a normal phytoplankton production could sustain a macrofauna of free swimming and drifting types in near-surface waters. But during the summer months the relatively cool water, upwelling in the eastern parts of the basin, was more strongly warmed and an excessive overproduction of phytoplankton, here with coccolithophorid blooms, poisoned the total water mass in the backreef basin. Macro-organisms, killed by such blooms, swiftly sank to the bottom and were shortly afterward covered by slowly sinking masses of dead coccolithophorids.

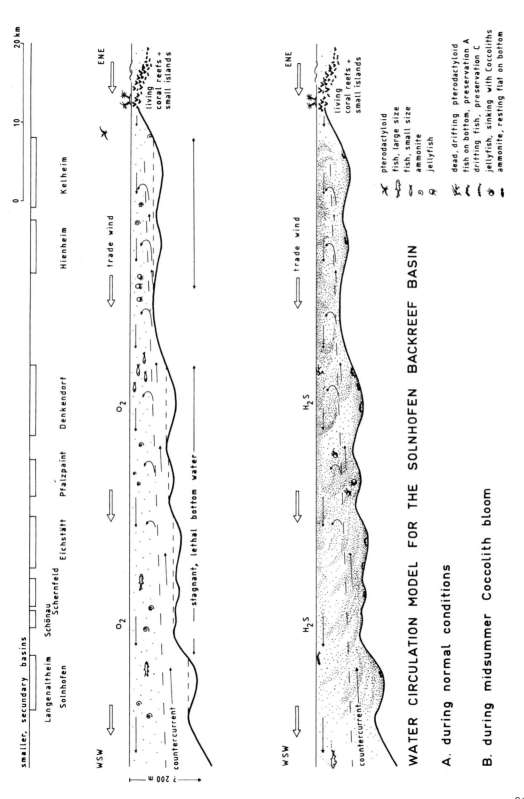

From this southwestern limit of the basin a countercurrent, bringing in deep and relatively cool ocean water, replenished the wind-driven surface water flowing out of the basin. This estuarine-like countercurrent system, with the easterly wind as the driving force, was favourable for the development of midsummer seawater blooms (Fig. 8).

The relatively cool and deep water overturned in the eastern parts of the backreef basin and brought inorganic nutrients, such as phosphates and nitrates, to the surface.

On its way out of the basin, this nutrient-rich surface water was warmed and could support high organic productivity. But especially during the summer months this high productivity could change into a seawater bloom.

At the onset of such midsummer coccolithophorid blooms, often occurring annually, nekton was killed and swiftly sank to the bottom. Shortly afterwards, the carcasses, lying on the bottom, were covered by slowly sinking suspensions, preferentially settling in the smaller, secondary basins of the pre-existing bottom relief. These suspensions soon condensed into oozes and sealed off the carcasses underneath from any aerobic decay.

But even after the micritic ooze was settled, the smaller secondary basins were lethal for all macro-organisms. Water was largely stagnant, probably of a high salinity, relatively cool and with an oxygen deficiency due to the decay of plankton. Within these smaller basins only very weak currents occurred now and again in the time interval between successive blooms. During such time intervals between blooms, one or more thin laminae of marly composition developed. These laminae represent the normal sedimentation in the backreef basin with normal organic productivity in near-surface waters.

From paleocontinental maps (SMITH et al., 1981; BERGGREN & HOLLISTER, 1974) it follows that during the Tithonian the Solnhofen area was situated more to the south at about 30° northern latitude. This means that the Solnhofen area was lying within, or – because polar icecaps were absent –, even south of the subtropical high pressure belt. Here rainfall was at a minimum and near-surface air masses were flowing to the southwest in the direction of the equator. This corresponds to our present-day's trade winds, in the northern hemisphere blowing all year long from the northeast or east.

Such a steady northeastern or eastern trade wind is considered to have been the driving force in the estuarine-like countercurrent system of the Solnhofen backreef basin.

Upwelling in the Solnhofen Limestone backreef basin possibly had an important climatological effect too. At the present-day, the hinterland of all regions of upwelling is comparatively dry and the aridity is extreme if the upwelling is strong and continuous.

This aridity is a result of a temperature inversion, precluding any precipitation in the coastal area (BRONGERSMA-SANDERS, 1969, 1971).

For the Solnhofen area this means that an almost rainless strip existed along the northwestern coast, comparable with the present-day's Namib desert along the coast of Southwest Africa.

Aridity in the hinterland is also indicated by fossil landplants, occasionally found in the Solnhofen Limestones or in coeval reef deposits. *Cycadopteris jurensis* (KURR), a type of seed-fern, possessed tough, leathery foliage with a heavy cuticular coating and overhanging, thickened rims of the leaves. The stomate cells were exclusively situated at the underside of the leaves, indicating extreme aridity (MEYER, 1974).

Another fossil plant, *Brachyphyllum nepos* SAPORTA, an Araucariacae belonging to the Coniferophyta, possessed rhombic, tough leathery leaves, arranged in spirals along the stems. Again the structure of the stomata is considered to be an adaptation to an arid climate (MUTSCHLER, 1927). Moreover, from *Brachyphyllum*, based on studies of cross-sections of the stems, it is known that this shrub-like Coniferophyte lived under hypersaline conditions, probably close to a coastline with salt marshes.

Other landplants, such as *Zamites moreaui* (BROGNIART) and *Sphenozamites latifolius* (BROGNIART), both belonging to the Bennettitales, are typical floral elements of the arid Indo-European region, the southern type flora of the Upper-Jurassic (VAKHRAMEEV, 1962).

From *Ginkgo* sp., recorded from the Solnhofen Limestones, it is still not certain whether this is not a marine algae instead of a landplant (BARTHEL, 1978). According to VAKHRAMEEV (1962) Ginkgophyta are infrequent in the arid Indo-European region, but predominate in the more humid, moderately warm climate of the Upper-Jurassic from the Siberian flora region. If the *"Ginkgo"* leaves of the Solnhofen Limestones are real Ginkgophyta, their occurrence could indicate that less arid circumstances possibly existed more landward from the extremely arid coastal belt postulated along the northwestern border of the Solnhofen backreef basin.

Another possibility here envisaged, is that less arid, semi-arid climatological conditions prevailed in mountain slopes far above sealevel. For instance in the mid-European landmass to the north, or in the Bohemian landmass to the east, ascending air masses could have caused semi-arid conditions. Such local, less arid conditions possibly even with small, temporarily rivulets and fresh-water pools, but well outside the vicinity of the Solnhofen backreef basin, are highly probable when we take into account the relatively rich and diversified insect fauna of the Solnhofen Limestones.

From the large size of the fossil insects, the conclusion has been drawn that they lived under warm, tropical circumstances. Their size often exceeds the size in modern, tropical insects (HANDLIRSCH, 1906–1908).

The diversification in the insect taphocoenose, indicating a variety of habitats and different ecological niches, stands in sharp contrast with the relatively poor, less diversified land flora of which the elements occur only as exceptional findings in the limestones.

This discrepancy between the richness and diversification in fossil insects, compared with the land flora, again can be explained when we accept the presence of an almost rainless zone between the Solnhofen backreef basin and landmasses further landward. Such an almost rainless belt with only a sparce vegetation, only incidentally would have supplied landplants into the marine backreef basin. But insects in most cases were capable of crossing such a zone of aridity.

That aridity prevailed in the direct surroundings of the Solnhofen basin may also be deduced from the rather frequently occurring dolomites in the coeval coral reefs. According to McKEE (1963) primary dolomites may be significant indicators of an arid paleoclimate.

If we accept that the autecology of the fossil corals in the reefs near the Solnhofen basin was identical with the autecology of modern reef corals, some conclusions may be drawn. Modern hermatypic corals cannot withstand strong departures from the normal marine salinity. Influx of fresh-water into the Solnhofen backreef basin was surely almost negligible.

The autecology of modern hermatypic corals also offers some information on the temperature near the seawater surface. For the coral reefs surrounding the Solnhofen basin, surface temperatures must have been relatively high, estimated between 26° and 28° C, with moreover only minor annual and daily variations.

Paleotemperatures on land in the vicinity of the Solnhofen backreef basin must have been relatively high also and with only minor annual and daily variation. Paleotemperatures on land are estimated at about 28° to 30° C for the mean annual temperature with an annual and daily variation of less than 5° C.

Together with the aridity, this means that the paleoclimate in the vicinity of the Solnhofen backreef basin was comparable with the arid, warm tropical desert climate of the KÖPPEN classification (BWh). But most probably this extremely arid paleoclimate was only present in a relatively narrow coastal belt and a less arid, semi-arid steppe climate (BS of the KÖPPEN classification) was present further landward of the arid coastal zone.

References Cited

BANZ, H.-U. (1969): Echinoidea aus Plattenkalken der Altmühlalb und ihre Biostratinomie. – Erlanger geol. Abh., **78**, 35 pp.

BARTHEL, K. W. (1964): Zur Entstehung der Solnhofener Plattenkalke (unteres Untertithon). – Mitt. Bayer. Staatssamml. Paläont. hist. Geol., **4**; 37–69.

BARTHEL, K. W. (1966): Concentric marks: Current indicators. – J. Sed. Petr., **36**: 1156–1162.

BARTHEL, K. W. (1970): On the deposition of the Solnhofen lithographic limestone (Lower Tithonian, Bavaria, Germany). – N. Jb. Geol. Paläont. Abh., **135**, 1: 1–18.

BARTHEL, K. W. (1972): The genesis of the Solnhofen lithographic limestone (Low. Tithonian): Further data and comments. – N. Jb. Geol. Paläont. Mh., **3:** 133–145.

BARTHEL, K. W. (1974): *Limulus*: a living fossil. Horseshoe crabs aid interpretation of an Upper Jurassic environment (Solnhofen). – Naturwissenschaften, **61:** 428–433.

BARTHEL, K. W. (1976): Coccolithen, Flugstaub und Gehalt an organischen Substanzen in Oberjura-Plattenkalken Bayerns und SE-Frankreichs. – Eclogae geol. Helv., **69,** 3: 627–639.

BARTHEL, K. W. (1977): A spur and groove system in Upper Jurassic coral reefs of Southern Germany. – Proc. 3. int. Coral Reef Symp. Univ. Miami, Florida, USA., 201–208.

BARTHEL, K. W. (1978): Solnhofen. – Ein Blick in die Erdgeschichte – 393 pp.; Thun (Ott Verlag).

BARTHEL, K. W. & SCHAIRER, G. (1977): Die Cephalopoden des Korallenkalks aus dem Oberen Jura von Laisacker bei Neuburg a. d. Donau. – Mitt. Bayer. Staatssamml. Paläont. hist. Geol., **17:** 103–113.

BERGGREN, W. A. & HOLLISTER, C. D., (1974): Paleogeography, Paleobiogeography and the History of Circulation in the Atlantic Ocean. in: Studies in Paleo-oceanography. – Soc. of Econ. Paleont. and Mineral., Spec. Publ., **20:** 126–186.

BERNARD, F. (1949): Remarques sur la biologie du Coccolithus fragilis Lohm., flagellé calcaire dominant du plancton mediterranéen. – Extr. travaux botaniques dédiés à R. Maire. Alger. Mém. Hors Sér. Soc. Hist. Nat. Afr. N., **2:** 21–28.

BERNARD, F. & LECAL-SCHLAUDER, J. (1953): Rôle des flagellés calcaires dans la sédimentation actuelle en Méditerranée. – Congr. Géol. Int. C. R., sect. 4: 11–24, Alger.

BRAARUD, T., GAARDER, K. R., & GRÖNTVELD, J. (1953): The phytoplankton of the North Sea and adjacent waters in May 1948. – Rapp. Proc. Verb. du Conseil, **133:** 1–87.

BRONGERSMA-SANDERS, M. (1957): Mass Mortality in the Sea. – Geol. Soc. Am., Mem. **67,** 1: 941–1010.

BRONGERSMA-SANDERS, M. (1965): Metals of Kupferschiefer supplied by normal sea water. – Geol. Rundschau, **55,** 2: 365–375.

BRONGERSMA-SANDERS, M. (1966): The fertility of the sea and its bearing on the origin of oil. – Advanc. Sci. (Bri. Ass. London), **23,** 107: 41–46.

BRONGERSMA-SANDERS, M. (1969): Permian wind and the occurrence of fish and metals in the Kupferschiefer and Marl Slate. – Proc. 15. Inter-Univ. Geol. Congr.; 61–71.

BRONGERSMA-SANDERS, M. (1971): Origin of major cyclicity of evaporites and bituminous rocks: an actualistic model. – Marine Geol., **11:** 123–144.

BRONGERSMA-SANDERS, M. & GROEN, P. (1970): Wind and water depth and their bearing on the circulation in evaporite basins. – Third symposium on Salt., Northern Ohio Geol. Soc., 3–7.

BUISONJÉ, P. H. DE (1972): Recurrent red tides, a possible origin of the Solnhofen Limestone. – Proc. Kon. Ned. Akad. Wetensch. (B), **75,** 2: 152–177.

FLÜGEL, E. & FRANZ, H. E. (1967): Elektronenmikroskopischer Nachweis von Coccolithen im Solnhofener Plattenkalk (Ober-Jura). – N. Jb. Geol. Abh., **127;** 245–263.

GALLOIS, R. W. (1976): Coccolith blooms in the Kimmeridge Clay and origin of North Sea oil. – Nature, **259;** 473–475.

GINSBURG, R. N. (1957): Early diagenesis and lithification of shallow-water carbonate sediments in South Florida. – Soc. Econ. Paleont. Min., Spec. Publ., **5:** 80–99.

GOLDRING, R. & SEILACHER, A. (1971): Limulid undertracks and their sedimentological implications. – N. Jb. Geol. Paläont., Abh., **137,** 3: 422–442.

GÜMBEL, C. W. VON (1891): Geognostische Beschreibung der Fränkischen Alb (Frankenalb). – 763 pp.; Cassel.

HANDLIRSCH, A. (1906–1908): Die fossilen Insekten und die Phylogenie der rezenten Formen. – Leipzig (Engelmann Verl.).

HÜCKEL, U. (1974): Vergleich des Mineralbestandes der Plattenkalke Solnhofens und des Libanon mit anderen Kalken. – N. Jb. Geol. Paläont., Abh., **145,** 2: 153–182.

JANICKE, V. (1969): Untersuchungen über den Biotop der Solnhofener Plattenkalke. – Mitt. Bayer. Staatssamml Paläont. hist. Geol., **9;** 117–181.

KEUPP, H. (1975): Der Solnhofener Plattenkalk – Ein neues Modell seiner Entstehung. – Jahresmitt. Naturhist. Ges. Nürnberg, 19–36.

KEUPP, H. (1976): Kalkiges Nannoplankton aus den Solnhofener Schichten (Unter-Tithon, Südliche Frankenalb). – N Jb. Geol. Paläont., Mh., **6,** 361–381.

KEUPP, H. (1977): Ultrafazies und Genese der Solnhofener Plattenkalke (Oberer Malm, Südliche Frankenalb). – Abh Naturhist. Ges. Nürnberg, **37,** 128 pp.

KUHN, O. (1977): Die Tierwelt des Solnhofener Schiefers. – Die Neue Brehm-Bücherei, **318,** 140 pp.

MALZ, H. (1976): Solnhofener Plattenkalk: Eine Welt in Stein. – Edited by Dr. T. KRESS, Solnhofer Aktien-Verein Maxberg, 109 pp.

MAYR, F. X. (1964): Die naturwissenschaftlichen Sammlungen der Philosophisch-theologischen Hochschule Eichstätt. – 400 Jahre Collegium Willibaldinum Eichstätt: 303–334.

MAYR, F. X. (1967): Paläontologie und Stratinomie der Plattenkalke der Altmühlalb. – Erlanger Geol. Abh., **67**, 40 pp.

MCKEE, E. D. (1963): Problems on the recognition of arid and of hot climates of the past. – In: Proc. of the NATO paleoclimates conference held at the Univ. of Newcastle-upon-Tyne., 367–377.

MEYER, R. K. F. (1974): Landpflanzen aus den Plattenkalken von Kelheim (Malm). – Geol. Bl. NO-Bayern, **24:** 200–210.

MUTSCHLER, O. (1927): Die Gymnospermen des Weißen Jura Zeta von Nusplingen. – Jber. Mitt. oberrh. geol. Verein, **16:** 25–50.

RIETSCHEL, S. (1976): *Archaeopteryx* – Tod und Einbettung. – Natur u. Museum, Frankfurt, **106**, 9; 280–286.

SCHLANGER, S. O. & DOUGLAS, R. G. (1974): The pelagic ooze-chalk-limestone transition and its implications for marine stratigraphy. – Spec. Publ. int. Ass. Sediment., **1:** 117–148.

SCHMIDT-KALER, H. (1979): Geologische Karte des Naturparks Altmühltal Südliche Frankenalb, 1:100 000. – Bayer. Geol. Landesamt München.

SEILACHER, A., ANDALIB, F., DIETL, G. & GOCHT, H. (1976): Preservational history of compressed Jurassic ammonites from Southern Germany. – N. Jb. Geol. Paläont., Abh., **152**, 3: 307–356.

SMITH, A. G., HURLEY, A. M., & BRIDEN, J. C. (1981): Phanerozoic paleocontinental world maps. – Cambridge Earth Sc. Series, Cambridge Univ. Press

STRAATEN, L. M. J. U. VAN (1971): Origin of Solnhofen Limestone. – Geologie en Mijnbouw, **50**, 1: 3–8.

VAKHRAMEEV, V. A. (1962): Jurassic floras of the Indo-European and Siberian botanical-geographical regions. – Colloque du Jurassique, Luxembourg, 1962

VIOHL, G. (1983): Forschungsprojekt "Solnhofener Plattenkalke". – Archaeopteryx, **1983**, 3–23.

WALTHER, J. (1904): Die Fauna der Solnhofener Plattenkalke, bionomisch betrachtet. – Jenaische Denkschr., **11** (E. Haeckel Festschr.): 133–214.

WELLNHOFER, P. (1970): Pterodactyloidea (Pterosauria) der Oberjura-Plattenkalke Süddeutschlands. – Abh. Bayer. Akad. Wiss., Math.-naturwiss. Kl., N. F., **141**, 133 pp.

ZEISS, A. (1975): Zur äthiologischen Deutung der großen Erlanger Limulusfährte. – Geol. Bl. NO-Bayern, **25**, 2–3: 95–99.

Author's address: Dr. PAUL H. DE BUISONJÉ, Hobbemakade 97, 1071 XS Amsterdam, Netherlands.

Philip J. Regal

Common Sense and Reconstructions of the Biology of Fossils: *Archaeopteryx* and Feathers

Abstract

The popularity of common sense interpretations of evolutionary questions traces back to an old tradition of pre-Darwinian philosophical and folk theories of causation and origins that is now part of our common sense. But such traditional thinking may contain flawed assumptions associated with essentialism and teleology. Ecology and comparative anatomy have shown that such common sense is not reliable.

We may tend to assume that a structure could not function in a certain way simply because it does not appear to be adapted for that role. But to take such a position may mean that we have slipped into essentialist thinking.

The often-heard idea that the origins of a structure must necessarily be linked to its presently assumed essential nature is not based on the empirical evidence of comparative anatomy. It is an attitude remaining from Aristotelian science.

Theories of the origin of feathers have long tended to be based on common sense, since there has been no direct evidence. Thus, it has been popular to assume that because feathers are essential to flight and heat retention in modern birds, then their origins must have involved one of these as causes. These ideas have had a life of their own despite considerable technical problems with each. The appeal of these ideas in the face of considerable difficulties appears to be an example of the attractiveness of common sense interpretation.

Studies on lizard scales show that they enlarge where they protect the animal from solar heat. On this basis a model of feather evolution can be proposed that avoids the physical and philosophical problems associated with the older models.

Once they had evolved for this they could easily have become coopted for heat-retention and flight.

If feathers did indeed first evolve as heat-shields then the advantages of the complex feather architecture over simple hairlike structures are understandable since width is advantageous in shading.

Zusammenfassung

Die Popularität von Interpretationen evolutionärer Fragen mittels des "gesunden Menschenverstandes" geht zurück auf eine alte Tradition vordarwinischer philosophischer und volkstümlicher Theorien über Ursachen und Entstehung, die nun ein Teil unseres "gesunden Menschenverstandes" sind. Aber solch traditionelles Denken kann fehlerhafte Annahmen enthalten, die mit Essentialismus und Teleologie zu tun haben. Ökologie und vergleichende Anatomie haben gezeigt, daß solch "gesunder Menschenverstand" nicht zuverlässig ist.

Wir können zur Annahme neigen, daß eine Struktur nicht in bestimmter Weise funktionieren konnte, einfach weil sie für diese Aufgabe nicht angepaßt erscheint. Aber eine solche Position kann bedeuten, daß wir in essentialistisches Denken abgeglitten sind.

Die häufig gehörte Vorstellung, daß die Ursprünge einer Struktur notwendigerweise mit seiner gegenwärtig angenommenen essentiellen Natur verbunden sind, basiert nicht auf dem empirischen Nachweis der vergleichenden Anatomie. Dies ist eine von der aristotelischen Wissenschaft überkommene Einstellung.

Theorien über den Ursprung der Federn neigten lange dazu, auf "gesunden Menschenverstand" gegründet zu sein, da es keinen direkten Beweis gab. Da Federn bei modernen Vögeln wesentlich für den Flug und die Wärmeisolierung sind, wurde allgemein angenommen, daß ihr Ursprung mit einer dieser Ursachen zusammenhängen mußte. Diese Ideen entwickelten ihr eigenes Leben, trotz beträchtlicher technischer Probleme mit jeder von ihnen. Der Reiz dieser Ideen angesichts erheblicher Schwierigkeiten erscheint als Beispiel für die Attraktivität von Interpretationen mit Hilfe des "gesunden Menschenverstandes".

Untersuchungen an Eidechsen-Schuppen zeigen, daß sie größer werden, wo sie das Tier vor Sonnenhitze schützen. Auf dieser Grundlage kann ein Modell für die Feder-Evolution vorgeschlagen werden, das die physikalischen und philosophischen Probleme, die mit den älteren Modellen verbunden waren, vermeidet.

Wenn sie erst einmal entwickelt waren, konnten sie leicht sowohl für die Wärmespeicherung als auch für den Flug verwendet werden.

Falls sich Federn tatsächlich zuerst als Hitzeschilder entwickelten, dann sind die Vorteile einer komplexen Feder-Architektur gegenüber einfachen haarartigen Strukturen verständlich, da die Breite beim Schattenspenden von Vorteil ist.

Introduction – "Essential functions" and evolutionary origins

It is very difficult to be able to determine how a dead structure was used in life simply by looking at it. Sometimes we can make very good assumptions about function by comparison with living forms when we have a good understanding of the biomechanics and ecology of these. This approach is particularly instructive if the structure in question is highly specialized and is limited in its mechanical potentials. But in any case there is no substitute for interpretations based on a direct study of the use of the structure by the living creature (GANS 1974).

The problem of interpretation is considerable when the structure that we are interested in is a feature of an ancient fossil with no very close relatives, and when the structure is so generalized that it could have been used in many ways, or when it is so odd that it has no close modern counterparts.

If the structure is ancestral to structures in modern forms then we may feel that this fact gives us a clue to guide our thinking. It may seem reasonable that the ancestral structure was developed by natural selection for the same "essential" survival functions that we see today in the modern forms. For many people this would seem only to be reasonable, good common sense. For example, to some persons it may seem as though the essential feature of the large human brain is to mediate complex social behavior. Thus, some argue that it is most reasonable to suppose that the progressive enlargement of brains in vertebrate evolution must be due to increasing sociality, even though there is little evidence for this and brains certainly have had nonsocial functions. Similarly, feathers are essential to bird flight and heat conservation. So, many persons like to argue that feathers must have evolved in conjunction with flight or heat conservation, despite a lack of empirical evidence, and even with the knowledge that feathers have other functions. I will argue that this sort of reasoning can be unsound.

Quite commonly, in discussions with scientists and laymen alike, one sees that the first few hypotheses that come to mind to explain the evolution of a feature are likely to be based on such common sense. Persons taking such positions seem to demand little empirical evidence and rather to depend almost entirely on the common sense logic, as though causes should be obvious from results and so the thesis needs little empirical support. On the other hand, they do demand considerable evidence for any hypothesis that paints a different sort of scenario. So the deciding factor is common sense.

Comparative anatomy contradicts the common sense approach

How safe, though, is it to depend on such common sense and to relax our standards of scientific rigor? Is it indeed reasonable to assume that a feature has evolved in conjunction with its present-day critical or essential function?

In fact, comparative anatomy teaches us that such thinking can be very misleading. Our hands, for example, are used for grasping and manipulation, feeding, fighting, and courtship. It is quite clear from comparative anatomy that none of these present functions were important for the early origins. The evidence is clear that differentiated distal portions of the forelimbs began as fins to help fishes move about in a liquid medium. Then they were used quite differently for locomotion and eventually for support on land. Grasping and manipulation came only late in evolution.

Our dentary jaw bones began as protective armor. The supporting elements, to which the dermal protective plates fused, began not as a chewing device but as gill arches – as support for the gill basket of the early jawless fishes.

Our ear bones began not as hearing devices but first as gill support elements and then as primary elements of the functional jaws of lower vertebrates.

Tails may be used by amphibians, reptiles and mammals for food storage, balance, and sometimes heat regulation. But they began as a portion of the body differentiated for propulsion in early chordates.

The swim bladder of bony fishes is used mostly as a hydrostatic organ, but it seems to have originated as a functional lung for respiration.

Examples are commonplace of shifts in the function of a structure over the course of evolution. Indeed, this is a general feature of evolution. Most features do not seem to have evolved de novo. Rather, cooptions have been common and structures have been reshaped for new roles – an arm becomes reshaped into a wing, and so on.

The nature of common sense

So the sorts of causes that seem attractive to many scientists and laymen are not based on views of nature that come from empirical study. Biology does not lead to the conclusion that a structure should have originated in connection with its contemporary functions. The notion seems rather to come from common sense thinking that is widespread in our culture. What is common sense, though? It is said to be that which is plain and obvious. Yet it differs from culture to culture. For some people it is obvious that dreams are only imaginary, while for other people it may seem obvious that dreams reveal a profoundly real world. To some tribes in New Guinea it is obvious that the cassowary is a form of human, while to Westerners it is obvious that the cassowary is a large bird. And so on.

Even when dealing with physical properties, common sense has often been wrong. Indeed, the major advances in science have been triumphs over common sense. It may seem that the sun and moon are about the same size, but it is now clear that the sun is much larger than the moon, no matter what our senses tell us is obvious. The earth does spin, though common sense once said that we should all fly off if the earth were spinning.

The physics before GALILEI and NEWTON was that of ARISTOTLE. The philosophy and science of PLATO and ARISTOTLE have been profoundly important sources of our Western common sense since our folk wisdom evolved its views of the world largely while pondering it through these filters.

In PLATO's view, behind the changing and variable material world are eternal, unchanging essences (in contrast to, for example, HERACLITUS who argued that reality is change). In ARISTOTLE's view things may come to exist because of the purposes that they serve. In his thinking and studies essences, functions, causes, and origins are closely related. In his system of causation things have material, formal, moving, and final causes that interact to bring them into being and make them what they are. Of these, the final cause would be the most important.

> "Clearly the first is that which we call the "Final" Cause – that for the sake of which the thing is formed – since that is the *logos* of the thing – its rational ground, and the *logos* is always the beginning for products of Nature as well as for those of Art." (Parts of Animals 639b)

There are three very closely related and precarious concepts here – teleological thinking, essentialist thinking, and typological thinking. But for brevity I believe we can think of some present problems as being mostly the results of essentialism.

Is behavior limited by biology or essentialism?

We often hear it said that a given structure could not have been used in such-and-such a manner because it is not "adapted" or specialized for that purpose. If good mechanical reasons are given then this is perfectly acceptable and we can agree or disagree with the evidence. But too often such statements are backed up only by the vague philosophical belief that creatures must be perfected for roles in nature, and must be as finely matched to these roles as a key is shaped to turn in a lock.

This view ignores the accumulated evidence of biology that the construction of creatures is adequate for survival, but nature does not reveal ideal perfection. The view also ignores how plastic the behavior of animals can be. One cannot simply look at the morphology of a fairly generalized animal and know what its habits and food are. If this were possible then there would be no need for ecological studies.

The student must learn through experience that animals do not conform to stereotypes. An early memorable experience for me was to find that the foxes on the California Channel Islands are almost entirely insectivorous. One expected from the carnivore anatomy, and from verbal and written accounts of "idealized" fox natural history, that foxes would have lived on the mice, lizards, and birds, but these are hard to catch there.

THULBORN suggested at this conference that *Archaeopteryx* did not eat insects or terrestrial food because its teeth were not specialized – so he argued that it must have been feeding on marine organisms and he reinterpreted much of its biology accordingly. This is not a necessary conclusion. Not only Channel Islands foxes but a number of animals eat insects and yet do not have dentitions specialized for insectivory.

Small bipedal kangaroo rats have feet and legs well adapted for hopping about on the desert sands. Yet ROSENZWEIG (1975) has caught them (*Dipodomys merriami*) in live-traps placed 25cm and even 60cm high in bushes. So some kangaroo rat species will confound our stereotypes and forage in bushes. This fact may be of interest to those who argue that the feet of *Archaeopteryx* were the same as those of any small bipedal dinosaur and so it was probably not climbing in trees.

One also learns that there are times when large strong bears live almost entirely on berries. Workers learned that hummingbirds cannot be kept only on nectar but that they must eat small arthropods, though their bills seemed clumsy for anything but nectarivory.

The list can go on and on. Animals have ranges of function that go beyond the stereotypes of them that we form in our minds. If we assume that they have essential natures or ideal behavior that they will not violate, we can be very wrong. The main rule that they follow is survival, however this is possible for them.

By no means am I suggesting that we completely abandon common sense and open the door to wild speculation of every sort. Instead I am suggesting that ecological and behavioral reconstructions of fossils be theories that are based on the best reasoning and evidence that is possible and that we do not become lax in the false security that common sense provides and use it as an instrument to try to push our interpretations beyond what the evidence allows. History shows that common sense assumptions merit particularly close scrutiny because they are so easy to take for granted – but this is not to say that they are always wrong.

Paleontology gives valuable insight into the past. But the power of resolution of evolutionary patterns is in a sense comparable to a compound microscope and not to an electron microscope. I think that it is possible that in trying to discover the boundaries of the discipline, as any healthy discipline should do, we may try to focus on fine detail and then discover that it is unresolvable. Then we may begin to conclude that we know very little about the past, when in fact we know a great deal at the level of resolution that is realistically possible.

Feathers

In short, it is largely because of our ancient Greek heritage that it seems reasonable that natural things have essential functions or natures, and that structures may have originated in connection with these. So when we try to decide how a primordial organ was used we must be very careful that we are equally demanding of all hypotheses and do not fall prey to the subtle vestiges of Western metaphysics and favor one over the others without any more reason than common sense.

For decades the speculation on the origin of feathers involved primarily two theories even though it was generally appreciated even by proponents that there were serious problems with each and no good evidence for either one. Even today one observes that these have great intuitive appeal – one encounters a certain amount of "wishful thinking" that there should be some way to have the theories make sense – despite their considerable problems. I attribute this appeal in the face of large difficulties and no positive evidence to a certain common sense logic. As discussed, this common sense logic is suspect.

Next I would like to outline the difficulties with these two theories and briefly to sketch an alternative. The alternative model was outlined in detail earlier, together with an extensive review of the literature, and it was also pointed out that any model for the evolution of feathers must in fact answer a wide range of difficult questions (REGAL 1975).

The Flight Hypothesis. Since feathers are wonderfully light and strong structures for flight, speculation has been that their origin involved aerodynamic selection pressures. But pennaceous feathers have an exceedingly complicated architecture and morphogenesis. It is hard to justify this complexity in terms of aerodynamic advantages in the intermediate stages. No one has proposed a series of transitional stages that would be aerodynamically advantageous at each step. Indeed, the intermediate stages that come to mind, flapping about in turbulent air, make me think of a half-built battleship at sea – unlikely to be of much use, likely to sink.

Some persons have suggested to me in conversation that I am taking a "gradualistic" stance on this question. But this is a misunderstanding and I should digress to clarify my position. I see no reason to suppose that feathers must have evolved at a slow or at a uniform gradual rate. They might have evolved rapidly. Rate is not an issue in my thinking on this subject. I do assume that any model for their evolution should try to outline and test its own assumptions. It should outline steps that make biological and mechanical sense for the entire transition. Each step should either provide a positive advantage to the organism or be neutral. If one proposes a disadvantageous condition at one stage then the author has the obligation to defend the reasonablness of doing so. There should be some reason stated for any element included in a model. Most gradualists and non-gradualists will agree on this.

I would not at all object to any model in which a large morphological change resulted from a small genetic and developmental change (I have even made proposals along these lines in earlier papers, e.g. REGAL 1977), as long as specific reasons were given to explain why we should take the particular suggestion seriously. Otherwise we are only being shown the loose folds of silk that cover the hat from which the illusionist pulls a rabbit. It is only in this spirit of asking for some completeness in model-building, so that we can examine each step in the reasoning and thus judge the soundness of the model, that I am critical of models that fail to explore their logical or biological implications.

Returning to the main discussion: FEDUCCIA has argued (at this conference) that the flight hypothesis should be resurrected. His main claim is, that whenever birds become flightless they lose the pennaceous structure of the feathers. But his interesting observation might suggest, if confirmed, only that the pennaceous structure is today maintained by aerodynamic selection pressures and it says nothing of origins.

Perhaps an analogy to illustrate the point of my criticism would be the swim bladder of bony fishes. This is often lost in bottom-feeding fishes since it is not needed for buoyancy. This demonstrates its hydrostatic function. But the evidence from comparative anatomy is that the swim bladder originated as a lung-like feature for gas exchange and that the hydrostatic function has been a secondary modification.

The detailed functional objections to the flight theory remain, moreover. Other vertebrates have membranous wings and this does seem like the most simple solution to aerodynamic problems. Pennaceous feathers are exceedingly complicated. No one has been able to present a complete model in which each transitional stage would be of particular advantage under aerodynamic selection pressures, let alone functionally possible. So most workers have assumed that pennaceous feathers must have evolved earlier, "in dry-dock" (to recall the half-built battleship analogy), for some other reason, and then were modified for flight later in evolution.

Heat Retention. Feathers, particularly down, are very good insulation and heat conservation is perhaps the most widespread critical function of body feathers for birds. So a popular theory for feather evolution has been that they originated for heat retention. (Authors often incorrectly equate "insulation" with heat retention, forgetting that insulation may function either to keep heat in or to keep it out.)

Briefly, the objections to this theory have been two. Hair is also good insulation and is a much simpler structure than feathers. So it is not clear why and how the complex structure of feathers would have been selected for by any "need" for insulation. Why not hair? as KEN PARKES (1966) put it.

Moreover, when looked at in detail the hypothesis has still more problems. Any need for heat conservation presupposes an endothermic animal without insulation. This would require a small animal with an exceptionally high metabolic rate and thus with exceptionally high food requirements. This is theoretically possible, though ecologically "clumsy", and it would be more reassuring if there were living animals resembling such an extravagant creature.

Why could endothermy and feathers not have evolved together?, some ask. This superficially appealing "moderate" position is not without problems. A little endothermy in an uninsulated animal would do little good, since small amounts of heat would be lost quickly and would not maintain elevated body temperatures. If the creatures depended to any degree on basking, then primordial feathers could have interfered with this function and their costs might well have outweighed any benefits. This is a potentially deceptive area of speculation that should not be entertained without some rather specific quantitative models to evaluate.

It would be physically (though perhaps not phylogenetically) easier to imagine larger warm-blooded animals with low rates of heat loss, because of their surface-to-mass ratios, gaining insulation as they evolved smaller body size, as MCNAB (1978) has suggested. In all such cases, though, the question remains: why not hair?

Heat Shielding. My own interest in feather evolution began while I watched lizards with enlarged raised scales basking at heat lamps in the laboratory. Once heated they would turn so that their scales seemed to act as small parasols to shade them. A series of experiments confirmed that the scales acted to reduce the rate of heat uptake at most angles and at no angle acted as a "solar furnace" that would concentrate heat uptake.

Other experiments showed that larger scales reduced heat uptake more than smaller scales did. It had previously been reported that both within genera and within species of basking lizards on three continents larger scales are found in hotter areas. The experiments provided a mechanism to understand this. In hotter areas the larger scales allowed the lizards to spend more time in the sun.

The width of a structure is very advantageous in providing shade (detailed discussion in REGAL 1975). So, given that the evolutionary sequence begins with a broad scale, width would be maintained even as the scale became putatively subdivided to provide a soft, flexible but tough covering for an active animal. (The answer to the question, why not hair?) The proposed intermediate steps and their functional advantages have been outlined previously.

The skin shedding system of lizards would seem not to allow the evolution of feathers in this phylogenetic line, so the lack of feathers on basking lizards may indicate nothing relevant to this issue. But there are nonpennaceous feather-like structures in a desert cactus *Mammalaria plumosa* and I showed that these function to shield the cactus from heat. Reportedly, some diurnal desert spiders also have feather-like structures (C. KRISTENSEN, personal communication).

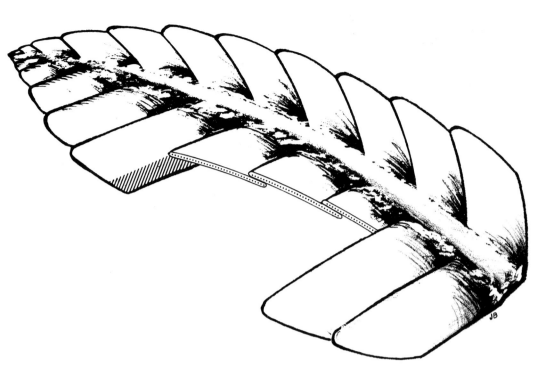

Fig. 1
An hypothetical intermediate stage between an enlarged scale and a feather (from REGAL 1975). The enlarged scale has become subdivided here, providing a broad heat-shield that is also flexible since it can bend from side to side and in other directions. The flexible covering would allow freedom of movement for an active animal. The lateral proto-barb, plate-like, structures (overlapping in cut-away protion) are held together not only at their bases but loosely at their edges by tiny (not visible here) proto-hooklet, Velcro-like, friction-producing structures. In the next stage, the proto-barbs would themselves become subdivided to add more flexibility while retaining the heat-shielding function of the protofeather. The width of a structure is important in providing shade (see theoretical reasons and experiments described in REGAL 1975). This hypothetical structure is very close to one that could easily be modified in details of strength and architecture to be coopted for heat retention or flight functions.

In this model of feather evolution, once the protofeathers became sufficiently enlarged and shaped for heat exclusion they easily could be coopted and refabricated for either flight or heat retention.

The loss of pennaceous structure in flightless birds may show that for an endothermic animal where the problem is largely heat retention, hair-like structures are indeed about as useful as the more complicated pennaceous feathers.

It is often assumed that since feathers are widely used to conserve heat today then their presence on *Archaeopteryx* is evidence that it was endothermic (see REGAL 1975, REGAL and GANS 1980). But feathers are not de facto evidence of endothermy since they might have been used for heat exclusion as easily as for heat retention. Even modern birds use feathers to protect themselves and young from the sun in some circumstances.

I suggested that endothermy might be expected in a small flying animal like *Archaeopteryx* because, 1) it could offset the high heat loss in a creature moving long distances through the air, and 2) the efficient repayment of oxygen debts would be of advantage in the context of the high energetic demands of powered flight, and this would take place most effectively at a high and stable body temperature. If *Archaeopteryx* was an energetic flyer or moved long distances then it was quite probably endothermic and the feathers had already become coopted for heat conservation.

I thank P. KITCHER and H. B. TORDOFF for comments on the manuscript.

References cited

GANS, C. (1974): Biomechanics: an approach to vertebrate biology. – 261 pp. Philadelphia (Lippincott).
MCNAB, B. K. (1978): The evolution of endothermy in the phylogeny of mammals.-Amer. Natur., **112**, 983: 1–21.
PARKES, K. C. (1966): Speculations on the origin of feathers. – The Living Bird **5**: 77–86.
REGAL, P. J. (1975) The evolutionary origin ot feathers. – Quart. Rev. Biol., **50**, 1: 35–66.
REGAL, P. J. (1977) Evolutionary loss of useless features: is it molecular noise suppression? – Amer. Natur., **111**: 123–133.
REGAL, P. J. and C. GANS. (1980): The revolution in thermal physiology: implications for dinosaurs. – In: A cold look at the warm-blooded dinosaurs, pp. 167–188, R. D. K. THOMAS and E. C. OLSON (eds.), AAAS Symposium **28**, Boulder, Colorado (Westview Press).
ROSENZWEIG, M., SMIGEL, B. and KRAFT, A. (1975): Patterns of food, space and diversity in desert rodents. – In: Rodents in desert environments, pp. 241–268, I. PRAKASH and P. K. GOSH (eds.), The Hague (W. V. JUNK).

Author's address: Prof. PHILIP J. REGAL, Museum of Natural Hisory, University of Minnesota, Minneapolis, Minnesota 55455, U. S. A.

Alan Feduccia

On Why the Dinosaur Lacked Feathers

Abstract

The hypothesis that feathers evolved initially as an insulatory mechanism for endothermic, ground-dwelling reptilian ancestors of birds is refuted on the basis of: 1) the structural complexity of feathers, which would represent gross "over-kill" for insulatory structures; 2) the aerodynamic design of feathers; 3) the fact that body or contour feathers are identical in basic structure to flight feathers; and 4) the fact that feathers become degenerate in flightless birds and lose their aerodynamic design, becoming very simplified and "hairlike" in structure. These facts provide indirect evidence to support the hypothesis that feathers evolved initially in an aerodynamic context and were then preadapted for insulating subsequently evolved endothermic birds.

Zusammenfassung

Die Hypothese, daß Federn ursprünglich als ein Isolationsmechanismus für endotherme, bodenlebende, reptilische Vogelvorläufer entstanden sind, wird zurückgewiesen. Dies wird begründet durch (1) die strukturelle Komplexität von Federn, welche für den Zweck des Wärmeschutzes allein unnötig wäre, (2) die aerodynamische Form von Federn, (3) die Tatsache, daß Körper- oder Konturfedern in ihrer Grundstruktur mit den Flugfedern identisch sind, und (4) die Tatsache, daß bei flugunfähigen Vögeln die Federn degenerieren und ihre aerodynamische Form verlieren, indem sie sehr einfach und "haarartig" werden..

Diese Tatsachen liefern einen indirekten Beweis für die Hypothese, daß Federn ursprünglich in einem aerodynamischen Zusammenhang entstanden und dann präadaptiert waren für eine Wärmeschutzfunktion bei den in der Folgezeit sich entwickelnden endothermen Vögeln.

It is commonly stated that without feathers in the specimen of *Archaeopteryx*, its avian affinities would never have been known. Indeed, it is true that aside from feathers and furcula, there are no features of *Archaeopteryx* that without question indicate avian status. Much of the hundred or so years of speculation concerning these remarkable specimens of reptile-birds have centered on their relationships to possible ancestral reptilian groups, and the nature of the paleobiological role of *Archaeopteryx*. Was *Archaeopteryx*, and other birds, derived directly from thecodonts or later from theropods? The fact that this question is still unanswered after a hundred years may well indicate that without additional fossils of more ancient birds we will never be able to establish for certainty the exact nature of the affinities of birds to ancestral reptile groups. Nevertheless, the presence of feathers in the specimens of *Archaeopteryx* has posed perhaps more interesting questions concerning the paleobiological role of *Archaeopteryx*, the function of feathers in *Archaeopteryx*, and the context in which feathers first evolved.

Basically two schools of thought have emerged to explain the origin of flight and feathers. Advocates of the arboreal school (see SAVILE, 1962; BOCK, 1965; PARKES, 1966; and FEDUCCIA, 1980 for summaries) have found little problem in explaining the evolutionary origin of feathers as offering an immediate aerodynamic advantage for the arboreal ancestors of birds. As SAVILE (1962) has shown, the slightest fringe of scale-like feather equivalents would offer an immediate advantage in "parachuting", which would represent a stage intermediate between "falling" and "gliding". On the other hand, advocates of the cursorial theory for the origin of avian flight tend to favor the argument that feathers evolved in endothermic, reptile ancestors of birds. For example, OSTROM (1974; 1979) envisions the "proavis" as a small, fast-running, bipedal coelurosaurian dinosaur that would be covered with contour feathers as an insulatory mechanism, that is, as an alternative to mammalian hair. Thus, in Ostrom's scenario (see also Bakker, 1975) feathers would have evolved first at a pre-flight stage in the context of endothermy, and would have later been preadaptive for flight. Ostrom's earth-bound theropod "pro-avis" or *Archaeopteryx* was unable to fly, and the feathers' primary function was insulation. Bob Bakker, carrying the case to an extreme, has illustrated a number of small theropods with feathered backs (see cover of Johns Hopkins Magazine, April, 1979). Obviously, then, the proponents of the idea of endothermy in dinosaurs, namely Ostrom and Bakker, have viewed *Archaeopteryx* in the context of a feathered coelurosaurian dinosaur, feathers having evolved in a context other than aerodynamic and a feathered "proavis" or *Archaeopteryx* at a pre-flight stage in avian evolution. Still another model for the evolutionary origin of feathers in a non-aerodynamic context is that of Regal (1975) who argues that feathers were derived from reptilian scales in the context of heat shields.

In the arboreal model, feathers first evolve in the context of flight, and are later preadaptive for insulation for endothermy, when true powered flight and high metabolic rates that characterize the currently extant class Aves evolved. Although the arguments have not always been stated as above, it is interesting to note that the two basic theories emerged almost at the same time, shortly after the discovery of *Archaeopteryx*, and the issue has been hotly debated until the present day. A cursorial origin for avian flight was first proposed by Williston in 1879, and the arboreal theory was proposed by Marsh in 1880.

I have earlier (Feduccia, 1980) summarized the now overwhelming evidence that *Archaeopteryx* had aerodynamically designed wings, the main critical evidence coming from the possession in *Archaeopteryx* of asymmetric primary feathers in the wings (Feduccia and Tordoff, 1979), a condition known only in flying birds. In addition, I presented evidence to show that feathers were designed for flight and not initially as an insulation for endothermic dinosaurs. Because these arguments have not been widely accepted (see Gould and Vrba, 1982; and Science, 1982, 216 (4551): 1212—1213), I would like here to restate and expand these basic arguments. It has been my belief that part of the misunderstanding is due to the lack of familiarity of the structural complexity of feathers by paleobiologists.

Feathers are features unique to birds, and there are no known intermediate structures between reptilian scales and feathers. Notwithstanding speculations on the nature of the elongated scales found in such forms as *Longisquama* (Sharov, 1970), *Cosesaurus* and *Scleromochlus* (see Bakker, 1975, and Martin, 1984) as being featherlike structures, there is simply no demonstratable evidence that they in fact are. In 1982 I examined the specimen of *Longisquama* in Moscow and could see no indication that the elongated scales were particularly feather-like. They are very interesting, highly modified and elongated reptilian scales, and are not incipient feathers. As an aside, it is interesting here to note that while most American workers (see Bakker, 1975, and Langston, 1981) have interpreted the fine markings on the slab of the pterosaur *Sordes pilosus* as "hair", I could not confirm this in my examination of the actual specimen in 1982. I would interpret the markings on the slab of *Sordes* as possibly being of plant origin, and I strongly urge paleobotanists to examine the specimen first hand. Paleontologists should be extremely cautious in proclaiming the identification of structures in fossils that cannot be established with absolute certainty. It is interesting to note here that although integumentary structures are commonly preserved in the Solnhofen limestone, for example, feathers in the specimens of *Archaeopteryx* and wing membranes in pterosaurs, the search for feathers in the Solnhofen dinosaur *Compsognathus* produced nothing (Ostrom, 1978: 115—116). Ostrom (op. cit.) stated that, "If the speculative question about feathered "coelurosaurs" can ever be answered, the Munich specimen of *Compsognathus* is the critical specimen to examine ... The reader can be sure that I made an exhaustive examination, ... but to no avail. If feathers had been present in *Compsognathus*, it is inconceivable to me that no evidence of them would be preserved ... But the fine-grained matrix shows nothing. Thus, I conclude that *Compsognathus* almost certainly was not feathered."

The fact that no feathers or feather-like structures are known in the fossil record except in association with non-avian fossils does not prove that they did not occur, but the main objection to the hypothesis that feathers evolved in endothermic reptiles as an insulatory mechanism is: why feathers? Feathers are by far the most complex derivatives of the integument to be found in any vertebrates, and their

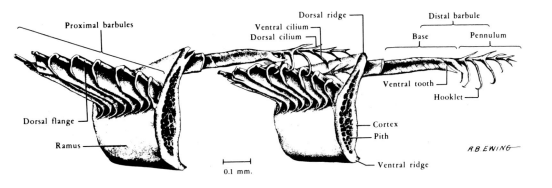

Fig. 1
Segments of two barbs from a contour feather of a chicken. The barbs are seen obliquely from the distal end to show interlocking of parts (from LUCAS and STETTENHEIM, 1972. Avian Anatomy: Integument, Part I, Fig. 170. U.S. Govn. Printing Office, Washington, D.C.).

embryology is still far from being understood. Yet, they would appear to offer no particularly great advantage over hair in terms of their insulatory properties. A typical feather consists of a rodlike shaft or rachis which gives rise on either side to a large number of branches, known as the barbs or rami, and these in turn have many still smaller branches, the barbules, both proximally and distally. The more distal barbules interlock with the proximal barbules of the adjacent barbs by means of microscopic hooklets that arise from the cilia (see Figure 1). This mechanism forms a coherent, nearly air-tight membrane known as the feather vane (RAWLES, 1960). Feathers are a near perfect aerodynamic design, and yet, when damaged, as is often the case, the barbules and hooklets permit the vane to come back together in its finest form. Feathers . . . "allow a mechanical and aerodynamic refinement never achieved by other means. Mechanically, feathers have great resilience, graded flexibility, and high strength to weight ratio. Aerodynamically, they provide smooth contours and streamlined sections; permit the development of slotted wings, which produce high lift at low speed; and allow birds to achieve laminar flow" (SAVILE, 1962: 113).

Body, or contour feathers, which according to the "feathered dinosaur" point of view evolved initially as insulatory structures and subsequently gave rise to the flight feathers, are virtually identical to the aerodynamically designed flight feathers, down to the minute detail. The body feathers are indeed miniature flight feathers (PARKES, 1966: 83). This fact alone would seem almost paradoxical if feathers evolved originally as an insulatory mechanism.

If feathers evolved initially as thermoregulatory structures, then the body or contour feathers of modern birds should be essentially similar to the original feather, that is, flight feathers should be secondary, and body feathers should be uniform in structure in both flying and flightless birds. In fact, this is not the case, and this single point provides the most critical argument against the hypothesis that feathers evolved initially in the context of thermoregulation. All known flightless birds evolved from flying ancestors, as attested to by the myriad of flight adaptations in the skeletons of these birds, among other features (see summary in FEDUCCIA, 1980). In the flightless ratites – the ostrich, emu, cassoway, emu and kiwi, – for which the feathers have value only in a thermoregulatory context, the feathers are very degenerate, are very loosely constructed and in fact resemble mammalian hair superficially. The feathers of the kiwi illustrate a stage at which the hooklets are strictly vestigial, remaining only as small spurs along the barbules. Even more telling, however, are the feathers of rails that have become flightless in very recent times. In the case of *Atlantisia rogersi*, a species known to be of very recent origin (OLSON, 1973) the flight feathers have lost their characteristic asymmetry (FEDUCCIA and TORDOFF, 1979), and, "Besides the remiges all the contour-feathers present markedly discontinuous rami, so that the body of the bird

Fig. 2
Distal barbules from identical places of the second primary flight feathers (counting from outward – in) of: the large, volant graynecked wood rail (*Eulabeornis cajaneus*), and the totally flightless Inaccessible Island Rail (*Atlantisia rogersi*). Note the complete disappearance of the hooklets of hamuli (H) and the near disappearance of the base (B) or basel lamella (modified after STRESEMANN, 1932).

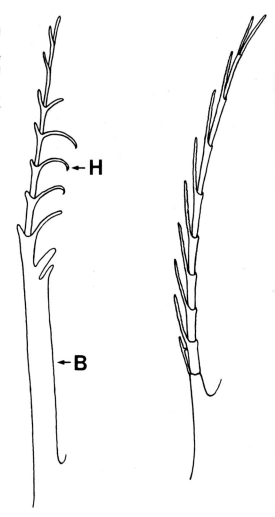

suggests a distinct appearance of hairiness" (LOWE, 1928: 104). But in addition, the flight feathers have lost the detail of feather structure that provides the aerodynamic design. The barbs show a discontinuity, the barbules exhibit a discontinuity, and there is a reduction in the breadth of the distal two-thirds of the base of the barbule. In addition, both the cilia and the hamuli or hooklets are ill-developed (Figure 2). As LOWE (1928: 107) put it, "... the barbules on the proximal side of the barb are ... discontinuous and must allow free passage of air between them, with consequent loss of resistance leading to an inability to fly."

Feathers represent an incredible structural complexity, by far the most complex of the integumentary derivatives of vertebrates. Their aerodynamic design is attested to not only through structural analysis, but is proven by the fact that they lose their structural integrity and become relatively simple or hairlike when flightlessness evolves. This fact indicates that feathers evolved as aerodynamically designed integumentary structures that were later preadapted for insulation in endotherms. That body or contour feathers are designed identically to flight feathers and also become structurally degenerate in flightless birds is further testimony to the hypothesis that feathers evolved in an aerodynamic context. Once present, feathers were perfectly preadapted for serving the dual functions of flight and insulation in later endotherms. To clothe a ground-dwelling, endothermic reptile with feathers for insulation is tantamount to insulating an ice truck with heat shields from the space shuttle.

References cited

Bakker, R.T. (1975): Dinosaur renaissance. – Sci. Amer., **232**(4): 58–78; San Francisco.
Bock, W.J. (1965): The role of adaptive mechanisms in the origin of the higher levels of organization. – Syst. Zool., **14**: 272–287; Washington, D.C.
Feduccia, A. (1980): The age of birds. – 196 pp. Cambridge (Mass.) and London (Harvard University Press).
Feduccia, A., & Tordoff, H.B. (1979): Feathers of *Archaeopteryx*: asymmetric vanes indicate aerodynamic function. – Science, **203**: 1021–1022; Washington, D.C.
Gould, S.J. & Vrba, E. (1982): Exadaptation – a missing term in the science of form. – Paleobiology, **8**(1): 4–15; Chicago.
Langston, W., Jr. (1981): Pterosaurs. – Sci. Amer., **244**(2): 122–136; San Francisco.
Lowe, P.R. (1928): A description of *Atlantisia rogersi*, the diminutive flightless rail of Inaccessible Island (South Atlantic), with some notes on flightless rails. – Ibis, 12th ser., **4**: 99–131; London.
Marsh, O.C. (1880): Odontornithes: a monograph on the extinct toothed birds of North America. – Report of the U.S. Geological Exploration of the Fortieth Parallel, **7**. Washington, D.C.
Martin, L.D. (1984): The origin of birds and of avian flight. – In: Current Ornithology, Vol. I; pp. 105–129; New York (Plenum Press).
Ostrom, J.H. (1974): *Archaeopteryx* and the origin of flight. – Quart. Rev. Biol., **49**: 27–47; Stony Brook (New York).
Ostrom, J.H. (1978): The osteology of *Compsognathus longipes* Wagner. – Zitteliana, **4**: 73–118; München.
Ostrom, J.H. (1979): Bird flight; how did it begin? – Amer. Sci., **67**: 46–56; New Haven.
Olson, S.L. (1973): Evolution of the rails of the South Atlantic islands (Aves: Rallidae). – Smithsonian Contr. Zool., **152**: 1–53; Washington, D.C.
Parkes, K.C. (1966): Speculations on the origin of feathers. – Living Bird, **5**: 77–86; Ithaca, (N.Y.).
Rawles, M.E. (1960): The integumentary system. – In: Biology and comparative physiology of birds, ed. A.J. Marshal, vol. I, pp. 189–240; New York (Academic Press).
Regal, P.J. (1975): The evolutionary origin of feathers. – Quart. Rev. Biol., **50**: 35–66; Stony Brook (New York).
Savile, D.B.O. (1962): Gliding and flight in the vertebrates. – Amer. Zool., **2**: 161–166; Lawrence, Kansas.
Sharov, A.G. (1970): An unusual reptile from the lower Triassic of Fergana. – Paleo. Z. hur., 127–130; Moscow.
Stresemann, E. (1932): La structure des remiges chez quelques rales physiologiquement apteres. – Alauda, **4**(1): 1–5; Paris.
Williston, S.W. (1879): Are birds derived from dinosaurs? – Kansas City Review of Sci., **3**: 457–460; Kansas City.

Author's address: Dr. Alan Feduccia, Dept. of Biology, Coker Hall 010A, The University of North Carolina, Chapel Hill, North Carolina 27514, U.S.A.

Richard A. Thulborn & Tim L. Hamley

A New Palaeoecological Role for *Archaeopteryx*

Abstract

Archaeopteryx is envisaged as an agile hunter that frequented shore-lines, pools and shallow waters. It probably subsisted on prey of moderate size, including small fishes and worms. The feathered forelimbs ("wings") may have been used as a canopy while *Archaeopteryx* foraged in water; its hunting techniques may have resembled those used by existing herons and egrets. Speculations on the natural history of *Archaeopteryx* prompt a new theory for the origin of avian flight: the first flying birds may have been aquatic forms, using their rudimentary flight apparatus to carry them from wave-crest to wave-crest.

Zusammenfassung

Archaeopteryx wird als flinker Jäger angesehen, der Küsten, Tümpel und seichte Gewässer aufsuchte. Er ernährte sich wahrscheinlich von Kleingetier, inklusive kleinen Fischen und Würmern. Die vorderen, gefiederten Gliedmaßen ("Flügel") wurden vielleicht während der Futtersuche im Wasser als Baldachin benutzt; diese Art der Futtersuche könnte wohl auch der gegenwärtiger Reiher ähnlich gewesen sein. Über die Naturgeschichte von *Archaeopteryx* angestellte Vermutungen gaben Anlaß zu einer neuen Theorie für den Ursprung des Vogelflugs: die ersten fliegenden Vögel könnten aquatisch gewesen sein und ihre Flügel dazu benutzt haben, sie von Wellenkamm zu Wellenkamm zu tragen.

Introduction

The most illuminating insights into the natural history of extinct organisms are often provided by trace fossils. Unfortunately no trace fossils can be attributed with certainty to *Archaeopteryx*. (Some Solnhofen tracks attributed to birds or bird-like reptiles (e.g. JAEKEL 1929) are now known to be the work of limulids). Consequently the natural history of *Archaeopteryx* must be investigated by finding analogues for the structural peculiarities of the body fossils. This procedure is risky because there are no universally accepted methods for identifying, selecting, and using analogues. In this paper two anatomical features will be selected as "primary indicators" to the natural history of *Archaeopteryx* – the dentition and the foot. The functions of both features are identified by reference to selected analogues (organic and inorganic). Study of the dentition should indicate the diet of *Archaeopteryx* and, hence, the environment in which it obtained its food. Study of the foot should indicate the animal's locomotor abilities and, hence, the environment through which it was accustomed to move. These preliminary findings provide a framework in which to appraise other structural peculiarities of the animal (furcula, forelimb, tail and plumage). The following speculations on the natural history of *Archaeopteryx* are not constrained by assumptions about the phylogenetic significance of the animal. On the contrary, those speculations may assist in the construction of a new phylogenetic scenario (*sensu* ELDREDGE 1979).

Primary Indicators of Habitat
Teeth, Jaws and Diet

There are 13 teeth in the upper jaw, and about 11 in the mandible. The teeth are widely spaced and have stud-like crowns which are basically conical in form; they are not sectorial, nor are they modified for crushing or puncturing the food. Analogous dentitions occur in sauropterygians, ichthyosaurs,

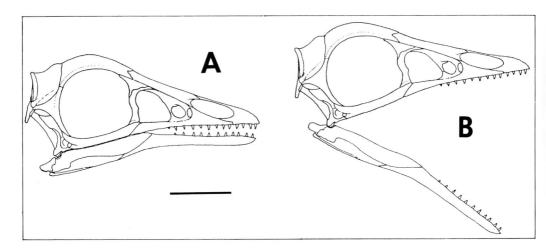

Fig. 1
Skull of *Archaeopteryx*. A, with jaws closed. B, showing gape necessary for the animal to swallow fish-like prey of estimated maximum size. Adapted from WELLNHOFER's (1974) reconstruction of the skull in the Eichstätt example. Scale bar indicates 1 cm.

crocodilians, some pterosaurs, odontocetes, and certain birds (Hesperornithiformes). In all these cases it seems that the teeth were adapted to apprehend struggling and/or slippery prey, such as fishes and cephalopods. There is no evidence of heterodonty in *Archaeopteryx*. The anterior teeth (both upper and lower) were probably food-obtaining teeth (as in most other vertebrates), but there are no posterior teeth ("cheek teeth") adapted for food-processing. The absence of specialized cheek teeth implies that food was not processed in the mouth, but that it was swallowed whole. Overall the jaws, with their small, widely spaced and rather stud-like teeth, appear to form a gripping device analogous to a pair of pliers.

The Eichstätt specimen shows the anterior half of the mandible downwardly deflected (Fig. 1). By manipulating WELLNHOFER's (1974) reconstruction of the Eichstätt skull (Fig. 1 A) it seems that upper and lower teeth would not have occluded when the jaws were closed. There remained a slight gap between upper and lower tooth rows, because the upper jaw is not so strongly deflexed as the mandible. However, this difference in the flexure of upper and lower jaws might be an artifact of preservation. Even if the jaws did close tightly along their entire length, few (if any) of the lower teeth would have occluded with the upper teeth. If the flexure in the jaws is not an artifact it may be analogous to that in a pair of curved forceps: it would allow upper and lower jaws to close more nearly in parallel; and it would ensure that the mandibular teeth bit upwards and backwards, tending to force the prey deeper into the predator's throat. This flexure might also have allowed an unobscured field of vision over the tip of the snout. It is conceivable that *Archaeopteryx* had binocular vision, as RUSSELL (1972) has suggested for ornithomimid theropods, though this possibility cannot be checked without evidence from an undistorted cranium.

From the structure of jaws and teeth it is possible to specify more closely the diet of *Archaeopteryx*, and the manner in which this animal dealt with its prey. First, the jaws and teeth appear to be those of an active hunter which sought prey agile enough or slippery enough to pose some threat of escape. Consequently it seems unlikely that *Archaeopteryx* was a scavenger (*contra* HECHT & TARSITANO 1982). There are no obvious adaptations (such as blade-like, serrated or sectorial teeth) for dismembering large prey or carrion, and it is difficult to envisage *Archaeopteryx* as a killer of bigger animals or as an habitual feeder upon large carcasses.

It is often suggested that *Archaeopteryx* was an insect eater, but the morphology of the teeth does not support this idea. Insectivores commonly have teeth adapted for crushing the exoskeleton of their prey; they tend to have closely packed and sometimes multi-cusped teeth which provide a large occlusal area. So, for example, insectivorous bats differ from carnivorous bats in having a distinctly larger tooth area in the occlusal plane (FREEMAN 1984). In *Archaeopteryx* the area of occlusion between upper and lower teeth must have been infinitesimal, no matter how the jaws are positioned and articulated.

Archaeopteryx probably took prey large enough to have been gripped safely between the jaws, but not so large that it could not have been swallowed whole. Such prey might have included fishes, worms and small tetrapods such as lizards. A diet so mixed would best be obtained in shallow water or along a shore-line, and not in an arboreal setting. DE BEER (1954: 46) considered that "the jaw movement in *Archaeopteryx* was restricted to snapping for insects or fish-catching". However, DE BEER rapidly abandoned the idea that *Archaeopteryx* might have been piscivorous and argued instead that the animal was an arboreal insectivore (1954, 1975). Indeed, it would seem difficult to reconcile fish-catching habits with an arboreal existence – especially if *Archaeopteryx* were incapable of strong powered flight. Nevertheless DE BEER's insight was probably correct: *Archaeopteryx* was adapted for seizing and retaining slippery or wriggling prey of moderate size, and such prey could well have included fishes. By contrast the *Archaeopteryx* dentition is not well suited for catching or holding insects, or for masticating them.

In WELLNHOFER's reconstruction of the Eichstätt skull (1974) the left and right jaw joints are about 10 mm apart. Evidently the animal could not have swallowed prey with a diameter much greater than about 9 mm. If the prey were approximately circular in cross-section (i. e. "worm-like") its maximum diameter could have been about 9 mm. If the prey were elliptical in cross-section (i. e. "fish-like") the figure of 9 mm would represent the maximum limit for transverse width; its maximum dorso-ventral depth could have been greater than 9 mm, limited only by the degree to which the floor of the throat was distensible in *Archaeopteryx*. Assume that an "average" fish would have an elliptical transverse section with maximum depth about 1,75 times maximum width: if the body width were 9 mm, then the body depth would be about 16 mm. In the ideal body design for a fast-swimming fish the length of the body is about 4,5 times its maximum diameter (ALEXANDER 1967). If the hypothetical fish-like prey approached such an ideal design it would have had a body length of about 7 cm. Apparently the Eichstätt example of *Archaeopteryx* (skull length 39 mm) could have swallowed an entire fish with a body length up to about 7 cm. Similar calculations for the Berlin example (skull length 52 mm) indicate a maximum limit of about 9,5 cm for the length of fish-like prey. Such feats of swallowing would not require an unusually large gape for the jaws (Fig. 1B). Among the contemporaries of *Archaeopteryx* are several types of small fishes (E. g. *Leptolepides, Propterus, Pleuropholis*) and worms (e. g. *Ctenoscolex*) which might represent potential prey. It is possible that *Archaeopteryx* exploited some of the smaller and abundant fishes which are known to have gathered in shoals (e. g. *Leptolepides sprattiformis*).

Foot, Hindlimb and Locomotion

The foot is similar to that in small theropod dinosaurs (coelurosaurs), as was emphasized by OSTROM (1974, 1976a). The hallux was at least partly reversed; it is quite short and emerges so high up the metatarsus that it could scarcely have touched the ground when the animal was in a normal (digitigrade) standing pose. From the size and location of the hallux it seems that the foot had limited grasping ability; it might have been quite unsuitable for perching, despite frequent assumptions to the contrary. OSTROM (1974) pointed out that the unguals of the foot are relatively straight and robust, with weak flexor tubercles; by contrast, birds with feet adapted for perching, or for grasping prey, have strongly curved unguals with prominent flexor tubercles. These features support OSTROM's (1974) view that *Archaeopteryx* was basically a ground-dwelling animal which was not adapted for perching in trees. In its hindlimb morphology *Archaeopteryx* resembles the small theropod *Compsognathus*, which was apparently an agile ground-based predator capable of catching and devouring lizards (OSTROM 1978); presumably *Archaeopteryx* would have been equally agile.

WELLNHOFER (1974) provides measurements of hindlimb bones in all examples of *Archaeopteryx*. By applying methods explained elsewhere (THULBORN 1984b; THULBORN & WADE 1984) it may be calculated that the "average" walking speed of the Eichstätt example was about 0,45 m/sec (1,6 km/hr); for the biggest example (London) the "average" walking speed may have been about 0,55 m/sec (2,0 km/hr). The maximal running speeds may have been as high as 2,17 m/sec (7,8 km/hr; Eichstätt example) and 2,77 m/s (10,0 km/hr; London example).

Other Anatomical Features

From the evidence of teeth, jaws, and hindlimb structure it is deduced that *Archaeopteryx* was not an arboreal insectivore; it was more probably an agile ground-based predator that exploited prey such as small fishes and worms. Such a diet implies that *Archaeopteryx* hunted along shorelines, through shallow waters and in pools. These preliminary findings provide a background against which to appraise some other structural peculiarities of the animal.

Furcula

OSTROM (1976b) suggested that the furcula might have served as a springy strut to maintain optimum spacing between the shoulder joints, and that it was not necessarily an indication that *Archaeopteryx* could fly in avian style. However, OLSON & FEDUCCIA (1979) considered that the furcula carried the attachment of a well-developed pectoralis muscle, which would have generated the power stroke of the "wing". The discovery of a furcula in various theropods – including allosaurids, caenagnathids (oviraptorids) and tyrannosaurids – dispels the idea that this element is a diagnostic component of the avian flight apparatus (BARSBOLD 1983; THULBORN 1984a). The furcula was in the first instance an adaptation of flightless predatory dinosaurs.

Forelimb

The forelimb is very large. It is much larger, in relation to the hindlimb, than in any theropod dinosaur. The manus, in particular, is relatively enormous; it represents about 40 % of the total length of the forelimb, and its digits would have touched the ground when the limb was even loosely extended. The phalanges are long and slender, and the slim sharp-edged claws are supported by ungual phalanges with prominent flexor tubercles. This large tridactyl manus, with its hook-like claws, has often been regarded as an adaptation for climbing, yet the form of the manus is nearly identical in theropods that have never been envisaged as climbers (e.g. *Deinonychus, Ornitholestes*). The functions of the manus were doubtless the same in *Archaeopteryx* as in those theropods: that is, the hands assisted the jaws in seizing prey and were probably important in holding and manipulating prey before it was swallowed. If the manus were adapted for climbing one might expect it to comprise more flexible digits (composed of shorter phalanges) terminating in thicker, blunter and mechanically stronger claws without sharp inferior edges – as in digits I–III of the pterosaur manus. The digits of the *Archaeopteryx* manus are in a general sense (though not precisely) analogous to the elongated third digit of the aye-aye *Daubentonia*: they are devices for obtaining food, and not for climbing.

Enlargement of the forelimb, and especially of the manus, was not primarily for the purpose of supporting the "wing" plumage. As OSTROM (1974) observed, if the forelimb were enlarged purely to provide anchorage for the feathers one would scarcely expect enlargement of the manus to such an extent that it became an "encumbrance". Evidently the forelimb was enlarged in the first instance for purposes of predation; only subsequently was it turned to use in supporting the "wing" feathers.

The manus could probably grasp as well (or as poorly) as the similar manus in theropods (see GALTON 1971 on *Syntarsus*). The innermost metacarpal is twisted along its length, and its distal articular surface is oblique to the long axis of the bone. During flexion the innermost digit would have swept slightly outwards, towards the other two digits, thus imparting some slight grasping ability to the manus. Overall the manus can be envisaged as a resilient and slightly flexible grappling hook. This adaptable manus would accommodate itself to awkwardly-shaped prey and would be resilient enough to act as a shock-absorber, damping down the struggles of the prey and reducing the risk of damage to the delicate hand bones. In modern birds heterocoelous centra endow the neck with great flexibility, so that the head can be darted accurately at prey which may be small, fast-moving or erratic in its movements. *Archaeopteryx* lacked heterocoelous centra and presumably had a less flexible neck; yet it may have succeeded in seizing equally elusive prey, by using its large hands as flexible grappling hooks which swept the prey towards, or even into, its jaws.

Tail

The tail is remarkably short by comparison with that in bipedal dinosaurs (both ornithopods and theropods). It represents between 88 % and 94 % of the maximum length of the hindlimb (both tail and hindlimb measured from the centre of the acetabulum, and regardless of which bones are or are not involved). Tail and hindlimb have similar proportions in the hesperornithiform bird *Baptornis*. Reduction of the tail probably did not impair the bipedal locomotor abilities of *Archaeopteryx*, for the pelvic bones were correspondingly re-modelled to accommodate the origins of hindlimb retractor muscles shifted forwards from the tail. This re-modelling is evident, for example, in the expanded anterior part of the ilium.

The principal effect of this reorganization in tail and pelvis would be to shift the centre of gravity forwards. That is, the counter-balancing and stabilising role of the tail was reduced, so that *Archaeopteryx* would have been a rather delicately balanced biped, with most of its body mass located in front of the hips. Even slight movements of the tail would generate instability, so that the animal could instantly lunge forwards and downwards in pursuit of its prey. In other words reduction of the tail made *Archaeopteryx* an inherently unstable biped: it was constantly poised to snatch its prey from ground-level.

Plumage

It has often been suggested that plumage originated for purposes of thermoregulation, either to conserve body heat (e.g. OSTROM 1974: DE BEER 1975) or to offer protection from solar radiation (e.g. REGAL 1975). In all probability the plumage could have served either function, depending on circumstances. Elsewhere it has been suggested that plumage originated in juvenile theropods (THULBORN 1984a), thus enabling them to compete actively with bigger animals that were inertial homoeotherms. This suggestion – that plumage originated as a juvenile adaptation and then persisted in smaller adult theropods and birds – does not imply that natal down is the original or "primitive" type of feather.

In *Archaeopteryx* it is likely that the general body plumage served for purposes of thermoregulation. According to the ideas outlined in this paper the plumage might also have served as waterproof insulation while *Archaeopteryx* splashed through shallow waters in pursuit of its prey.

The function(s) of the tail plumage cannot be ascertained with any great degree of confidence. The feathered tail might have played a role in locomotion, as a device to control stability (as suggested above), or perhaps as "a braking mechanism in a fleet-footed, bipedal animal" (OSTROM 1974: 41). Perhaps the tail was used in display. It is also possible that the tail was used in foraging and hunting – to disrupt the animal's body outline or to flush out prey (by being flicked from side to side or up and down, as in some modern birds). However, these are only guesses as to the possible use(s) of the tail. In promoting the idea that *Archaeopteryx* was a terrestrial predator OSTROM (1974) pointed out that long tail plumage occurs in some modern birds that seek their prey on the ground (e.g. the road runner, *Geococcyx californianus*, and the secretary bird, *Sagittarius serpentarius*). It may be added that rather long and stiff rectrices are also found in a number of piscivorous birds, such as the darter or snake bird (*Anhinga anhinga*) and the shags and cormorants (*Phalacrocorax* spp.).

The feathered "wing" has usually been regarded as an adaptation for aerial locomotion. FEDUCCIA & TORDOFF (1979) reinforced this widely held view by claiming that asymmetry of the "remiges" was an unmistakable indicator of aerodynamic function. This is not necessarily true. The degree of asymmetry of the feathers might instead be correlated with their size, their spatial arrangement and their mobility.

That is, a series of feathers or feather-like structures can be used in any of numerous (hypothetical) ways to construct a wing, and some patterns of wing construction will be more economical than others. In the avian wing the shafts of the feathers do not radiate from a single focus; they are strung out in series. Nor are the feathers uniform in size (transverse width). Moreover, the feathers in different locations (distal or proximal) are extended to different degrees when the whole wing is extended. So, for example, a distal primary might divaricate much more from the sagittal plane of the body than would a proximal secondary. With these basic constructional and mechanical limitations the differing degrees of asymmetry from distal primary (strongly asymmetrical) to proximal secondary (close to symmetrical) may reflect nothing more than an economical method of wing design. Such a wing design might ensure (a) that there would be no excessive (and functionless) overlap of one feather on to another, and (b) that there would be no excessive development of slots (gaps) between the feathers. Thus the avian pattern of wing structure may reflect nothing more than economy of design using feathers which show a proximo-distal gradient in size and mobility. The primary advantages of that wing structure could have lain in the following attributes: its economy (without excessive imbrication of feathers), its continuity (without slots), its lightness, its modifiable shape, and its easy repair and maintenance. The aerodynamic and hydrodynamic advantages of such a wing might have been exploited only secondarily.

The idea that *Archaeopteryx* was adapted for aerial locomotion is founded on the presence of avian characters in the animal. An analysis presented elsewhere (THULBORN 1984a) demonstrates that such characters are very few and that *Archaeopteryx* is no more bird-like in its osteology than are theropods such as *Tyrannosaurus* and *Allosaurus*. In terms of its skeletal anatomy *Archaeopteryx* is nowhere near attaining the integrated system of coadaptations that forms the avian flight mechanism. Consequently it is rather doubtful that the animal was adapted for aerial locomotion. These findings raise an obvious question: what could have been the function(s) of a feathered "wing" in an animal incapable of aerial locomotion? OSTROM's answer to this question (1974) was that the feathered "wings" enhanced the prey-catching function of the hands – that their large surfaces were used to trap insects. The answer presented here is slightly different: the "wings" were indeed used to enhance the prey-catching function of the hands (and jaws) – but they were used in the form of a canopy or "sunshade" to reveal (or even to attract) aquatic prey. In short, *Archaeopteryx* may have hunted and foraged in much the manner of the existing herons and egrets (Ardeidae). These birds employ many styles of fishing, a number of which involve the use of the wings (see, for example, MEYERRIECKS 1972; KUSHLAN 1976). In some cases these birds may use the technique of "wing-flicking"; this presumably startles prey into activity, thereby enabling the bird to seize animals not previously disturbed by its approach. Sometimes one or both of the wings will be held out laterally to cast shadows on the water; or sometimes the wings are brought forwards to form a canopy beneath which the bird seeks its prey – a technique developed to a remarkable extreme by the African Black Heron, *Egretta (Melanophoyx) ardesiaca* (see, for example, MILSTEIN & HUNTER 1974). On some occasions herons stand upright with wings extended, while on other occasions they may crouch so low that the tips of the wings enter the water. The advantages of under-wing or canopy feeding are several: startled fishes may flee into the shade of the canopy, where they can be seized; the shadow of the wings eliminates reflections and surface glare, thus providing a clear field of view for hunting; side-lighting may reveal fishes in the waters surrounding the shadowed area; against the dark background of the canopy any striking movement of the predator's head will be invisible to the prey. It is also possible that a low crouching posture, with wings extended, allows a heron to approach more closely to its prey when the water is clear (e.g. RECHER & RECHER 1972). Today herons and egrets use these and other techniques in fresh waters, in tidal pools, and on reefs and shores.

In summary, *Archaeopteryx* may have resembled the existing herons and egrets in some aspects of its foraging behaviour (Fig. 2). That is, *Archaeopteryx* may have stalked through shallow waters and pools, perhaps stirring up potential prey with its feet (and maybe even with flicking movements of its tail). The "wings" may have been extended as a canopy or "sunshade" to assist in revealing aquatic prey, or perhaps even in attracting such prey. Once its prey was located, *Archaeopteryx* might have made a swift forwards lunge to seize it with the jaws. The reduced tail probably ensured that *Archaeopteryx* was perfectly poised for such a lunge. Alternatively *Archaeopteryx* might have swept its clawed hands

Fig. 2
Archaeopteryx foraging, with wings extended as a canopy. For clarity the animal is shown in high-standing posture in very shallow water. It might conceivably have used a lower (crouching) posture and might have foraged in deeper water.

inwards from the edge of the canopy, attempting to grasp the prey or to drive it towards the jaws. It is also conceivable that jaws and hands were used in concert to snatch the prey. Subsequently the hands would be used to manipulate and orient the prey before it was swallowed whole. In plunging into the shallows in pursuit of its prey *Archaeopteryx* would have been protected to some extent by its gastralia and its waterproof plumage.

[R.A.T.]

Speculations on the Origin of Avian Flight

Currently there are two main lines of thought on the origin of powered avian flight: (1) that flight was originated by arboreal creatures, perhaps via stages of parachuting and gliding (e.g. HEILMANN 1926; DE BEER 1954), and (2) that flight was originated by ground-based creatures, perhaps via the flapping of feathered forelimbs to assist in the capture of prey (e.g. OSTROM 1974; CAPLE et al. 1983). The preceding suggestions about the biology of *Archaeopteryx* prompt some quite different speculations on the origin of flight. For the sake of brevity those speculations are presented as a sequence of hypothetical phylogenetic stages:

Stage 1. Terrestrial stem-group birds (theropod dinosaurs), with a framework of avian characteristics developed initially for purposes of predation (e.g. tridactyl manus, furcula, reversed hallux). Plumage developed as insulating blanket in juveniles, thereby allowing them to match activities of adults that were inertial homoeotherms.

Stage 2. A shift to foraging in shallow water (*Archaeopteryx*). Body plumage serves as waterproof insulation; "wings" serve as canopy for hunting in the fashion of existing herons and egrets. Asymmetry of feathers a constructional requirement of "wing" formation.

Stage 3. A shift to swimming and feeding at (or near) the surface of near-shore waters. Body plumage continues to serve as waterproof insulation and might perhaps enhance buoyancy. Asymmetry of feathers anticipates hydrodynamic requirements. This stage is not entirely speculative since it is known that some early stem-group birds were capable of swimming (COOMBS 1980).

Stage 4. Flapping of feathered forelimbs to assist in carrying the animal from wave-crest to wave-crest; updraughts from wave-fronts may assist in providing lift. This innovation greatly extends the animal's feeding range, and might allow it to descend on unsuspecting prey. Precise aerodynamic control is not required: these first tentative fliers can splash down safely into the water. Asymmetry of feathers meets aerodynamic requirements.

Stage 5. Continued improvements in flapping ability, wing structure and aerodynamic control permit the first sustained flights over water (e.g. ?Enantiornithes, ?*Ichthyornis*).

Stage 6a. Shift to foot-propelled diving after prey (e.g. Hesperornithiformes), and...

Stage 6b. Shift to wing-propelled diving after prey (e.g. penguins, auks), and...

Stage 6c. Return to littoral environments and shore-line foraging (e.g. Charadriiformes), and...

Stage 6d. Return to terrestrial environments.

Stage 2 (*Archaeopteryx*) is dated as Late Jurassic, and stage 5 (appearance of the first sustained fliers) would be dated as Late Cretaceous. Stages 6a to 6d encompass an adaptive radiation from stage 5, though it is possible that some lines of this radiation emerged directly from stage 4, before the advent of sustained fliers.

This sequence of stages is hypothetical, but it does have certain points in its favour. First, this hypothesis accords with the commonly remarked fact that most Cretaceous birds are maritime or littoral types. Second, this hypothesis might explain why some apparently "experimental" flying birds (Enantiornithes) are of such relatively late stratigraphic occurrence (Upper Cretaceous). Third, all the changes envisaged might have conferred immediate benefits on the animals involved (i.e. there are no "inadaptive stages"). Fourth, the earliest flying birds, with rudimentary flight apparatus, would have performed their aerial exploits without fear of injury; they would simply have splashed down safely into the water. The first poorly controlled fliers might have been at much greater risk over dry land or among trees. And, last, the emergence of the first flying birds in the Late Cretaceous might account for the demise of pterosaurs shortly thereafter, perhaps as a consequence of direct competition.

In summary, we suggest that the first avian fliers did not launch themselves from the trees; nor did they leap from the ground. We suggest, instead, that they took off from the waves.

[T. L. H. & R. A. T., with priority determined by toss of a coin]

Acknowledgements

DOUGLAS DOW and JIRO KIKKAWA helped with information on bird behaviour, and INGE MCDOWALL translated the Abstract into German.

References Cited

ALEXANDER, R. McN. (1967): Functional Design in Fishes. – 160 pp. London (Hutchinson).
BARSBOLD, R. (1983): Khishchnie dinozavri Mela Mongolii [Carnivorous dinosaurs from the Cretaceous of Mongolia]. – Trudy Sovetsko-Mongolskaya Paleont. Eksped. [Trans. Joint Soviet-Mongolian Palaeont. Exped.], **19**: 1–120; Moscow.

Beer, G. R. de (1954): *Archaeopteryx lithographica*, a study based upon the British Museum specimen. – 68 pp. London (British Museum [Natural History]).

Beer, G. R. de (1975): The evolution of flying and flightless birds. – In Head, J. J. (Ed.), Oxford Biology Readers, London (Oxford University Press); **68**: 1–16.

Caple, G., Balda, R. P. & Willis, W. R. (1983): The physics of leaping animals and the evolution of preflight. – Amer. nat., **121**: 455–476; Chicago.

Coombs, W. P. (1980): Swimming ability of carnivorous dinosaurs. – Science, **207**: 1198–1200; Washington.

Eldredge, N. (1979): Cladism and common sense. – In Cracraft, J. & Eldredge, N. (Eds), Phylogenetic analysis and paleontology, New York (Columbia University Press): 165–198.

Feduccia, A. & Tordoff, H. B. (1979): Feathers of *Archaeopteryx*: asymmetric vanes indicate aerodynamic function. – Science, **203**: 1021–1022; Washington.

Freeman, P. W. (1984): Functional cranial analysis of large animalivorous bats (Microchiroptera). – Biol. J. Linn. Soc., **21**, 4: 387–408; London.

Galton, P. M. (1971): Manus movements of the coelurosaurian dinosaur *Syntarsus* and opposability of the theropod hallux. – Arnoldia (Rhodesia), **5**, 15: 1–8; Salisbury.

Hecht, M. K. & Tarsitano, S. (1982): The paleobiology and phylogenetic position of *Archaeopteryx*. – Geobios, Mem. spec., **6**: 141–149; Lyon.

Heilmann, G. (1926): The origin of birds. – 208 pp. London (Witherby).

Jaekel, O. (1929): Die Spur eines neuen Urvogels (*Protornis bavarica*) und deren Bedeutung für die Urgeschichte der Vögel. – Paläont. Z., **11**: 201–238; Stuttgart.

Kushlan, J. A. (1976): Feeding behavior of North American herons. – Auk, **93**: 86–94; Lawrence (Kansas).

Meyerriecks, A. J. (1972): Diversity typifies heron feeding. – Natural Hist., **71**, 6: 48–59; New York.

Milstein, P. le S. & Hunter, H. C. (1974): The spectacular Black Heron. – Bokmakierie, **26**, 4: 93–97; Johannesburg.

Olson, S. L. & Feduccia, A. (1979): Flight capability and the pectoral girdle of *Archaeopteryx*. – Nature, **278**: 247–248; London.

Ostrom, J. H. (1974): *Archaeopteryx* and the origin of flight. – Q. Rev. Biol., **49**: 27–47; Stony Brook (New York).

Ostrom, J. H. (1976a): *Archaeopteryx* and the origin of birds. – Biol. J. Linn. Soc., **8**, 2: 91–182; London.

Ostrom, J. H. (1976b): Some hypothetical anatomical stages in the evolution of avian flight. – Smithsonian Contribs Paleobiol., **27**: 1–21; Washington.

Ostrom, J. H. (1978): The osteology of *Compsognathus longipes* Wagner. – Zitteliana, **4**: 73–118; Munich.

Recher, H. F. & Recher, J. A. (1972): The foraging behaviour of the Reef Heron. – Emu, **72**: 85–90; Melbourne.

Regal, P. J. (1975): The evolutionary origin of feathers. – Q. Rev. Biol., **50**: 35–66; Stony Brook (New York).

Russell, D. A. (1972): Ostrich dinosaurs from the Late Cretaceous of Western Canada. – Can. J. Earth Sci., **9**: 375–402; Ottawa.

Thulborn, R. A. (1984a): The avian relationships of *Archaeopteryx*, and the origin of birds. – Zool. J. Linn. Soc., **82**: 119–158; London.

Thulborn, R. A. (1984b): Preferred gaits of bipedal dinosaurs. – Alcheringa, **8**: 243–252; Adelaide.

Thulborn, R. A. & Wade, M. (1984): Dinosaur trackways in the Winton Formation (mid-Cretaceous) of Queensland. – Mem. Qd Mus., **21**, 2: 413–517; Brisbane.

Wellnhofer, P. (1974): Das fünfte Skelettexemplar von *Archaeopteryx*. – Palaeontographica, A, **147**, 4–6: 169–216; Stuttgart.

Authors' address: Dr. Richard A. Thulborn and Mr. Tim L. Hamley, Department of Zoology, University of Queensland, St. Lucia, Queensland 4067, Australia.

D. W. Yalden

Forelimb Function in *Archaeopteryx*

Abstract

Ostrom has argued forcefully for the view that *Archaeopteryx* was a terrestrial, cursorial, bipedal, predator, in opposition to the view, more generally held, that it was an arboreal predator. Ostrom's interpretation of the ungual phalanges and their horny claws is an important part of this argument.

The negligible flexor tubercle and shallow curvature of the ungual phalanges of the pes certainly seem to indicate a terrestrial function, as Ostrom argued. The highly curved, very sharp ungual phalanges and claws of the manus, which Ostrom suggested as indicative of a predatory function, seem better explained as indicative of a climbing function. So too does the orientation of these claws.

If *Archaeopteryx* climbed up tree trunks in the manner of a squirrel, it would require claws which were orientated at 90° to the surface of the wing – apparently it has. It would grasp the trunk using, especially, large pectoralis muscles – these too it apparently had. If it were invariably to climb upwards and fly (or glide) downwards, its hind limbs would not be specialized in the same way. Moreover, this would be commensurate with elongated fore limbs as compared to hind limbs. This too accords with what we find.

Zusammenfassung

Ostrom hat nachdrücklich die Ansicht vertreten, daß *Archaeopteryx* ein terrestrischer, biped laufender Räuber war, im Gegensatz zur verbreiteten Meinung, daß er ein arboricoler Räuber war. Ostrom's Interpretation der Krallenglieder und ihrer Hornkrallen ist ein wichtiger Teil seines Arguments.

Das zu vernachlässigende Tuberculum flexoris und die schwache Krümmung der Krallenglieder des Fußes scheinen sicher eine terrestrische Funktion anzuzeigen, wie Ostrom meint. Die stark gekrümmten, sehr scharfen Krallenglieder und Krallen der Hand, die nach Ostrom eine räuberische Funktion anzeigen, scheinen sich besser als Indiz für eine Kletterfunktion erklären zu lassen.

Falls *Archaeopteryx* Baumstämme nach Art eines Eichhörnchens hinaufkletterte, benötigte er Krallen, die unter einem Winkel von 90° zur Flügelfläche orientiert waren. Dies war offenbar der Fall. Er würde sich am Stamm ankrallen unter Einsatz großer Pectoralis-Muskeln, die er offenbar auch besaß. Wenn *Archaeopteryx* stets aufwärts kletterte und abwärts flog (oder glitt) würden seine Hinterbeine nicht in der gleichen Weise spezialisiert sein. Darüberhinaus würde dies in Einklang stehen mit gegenüber den Hinterbeinen verlängerten Vordergliedmaßen. Auch dies entspricht dem, was wir finden.

Introduction

In reformulating the theory that birds originated as terrestrial, cursorial, predators, Ostrom (1974) drew attention to the nature of the ungual phalanges and claws of *Archaeopteryx*. As he noted, the claws of the hand are highly curved, sharp and match those of various predators. Conversely, the phalanges of the feet seemed to him less highly curved, the claws less sharp, and he noted that the flexor tubercles, which are well developed on the equivalent phalanges of both predatory and perching birds, were lacking. He drew the conclusion that the feet were those of a cursorial bird, whereas the hands were those of a predator.

As someone who much prefers the concept of an arboreal ancestor for birds in general, and envisages *Archaeopteryx* as an arboreal animal, I found these arguments quite convincing, and therefore disturbing. They prompted a re-examination of claws of birds and mammals, and some further thoughts on forelimb function in *Archaeopteryx*.

Comparative Anatomy

Claws. OSTROM (1974) emphasized the sharpness of the claws in *Archaeopteryx*, particularly those of the manus. Although he depicted the ungual phalanges of several climbing birds as well as those of predators, he compared these primarily with the equivalent hind toe bones of *Archaeopteryx*, in emphasizing the less curved nature of the latter. For the manus, he drew attention to the "extreme curvature and the needle-like point" of the horny sheath of the claw (OSTROM 1974, fig. 10), and concluded by comparison with bipedal predatory theropods that *Archaeopteryx* too used its forelimbs for grasping prey.

From an examination of a moderately wide (but by no means exhaustive) range of bird and mammal claws, I would suggest that "needle-like points" to the claws are **not** characteristic of predators, but are characteristic of trunk-climbing vertebrates. Predators have conical, tapered claws, (Fig. 1, *Asio, Accipiter*). They are, indeed, highly curved in lateral view, but no more so than those of some climbers. They taper in "dorsal" (extensor) view, indicating their strength, and the fact that muscular pressure, applied at the broad base of the claw, is used to drive the tip into the prey. The tip, however, is not "needle-point" – it would be too fragile if it were.

By contrast, woodpeckers, Piciformes, do have needle-like points to their claws (Fig. 1. *Dendrocopus*); not only are the claws curved, but in extensor view they are extremely narrow. The ungual phalanges are so narrow that they bear lateral grooves where the horny claws are attached to them; the horny claws are thickened around the dorsal curve, and this thickening extends distally to give the needle-point to the claw.

The claws of swifts (Fig. 1, *Apus*) are also sharp, and highly curved, but in extensor view they are more tapered, not laterally compressed, and the points seem less "needle sharp". However, *Apus melba* is a cliffdwelling species; I was intrigued to note that in a tree-swift (*Hemiprocne*) of S. E. Asia, the claws were narrow, like those of a woodpecker, and very sharp.

Mammalian claws are constructed rather differently from those of birds; in particular, the base of the ungual phalanx is hooded, the hood overlying the base of the claw. This tends to obscure similarities of functional morphology. Trunk climbing mammals (Fig. 1, *Sciurus, Cynocephalus*) share with woodpeckers the laterally compressed form of both phalanges and their claws; moreover, the claws have needle-sharp points. The most extreme cases of both lateral compression and needle-sharp claws are seen in the bats (Fig. 1, *Pteropus, Hipposideros*). Moreover, these thin claws are thickened dorsally, round the outside of the curve, and this thickening is extended to produce the point, as in the woodpeckers.

The close resemblance of the claws of *Archaeopteryx* to this format must be obvious to anyone who has studied the magnified photographs presented by OSTROM (1974) or the photographs and diagrams in WELLNHOFER (1974). The ungual phalanges were evidently very thin – they too show the depressions laterally where the horny claws attached, and seem, as preserved, to have no great thickness. The claws show the dorsal thickening seen in woodpecker and bat claws, and this thickening extends at the tip of the claw to provide the sharp point. Moreover, although OSTROM (1974) would have regarded this as a

Fig. 1
Scale drawings of the ungual phalanges (heavy line) and horny claw (thin line) of a range of birds and mammals, showing lateral and extensor views. *Archaeopteryx* claws (after WELLNHOFER 1974, fig. 13) are shown for comparison. The selection includes a perching bird which forages on the ground (magpie, *Pica pica*); two avian predators which use their feet to grasp prey (long-eared owl, *Asio otus*; sparrowhawk, *Accipiter nisus*); a cliff nesting bird (alpine swift, *Apus melba*); a trunk climbing woodpecker (lesser spotted woodpecker, *Dendrocopus minor*, and (claw only, from a study skin) white-backed woodpecker, *D. leucotus*); two trunk climbing mammals (grey squirrel, *Sciurus carolinensis*; cobego, *Cynocephalus volans*); two bats (fruit bat, *Pteropus* sp., leaf nosed bat *Hipposideros* sp.) and a ground-dwelling game bird (grey partridge, *Perdix perdix*).

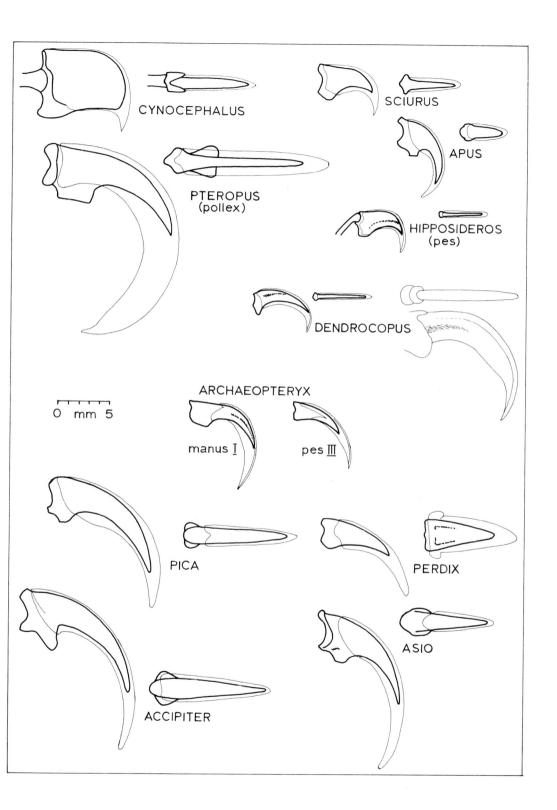

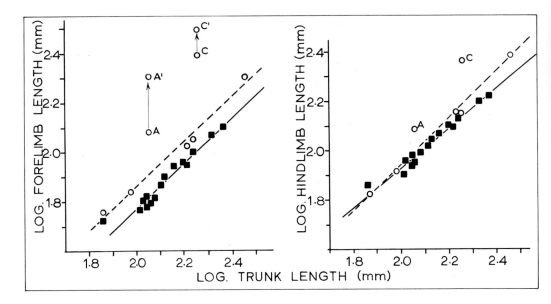

Fig. 2
Relative limb lengths for *Archaeopteryx* and *Cynocephalus* plotted on the graphs (from THORINGTON & HEANEY, 1981) of limb lengths in squirrels. The lower points for *Archaeopteryx* and *Cynocephalus* are those calculated (as by THORINGTON & HEANEY) with forelimb length as (length humerus + length radius); the upper points are plotted with the length of the longest digit added.

description of the claws of the hand, it is clear that it also applies to the claws of the foot (vide WELLNHOFER 1974, pl 22, fig. 5).

OSTROM (1974) argued that the ungual phalanges of the pes had a shallow curvature, and matched those of ground-living birds such as game birds. Such phalanges, and their claws, are very broad in extensor view; while all the equivalent claws of *Archaeopteryx* are preserved in lateral view, none suggest any great lateral thickness.

Claw Orientation. I am grateful to THULBORN & HAMLEY (1982) for pointing out that the claws of the hand in *Archaeopteryx* have toppled or twisted to lie on their sides; they pointed this out in respect of the Berlin specimen (their Fig. 4), but it is also evident in the Eichstätt specimen (WELLNHOFER, 1974, pl. 22, Fig. 3). The claws evidently, in life, pointed ventrally from the wing surface. This does not preclude them from functioning, as OSTROM (1974) argued, in predation, but any other orientation would preclude them from functioning in tree climbing. The crossing of the second and third fingers, seen in the Berlin specimen (both wings) and the Eichstätt specimen (right wing, in particular), has been regarded by previous authors (up to and including YALDEN, 1984) as the natural condition. The arguments given here, and the Maxberg specimen (WELLNHOFER, this volume, Fig. 1) make it clear that this was not the case.

Forelimb Proportions. Elongated forelimbs are characteristic of gliding and flying vertebrates, as OSTROM (1974) acknowledged. THORINGTON & HEANEY (1981) have recently documented this by comparing gliding squirrels with their non-gliding relatives. In squirrels, the hind limbs are not elongated in the gliding forms, by comparison with the non-gliding forms, but they are in *Cynocephalus*. In its limb proportions (expressed in relation to trunk length), *Archaeopteryx* has more elongate limb bones than flying squirrels, but its limb bones (humerus + radius, femur + tibia – as considered by THORINGTON & HEANEY, 1981) are less elongate than those of *Cynocephalus* (Fig. 2). If the length of the longest finger is added to the forelimb measurement, then both *Cynocephalus* and *Archaeopteryx* have forelimbs that are 1.7 times their trunk length.

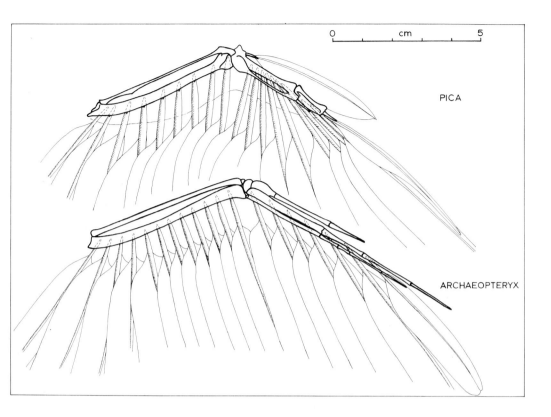

Fig. 3
Scale drawings of the skeleton of the forearm of magpie *Pica* and *Archaeopteryx*, to show the insertion of the remiges. The skeleton is drawn in ventral view; the remiges, which insert dorsally on the bones of the hand in *Pica*, and presumably did also in *Archaeopteryx*, are shown in dotted lines where they pass behind the bones.

Feather Insertions and "Free Fingers". It seems to be generally accepted that *Archaeopteryx* had "three free, clawed, fingers". HURST (1895) certainly argued that the evident fingers (1, 2, and 3, without wishing to prejudge the issue of their homology) did not support the primary feathers, but then he reasoned that fingers 4 and 5 of the pentadactyl hand were still present in *Archaeopteryx*, in a plane in the rock below that of the feathers. (He was arguing primarily from and about the Berlin specimen). The fact that, in the London specimen, the wings have remained intact, despite the disintegration of the skeleton of the hand, has also been taken to imply that the flight feathers had at best only a slight attachment to those bones. OSTROM (1974) has pointed out, too, that the ulna of *Archaeopteryx* shows no sign of the nodes to which the secondary feathers attach; this might be further evidence that the connection between the wings and forearm skeleton was only slight, though it does not impinge directly on the freedom of the fingers.

In the wing of a modern bird (Fig. 3, *Pica*), the secondaries are strongly attached to the ulna. They may attach above quite pronounced nodes, or tubercles, as they do in *Pica*, and also (in my reference collection) in *Vanellus, Cuculus,* and *Picus*, to take examples from four orders. On the other hand, the nodes may be very slight, just discernible if a finger-nail is scraped down the ulna, or quite absent; again, to quote a few examples, this is the case in my reference skeletons of *Lagopus, Gallinula, Fulmarus, Falco, Strix* and *Asio.* Four skeletons of Sparrowhak, *Accipiter nisus,* are particularly instructive; three have faint nodes and the fourth, a 2 year-old male, has no evident nodes. I conclude that the absence of

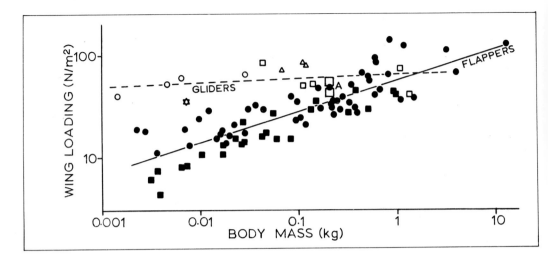

Fig. 4
Graph of wing loading against body mass (on logarithmic scales) for gliding and flapping-flight vertebrates (from RAYNER, 1981), with *Archaeopteryx* added, to emphasize its intermediate placement. Mass of *Archaeopteryx* 250 g, wing area strictly 479 cm^2 (upper point) or 668 cm^2 if the tail area and a more generous body strip estimate is used (lower point) (from YALDEN 1971, 1984).

strong nodes on the ulna of *Archaeopteryx* is not a significant feature, and that it too had the secondary feathers strongly attached to the ulna. Moreover, it should be noted that the feathers do not attach directly to the nodes but pass dorsally past them, to insert on the dorsal side of the ulna.

Turning to the primary feathers, these attach, in a modern bird, to the dorsal surface of metacarpal 2, and to the dorsal surface of the two phalanges of digit 2; the much slenderer shaft of metacarpal 3, fused into the composite carpo-metacarpus, is scarcely involved in this feather attachment, though it does underlie the bases of the shafts of the primaries (Fig. 3). In a passeriform, such as the magpie *Pica*, there are nine full and one reduced primary; of these, seven attach to metacarpal 3, two to the proximal phalanx of digit 2, and the reduced 10th primary attaches to the distal phalanx. Though there has been some argument about the precise number of primaries in *Archaeopteryx*, most authors who have examined the Berlin specimen (with much the best preserved plumage) recognise 7; that seemed to me, interpreting photographs, to be correct (Fig. 3). Of these, it appears that four attached to metacarpal 2, two attached to the proximal phalanx of digit 2 and the outermost one attached to the base of the middle phalanx of digit 2. This arrangement not only accords with the likely homology of the bones with those in a modern bird, but also produces a convincing geometric arrangement, with the feather shafts almost parallel, as they indeed appear to be in the Berlin specimen. It follows that the second finger was not "free" – only its terminal, clawed, phalanx was. Further, the third finger must have underlain the bases of the primary feathers, as it does in modern birds, and it could not have been "free" either; but its claw presumably projected from the ventral surface of the wing. Except that he counted twelve primaries, the reconstruction (Fig. 3) resembles that by HEILMANN (1926) quite closely. RIETSCHEL (this volume) demonstrates clearly that there were in fact 10 or 11 primary feathers, but the "extra" ones seem to insert onto metacarpal 2, leaving the three other feathers to insert, as shown in Fig. 3, on the phalanges.

Discussion

The analogies and homologies discussed here suggest to me that *Archaeopteryx* was a trunk-climbing vertebrate, and further suggest a moderately close analogy with the cobego *Cynocephalus*. The curved claws, their orientation, the elongated limbs, and the general anatomy of the hand all conform to that analogy at least as well as to Ostrom's predatory analogy, and the sharpness of the claws fits a tree climbing analogy better. OSTROM (1974) argued that a "fly-swatting" predator would need powerful pectoralis muscles, and thus explained the development of the downstroke muscles in birds. Equally, an animal hugging a trunk in the manner of a squirrel or cobego would need powerful pectoralis muscles. OSTROM further questioned the arboreality of *Archaeopteryx* on the grounds that the claw of the hallux is relatively short; perching birds, as he noted, tend to have long hallucial claws. However, this is not true of trunk climbers – Piciformes, *Sitta, Certhia* and similar birds tend to have claws of more nearly equal length on their toes.

I do not wish to pursue the analogy with *Cynocephalus* too far; I am certainly not suggesting that *Archaeopteryx* was at the gliding stage of flight evolution, as is that mammal. The highly asymmetrical primary feathers (FEDUCCIA & TORDOFF, 1979), the fact that primaries and secondaries are differentiated, their curved shafts (R. A. NORBERG, this volume), the stout furcula (OLSON & FEDUCCIA, 1979), the prominent deltopectoral crests on the humerus (YALDEN 1971), and the depth of the furcula, projecting 6mm below (ventrally to) the coracoid plate (YALDEN, 1984) all indicate that flapping flight was possible, and that the pectoralis muscles were well developed.

RAYNER (1981) has argued the energetic advantages of arborealty and gliding, as well as discussed the stages of evolution of flapping flight. His graphs of wing loading against mass for gliders and flappers converge towards a body mass of 1 kg, but the scatter of points round both lines is generous. If *Archaeopteryx* had a mass of 250 g and a wing area of 479 cm^2 (YALDEN, 1971, 1984), its wing loading would have been 51 N/m^2, a figure nicely intermediate between flappers and gliders of that mass (Fig. 4). Given the morphology of its rib cage, *Archaeopteryx* was not capable of sustained flapping flight; as in so much of its anatomy, so in its habits, it seems to have been at an intermediate stage.

Acknowledgements

I thank Dr. M. V. HOUNSOME, Keeper of Zoology, Manchester Museum, for access to specimens in his care, and Mrs. PAMELA HALL for typing the paper. I also wish to thank the University of Manchester for a staff travel grant which helped me to attend the Conference.

References Cited

FEDUCCIA, A. & TORDOFF, H. B. (1979): Feathers of *Archaeopteryx*: asymmetric vanes indicate aerodynamic function. – Science **203**: 1021–1022; New York.
HEILMANN, G. (1926): The origin of birds. London (Witherby).
HURST, C. H. (1895): The structure and habits of *Archaeopteryx*. – Stud. Biol. Dept. Owens Coll. **3**: 267–287; Manchester.
NORBERG, R. A. (1985): Function of vane asymmetry and shaft curvature in bird flight feathers: inferences on flight ability of *Archaeopteryx*. – This volume.
OLSON, S. L. & FEDUCCIA, A. (1979): Flight capability and the pectoral girdle of *Archaeopteryx*. – Nature **278**: 247–248; London.
OSTROM, J. H. (1974): *Archaeopteryx* and the origin of flight. – Quart. Rev. Biol. **49**: 27–47; Stony Brook, New York
RAYNER, J. M. V. (1981): Flight adaptations in vertebrates. – Symp. Zool. Soc. Lond. **48**: 137–172; London.
RIETSCHEL, S. (1985): Feathers and wings of *Archaeopteryx*, and the question of her flight ability. – This volume.
THORINGTON, R. W. & HEANEY, L. R. (1981): Body proportions and gliding adaptations of flying squirrels (Petauristinae). – J. Mammal. **62**: 101–114; Lawrence, Kansas.
THULBORN, R. A. & HAMLEY, T. L. (1982): The reptilian relationships of *Archaeopteryx*. – Aust. J. Zool. **30**: 611–634. Melbourne.
WELLNHOFER, P. (1974): Das fünfte Skelettexemplar von *Archaeopteryx*. – Palaeontographica, A, **147**: 216; Stuttgart.
WELLNHOFER, P. (1985): Remarks on the digit and pubis problems of *Archaeopteryx*. – This volume.
YALDEN, D. W. (1971): The flying ability of *Archaeopteryx* – Ibis, **114**: 349–356; London.
YALDEN, D. W. (1984): What size was *Archaeopteryx*? – Zool. J. Linn. Soc., **82**: 177–188; London.

Author's address: Dr. D. W. YALDEN, Department of Zoology, University of Manchester, Williamson Building, Manchester M13 9PL, United Kingdom

Allen Peterson

The Locomotor Adaptations of *Archaeopteryx*: Glider or Cursor?

Abstract

The modern avian features of *Archaeopteryx* (flight feathers, wing structure, furcula) indicate that the opportunity for extensive selective morphological change through natural selection existed. Therefore it may not be appropriate to assume that all the reptilian characteristics of *Archaeopteryx* were unused vestiges. If one assumes that the reptilian characteristics of *Archaeopteryx* were in fact useful structures that were selectively retained, a different perspective of the evolution of avian flight is obtained.

The relatively long limbs and neck of *Archaeopteryx* are not traits of an arboreal leaping organism, rather they indicate a cursorial existence. In addition, arboreal gliders tend not to manuever in flight, however, the elliptical wings and long tail of *Archaeopteryx* are adaptations for manueverability. The undeveloped pectoral anatomy of *Archaeopteryx* in the presence of well developed, lift producing wings suggests that wing development may not have initially occurred to enhance aerial capabilities.

If wing and feather development first occurred to improve the braking and turning abilities of a cursorial predator, the lag between wing and pectoral development is explained.

Zusammenfassung

Gewisse Züge, die *Archaeopteryx* mit heutigen Vögeln gemeinsam hatte (Flugfedern, Flügelbau, Furcula) zeigen, daß die Möglichkeit für morphologische Veränderungen durch natürliche Auslese existierte. Es könnte daher angenommen werden, daß alle reptilartigen Merkmale von *Archaeopteryx* nicht verkümmerte Glieder waren. Nimmt man an, daß diese Merkmale tatsächlich brauchbare Strukturen waren, die durch natürliche Auslese beibehalten wurden, so ergibt sich eine neue Perspektive der Evolution des Vogelflugs.

Die verhältnismäßig langen Glieder sowie der lange Hals von *Archaeopteryx* sind nicht die Merkmale eines arboreal springenden Vogels, sondern weisen auf die eines Laufvogels hin. Außerdem machen arboreale Gleiter im allgemeinen keine Manöver während des Fluges; die elliptischen Flügel und der lange Schwanz von *Archaeopteryx* waren jedoch Anpassungen an bessere Manövrierfähigkeit. Wir können auf Grund der unentwickelten Brustanatomie bei gleichzeitigem Vorhandensein auftriebserzeugender Flügel annehmen, daß die Flügelentwicklung ursprünglich nicht stattgefunden hat, um die Flugfähigkeit zu fördern.

Nehmen wir an, daß Feder- und Flügelentwicklung zuerst stattfand um die Halt- und Lenkungsfähigkeit eines "cursorialen" Raubvogels zu fördern, so ist das Zurückbleiben der Brust – gegenüber der Flügelanatomie zu erklären.

Introduction

There recently has been a radical change in our perception of the Mesozoic reptiles from lumbering, evolutionary misfits to swift, highly advanced and extremely successful organisms. This change in perception appears now to be focusing on *Archaeopteryx lithographica* and the evolution of flight. In the 1970's, OSTROM challenged the traditional, arboreal theory in a series of papers (1974, 1976a, 1976b, 1978, 1979) wherein he proposed that *Archaeopteryx* was a coelurosaurian, cursorial predator not unlike *Deinonychus* or *Ornitholestes*. He also speculated that contour feathers, originally for heat conservation of endotherms, were selectively modified at the wingtips to serve as elongated swatters to knock insects out of the air. CAPLE, WILLIS and BALDA (1983) developed a detailed physical model of a cursorial evolution of flight, however they rejected OSTROM's insect swatter idea and pictured wing development as the means of controlling body orientation during a leap to capture, in the jaws, flying insects.

The purpose of this paper is to discuss some of the locomotor adaptations possessed and not possessed by *Archaeopteryx* and compare them with other vertebrates, primarily arboreal leapers and terrestrial cursors for clues to the origin of avian flight. I also comment upon CAPLE's model of a cursorial proavis. Many of the ideas presented are gleaned from an earlier, unpublished version of this paper which was written in 1979.

Many earlier writers, including ROMER (1959) saw *Archaeopteryx* as a feeble evolutionary link to the modern birds with almost as many vestiges as useful adaptations. This viewpoint may have biased their conclusions about the evolution and lifestyle of the early birds. In this paper, I do not assume that *Archaeopteryx* was a "clumsy" or "primitive" life form nor that its reptilian features were simply unused vestiges. Instead, I assume that in order to have survived the highly competitive Mesozic reptilian radiation, *Archaeopteryx* and its proavian ancestors must have each been successful end products in their own right. I assume that they were highly specialized reptilian predators that lost structures they did not use, evolved structures they needed and did not evolve structures they did not need. This perspective leads me to radically different conclusions than those of some earlier writers.

The Body Form

Postulating upon the origin of flight via examination of *Archaeopteryx* is admittedly risky. One is forced to assume *a priori* that *Archaeopteryx* was among other things, a mainline descendent of a coelurosaurian proavis that had undergone no major changes from adaptive radiation or reduction of vestiges. However, due to its many similarities with other coelurosaurian genera, this seems a defensible assumption. OSTROM (1976) also defends this assumption and details the coelurosaurian traits of *Archaeopteryx*.

In addition to the conditions in *Archaeopteryx* listed by OSTROM (1974) which support the cursorial theory, there are other aspects of the general body form of *Archaeopteryx* which support the cursorial but not the arboreal theory. The selective pressures upon vertebrate arboreal leapers are markedly different from those exerted upon terrestrial cursors. Arboreal leapers, as typified by the genus *Sciurus* undergo a lengthening of the trunk and a shortening of the limbs (GUNDERSON 1976). This trend is also clearly seen in the arboreal gliders like *Glaucomys* and *Draco*. The limb to trunk ratio in these species approximates 1 : 2, depending on the degree of arboreal habit. A trend toward a short, sturdy neck is also seen in these arboreal leapers and gliders. Terrestrial cursors however, are exposed to radically different selective pressures. They generally experience a lengthening of the limbs and fusion or loss of the metatarsals and metacarpals to increase running speed. The limb to trunk ratio in these species approximates 1 : 1. They also tend to possess elongated necks to shift the center of mass forward during acceleration (GUNDERSON 1976).

Proponents of the arboreal theory hold that *Archaeopteryx*'s ancestors radiated into arboreal habitats and evolved through leaping, parachuting, and gliding stages to powered flight (BOCK 1965, 1969). However, *Archaeopteryx* does not exhibit the trunk elongation or limb and neck shortening seen in arboreal leapers and gliders. It has, however, retained all the above noted cursorial adaptations in addition to those presented by OSTROM (1974). It's limb to trunk ratio is nearly 1 : 1. This supports the hypothesis that *Archaeopteryx* was, and evolved as, a cursorial organism. It is interesting to note that ROMER's (1959) sketch of an arboreal proavis had an elongated trunk, reduced hindlimbs, and virtually non-existent neck, all good arboreal leaping adaptations, but none of which were possessed by *Archaeopteryx*, which the arborealists place at the gliding stage (BOCK 1965, 1969).

The Pectoral Structure

Another aspect of the morphology of *Archaeopteryx* usually cited as proof of its arboreal, gliding existence is its lack of flight adaptations to the pectoral anatomy. Because *Archaeopteryx* had no such adaptations for powered flight, it was assumed to be a glider and hence arboreal, since level ground is a poor place from which a glider can launch itself. However, I suggest that the lack of pectoral development may be evidence against an arboreal evolution of flight and may in fact be evidence of a cursorial evolution.

If *Archaeopteryx* and its ancestors had been arboreal leapers, with wing and feather development both distally and proximally to increase lift and extend the glide path, then force, both physical and selective would have been exerted upon the pectoral anatomy concurrent with wing and feather development. These forces would have been greater than those exerted upon modern gliders (which have few pectoral modifications), because the gliding planes in modern gliders are proximal to the body. Distal lifting surfaces exert a relatively greater force upon the pectoral skeleton than proximal ones. This force is exerted upon an airborne organism regardless of whether it flaps or merely glides. This may be experienced by leaping off a tall building with a pair of wings strapped to the arms. One's shoulders will dislocate even if one wishes only to glide.

Therefore, if proaves had evolved as arboreal leapers, one would expect the pectoral region to have developed strength and rigidity with the development of the wings and feathers. Hence one would expect *Archaeopteryx* to exhibit modifications of the pectoral anatomy proportionate to modifications of the wings. This however was not the case. Pectoral development in *Archaeopteryx* lagged far behind wing and feather development. In addition, there was no loss of flexibility (OSTROM 1976). This leads one to suspect that either the wings did not evolve to provide lift for an airborne organism, or that there were few lifting stresses upon the pectoral region during wing development, or both. Well developed wings and undeveloped pectoral structures might be seen in a predatory, cursorial organism that used wings not to enhance aerial capabilities but to assist terrestrial agility during pursuit of prey. Furthermore, *Archaeopteryx* did have a robust furcula which served as a transverse spacer to provide proper separation of the shoulder joints (OSTROM 1976). The question then must be asked, if *Archaeopteryx* evolved as an arboreal leaper and glider, why did only one bone in the entire pectoral skeleton become stronger while the rest remained unchanged? The answer may be that the furcula did not evolve as a response to lifting stresses. It may have been a completed response to stresses generated by the purely terrestrial wing functions of braking and turning. In fact, OSTROM (1976) and YALDEN (1970) both noted the presence of a sturdy furcula in an essentially non-avian pectoral girdle and OSTROM suggested non-avian functions, "perhaps predation".

The Wing and Tail Shape of *Archaeopteryx*

A third weakness of the arboreal theory is the shape of *Archaeopteryx*'s wings and tail. The arboreal theory places *Archaeopteryx* at the gliding stage. However *Archaeopteryx* did not have rectangular or pointed wings or a broad rounded tail, both of which are primary adaptations of a gliding or soaring bird (PASQUIER 1977). Instead, it had an elliptical or rounded wing with the middle primaries being the longest feathers (SAVILE 1957a). This is a highly maneuverable (SAVILE 1957b) type used by forest and brushland dwellers like the ruffed grouse (*Bonasa umbellus*) and COOPER's hawk (*Accipiter cooperi*) to execute rapid twisting and turning movements. Also, the elliptical wing, because of its rounded nature, is not an efficient type for gliding (PASQUIER 1977). The tail was long and narrow, which is an adaptation for maneuverability (PASQUIER 1977) typified by the accipitrine hawks. However, gliding is essentially a straight line affair, there is little in-flight maneuvering (GUNDERSON 1976). Therefore, if *Archaeopteryx* was an arboreal, non-powered glider, it achieved this condition without two primary gliding adaptations, the pointed wing and the broad rounded tail and instead evolved two structures it did not need, an elliptical wing and a long, narrow tail. This does not agree with, among other biological laws, the premise of BOCK and VON WAHLERT (1965) that a close correlation exists between the morphology and the function of a structure. It might be argued that the long tail is simply a primitive trait which had not yet been lost. However, the "appreciably advanced" wing (SAVILE 1957b) indicates that ample opportunity for selective morphological change existed. Assuming that the law of natural selection applied to both ends of proavis, one might also conclude that the long tail was retained in *Archaeopteryx* for a reason.

The Cursorial Theory of CAPLE et al.

The recent work by CAPLE et al. (1983) must be considered significant. When combined with the recent works of OSTROM, it presents a strong argument in favor of a cursorial evolution of flight. To their theory, I would like to propose one modification and one addition.

CAPLE et al. have proposed that proavis was a cursorial reptile which leapt to capture in the jaws, flying insects. They postulate that flight feathers evolved at the distal portion of the wings to control body orientation during the leap and in the latter stages, to extend the leap. Their theory, by itself, appears beyond any significant criticism. However, in examining the pectoral area of *Archaeopteryx*, the same question regarding its undeveloped nature in the presence of well-developed, lift producing wings arises. If the ancestors of *Archaeopteryx* evolved as terrestrial leaping organisms, stresses both physical and selective upon the pectoral anatomy would have increased with the lift capabilities of the wing. Therefore, by the time wing development was completed, one might expect to see some pectoral modifications.

CAPLE's theory of a cursorial proavis, with the legs as the primary locomotor source, explains much of the gap between wing and pectoral development in *Archaeopteryx*. I propose a modification which may explain even more. Like OSTROM, I feel that the path to flight began with a small, bipedal, reptile (coelurosaur) which had body contour feathers. It was an active, cursorial predator which lived by running down a variety of small terrestrial organisms. Since agility is a desired trait in cursorial predators, proavis would be under selective pressures for elongation of tail and distal wing contour feathers to assist in braking and turning. This differs from NOPSCA's (1907) cursorial theory in that the wings did not develop as "propellers" to increase running speed. It also differs from the theory of CAPLE et al. in that wing development did not first occur to aid airborne movements. Rather it and tail feather development began strictly to aid the agility of a purely terrestrial form and in the earlier stages resulted in few stresses to the pectoral anatomy. Because the wings and tail were used from their onset to improve maneuverability, they would evolve towards elliptical and long forms, respectively, as is ultimately seen in *Archaeopteryx*.

Initially, this proavis ran with the proto-wings and tail extended and they were "flapped" only during braking and turning. This type of proavis in pursuit of prey would somewhat resemble a modern day road-runner (*Geococcyx californianus*) with clipped wings chasing a lizard. Through all these early stages, little development of the bony structure of the pectoral skeleton was necessary, because the animal's weight was still being borne by its legs. In order to retain the ability to execute the rapid twisting movements necessary for its cursorial lifestyle, the developing proavis did not evolve the rigidity in the pectoral and thoracic areas as is seen in modern birds. Structures such as uncinate processes were not selected for because the strength they supplied to the skeleton was not necessary and the rigidity they imposed would be detrimental. Gastralia, in contrast, were retained (OSTROM 1976). This may be because they were not anchored to the vertebral column, they "floated" in the tissue. Hence they provided strength without imposing rigidity.

At some point after feather elongation began, the wings and tail may even have been beaten in alternating strokes during a run to improve the stability as well as the agility of the running organism. I propose that it is after rudimentary flight structures and possibly motions evolved on the ground that proavis began the leaping sequence so well presented by CAPLE et. al. This delay between wing development and actual take-off may explain the gap between pectoral and wing development.

I also propose an addition to CAPLE's theory that the proavis leapt after sizeable flying insects that were too large to be killed with a single snap or swallowed whole. I cite two aspects of the anatomy of *Archaeopteryx* in support of this concept. First, the facial anatomy of *Archaeopteryx* does not indicate small prey capture. Airborne organisms which capture small airborne insects tend to have short, broad bills or muzzles with wide gapes to maximize the surface area of the capturing surface, the open mouth. Modern birds exhibiting this trait include the swallows (Hirundinidae), swifts (Apodidae) and nightjars (Caprimulgidae). Birds which hunt by seizing larger prey and bringing it down for the kill tend to retain the generally pointed and narrow bill shape. This is seen in the modern bee-eaters (*Merops* spp.) jacamars (*Galbula* spp.) and in certain terrestrial birds, including the road-runner (PERRINS and HARRISON 1979). The snout of *Archaeopteryx* is sharply pointed, not unlike the theropod predators *Velociraptor* and *Ornitholestes* (OSTROM 1976).

Secondly, since a terrestrial *Archaeopteryx* by definition did not use its claws and hands for climbing about in trees, they must have served some other function, or be lost via the law of use and disuse. An *Archaeopteryx* leaping after large flying prey could have used its claws to grasp the prey after it was seized in the jaws and to assist in killing and eating the captured prey on the ground.

A final thought; the presence of scales on the face of *Archaeopteryx* may not then have been simply a reptilian vestige. The absence of feathers may have been a sanitary adaptation similar to that seen in present day vultures (Cathartidae) to prevent the soiling of feathers in an area not easily preened.

References Cited

BOCK, W.J. (1965): The role of adaptive mechanisms in the origins of higher levels of organization. – Syst. Zool. **14**: 272–287.
BOCK, W.J. and VON WAHLERT, G. (1965): Adaption and the form-function complex. – Evolution **19**: 269–299.
BOCK, W.J. (1969): The origin and radiation of birds. – Ann. N.Y. Acad. Sci. **167**: 147–155.
CAPLE, G., BALDA, R.P. and WILLIS, W.R. (1983): The physics of leaping animals and the evolution of preflight. – Am. Naturalist **121** (4): 455–476.
GOULD, S.J. (1977): The tell-tale wishbone. – Nat. History **86**: 26–38.
GUNDERSON, H.L. (1976): Mammalogy. – 483 pp. New York, St. Louis. McGraw-Hill Book Co.
NOPSCA, F. (1907): Ideas on the origin of flight. – Proc. Nat. Zool. Soc. London. pp. 223–226.
OSTROM, J.H. (1974): *Archaeopteryx* and the origin of flight. – Q. Rev. Biol. **49**: 27–47.
OSTROM, J.H. (1976a): *Archaeopteryx* and the origin of birds. – Biol. J. Linn. Soc. **8**: 91–182.
OSTROM, J.H. (1976b): Some hypothetical anatomical stages in the evolution of avian flight. – Smithson. Contrib. Paleobiol. **27**: 1–21.
OSTROM, J.H. (1978): A new look at dinosaurs. – Nat. Geographic **156** (8): 168. August, 1978.
OSTROM, J.H. (1979): Bird flight: how did it begin? – Am. Scientist **67**: 46–56.
PASQUIER, R.F. (1977): Watching birds: an introduction to ornithology. – 301 pp. Boston. Houghton Mifflin Co.
PERRINS, C. and HARRISON, C.J.O. (1979): Birds: their life, their ways, their world. – 411 pp. Pleasantville (N.Y.) Readers Digest Asoc. Inc., Original 1976 edition by H.N. Abrams, Inc., N.Y.
ROMER, A.S. (1959): The Vertebrate Story. Fourth Edition. – 437 pp. Chicago, Univ. of Chicago Press.
SAVILE, D.B.O. (1957a): Adaptive evolution in the avian wing. – Evolution **11**: 212–224.
SAVILE, D.B.O. (1957b): Primaries of *Archaeopteryx*. – Auk **74**: 99–101.
YALDEN, D.W. (1970): The flying ability of *Archaeopteryx*. – Ibis, **113**: 349–356.

Author's address: ALLEN PETERSON, RD 1, Ford Road, Owego, NY 13827, U.S.A.

Michael E. Howgate

Problems of the Osteology of *Archaeopteryx*
Is the Eichstätt Specimen a Distinct Genus?

Abstract

Differences in osteology between the Eichstätt specimen and other specimens of *Archaeopteryx* are unlikely to be due to the supposed juvenile status of the Eichstätt specimen. Neither the 'vertically' directed pubis of the Eichstätt specimen nor the 'opisthopubicly' directed pubis of the Berlin specimen need to be reoriented. Both are only slightly distorted, probably as a result of diagenetic compaction, and lie in approximately their 'life' positions. The differences in size, limbic proportions, shape of the ischium, and relative position of the pubes can therefore be regarded as of phylogenetic significance and indicate that the Eichstätt specimen is probably a distinct genus. It is proposed that the name *Jurapteryx* designate the new genus based on the Eichstätt specimen.

Zusammenfassung

Es ist unwahrscheinlich, daß die osteologischen Unterschiede zwischen dem Eichstätter und den anderen Exemplaren von *Archaeopteryx* auf den angenommenen juvenilen Zustand des Eichstätter Exemplares zurückzuführen sind. Weder das "senkrecht" orientierte Pubis des Eichstätter Exemplares, noch das nach "hinten" gerichtete Pubis des Berliner Exemplares brauchen neu orientiert zu werden. Beide sind nur leicht verdreht, wahrscheinlich aufgrund diagenetischer Kompaktion, und liegen annähernd in ihrer "Lebens"-Stellung. Die Verschiedenheiten in Größe, Gliedmaßen-Proportionen, Form des Ischiums und relativer Lage der Pubes können deshalb als phylogenetisch signifikant angesehen werden. Sie zeigen, daß das Eichstätter Exemplar wahrscheinlich zu einer anderen Gattung gehört. Für diese neue Gattung wird der Name *Jurapteryx* vorgeschlagen.

Introduction

Palaeontological opinion since DE BEER (1954) has placed all the known specimens of *Archaeopteryx* into a single monotypic species, *Archaeopteryx lithographica* VON MEYER. Elsewhere (HOWGATE 1984a) I have proposed that the Eichstätt specimen is distinct from the other (London, Berlin, Teyler and Maxberg) specimens at the specific, if not at the generic level. This was mainly based on the evidence of the teeth and a reinterpretation of the size and allometric differences previously (WELLNHOFER, 1974; OSTROM, 1976) regarded as indicative of the juvenile status of the specimen. However, I left open the question of two more problematic differences, the absence of a furcula in the Eichstätt specimen and the vertical orientation of the pubes, which would indicate that the Eichstätt specimen was distinct from the other specimens at least at the generic level. The problem of deciding which characters are of generic rather than specific value is, of course, a subjective one.

Previously I tried to weigh the evidence of the tooth difference between the Eichstätt and London/Berlin specimens in terms of four competing hypotheses, sexual dimorphism, ontogenetic difference, phylogenetic difference, and polymorphism (HOWGATE 1984a). While it is possible to regard almost any differences in osteology as merely extensive polymorphism, so long as breeding criteria cannot be applied, it becomes impractical when dealing with fossils to lump together specimens which diverge distinctly in osteology. Similarly with sexual and ontogenetic differences, small differences in relative size, development of cranial ornamentation and pelvic structure can be accomodated within these hypotheses. However differences in food gathering apparatus (teeth), locomotion (absence of furcula, longer hind limbs), and balance (orientation of the pelvis) cannot be explained in terms of the above

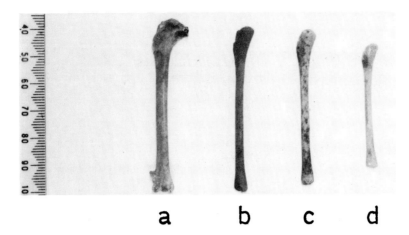

Fig. 1
Right humeri of Audubon's shearwater (*Puffinus l'herminieri*) in dorsal aspect; a) adult b,c) larger juveniles (just about fledging), d) small juvenile.

hypotheses except by introducing a degree of sexual and/or ontogenetic diversification which would be considered as unwarrented were any other fossil than *Archaeopteryx* being considered.

While many paleontologists may feel that it is better to err on the side of caution, I consider that a more parsimonious interpretation of the osteology is that those differences noted between the Eichstätt and other specimens attributed to *Archaeopteryx* are good phylogenetic characters. The osteology points to the adult status of the Eichstätt specimen, so why assume a juvenile condition? The pubes of the Eichstätt and Berlin specimens do not appear to be disarticulated, so why try to reorient them?

The juvenile(?) status of the Eichstätt specimen

The juvenile condition of the Eichstätt specimen was assumed by WELLNHOFER (1974) and OSTROM (1976) on account of its smaller size. Further anatomical differences, most notably the absence of an ossified furcula, were then interpreted in the light of this assumption. OSTROM (1976) noted that the absence of this bone may be an artefact of preservation (though this is unlikely, given the perfect articulation of the specimen and the detail preserved on the almost complete counter-slab) but concluded that "a more likely explanation is that, because of the small size and probable immaturity of the specimen, the furcula had not yet ossified" (p. 139)

Even assuming that ossification was not of the avian pattern, the absence of such a prominent bone as the furcula, while the rest of the skeleton appears well ossified, would be unusual to say the least. Particularly as the furcula is one of the first bones to ossify in the chick, "the process being completed by 14–16 days in the chick embryo" (LANSDOWN, 1968). Not only would one expect a furcula to be present in a two-thirds grown chick, as the Eichstätt specimen is assumed to be, but one would expect that other bones which ossify later would betray the juvenile status of the specimen – such is not the case. The Eichstätt specimen appears to be as well ossified as the other larger specimens, all the bones having a smooth finish of periosteum, even the epiphyses of the limbs and the finest tips of the most delicate bones (see WELLNHOFER 1974 Fig. 7). Whereas typical juvenile and fledgling birds have a porous and distinctly unfinished appearance to the termini of the limb-bones which lack the detailed sculpturing of the adult. As I have previously explained (HOWGATE 1984a), such a late ossification as the ascending process of the calcaneum (MARTIN, STEWART & WHETSTONE, 1980) is a prominent bone in the Eichstätt specimen, and the gastralia and foot of the pubes appear to be better ossified in the 'juvenile' Eichstätt specimen than in the 'adult' Berlin specimen. A disproportionate increase in the ossification of the gastralia is also an indication of adulthood among extant crocodilians.

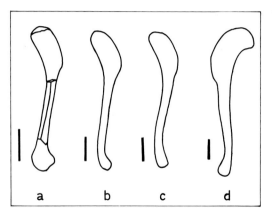

Fig. 2
Right humeri of specimens attributed to *Archaeopteryx*; a) Eichstätt (after WELLNHOFER 1974) in dorsal view, b) Berlin (after OSTROM 1976) in dorsal view, c) Maxberg (redrawn from HELLER 1959) in dorsal view, d) London (after HARRISON & WALKER 1973) in palmar view. Scale bars = 1 cm.

Fig. 3
Restorations of the pelvis of *Archaeopteryx* a,b) Berlin specimen (after HEILMANN 1926); c) Eichstätt specimen (after WELLNHOFER 1974); d) Composite restoration after OSTROM (1975); e) Composite restoration after WALKER (1980).

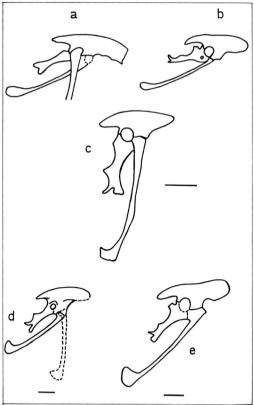

An examination of the humeri of the specimens ascribed to *Archaeopteryx* and comparison with a growth-series from sub-fossil bird material will illustrate the lack of juvenile features in the Eichstätt material. A series of humeri of the colonial burrow-nesting Shearwater (*Puffinus l'herminieri*) provided ideal material for study as the chick remains in the nest right up to fledging and so a full growth series could be obtained. Also the underground nesting habit provides a high proportion of undamaged material. The humerus of a growing hatchling (Fig. 1a-c) is not only smaller than adult humerus (Fig. 1d) but the extremities are barely differentiated from the shaft. As the hatchling increases in size there is no indication that the termini of the humeri are becoming further ossified. Even when the chick has become fully fledged, by which time it has reached full adult size, it is noticeable that particularly the distal epiphysis is totally unossified. However once this terminal size has been reached there is further ossification of the epiphyses to the adult condition.

On examing the humeri of the Eichstätt, Berlin, Maxberg, and London specimens (Fig. 2a–d) it is noticeable that the proximal articulations of all the specimens appear equally well ossified i.e. all are equally 'adult' in appearance. However the distal articulation of the Eichstätt specimen appears distinctly more 'adult' than the other putatively more 'adults' specimens. This is particularly obvious in restorations of the Berlin and London humeri in which the distal epiphyses are dotted in as the outline is unclear and presumably cartilaginous. The Eichstätt humerus has in contrast left a distinct impression of the distal epiphysis which compares favourably with the 'adult' Shearwater material.

As the evidence suggests that the Eichstätt specimen is a good adult, then the absence of a furcula must be considered an artefact of preservation (an unlikely assumption as there is no sign of disruption in the vicinity and no unidentified fragments of bone), or a factor of phylogenetic significance. If the latter is the case then the Eichstätt specimen may represent a phylogenetic precursor to the larger *Archaeopteryx* specimens, less well adapted to flight as it lacked an ossified furcula. Alternatively it could be considered a later 'degeneration' along an Archaeopterygiform side-branch reverting to a more terrestrial habit among the lagoonal islands of the Solenhofen limestone sea.

The pelvis of *Archaeopteryx* and 'reorientation'

If all the specimens of *Archaeopteryx* are regarded as representatives of a monotypic genus then the preserved orientation of the pubes and ischia of the Eichstätt specimen, which is noticeably different from that of the Berlin specimen, presents a problem. OSTROM (1975, 76) proposed that the orientation of the pubis in the Eichstätt specimen, which is almost verical or only slightly opisthopubic and hence quite theropod looking, was the correct life orientation for all the specimens (Fig. 3c). He based this contention on the orientation of the single broken pubis of the Teyler specimen which is preserved at right angles to a disconnected series of vertebrae (OSTROM 1972 Fig.1); and claimed further support for his thesis from the even more disrupted Maxberg skeleton (OSTROM 1976).

In proposing that the pubis was oriented vertically in life, OSTROM was faced with the distinctly opisthopubic orientation of the pubis of the well preserved and fully articulated Berlin specimen (Fig. 3a–b), which was previously considered to be the life orientation. OSTROM's answer was to claim that the pubes of the Berlin specimen had suffered post-mortem disarticulation and rotation and needed to be restored to a 'vertical' position (Fig. 3d). The disarticulation on the right side of the pelvis, OSTROM maintained, took place along a fracture between the right pubis and ilium (1976, Fig. 7A) leaving a triangular space between the two bones, HEILMANN's (1926) "indeterminable mass 'x'" (Fig. 3a, dotted area), filled with fine calcite crystals.

OSTROM further proposed that the reorientation of the pubes was not accompanied by a similar rotation of the ischia – necessary if the orientation of both the pubes and ischia as seen in the Eichstätt specimen was to be considered the putative life orientation. Instead, because there is no sign of the ischia paralleling the pubes in the Teyler specimen, OSTROM assumed that the natural orientation of the ischia was that preserved in the Berlin specimen. Thus OSTROM's *Archaeopteryx* pelvis is a composite of the best (or worst) of both worlds – ischia as oriented in the Berlin specimen and pubes as oriented in the Eichstätt specimen.

A close examination of HEILMANN's indeterminable mass (Fig.4) which provides OSTROM with a suitable triangular area to accomodate his reorientation, indicates that it is composed of damaged bone; puncture marks presumably made by a preparators needle can be discerned. The lower margin of the disputed area is clearly a broken edge of the surface periosteum and the antero-dorsal extension of the pubis was presumably coterminous with the 'indeterminable mass' as indicated by WALKER (1980, Fig. 3e). The upper margin is a thin sliver of smooth periosteal bone which is continuous with the pubis and abutts against or slightly overlaps the ventral margin of the ilium. A small lobate flange at the antero-dorsal tip of this thin sliver of the pubis (a in Fig.4) definitely does overlap the ilium.

Criticism of OSTROM's 'theropod-looking' pelvis theory has come from two quarters: WALKER (1980) has suggested that it is the pelvis of the Eichstätt specimen that has suffered post-mortem dislocation and rotation from an originally opisthopubic orientation, and that the life orientation was as displayed by the Berlin specimen. TARSITANO and HECHT however, (1980) consider that "the Eichstätt specimen reveals a pelvis too altered to allow a reliable interpretation" (p.168). The latter authors go on to ascribe an opisthopubic orientation to the pelvis of *Archaeopteryx* on the basis of the orientation of the distal heads of the conjoined pubes as seen in the London specimen. This appears to be good confirmatory evidence that the larger *Archaeopteryx* specimens had an opisthopubic pelvis. However there is no evidence to suggest any but minor distortion of the pelvis of the Eichstätt specimen, in particular there is no indication that there has been any differential movement between the ischia and pubes as would be necessary to accommodate OSTROM's hypothesis.

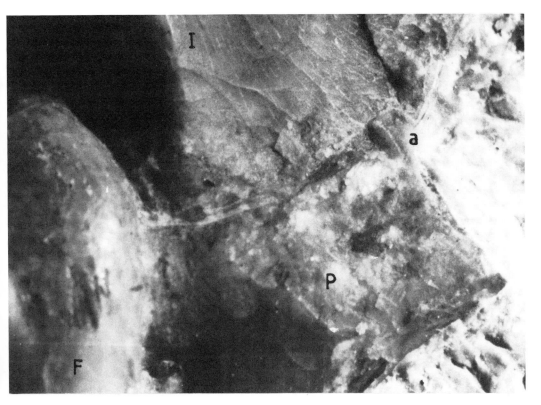

Fig. 4
Berlin specimen of *Archaeopteryx lithographica* – detail of contact between right pubis and ilium, HEILMANN's "indeterminate area 'x'". F = femur, I = Ilium, P = pubis, a = lobate flange.

WALKER (1980) considers the possibility that the pelvis of the Eichstätt specimen might be in its life orientation, but dismisses this conjecture as unlikely as it would mean that by some quirk of fate palaeontologists had actually caught a 'saurischian' pelvis (Eichstätt) evolving into a more bird-like pelvis (Berlin and London) within the same species and at the same horizon. This embarrassment of evidence he counterposed with a hypothetical post-mortem backwards rotation of the pubes and ischia of the Eichstätt specimen (Fig. 3e) for which no evidence is advanced. WALKER does make the observation that the pubes and ischia of the left side are 3 mm closer together than their right side counterparts and, I believe correctly, surmises that this is due to post mortem compression between the opposing femora during diagenesis (see WELLNHOFER 1974 Pl. 23; WALKER 1980 Pl. 1). However, he is reduced to hypothesising 'juvenile' cartilaginous tracts between the bones of the pelvis so as to circumvent the precise fit of WELLNHOFER's (1974, Fig. 10) restoration. He is thus able to reorient his composite Berlin looking pubis and Eichstätt looking ischium into an opisthopubic orientation (WALKER, 1980 Fig. 3b).

Diagenesis and the distortion of the pelvis

Osteologically the Eichstätt and Berlin specimens are the most perfectly preserved skeletons one could hope to find, every bone being in perfect articulation. It seems unnecessary to try and adduce evidence from incomplete and disarticulated specimens in order to reorient 'intransigent' elements of the anatomy of such complete specimens. Against both 'reorientation' theories the following three points may be noted:

1) The heads of the right and left femora are within the circuit of their respective acetabulae, even when these have been skewed out of their three-dimensional orientation. This even appears to be the case with the left femur of the Berlin specimen (OSTROM 1976 Fig. 7a,b). In the Eichstätt specimen the only displacement is that of the heads of the opposing femora which do not lie directly opposite each other (WELLNHOFER 1974 Pl.10A).

2) There is only slight disruption of the most posterior gastralia with no indication of any anteriorly or posteriorly directed thrust which supposedly disarticulated the pelvis and swung the pubes (and ischia) through 40–50° into or away from the gastralia.

3) There is no indication of anything which might have caused such selective disruption and no hypothesis has ever been but forward to account for it.

As I have previously suggested (HOWGATE 1984a) the distortion seen in the pelvis of the Berlin and Eichstätt specimens can be most parsimoniously explained as reorientation of the various component bones of the pelvis during the normal process of diagenetic compaction. As can be seen in the dorsal exposure of the conjoined pubes in the London specimen (HEILMANN, 1926 Fig. 9; TARSITANO & HECHT, 1980 Fig. 7), they consist of two diverging, almost cylindrical, rami joined distally at a long symphysis. This produces a strong springy bone, similar mechanically to the wishbone (furcula) of modern birds. The ilium and ischium are in contrast plate-like elements which would not distort significantly under compaction.

What appears to have happened is that under vertical compaction, the conjoined pubes, which were oriented vertically in the Eichstätt and Berlin specimens, initially resisted the stress imposed. A large amount of back-pressure built up at the weakest point which appears to have been the junction of the pubis with the ilium and ischium at the acetabulum. At first any reorientation of the pubes would be resisted by the vertical pressure on the laterally oriented plate-like ilia and ischia. However, once the back-pressure of the pubic 'spring' was enough to overcome the strength of the pelvis, one or both of the pubes would spring free and reorientation of the conjoined pubis would take place so as to offset the vertical pressure. In doing so the rami would rotate about their symphysial axis so as to present the divergent aspect seen in the Berlin, Eichstätt, and Maxberg specimens. In the Berlin specimen the left pubis has skewed posteriorly, though it is impossible to say where the pelvis 'gave'. The dorsal displacement of the ilium suggests that the sacrum disarticulated but wether the pubis and ischium remained joined cannot be ascertained. The head of the right pubis appears to have, in compensation for the posterior movement of the left pubic ramus, slightly over-ridden the right ilium.

In the Eichstätt specimen the position is somewhat different. It can be clearly seen that both pubes have sprung free at the acetabulum. The right pubis which has sprung completely free now lies anterior to its contact with the ilium, while the head of the left pubis has moved posteriorly only slightly into its acetabulum. There is no indication of any other distortion or twisting of the pelvis, the crushed appearance of the ilium being explicable as purely diagenetic compaction around the sacrum. It thus appears more parsimonious to consider the differing orientations of the pelves of the Berlin and Eichstätt specimens as being approximately in their corresponding 'life' orientations.

The Eichstätt specimen – a new genus

It is difficult to determine which distinctive features or degrees of difference in orientation observed in the osteology of the Eichstätt specimen should be regarded as of generic rather than specific significance. The taxonomy of extant birds is of little value in this problem due to the very fine resolution of specific differences within a basic avian bauplan on which it is based. The problem with *Archaeopteryx* is to differentiate degrees of significance at a much more ancestral level where there is a mosaic of 'class' characters to be interpreted in terms of specific and generic differences.

However, the differences between the Berlin/London (plus the other larger specimens) and the Eichstätt specimens can be considered as falling into two categories. Such characters as differences in the teeth, shape of the ischium, limbic proportions, and overall size may be regarded as of specific significance, while other differences which are probably associated with a relative ecological shift, such as from a more terrestrial to a more arboreal habitat, may be considered as of generic significance. Characters which could be considered as falling into the category are:

1) The absence of a furcula, which is present as a robust bone in the London and Maxberg specimens, and which should be preserved at least as an impression on the near perfect counterslab, its absence being presumed to be due to lack of ossification of the structure in the adult.

2) The ischia and pubes of both the Berlin and the Eichstätt specimens are in approximately natural articulation, both having undergone similar distortion during compaction. The pubes of the Eichstätt specimen being almost vertical/slightly opisthopubic while the pubes of the Berlin specimen are distinctly opisthopubic. In both cases the ischia parallel the pubes.

Taxonomy and Diagnosis

Subclass Archaeornithes
Order Archaeopterygiformes
Genus *Jurapteryx* gen. nov.

Derivation of name: from Jura a common name for the Jurassic Alb where the specimen was discovered, and the name of the museum where the specimen is housed, (also indicative of the Jurassic age of the specimen) and 'pteryx' (Greek) a wing.

Type species: *Jurapteryx recurva* (HOWGATE 1984).

Synonymy: described by WELLNHOFER (1974) as a juvenile specimen of *Archaeopteryx lithographica* and HOWGATE (1984) as a distinct species *Archaeopteryx recurva*.

Holotype: an almost complete skeleton (part and counterpart) with indistinct feather impressions – monographed by WELLNHOFER (1974). Jura Museum, Eichstätt, West Germany.

Horizon: Upper Jurassic (Tithonian), Malm Zeta 2b.

Type locality: Petershöhe, near Eichstätt, Bavaria, West Germany.

Diagnosis. Archaeopterygiform bird similar to but two-thirds the size of *A. lithographica* and with the skeleton disproportionately more gracile. Teeth smooth without ornament and slightly recurved, furcula absent? and forelimbs shorter in proportion to hindlimbs than in *A. lithographica*. Pubes and ischia sub-parallel and oriented approximately vertical, ventral margin of ischia markedly concave, lower terminal prong robust and squared off. Metatarsals unfused.

Evolutionary and Ecological significance.

WALKER (1980) expressed the fear that very rapid evolution would have to be invoked if two specimens from the same stratum, regarded as conspecific, were to exhibit a marked divergence in pubic orientation. To overcome this problem, he not only proposed disruption of the pelvis but also hypothesised that this rotation from 'vertical' to opisthopubic took place gradually during the late Triassic or early Jurassic. This may be the case, but a stage of rapid evolution due to a dramatic change in the direction of selection pressure commensurate with the invasion of a new ecological niche, cannot be ruled out. What can be adduced from the evidence however is that the Eichstätt specimen was ecologically divergent from *A. lithographica*, (not only in terms of diet as is evident from the difference in tooth structure [HOWGATE 1984a,b]). The absence of an ossified furcula and the proportionately larger pes indicate that the Eichstätt specimen was more cursorial and less well adapted to flight than its contemporary. Although the absence of a furcula on its own may not be regarded as sufficient evidence on which to reconstruct a terrestrial habit for the Eichstätt specimen, as budgerigars, which are good fliers, also lack an ossified furcula. A tendency to reduction of the pes is seen in the one extant terrestrial biped which has taken to the trees, the tree-kangaroo *Dendrolargus*, which also does not need an opposable digit in order to be a passable climber. Posterior deflection of the pubis in *A. lithographica* may be associated with the need for more precise balance and the realignment of the vicera posteriorly over the centre of gravity.

The Solnhofen specimens of the first birds may not represent *in situ* evolution from terrestrial carnivore to arboreal insectivore along the main line of bird evolution. The Eichstätt specimen may, indeed, represent a return to a terrestrial habit, on one of the islands fringing the Solnhofen sea, from a previously flying stock. However it provides as good an intercallary type (*sensu* Huxley) as the other specimens in indicating the ecological and evolutionary pathway which birds must have travelled from their archosaurian ancestors.

Acknowledgements

I would like to thank Dr. G. VIOHL, Jura Museum, Eichstätt; Dr. A. CHARIG, British Museum (Natural History), London and Dr. H. JAEGER, Museum für Naturkunde, Berlin for permission to examine the specimens of *Archaeopteryx* in their care, and Mr. G. COWLES of the British Museum (Natural History), Tring, for his help in my examination of the Shearwater material.

References Cited

DE BEER, G. R. (1954): *Archaeopteryx lithographica*; a study based on the British Museum specimen. – 68 pp. London (British Museum (Natural History)).
HARRISON, C. J. O. & WALKER C. A. (1973): *Wyleyia*: a new bird humerus from the lower Cretaceous of England. – Palaeontology, **16**: 721–728; London.
HEILMANN, G. (1926): The origin of birds. – 208 pp. London (Witherby).
HELLER, F. (1959): Ein dritter *Archaeopteryx*-Fund aus den Solenhofener Plattenkalten von Langenaltheim/Mfr. – Erlanger Geol. Abhandl., **31**: 3–25; Erlangen.
HOWGATE, M. E. (1984a): The teeth of *Archaeopteryx* and a reinterpretation of the Eichstätt specimen. – Zool. J. Linn. Soc., **82**: 159–175; London and New York.
HOWGATE, M. E. (1984b): On the supposed difference between the teeth of the London and Berlin specimens of *Archaeopteryx lithographica*. – N. Jb. Geol. Palaont. Mh., **1984** (11): 654–660; Stuttgart.
LANSDOWN, A. B. G. (1968): The origin and early development of the clavicle in the quail (*Coturnix c. japonica*). – J. Zool., **156**: 307–312; London.
MARTIN, L. D., STEWART, J. D. & WHETSTONE, K. N. (1980): The origin of birds: structure of the tarsus and teeth. – Auk, **97**: 86–93;
OSTROM, J. H. (1972): Description of the *Archaeopteryx* specimen in the Teyler Museum, Haarlem. – Proc. Koninkl. Nederl. Akad. Wet., B, **75**, 4: 286–305; Amsterdam.
OSTROM, J. H. (1973): The ancestry of birds. – Nature, **242**: 136; London.
OSTROM, J. H. (1975): On the origin of *Archaeopteryx* and the ancestry of birds. – Proc. Centre Nat. Rech. Sci., Colloq. Internat., **218**: 519–532; Paris.
OSTROM, J. H. (1976): *Archaeopteryx* and the origin of birds. – Biol. J. Linn. Soc., **8**: 91–182: London and New York.
TARSITANO, S. & HECHT, M. (1980): A reconsideration of the reptilian relationships of *Archaeopteryx*. – Zool. J. Linn. Soc., **69**, 2: 149–182; London and New York
WALKER, A. D. (1980): The pelvis of *Archaeopteryx*. – Geol. Mag., **117**: 595–600; Cambridge.
WELLNHOFER, P. (1974): Das fünfte Skelettexemplar von *Archaeopteryx*. – Palaeontographica, A. **147**, 4–6: 169–216; Stuttgart.

Author's address: Mr. MICHAEL E. HOWGATE, Dept. of Zoology, University College London, Gower Street, London WC1 6BT, United Kingdom

Peter Wellnhofer

Remarks on the Digit and Pubis Problems of *Archaeopteryx*

Abstract

The phalangeal formula of the manus of *Archaeopteryx* is shown to be 2.3.4. This is the case in all three specimens with preserved hand skeletons, the Berlin, "Maxberg", and Eichstätt specimens. The Triassic thecodont *Megalancosaurus* with a phalangeal formula of the manus 2.2.3.3.3. is thus unlikely to be ancestral to *Archaeopteryx* and birds as suggested elsewhere, neither when its fingers are numbers I, II, III, nor II, III and IV.

Based on the London, Berlin, "Maxberg", and Eichstätt specimens a new reconstruction of the pelvis of *Archaeopteryx* is given, suggesting a backward orientation of the pubis relative to the long axis of the ilium at an angle of 110°. The pelvis of *Archaeopteryx* is not birdlike. Its functional significance is discussed. The backward rotation of the pubis during avian evolution is related to the changing pelvic musculature. This is probably due to the moving of the centre of gravity at the same time the vertebral tail was reduced. Since contemporary theropods show a propubic condition, *Archaeopteryx* must have already gone a long way in its evolution in the Upper Jurassic, originating in early, as yet unknown theropods.

Zusammenfassung

Die Phalangenformel der *Archaeopteryx*-Hand lautet nach den Befunden am Berliner, "Maxberg"- und Eichstätter Exemplar 2.3.4.. Der Trias-Thecodontier *Megalancosaurus* (CALZAVARA, MUSCIO & WILD 1981) kommt wegen seiner Hand-Phalangenformel 2.2.3.3.3. als Vorläuferform von *Archaeopteryx*, wie von TARSITANO & HECHT (1980) angenommen, kaum in Betracht, und zwar weder wenn die Finger von *Archaeopteryx* als zweiter, dritter und vierter, noch wenn sie als erster, zweiter und dritter anzusprechen sind.

Auf der Grundlage des Londoner, Berliner, "Maxberg"- und Eichstätter Exemplares wird eine erneute Rekonstruktion des *Archaeopteryx*-Beckens vorgelegt, bei welcher das Pubis um 110° gegenüber der Ilium-Längsachse nach hinten orientiert ist. Das Becken ist nicht vogelartig. Seine funktionelle Bedeutung wird diskutiert. Die Rotation des Pubis während der Evolution der Vögel steht in Beziehung zur Änderung der Beckenmuskulatur, diese wiederum mit einer wahrscheinlichen Verlagerung des Körperschwerpunktes, während zugleich der lange Wirbelschwanz reduziert wird.

Da mit *Archaeopteryx* zeitgleiche Theropoden ein nach vorne orientiertes Pubis haben, muß *Archaeopteryx* im Oberen Jura schon einen längeren stammesgeschichtlichen Weg zurückgelegt haben, hervorgegangen aus frühen, noch unbekannten Theropoden.

Introduction

Several recent publications deal with certain aspects of the skeletal anatomy of *Archaeopteryx*. Since some authors have reached conclusions contrary to those published in my 1974 paper on the Eichstätt specimen, I would like to clarify my results on the basis of all five specimens as far as they are relevant for the problems involved.

In this paper I want to illuminate just two aspects of the skeletal anatomy of *Archaeopteryx*, firstly the digit problem, and secondly the pubis problem.

The digit problem of *Archaeopteryx*

TARSITANO & HECHT (1980) have questioned the conventional count of the fingers of *Archaeopteryx*. They pointed out that the three fingers of *Archaeopteryx* were not homologous with the three digits of the theropods, that is, the digits I, II and III. They identified the hand digits of *Archaeopteryx* as numbers II, III and IV by means of embryological evidence in modern birds (HINCHLIFFE 1977).

Furthermore, these authors considered the first two phalanges of the outermost digit as possibly being the broken fragments of originally o n e phalanx. From this would follow the phalangeal formula 2.3.3. instead of the generally accepted formula 2.3.4. (see also HECHT & TARSITANO 1982: 144 and 1984: 588).

According to OSTROM (1976: 111) there is good evidence for the conservative count of the digits of the hand in both theropods and *Archaeopteryx*. Today, however, it seems that the hand digits of modern birds can be generally taken for the 2nd, 3rd, and 4th fingers (HERZOG 1968: 41), the first and fifth digits having been lost (HINCHLIFFE 1985).

In this situation THULBORN & HAMLEY (1982: 617) have offered a compromise between these two contradictory views. As a criterion for the finger count they suggested the relative sizes of the metacarpals, regarding the axial or third digit as being the generally dominant one in pentadactyl tetrapods. Their conclusion is therefore that the digits of the hand are numbers II, III and IV in theropods, *Archaeopteryx*, and neornithiform birds as well.

At first, this discussion seems to be a more theoretical one, but the problem is important for the interpretation of the relationships of *Archaeopteryx*. One point, however, can be absolutely clarified, that is, the number of phalanges, – the phalangeal formula. The number of phalanges in the topographic third digit is definitely four in all three, i. e. the Berlin, "Maxberg", and Eichstätt specimens (Fig. 1). The first two phalanges of this digit show well developed articulations to each other (WELLNHOFER 1974: 194; OSTROM 1976: 111; THULBORN & HAMLEY 1982: 618; HOWGATE 1983: 644, 1984: 104; PAUL 1984).

HECHT & TARSITANO (1982: 144), thought they recognized a break in the proximal phalanx of the topographic third digit in both hands of the Berlin specimen. Their measurements indicated that the break was not symmetrical in both the left and the right digit. There is a peculiar "flange" on the proximal phalanx of the middle left finger of that specimen which, according to TARSITANO & HECHT (1980: 162), could have caused the break of the proximal phalanx of this finger into two pieces (Fig. 1). This "flange" is neither present in the middle finger of the right side nor is it met with in any other *Archaeopteryx* specimens. THULBORN & HAMLEY (1982: 619) took this structure to be an "artifact of preservation". It seems to me to be a natural, but pathological outgrowth of the bone, maybe due to an injury during life, an opinion also expressed by HOWGATE (1984).

Furthermore, HECHT & TARSITANO (1982) argued, with reference to HELLER (1959: 17), that in the "Maxberg" specimen the proximal phalanx of the last digit was one piece in one manus and two pieces in the other manus, whereas Heller himself simply wrote: "Der dritte Finger wird durch das Vorhandensein von 4 Phalangen charakterisiert" ("The third digit is characterized by the presence of four phalanges"). Evidently this is the case in the right manus (Fig. 1). In the left manus the outer finger is hidden by the middle digit and the drawing by HELLER (plates 7 and 15) is rather schematic. I am convinced that a restudy of this specimen, which is badly needed, would reveal the presence of four phalanges in the topographic third digit of the left hand also.

In the right manus of the Eichstätt specimen the outermost digit is also composed of four phalanges, although documented in part by impressions only. In the left hand this finger is only partly preserved. The hand skeleton is not complete in the London and Haarlem specimens.

The phalangeal formula and its interpretation suggested by TARSITANO & HECHT (1980) were discussed at some length by THULBORN & HAMLEY (1982) and HOWGATE (1983).

The discussions about the number of phalanges in the outer digit of the hand of *Archaeopteryx* could be regarded as not being very important were it not for the phylogenetic conclusions which have been based upon it. So, a possible evolutionary relation to the Upper Triassic *Megalancosaurus* (CALZAVARA,

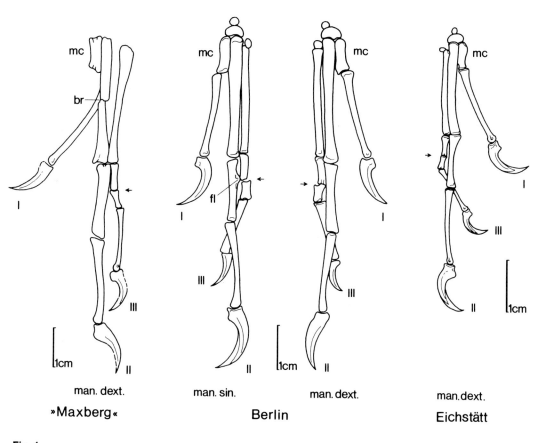

Fig. 1
The hand skeletons of the "Maxberg", Berlin and Eichstätt *Archaeopteryx* specimens. The fingers are counted conventionally as being the 1st, 2nd, and 3rd. Arrows indicate the joints between the first and second phalanx of digit III. Abbreviations: br break of the metacarpal II in the "Maxberg" specimen, fl flange on the proximal phalanx of left digit II in the Berlin specimen, mc metacarpals.

MUSCIO & WILD 1981) was suggested, with the phalangeal formula of the manus 2.2.3.3.3. According to HECHT & TARSITANO (1982: 145), this small thecodont is a form "with an avian-like skull and pelvis, and large forelimbs ... from which the archaeopterygian forelimb can be derived".

Since the phalangeal formula of *Archaeopteryx* can be shown to be 2.3.4., a *Megalancosaurus*-type archosaur is unlikely to be ancestral to *Archaeopteryx*, neither if the count of the digits of the hand is 1st, 2nd, 3rd, nor if it is taken as being 2nd, 3rd, and 4th. In both cases, during evolution towards *Archaeopteryx*, there would have been an increase in the number of phalanges rather than a reduction to be expected.

The pubis problem of *Archaeopteryx*

The second problem – the pubis problem – has been discussed in recent years, focusing mainly on the question of whether the pubis of *Archaeopteryx* was oriented backwards and thus bird-like, or more vertically or downwards suggesting a more dinosaurian appearance.

According to OWEN's first description of the London specimen (1863), its pubes were not yet prepared,

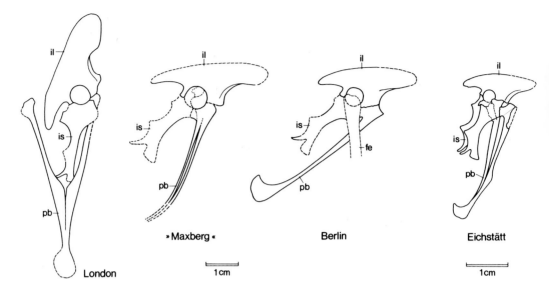

Fig. 2
The pelves of the London, "Maxberg", Berlin, and Eichstätt specimens as preserved. The "Maxberg" pelvis is drawn after X-ray photographs (Fig. 3). The drawing of the Eichstätt pelvis is combined from the main slab and the counterpart slab.

and thus unknown. Only in 1917 were the pubes revealed by Petronievics & Woodward. Later Petronievics (1921), argued that the right pubis was in its natural position with the ilium and the ischium, whereas only the left pubis was displaced.

However, it is obvious to me that the pelvic elements, – sacrum, ilium, ischium, and the pubes – in the London specimen do not have natural contact, but are separated. This also was the opinion of De Beer (1954: 27).

The right ilium and the right ischium are exposed laterally (Fig. 2). The ischium appears to have rotated caudally to a position almost parallel to the long axis of the iliac blade. This, however, has been regarded as the natural position by Tarsitano & Hecht (1980: 168). The left ilium shows its medial surface and is partly hidden. Apparently the ilia were connected to the sacrum by cartilage only.

In front of the acetabulum a broad facet for the pubis is formed by two projections, called "anteroacetabular apophysis" and "iliac spine" by De Beer (1954), and by others, the "pubic peduncle". Both pubes, fused in a ventral symphysis, are exposed in posterior aspect. They too, have been rotated backwards. It is remarkable that the contacts between the sacrum, ilium, ischium, and pubes were relatively weak, whereas both pubes remained in a firm symphyseal contact. This observation is of some importance with regards to the other *Archaeopteryx* specimens, and to the restoration of the pelvis of *Archaeopteryx* in general.

Distally there is the typical expansion, – the foot or hook of the pubic symphysis. It is partly covered by a mass of calcite in the London specimen indicating a cartilaginous extension of the pubic foot similar to the structure found also in the Eichstätt specimen (Wellnhofer 1974: 197, Abb. 10), and called symphyseal cartilage.

The proximal distance between the free rami of the pubes is a direct indication of the breadth of the pelvis (Fig. 4). According to the angle of the posteriorly inclined articular facets to the ilium, both Walker (1980) and Tarsitano & Hecht (1980) suggested a backward orientation of 135° and 160° respectively, relative to the long axis of the ilium. I presume, however, that the proximal articular expansion of the pubes has been twisted into the bedding plane due to sedimentary pressure and compaction. Therefore, the inclination of the articular facets of the pubes can not be determined exactly. In my opinion, the orientation of the pubes can not be restored on the basis of the London specimen.

Originally, very little was known about the pelvic girdle in the Berlin specimen also (DAMES 1884). Only after further preparation could DAMES (1897) show that the pelvis was complete. In his opinion the pelvic bones were in their natural position suggesting that the pubis had a backward orientation of 135° relative to the long axis of the ilium.

The right pubis is exposed laterally. The right femur is still articulated in the acetabulum. The right ilium is complete except for its anterior portion. The left ilium is displaced dorsally indicating that there was no firm contact with the sacrum. Both ischium and pubis appear to have been rotated caudally, relative to the ilium. There is no natural contact between the pubis and the ilium. The pubis has not only been rotated backwards but has, additionally, been moved dorsally, sliding over the pubic peduncle.

WALKER (1977; 1980) regarded the proximal anterior projection of the right pubis as being a bird-like pectineal process. When the pubis is rotated forwards to a position displayed in both the "Maxberg" and the Eichstätt specimens, this projection is turned back, below, and in contact with the pubic peduncle of the ilium. Therefore, the contact line between the ilium and the pubis is not a real suture but the proximal margin of the processus ischiadicus of the pubis, pushed over the ilium, and probably reduced in size due to preparation.

Therefore, I agree with OSTROM (1972; 1973; 1974; 1975; 1976), considering the Berlin pubis as having been rotated backwards – post mortem.

Furthermore, no other specimen of *Archaeopteryx* shows a pronounced pectineal process on the pubis, although WALKER (1980), suggested one in both the London and Eichstätt specimens. In modern birds the pectineal process, if present, serves as the origin of the M. ambiens, and is usually a part of the ilium. Only in ratite birds does this process arise from the pubis (STRESEMANN 1927). So, neither would it be necessary to locate the origin of the ambiens muscle of *Archaeopteryx* on the pubis, nor to suppose a pronounced pectineal process at all.

PETRONIEVICS (1921), suggested that the pubes of the Berlin specimen were not united in a symphysis. This was one of the reasons that led him to separate the London and the Berlin specimens generically into *Archaeopteryx* and *Archaeornis*. The right pubis of the Berlin specimen has in its distal portion an expansion pointing to an "apron", which is also present in the London and Eichstätt specimens. A detailed analysis about the position, preservation, and interpretation of the Berlin pubis was published by OSTROM (1976: 104 ff.).

In his description of the third *Archaeopteryx* specimen, which I have called the "Maxberg" specimen (WELLNHOFER 1974), HELLER (1959: 17) recognized only the anterior part of the ilium and the acetabulum, mainly based on X-ray photographs. He identified the parallel structures ventral to the ilium as being the tarsometatarsus (HELLER 1959: Taf. 14, 15). But they are in fact the pubes, exactly where they are expected to be, in natural contact with the ilium, a fact which I have already noted (WELLNHOFER 1974), and which was discovered by OSTROM (1976) independently.

On the surface of the slab only the proximal portions of the pubes are exposed, the right one showing suture lines between the ilium and very clearly with the ischium. Distally the pubes are completely covered by matrix and are crossed by the right tibia.

At the posteroventral border of the acetabulum the ischium shows its natural contact with the ilium and the pubis. This connection can also be observed in the X-ray photographs. The suture line between the pubis and the ischium lies below the middle of the acetabulum (WELLNHOFER 1974: 210).

The distal extremity of the right ischium is exposed, but split. The interior trabecular structure of the bone indicates a bifurcation as in the London, Berlin, and Eichstätt specimens.

Professor STÜRMER, Erlangen, very kindly provided me with several unpublished X-ray photographs which he took of the "Maxberg" specimen in the late fifties. Amongst these are also X-rays of the pelvic

 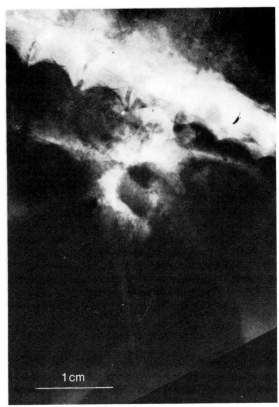

Fig. 3
Opposite X-ray photographs of the pelvic region of the "Maxberg" specimen. The sacrals are displaced. Clearly shown is the circular acetabulum. The parallel downwardly directed structures are the rami of the pubes crossed by the right tibia. The pubic "foot" lies outside the film. Traces of the ischia show up very faintly. Photos by Prof. STÜRMER, Erlangen.

region (Fig. 3). One can recognize the ventral border of the ilium, the acetabulum, and both rami of the pubes clearly, showing their proximal expansion with the contact with the ilium and the ischium. The distal extremity – the pubic foot – lies outside the film. The close parallel position of the pubes indicates only minor displacement, and that the pubes probably were firmly linked in a ventral symphysis.

The concave anterior margin of the ischium shows up faintly, and marks its natural position with regards to the suture lines already indicated.

From the X-ray photographs it becomes apparent that the sacrum was separated from the ilia – post mortem – as was the case in the London, Berlin, and Eichstätt specimens.

Generally, the orientation of the pubes in the "Maxberg" specimen seems to be more or less natural. The backward inclination is about 115° relative to the long axis of the ilium. This is about the same as in the Eichstätt specimen (WELLNHOFER 1974: 197; OSTROM 1976: 107).

Unfortunately it has not been possible in recent years to study the "Maxberg" specimen in detail or to carry out further preparations. After having removed the slabs from the Solnhofen museum at the Maxberg in 1974, the owner has not given access to his specimen. So, at the moment, it is not available to the scientific community, and there is no law in Bavaria to change this situation.

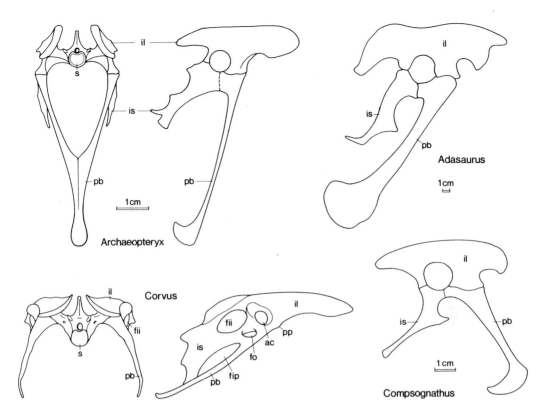

Fig. 4
Restoration of the pelvis of *Archaeopteryx* in anterior and right lateral views based on the size of the London specimen. For comparison the pelves of a modern bird (*Corvus* sp.) in anterior and right lateral views, and of the theropods *Compsognathus longipes* WAGNER (after OSTROM 1978) and *Adasaurus mongoliensis* BARSBOLD (after BARSBOLD 1983). Abbreviations: ac acetabulum, fii fenestra ilio-ischiadica, fip fenestra ischio-pubica, fo foramen obturatum, il ilium, is ischium, pb pubis, pp processus pectinealis, s sacrum.

In the Haarlem specimen only the distal extremity of the left pubis and a presumed fragment of the right pubic shaft is preserved (OSTROM 1972: 298). Imprints of the shaft indicate the long axis of the pubes, but it is not possible to determine the orientation of the Haarlem pubis exactly, although OSTROM (1972), has suggested an "orientation of the pubes at nearly right angles to the trace of the vertebral column". Since in all other specimens the pelvis was separated from the vertebral column – post mortem – the angle between the pubis and the vertebral column is not of much value with regards to the original orientation of the pubis.

But there is another reference line: the line which is marked by the ventral limit of the gastralia. The angle between the femur and this line in the Haarlem specimen is the same as in the Berlin specimen – almost perpendicular. The Haarlem pubis runs parallel to the femur, whereas the right pubis of the Berlin specimen has an angle of 45° relative to the right femur. This is evidence that the pubes of the Haarlem specimen were oriented rather ventrally, although no precise angle, relative to the long axis of the ilium, can be given.

As far as the Eichstätt specimen is concerned I would like to refer to my original description of the pelvis (WELLNHOFER 1974: 195 ff.). Here I want to present a drawing of the pelvis combined from the main slab and the counterpart slab (Fig. 2). According to HECHT & TARSITANO (1982: 144), the "Eichstätt specimen's pelvis is so badly damaged as to make restoration unreliable". In this point I disagree. All elements of the pelvis are preserved, the ilium, both ischia, and both pubes, although separated and only partly preserved on the main slab and the counterpart. Therefore, a restoration of the pelvis of *Archaeopteryx* on the basis of the Eichstätt specimen is more reliable than one based on the London example only.

The tree elements, ilium, ischium, and the pubes are not exactly in their natural contact with each other, but their displacement is only minimal. The shaft of the left pubis was shifted backwards, the right shaft forwards, relative to the pubic peduncle of the ilium.

Both pubes were firmly united in a distal symphysis as in all other specimens. Also, this connection was not separated after death. The pubes in this specimen appear to have been a fork-like structure, the free ends of which were twisted into the bedding plane during compaction of the sediment. This was also the case in the London and the Berlin specimens (OSTROM 1976: 107).

The suture line between the pubis and the ischium which I indicated in my restoration of the Eichstätt pelvis (WELLNHOFER 1974: Abb. 10c), was criticized by WALKER (1980: 596), as "clearly a minor crack" in the pubis. Today, I agree with this observation. The part I took as belonging to the right ischium, must then be included in the right pubis which is in agreement with the situation in the "Maxberg" specimen.

In the Eichstätt specimen the pubo-ischiadic contact also lies below the middle of the acetabulum.

The anterior corner of the proximal end of the Eichstätt pubis regarded by WALKER (1980: Fig. 2b) as being a "pectineal process" will be shifted below the pubic peduncle of the ilium, if properly articulated.

My new reconstruction of the pelvis of *Archaeopteryx* (Fig. 4) is based on the size of the London specimen with the orientations of the ischium and the pubis derived from the other specimens, particularly the "Maxberg" and Eichstätt specimens. Compared to my former restoration (WELLNHOFER 1974), the orientation of the ischium and the position of the pubo-ischiadic suture has changed slightly. Not changed is the orientation of the pubis relative to the long axis of the ilium with an angle of approximately 110°. This is a little more than the 100° estimated by OSTROM (1976: 107), but considerably less than the 135° and 160° suggested by WALKER (1980: 598) and TARSITANO & HECHT (1980) respectively.

This downward, but nevertheless opisthopubic inclination of the pubis of *Archaeopteryx* is taken directly from the preservation of the "Maxberg" and Eichstätt specimens, indirectly also from the Haarlem specimen, because in those three specimens post-mortem displacement appears to have been at its least.

Conclusions

The backward inclination of the pubis of *Archaeopteryx* has been taken as being bird-like, and to be evidence for an intermediate stage between reptiles and birds. Of course, other archosaurs, like the ornithischians, also have an opisthopubic pelvis. BARSBOLD (1983) has shown that even some dromaeosaurid theropods had an opisthopubic condition: *Adasaurus* and *Segnosaurus* from the Upper Cretaceous of Mongolia. During the Mesozoic different archosaurian groups featured a trend for a backward rotation of the pubis and not only in the ancestral stock of the birds (THULBORN & HAMLEY 1982). A backwards inclined pubis means no more than a downward directed pubis as far as the relationships of *Archaeopteryx* are concerned.

The only bird-like pelvic feature is the elongation of the preacetabular portion of the ilium. The difference between the pelvis of *Archaeopteryx* and modern birds, including the Cretaceous birds, is in fact considerable and becomes apparent if we compare the front views (Fig. 4). Whereas the pelvis is open ventrally in modern birds, the pelvis of *Archaeopteryx* is closed by a long pubic symphyis. (Only in *Struthio* are the pubes in contact distally.) The aperture of the pelvis in *Archaeopteryx* was relatively narrow, leaving just a small passage way for the eggs.

What is the significance of this construction of the pelvis? We must suppose functional reasons related mainly to the pelvic musculature of the hind legs. In *Archaeopteryx* a long vertebral tail is present. Therefore the centre of gravity must have been different from the "tail-less" modern birds. Accordingly, the geometrical courses of lines of muscle action between the pelvis and the hind legs must have been different. Perhaps there even existed an indirect correlation between the length of the tail and the orientation of the pubis within avian evolution.

The long, ventrally closed pubes may have served another purpose as well. In most modern birds the viscera are supported by the large sternum. There was no ossified sternum in *Archaeopteryx*. The gastralia provided only minor support. Of greater importance were the long, distally fused pubes which could only function as a supporting device for the viscera if they were oriented more or less downwards.

Theoretically, it could be expected that avian evolution from the *Archaeopteryx* level to modern birds led to the improvement of flight, and corresponded with the ossification and the enlargement of the sternum.

Simultaneously the tail could be reduced, thus altering the centre of gravity of the bird's body and altering the pelvic musculature and the areas of muscle origins on the pelvic elements. Therefore the pubis started to rotate backwards, and separated to leave the pelvis open ventrally. The function of supporting the viscera was increasingly taken over by the sternum which in turn displaced the gastralia.

In conclusion, the pelvis of *Archaeopteryx* is not bird-like at all. There is, however, a trend of ornithization, compared to the coelurosaurian pelvis, by the enlargement of the preacetabular portion of the ilium and by the elongation of the pubes. The same trend occurred in other theropods as well, without any closer relationship to the ancestral stock of *Archaeopteryx*, for example in the Upper Cretaceous *Adasaurus* and *Segnosaurus* (BARSBOLD 1983).

Since contemporaneous coelurosaurs like the Upper Jurassic *Compsognathus* had a propubic pelvis, *Archaeopteryx* must have already gone a long way in its evolution, originating in early, yet unknown theropods.

Acknowledgements

I am very much indepted to Prof. Dr. Dr. h.c. WILHELM STÜRMER, Erlangen, for providing me with X-ray photographs of the "Maxberg" specimen, to Prof. Dr. JOHN H. OSTROM, Yale Peabody Museum, New Haven, for reviewing the manuscript and his valuable suggestions, and to Mrs. BERYL HÖFLING, München, for improving the English text. My sincere thanks to all.

References Cited

BARSBOLD, R. (1983): Carnivorous dinosaurs from the Cretaceous of Mongolia. – Joint Soviet-Mongolian Palaeont. Exp., Trans., **19**: 120 pp., 30 figs.; Moskau (russian).
CALZAVARA, M., MUSCIO, G. & WILD, R. (1981): *Megalancosaurus preonensis* n. g., n. sp., a new reptile from the Norian of Friuli, Italy. – Gortania, Atti Mus. Friul. Storia Natur., **2** (1980): 49–64, 5 figs.; Udine.
DAMES, W. (1884): Ueber *Archaeopteryx*. – Palaeont. Abh., **2** (3): 119–196, 5 Abb., Taf. 15; Berlin.
DAMES, W. (1897): Über Brustbein, Schulter- und Beckengürtel der *Archaeopteryx*. – Sitz.-Ber. Kgl. Preuss. Akad. Wiss., **38**: 818–834; Berlin.
DE BEER, G. (1954): *Archaeopteryx lithographica*. – Brit. Mus. (Nat. Hist.) London: 1–68, 9 figs., 16 pl.; London.
HELLER, F. (1959): Ein dritter *Archaeopteryx*-Fund aus den Solnhofener Plattenkalken von Langenaltheim/Mfr. – Erlanger Geol. Abh., **31**: 1–25, 2 Abb., 15 Taf.; Erlangen.
HECHT, M.K. & TARSITANO, S. (1982): The Paleobiology and Phylogenetic Position of *Archaeopteryx*. – Geobios, Mém. Spec., **6**: 141–149, 2 tabl.; Lyon.
HECHT, M.K. & TARSITANO, S. (1984): *Archaeopteryx* palaeontological myopia. – Nature, **309**: 588; London.
HERZOG, K. (1968): Anatomie und Flugbiologie der Vögel. – 180 S., 100 Abb., 9 Tab.; Stuttgart (G. Fischer).
HINCHLIFFE, J.R. (1977): The chondrogenetic pattern in chick limb morphogenesis: a problem of development and evolution. – In: Vertebrate limb and somite morphogenesis (eds. EDE, D.A., HINCHLIFFE, J.R. & BALLS, M.), pp. 293–309; Cambridge.

HINCHLIFFE J. R. (1985): 'One, two, three' or 'Two, three, four': An Embryologist's View of the Homologies of the Digits and Carpus of Modern Birds. – This volume.
HOWGATE, M. (1983): *Archaeopteryx* – no new finds after all. – Nature, **306** : 644; London.
HOWGATE, M. (1984): *Archaeopteryx*'s morphology. – Nature, **310**: 104; London.
HOWGATE, M. (1985): Problems of the Osteology of *Archaeopteryx*, Is the Eichstätt Specimen a Distinct Genus? – This volume.
OSTROM, J. H. (1972): Description of the *Archaeopteryx* Specimen in the Teyler Museum, Haarlem. – Proc. Koninkl. Nederl. Akad. Wet., (B) **75**, 4: 289–305, 1 fig., 6 pl.; Amsterdam.
OSTROM, J. H. (1973): The ancestry of birds. – Nature, **242**: 136; London.
OSTROM, J. H. (1974): On the origin of *Archaeopteryx* and the ancestry of birds. – Coll. int. C. N. R. S., **218**: 519–532, 6 figs.; Paris.
OSTROM, J. H. (1975): The origin of birds. – Ann. Rev. Earth and Planet. Sci., **3**: 55–77, 9 figs.;
OSTROM, J. H. (1976): *Archaeopteryx* and the origin of birds. – Biol. J. Linn. Soc., **8** (2): 91–182, 36 figs.; London.
OSTROM, J. H. (1978): The osteology of *Compsognathus longipes* WAGNER. – Zitteliana, **4**: 73–118, pl. 7–14; München.
OWEN, R. (1863): On the *Archaeopteryx* of VON MEYER, with a description of the Fossil Remains of a Long-tailed species, from the Lithographic Stone of Solenhofen. – Phil. Trans., **153**: 33–47, pl. 1–4; London.
PAUL, G. S. (1984): The hand of *Archaeopteryx*. – Nature, **310**: 732; London.
PETRONIEVICS, B. (1921): Ueber das Becken, den Schultergürtel und einige andere Teile der Londoner *Archaeopteryx*. – 31 S., 2 Taf.; Genf (Georg & Co.).
PETRONIEVICS, B. & WOODWARD, A. S. (1917): On the Pectoral and Pelvis Arches of the British Museum Specimen of *Archaeopteryx*. – Proc. Zool. Soc. London, **1917**: 1–6, pl. I; London.
STRESEMANN, E. (1927): Aves. – In Handb. Zool. (Hrsg. T. KRUMBACH), **7** (2), Lfg. 1: 1–112; Berlin und Leipzig (W. de Gruyter & Co.).
TARSITANO, S. & HECHT, M. K. (1980): A reconsideration of the reptilian relationships of *Archaeopteryx*. – Zool. J. Linn. Soc., **69** (2): 149–182, 9 figs.; London.
THULBORN, R. A. & HAMLEY, T. L. (1982): The Reptilian Relationships of *Archaeopteryx*. – Austr. J. Zool., **30**: 611–534, 6 figs.; Melbourne.
WALKER, A. D. (1977): Evolution of the pelvis in birds and dinosaurs. – Probl. Vert. Evolution: Linn. Soc. Symp. Ser. No. **4**: 319–358, 14 figs.; London.
WALKER, A. D. (1980): The pelvis of *Archaeopteryx*. – Geol. Mag., **117** (6): 595–600, 3 figs., pl. 1–2; London.
WELLNHOFER, P. (1974): Das fünfte Skelettexemplar von *Archaeopteryx*. – Palaeontographica, A. **147**: 169–216, 13 Abb., Taf. 20–23; Stuttgart.

Author's address: Dr. Peter Wellnhofer, Bayerische Staatssammlung für Paläontologie und historische Geologie, Richard-Wagner-Straße 10, 8000 München 2, Federal Republic of Germany.

Alick Walker

The Braincase of *Archaeopteryx*

Abstract

A re-interpretation of the otic region of the London specimen is given. The 'quadrate cotyle' of WHETSTONE (1983) is considered to be the smooth floor of the entrance to the posterior tympanic recess. The London braincase shows no sign of a squamosal or of an articulation for the quadrate. From comparison with the Eichstätt specimen it is concluded that the quadrate in *Archaeopteryx* was of 'normal' archosaurian type, articulating by a single head with the squamosal. The otic capsule is of a primitive, basically avian type from which those of later birds can readily be derived; in comparison, ratites are more specialized than carinates. Juvenile features of the London and Eichstätt skulls are pointed out. A possible method of enclosure of the superior tympanic recess by the squamosal is outlined. Some comments are made on bird-crocodile relationships.

Zusammenfassung

Es wird eine Neu-Interpretation der Ohrregion des Londoner Exemplares gegeben. Der "quadrate cotyle" von WHETSTONE (1983) wird als der glatte Boden der Öffnung zur hinteren tympanischen Nische gedeutet. Die Londoner Gehirnkapsel zeigt keine Anzeichen eines Squamosums oder einer Artikulation für das Quadratum. Aufgrund eines Vergleichs mit dem Eichstätter Exemplar wird geschlossen, daß das Quadratum bei *Archaeopteryx* vom "normalen" Archosauriertypus war, indem es mit einem einzigen Gelenkkopf mit dem Squamosum artikulierte. Die Ohrkapsel ist von einem primitiven, im wesentlichen vogelartigen Typus, von welchem diejenigen der späteren Vögel leicht abgeleitet werden können. Im Vergleich dazu sind die Ratiten mehr spezialisiert als die Carinaten. Juvenile Merkmale des Londoner und des Eichstätter Schädels werden dargelegt. Eine mögliche Art der Umschließung der oberen tympanischen Nische durch das Squamosum wird aufgezeigt. Es folgen einige Bemerkungen zu den Beziehungen zwischen Vögeln und Krokodilen.

Introduction

WHETSTONE (1983) has described the results of new preparation of the London specimen of *Archaeopteryx*, and has presented an interpretation of the braincase. While agreeing to a large extent with this, I believe that he has misunderstood the otic region and as a result has identified a quadrate articulation where none exists. In fact there is no quadrate articulation on this braincase. Comparative study of the London and Eichstätt specimens demonstrates that the quadrate articulation in *Archaeopteryx* was with the squamosal and was of "normal" archosaurian type. The bone originally lying close to the braincase of the London specimen (DE BEER, 1954, Pl. VII) is almost certainly the right quadrate. It has a single head.

Citation here of various bird groups as examples has been governed by what I have been able to collect myself and what has been available to me. I have found it particularly difficult to obtain juvenile birds at the right stage to display the definitive suture pattern in the braincase before it is obliterated by fusion. I have attempted to obtain examples of the groups usually thought to be most primitive, but coverage is patchy and no significance should be attached to omission of particular groups. Results suggest that otic capsule pattern is constant within groups and should be a useful aid to classification. (In this paper 'juvenile' is not used in the strict ornithological sense).

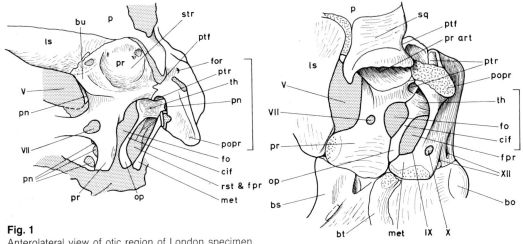

Fig. 1
Anterolateral view of otic region of London specimen. Abbreviations as in ROMER (1956), with the addition of: aps, alaparasphenoid; bt, basitemporal; bu, prootic-laterosphenoid buttress; cif, crista interfenestralis; eov, foramen for external occipital vein; ep, epiotic; fm, margin of foramen magnum; fo, fenestra ovalis; for, foramen; fpr, fenestra pseudorotunda; met, metotic; mp, mammillary process; pn, pneumatic cavity; pr art, prootic articulation for quadrate; ptr, posterior tympanic recess; rst, recessus scalae tympani; sq art, squamosal articulation for quadrate; str, superior tympanic recess; th, threshold to posterior tympanic recess. Scale length 5 mm.

Fig. 2
Anteroventrolateral view of otic region of juvenile penguin 'No. 3A', *Eudyptes chrysolophus*, Coll. by British Antarctic Survey. (Now B.M.N.H. S/1984. 94.1), age from hatching 21 days. In the fresh skull a little cartilage intervenes between prootic and metotic. Dried cartilage: irregular dashed pattern. Scale length 5 mm.

The otic capsule of the London specimen

The most important point to be decided is the nature of the smooth area ('th' in Fig. 1) which WHETSTONE (1983, Fig. 8 and p. 447) designated the 'facies articularis pro quadrato'. In its present orientation this area is concave anteroposteriorly and is gently arched at right angles to this. Running downwards and forwards from this area is a slender bar of bone which curves downwards and broadens at its lower end. Behind the bar is a deep concavity bounded behind by a thin sheet of bone. It is evident that these bones have been crushed upwards and inwards as a unit, since the lower surface of the paroccipital process is extensively cracked and pressed inwards (Fig. 6). One could select several species of juvenile bird skull for comparison, but the most striking one available to me for general resemblance is that of a specimen of *Eudyptes chrysolophus* (Sphenisciformes) (Fig. 2). Without labouring the point, it seems to me obvious that the slender bar of bone in *Archaeopteryx* corresponds to the crista interfenestralis (SÄVE-SÖDERBERGH, 1947, p. 513) of the opisthotic of the penguin skull. This terminates dorsally in an expansion closely similar to that of *Archaeopteryx*. The expansion differs in being traversed by a ridge running anterodorsally, which dies out before reaching the prootic/opisthotic suture. As usual in birds, this suture in the penguin runs posterodorsally from the upper margin of the fenestra ovalis into a cavity within the paroccipital process. The latter is probably formed largely by ossification in the metotic cartilage (STRESEMANN, 1927–34, p. 50), here termed 'metotic' for brevity, and the cavity is the posterior tympanic recess (PYCRAFT, 1902, p. 282) ('antrum pneumaticum centrale' in the terminology used by WHETSTONE). (The area labelled 'met' on Figs. 1–3 and 6 may in fact result from ossification of the lateral region of the basal plate of the embryo, rather than of the metotic cartilage). A slightly older skull of the penguin *Pygoscelis papua* (Fig. 3a) presents a more exact comparison with *Archaeopteryx* in that the threshold of the posterior tympanic recess formed by the opisthotic is smooth, concave from

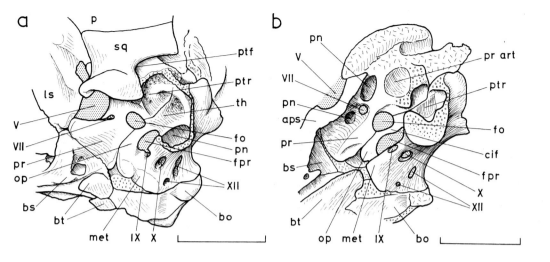

Fig. 3
(a). Ventrolateral view of otic region of juvenile *Pygoscelis papua*, B.M.N.H. 1898.7.1.1. Part of paroccipital process removed to show interior of posterior tympanic recess. In spite of appearances, the opisthotic almost certainly does not reach the lower end of the cochlear recess. Scale length 10 mm. (b) Ventrolateral view of otic region of juvenile *Larus* sp. (still with egg-caruncle). Scale length 5 mm.

anterodorsal to posteroventral, and slightly convex in a direction at right angles to this. In the penguin, as in *Archaeopteryx*, the concave area ends at a sharp edge posteroventrally, giving way to a deeper pneumatic space within the paroccipital process.

These features of the crista interfenestralis and threshold or floor of the posterior tympanic recess can be matched in other juvenile bird skulls. Specimens of the petrel *Daption capensis* (Procellariiformes) if anything match the 'articular area' of *Archaeopteryx* even more closely, but are rather small and, having the area set more deeply within the mouth of the recess, are difficult to illustrate. A young gull, *Larus* sp. (Charadriiformes) (Fig. 3b) is figured for general comparison. Specimens of the eider, *Somateria mollissima* (Anseriformes) and *Gallus gallus* (Galliformes) are basically similar, but in the galliforms studied a stronger ridge crosses the threshold and in both groups the lower end of the opisthotic differs from that of *Archaeopteryx*, owing to the fact that this bone takes up a smaller share of the cochlear recess. In all these Recent genera the deeper posteroventral pneumatic cavity within the paroccipital process is present, although its exact configuration varies from group to group.

It is concluded, therefore, that the apparent articular area in *Archaeopteryx* is actually the smooth threshold of the entrance to the posterior tympanic recess. Its curvature has probably been exaggerated by crushing against the prootic. It is, in addition, too deeply set, even allowing for crushing, to be a quadrate articular area. The slender bar of bone running down from this area is thus the crista interfenestralis, with the foramen perilymphaticum lying medial to it. Fenestra ovalis and fenestra pseudorotunda are situated in front of and behind the crista respectively. The deep recess behind the crista is the recessus scalae tympani, which WHETSTONE has identified as the 'antrum pneumaticum

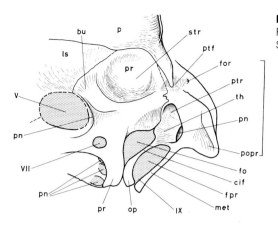

Fig. 4
Restoration of the otic region of the London specimen. Scale length 5 mm.

centrale'. The close resemblance of the thin sheet of bone behind this cavity to the metotic buttress of the juvenile *Eudyptes, Daption* and *Larus* leaves no doubt of the correct identification. The foramen perilymphaticum should open backwards and downwards into the recessus, but is blocked by matrix. A short section of the medial border of the foramen has been exposed at the dorsal end of the recessus. Behind the metotic buttress the foramina for the vagus and hypoglossal nerves occur in a depression (Fig. 6), contrary to the usual situation in birds, which suggests that the anteroposterior dimension of the recessus has been reduced by crushing. Nevertheless, the anterior surface of the buttress develops a broad forward projection centrally at depth in the recessus, implying that the metotic foramen was originally elongate-reniform. The glossopharyngeal nerve may have exited from the lower end of the recessus scalae tympani as shown in Fig. 4. In the juvenile penguins and other birds nerve IX progressively notches the anterior surface of the buttress at its lower end as growth proceeds. A similar notch occurs in the adult *Fulmarus*, or may be converted into a foramen. In *Rhea* and *Struthio*, however, IX exits posteriorly with X. The position of the vagus foramen supports the identification of the recessus scalae tympani. Removal of matrix from this large foramen should provide information on the direction which it follows; one would expect it to pass into the deepest part of the recessus (or, more precisely, into the metotic foramen), though the actual connection cannot be exposed because of the compression which the cavity has suffered.

Turning now to the area in front of the crista interfenestralis, it is clear that the projection which WHETSTONE has identified as the 'remnant of the interfenestral process' (i.e., the crista) cannot be this structure. The 'arcuate posterior margin' of the prootic above this projection, noted by WHETSTONE (p.447), is in fact the boundary of an area of the prootic which has been bent inwards by the pressure of the threshold area of the opisthotic (WHETSTONE's 'cotyle') against it. The middle part of this curved edge is a ridge running up and back to die out on the lower part of the strong ridge of prootic which contacts the paroccipital process. However, the area medial to the ridge, from which more matrix could be removed, has certainly been bent inwards. Thus the projection referred to above is actually the beginning of the dorsal margin of the fenestra ovalis (WHETSTONE's 'fenestra pseudorotunda'), which formerly curved round to join the anterior edge of the threshold area of the opisthotic. The portion of the prootic which WHETSTONE has regarded as the anterodorsal part of the cotyle is the continuation of the threshold on to the prootic, as in the bird skulls already cited.

To restore the otic region (Fig. 4), opisthotic plus metotic must be imagined as pulled downwards and outwards at the upper end, inwards at the lower end, and rotated clockwise as a whole so that the lower portion of the opisthotic just touches the prootic. The upper end of the crista interfenestralis and the anteroventral edge of the opisthotic threshold are slightly damaged; nevertheless, it seems that there was a notch at the prootic/opisthotic contact at the top of the fenestra ovalis, as in the juvenile penguin figured and in other juvenile birds. This notch disappears with increasing ossification (Fig. 3). The anterior margin of the opisthotic part of the threshold evidently corresponds closely to the original

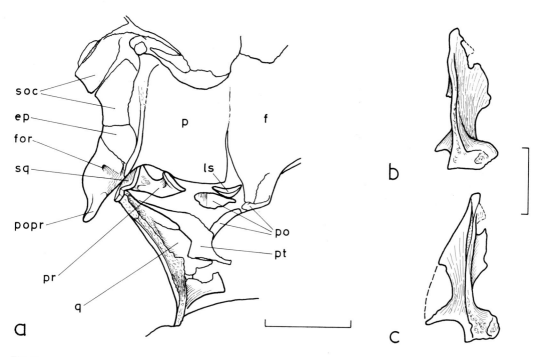

Fig. 5
(a) Posterior portion of skull of Eichstätt specimen. The epiotic appears to be narrower than in the London skull. Scale length 5 mm. (b) posterior and (c) posteromedial views of right quadrate of London specimen. Scale length 5 mm.

contact with the prootic which, as in Recent birds, ran up and back into the mouth of the posterior tympanic recess. The precise nature of the transition to the preserved part of the prootic/opisthotic suture at the posterior end of the pootic ridge (figured in ventrolateral view by WHETSTONE, 1983, Fig. 7) is uncertain. It is assumed here that, as in recent juvenile birds (Figs. 3a, b), the suture ran straight up and back and then turned abruptly dorsally to join the preserved section below the prootic ridge. In the modern bird, however, this change of direction takes place well back within the posterior tympanic recess, whereas in *Archaeopteryx* it would originally have been visible just in front of the entrance to this cavity. The entrance itself arches more dorsally in *Archaeopteryx* and the smooth threshold area would originally have been exposed to a greater extent in front of it. Acquisition of a contact by the quadrate head with the rear end of the prootic ridge in typical birds (Fig. 7b) has thus resulted in the greater occlusion of the entrance to the posterior tympanic recess. When opisthotic plus metotic are returned to their presumed original position, the orientations of the curvatures of the saddle-shaped threshold area correspond closely to those in the juvenile *Pygoscelis* or *Daption* skulls, with the axis of the 'trough' running down into the upper margin of the recessus scalae tympani.

Below the fenestra ovalis one would expect to find the tubular cochlear recess, and there is no reason to doubt that the lower part of the prootic and the downward-curving part of the opisthotic together made up this structure in the London cranium. How much they contributed medially cannot be determined, but the restored position of the opisthotic takes this bone as far ventrally as the prootic and suggests that the capsule was more primitive than that of any Recent bird group that I have studied, and similar in this respect to that of the Triassic crocodilomorph *Sphenosuchus* (WALKER, 1972). It seems likely also, from the observed structure, that the opisthotic in *Archaeopteryx* formed a complete loop around the foramen perilymphaticum as in *Sphenosuchus*, crocodiles and lizards, although this cannot definitely be asserted in the present state of the specimen. I have only observed this complete perilymphatic loop of the opisthotic in the Galliformes amongst the Recent birds studied; in other groups, including ratites, metotic or 'exoccipital' intervenes medially to meet prootic and break the continuity. Because of its position behind the opisthotic, the lower end of the metotic almost certainly did not enter the lower end of the cochlear recess (even supposing ossification to have been incomplete), as it does to a varying extent in all Recent bird groups studied (including ratites), with the exception of the Anseriformes, in which the recess is largely made up of prootic. Prootic and opisthotic appear to have curved apart at the lower end of the recess, at the side of the presumed lagenar region. This is usually the case in living carinates (I have not been able to study young enough specimens of ratites) in early stages except that metotic may replace opisthotic, and is also seen in *Sphenosuchus* and typical adult crocodiles.

The quadrate articulation

The London braincase thus displays no sign of an articulation for the quadrate or of a squamosal. For elucidation of this problem we must turn to the Eichstätt specimen (Fig. 5). Here the head of the quadrate is in contact with a piece of bone which thins medially and is itself in contact on its anterior face with a curved bone. WELLNHOFER (1974, Fig. 5B) identified these two elements as prootic and (with doubt) laterosphenoid respectively. I identified the anterior of the two as prootic (WALKER, 1980) and this has been confirmed by the preparation of the London braincase. WHETSTONE (1983) divides the posterior element into two, adding one part to the prootic (without a break between them) and regarding the other as possibly part of the head of the quadrate. I believe that this is inadmissible, since there is a clear break in the position shown by WELLNHOFER, and I agree with him that the posterior piece is all one element. The prootic ridge in the London cranium terminates smoothly posteriorly against the paroccipital, with no possibility of there having been a quadrate socket on it. Although the fragment in contact with the quadrate in the Eichstätt skull does resemble the rear part of the prootic of the London specimen, comparison of the anterior parts of the prootics of the two skulls shows that about one-third of the bone is missing at the rear of the Eichstätt prootic. The fragment in contact with the quadrate in the Eichstätt skull thickens laterally and shows no dip in its upper edge for the post-temporal fossa. In contrast, the London prootic thins out against the paroccipital, and has a slight dip in its upper surface, below the parietal tongue, which marks the lower border of the post-temporal fossa. I conclude, therefore, that the posterior bone in the Eichstätt skull is the inner part of the squamosal, driven medially by the lateral rotation of the foot of the quadrate, and that the anterior two-thirds of the prootic has been broken off and displaced upwards. The posterior surface of this squamosal fragment is slightly convex, agreeing with the slight concavity of the anterior surface of the paroccipital process lateral to it. The counterpart slab shows an indistinct area of bone between quadrate head and paroccipital process, which presumably represents the continuation of the squamosal, and the remainder of the bone would thus lie at a deeper level in the counterslab.

According to this interpretation, *Archaeopteryx* had a quadrate articulation of 'normal' archosaurian type, single-headed, and essentially with the squamosal. In thecodontians the quadrate head may also contact the paroccipital process to some extent (PRICE, 1946; WALKER, 1961; CHATTERJEE, 1978) so it is possible that this was the case in *Archaeopteryx*.

The post-temporal fossa in *Archaeopteryx* was a small aperture below the lateral tongue of the parietal. The dip in the upper margin of the prootic below this tongue is continued posteriorly by a shallow groove on the paroccipital (more pronounced in the Eichstätt skull), ending in a foramen (Figs. 1, 5, 6). The fossa no doubt transmitted the occipital or cervical artery (OELRICH, 1956; ALBRECHT, 1967), the avian homologue of which appears to be the ramus occipitalis of the external ophthalmic (stapedial) artery (SUSCHKIN, 1899; SAIFF, 1974), and the terminal foramen was probably for a branch of this. The bony tube which crosses the upper tympanic recess in some Recent bird skulls (e.g. *Gavia, Alca, Somateria*) terminates in a foramen or group of foramina on the occiput or dorsally just in front of it, and probably carries the occipital artery. In juvenile birds this vessel emerges in a groove beneath the sinus at the posterior margin of the squamosal. For this reason I labelled these foramina 'post-temporal fossa' (1972, Fig. 5d). In reptiles the fossa may also transmit the vena capitis dorsalis (BRUNER, 1907; O'DONOGHUE, 1921), but I have not been able to determine whether a vein drains the adjacent area in birds.

It is presumed that in *Archaeopteryx*, as in some archosaurs and birds (e.g. *Sula*) the squamosal sent a process inwards to overlap the tongue of the parietal (Figs. 6, 7a), and that this is the portion of the

Fig. 6
Occipital view of the London braincase. Squamosal restored and approximate position of midline indicated in broken lines. Scale length 5 mm.

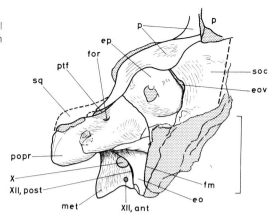

squamosal now in contact with the broken rear end of the prootic in the Eichstätt specimen. The squamosal would thus have projected above the paroccipital process (Fig. 6) and extended down its distal anterior surface for an unknown distance. It probably also had an anterior contact with the postorbital, since there are areas of bone and impression on both Eichstätt slabs indicating the presence of such a bone (Fig. 5a), and it is one of the few reasonably certain things about the Berlin skull. The postorbital seems to have had a descending process to meet the jugal, so that the suspensorium would have been effectively braced.

The parietal tongue may have been pressed down a little in the London skull and, since the saddle below it extends laterally beyond the parietal tip, the squamosal probably entered the post-temporal fossa laterally (Fig. 6). In thecodontians the squamosal commonly enters the lateral margin of the fossa. In modern birds the fossa lies entirely behind the squamosal (Figs. 2, 3a) because of the change in position of the squamoso-parietal contact. In juvenile carinates a sinus at the rear of the bone exposes the prootic/opisthotic suture running posterolaterally, crossed at right angles by a groove for the occipital artery. The resemblance to reptiles confirms the identification of the fossa, but in the bird forward migration of the squamosal has exposed the original floor on the occiput.

The quadrate

It seems evident that this was more reptilian in *Archaeopteryx* than has hitherto been supposed. The bone formerly lying adjacent to the braincase of the London specimen (DE BEER, 1954, Pl. VII), and recently removed by Mr. WHYBROW, acquires a new significance in this context. It seems inescapable that this bone (Fig. 5b, c) is the r i g h t q u a d r a t e, being extremely like the corresponding element in archosaurs, with its two wings departing from the shaft approximately at right angles to each other and a lateral embayment in the region of the quadrate foramen. One must allow, however, that the lower articular surface has been lost (a condyle-like fragment close to the London braincase might be the missing portion, but this is doubtful). WHETSTONE (1983, p. 448) states that this bone shows no resemblance to the Eichstätt quadrate. However, there is a discontinuity with a finished anterior edge running down the area labelled 'q' by WELLNHOFER (1974, Fig. 5B), and this sweeps down to the anteroventral corner in a manner closely similar to the presumed pterygoid wing of the London bone (Figs. 5a, c, herein). Interpretation of adjacent areas is shown in Fig. 5a. The lateral wing, which may have articulated with the squamosal, would thus project into the matrix of the Eichstätt counterslab. The inner surface of the lower end of the quadratojugal is concave, corresponding to the convexity seen laterally in the London quadrate. Even allowing for loss of bone on breakage it does seem, however, that the Eichstätt quadrate is less extensive laterally than the London one. The latter may include some quadratojugal, but there is no indication of a suture. The quadrates compare very closely and are in correct proportion, but present certain differences which may be specific or even generic.

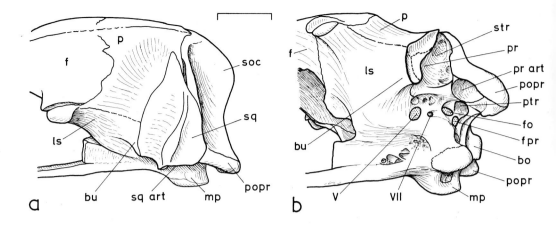

Fig. 7
Rear part of skull of sub-adult *Sula bassana* in (a) dorsolateral view, and (b) anteroventrolateral view with squamosal removed. Scale length 10 mm.

Other points

I differ from WHETSTONE in the interpretation of certain other details of the London skull, and from both WHETSTONE and WELLNHOFER in the interpretation of the Eichstätt skull, notably as concerns the mandible. Some points may be briefly mentioned. The foramen labelled 'V?' by WHETSTONE (London skull) is almost certainly pneumatic. A pneumatic foramen in just this position occurs in the juvenile eider *Somateria mollissima* but not in the adult, presumably because of backward growth of the 'alaparasphenoid' (Jollie, 1957, p. 406). This ossification could have been present originally in *Archaeopteryx*, since its removal when not firmly sutured to the prootic leaves remarkably little trace on the bone. It may be represented by a thin sheet of bone ventrolateral to the prootic. It seems to me that the lower part of the laterosphenoid has been bent medially and that this originally joined on to the prootic in front of the trigeminal foramen, which is very large. The uppermost of the three pneumatic cavities entering the anterior wall of the cochlear recess is doubtfully separate from the one below. In many juvenile bird skulls pneumatization begins with two depressions in this area, and a stronger central trabecula may persist into the adult. It is interesting that in *Sphenosuchus* two small cavities on the prootic occur in the same position, i.e. along the course of the palatine nerve. The foramen for the external occipital vein (Fig. 6) enters about half-way up the contact between supraoccipital and epiotic, and there seems to have been a narrow slit above it, as in juvenile bird skulls. This vein arches over the anterior vertical semicircular canal on the inner wall of the avian cranium, and I believe that this canal can be seen within the bone of the epiotic of the London skull, immediately lateral to the contact with the supraoccipital.

Are the *Archaeopteryx* specimens juveniles?

A number of features in the skulls suggest the possibility that even the London specimen may not be an adult. These are: (1) very large trigeminal foramen; (2) fenestra ovalis pointed at the lower end and probably notched at the upper end; (3) extremely slender crista interfenestralis; (4) cochlear recess notched laterally at its base; (5) lack of a notch for nerve IX; (6) upper part of supraoccipital/epiotic suture apparently a slit; (7) lack of irregular surfaces at squamosal/paroccipital contact and at presumed overlap on parietal tongue; (8) same on prootic for possible alaparasphenoid; (9) gently curving, smooth, slightly overlapping contacts between frontals and parietals, and the way the dermal bones have slid over each other during fossilization. Many of these features are typical of juvenile birds; some probably also occur in juvenile reptiles, but I have not had the opportunity to compare these. (4) is seen in adult crocodiles. A rough estimate of the stage of development of the London specimen in comparison to *Eudyptes*, *Daption* and *Somateria* suggests that it is about 50–70 % of full skull length. The ratio of the sizes of the London and Eichstätt specimens, based on the most reliable postcranial parameters (WELLNHOFER, 1974, Table 1), averages at 1.7, whereas certain skull dimensions give a ratio of 1.2. Using the latter, and taking snout to supraoccipital length for the Eichstätt skull as 41 mm, the London skull would have been

49 mm long, and its estimated adult length as from about 70 to 100 mm. The implication of this suggestion would be that *Archaeopteryx* continued to increase substantially in size after fledging, unlike the young of most modern birds, with the notable exception of the Galliformes.

The quadrate articulation in relation to the squamosal

In *Archaeopteryx* the superior tympanic recess (PYCRAFT, 1902, p. 281), or 'antrum pneumaticum dorsale', is well developed as a depression on the upper surface of the prootic but, as noted by WHETSTONE (1983, p. 451), there is no sign that the squamosal covered this region. Indeed, if the squamosal was of 'normal' archosaurian type, as suggested here, it could hardly have done so. Nevertheless, the recess may still have been pneumatic, since in modern birds pneumatic diverticula frequently ramify amongst the jaw muscles and lie against the bones (BIGNON, 1889). The appearance certainly suggests pneumaticity, but it may have been partly or entirely a muscle origin area, as it is in some Recent birds (PYCRAFT, 1898; SAIFF, 1974). In typical birds the squamosal has moved forwards as a whole compared with its position in reptiles, and this appears to be connected with the forward movement of the squamosal articulation of the quadrate head. It is instructive to compare the skull of *Archaeopteryx* with that of the gannet, *Sula bassana* (Pelecaniformes), purely as a morphological comparison, since each has a large superior tympanic recess and a strong posterior ridge on the prootic, which in *Sula* bears a socket for the quadrate but in *Archaeopteryx* is quite smooth (Figs 4,7) *Sula* also has a short squamosal, an apparently primitive feature. In *Archaeopteryx* the prootic-laterosphenoid buttress is noticeably inset compared to the paroccipital process, whereas in *Sula* it projects to an equal extent (this is in a late juvenile specimen – in the adult the buttress disappears, the squamosal socket being actually some distance behind its former position). Starting from *Archaeopteryx*, as the braincase (particularly the area surrounding the cerebral hemispheres) became relatively expanded, the buttress could eventually have contacted the anterior part of the squamosal in the postorbital/squamosal bar. At the same time, as the socket for the quadrate moved forwards below the squamosal, the anterior margin of the medial process of the latter (which overlapped the parietal tongue) became drawn forwards, the bone expanding like a fan about its inner point as pivot, so that it eventually became a triangle covering the superior tympanic recess. In this way the rear part of the supratemporal opening would have been obliterated. Later, the postorbital would have been lost. A process such as this would account for the fact that in several carinate bird groups usually regarded as primitive the squamosal socket is a short distance behind the buttress and does not extend on to it. Otherwise the apparently fortuitous coincidence of squamosal socket and buttress is hard to explain. Jaw muscles could still have passed into the superior tympanic recess and on to the skull roof during these changes, in fact closure of the 'lid' of the upper tympanic recess was probably in part due to increase in the importance of muscle origins on the squamosal's upper surface.

Relative enlargement of the braincase probably also largely accounts for the acquisition of a contact of the quadrate head with the rear end of the prootic ridge, although it seems likely that changes in streptostyly and kinesis were also involved. One would expect that there would have been an intermediate stage during which the quadrate head contacted both paroccipital and prootic, but it is not clear whether this was necessarily so. In ratites there is such a contact, but whether this is truly primitive is doubtful, since the articulation as a whole is placed well forward and the paroccipital contact is on an inner extension of the head and this appears to be a specialized arrangement. In detail the otic capsule too is highly specialized in the ratites I have studied (*Struthio, Rhea, Dromaius, Casuarius*). Freeing of the paroccipital process from its role as a quadrate support seems to have been a major factor in the upward expansion of the metotic into this process at the expense of the opisthotic. In *Archaeopteryx* it seems probable that the process was still formed by the opisthotic.

Archaeopteryx in relation to modern birds

The analysis of the braincase of *Archaeopteryx* presented here differs considerably from that of WHETSTONE (1983), particularly as concerns the otic capsule and quadrate articulation. Consequently my view of its phylogenetic position also differs from his. It seems to me that there is nothing in the structure of those regions which I have particularly studied which debars *Archaeopteryx* from the direct ancestry of modern birds. Two of the three typical pneumatic recesses, superior and posterior, are already present, and such differences as they exhibit from their homologues in the modern bird skull are readily explicable on the basis of the reptilian nature of the quadrate and squamosal. As noted by WHETSTONE, there is good reason to think that the anterior tympanic recess was also developed. The prootic is more deeply inserted into the braincase in the modern bird skull, but this can be accounted for by changes in relative brain size and proportions. The otic capsule details, as far as can be observed, correspond exactly to the expected, primitive pattern, the opisthotic probably forming a complete perilymphatic loop and the metotic being excluded from the cochlear recess. It should be stressed, however, that the otic capsule of *Archaeopteryx* is extremely avian; in comparison, ratites are more specialized than carinates. In ratites the opisthotic does not extend further than the crista interfenestralis and so does not enter the cochlear recess. The latter is formed by prootic and metotic and the basioccipital does not enter its base as it does in carinates.

The relationships between birds and crocodiles

I have suggested (WALKER, 1972; 1974; 1977) that birds and crocodiles share a common ancestry above the thecodontian level and this view has been supported by WHETSTONE & MARTIN (1979; 1981) and MARTIN (1983). However, several lines of evidence have since combined to throw considerable doubt on this hypothesis.

(1) It is highly probable that the forward movement of the quadrate head, leading to a prootic contact in both groups, was a parallel development. In recent crocodiles the stapedial artery (strictly, the temporoorbital artery) passes forwards in a canal dorsomedial to the head of the quadrate (HOCHSTETTER, 1906), whereas in birds the stapedial (or external ophthalmic) artery passes forwards below the quadrate articulation in the usual reptilian position. Previously it had seemed probable to me that this part of the course of the artery in 'true' crocodiles was a by-pass, owing to the fusion of quadrate and pterygoid to the braincase. However, the method of enclosure by moving forward of the quadrate head has since proved relatively simple to understand. Recent unpublished work on the skull of the Upper Triassic thecodontian *Stagonolepis* (WALKER, 1961) has clarified this process. The similar, albeit more primitive, build of the quadrate articular region in the Upper Triassic crocodilomorph *Sphenosuchus* to that of recent crocodiles leaves little doubt that the stapedial artery in this reptile passed forwards in a similar manner. On the other hand, there is no reason to doubt that the avian 'external ophthalmic artery' is the homologue of the reptilian stapedial artery. All the anatomical and embryological evidence supports this conclusion (HAFFERL, 1921; HUGHES, 1934; STELLBOGEN, 1930; STRESEMANN, 1927—34; TWINING, 1906).

Thus, if there has been a common ancestral stock for birds and crocodiles, the quadrate head must have been in the 'normal' archosaurian position. Characters associated with the forward position of the quadrate head: formation of a post-quadrate tympanic cavity and an external auditory meatus, protection of the tympanic membrane, etc, therefore developed in each group as a result of parallel evolution. This probability tends to throw doubt on the value of details of the otic capsule, which were believed to be part of a phase of general improvement in auditory structures in the common ancestral stock (WALKER, 1972).

(2) WHETSTONE & MARTIN (1979, 1981) argued that birds and crocodiles share otic specializations not present in early thecodontians or in dinosaurs. The latter, in their estimation, were more like *Sphenodon* in otic structure. They apparently visualize a sudden shift of the foramen perilymphaticum to a morphologically different, more lateral position to give the bird or crocodile condition. (It should be noted that the subcapsular process does not, as they state, 'divide the metotic fissure'). It seems more probable, however, that the inner margin of the foramen perilymphaticum (which faces posteromedially in *Sphenodon* and *Euparkeria:* CRUICKSHANK, 1971, and personal observations), previously cartilaginous, became ossified and the foramen itself rotated laterally. At the same time a subcapsular process (SHIINO, 1914) added bone to the lateral edge of the exoccipital behind the metotic foramen, thus extending the perilymphatic sac laterally and leading to the formation of a fenestra pseudorotunda and a secondary tympanic membrane (see also BAIRD, 1960, p. 941). At some stage in this process the vagus nerve became enclosed either within, or behind, the subcapsular portion of the exoccipital, and thus (in 'true' crocodiles) it now runs back posterolaterally behind a strong 'exoccipital' ridge. Medially, however, its exit still lies within the metotic foramen. Further preparation of a skull of the thecodontian *Stagonolepis* has revealed an otic structure transitional to that of crocodiles, with a foramen perilymphaticum surrounded by a loop of opisthotic. An exoccipital ridge probably indicates some modest development of a subcapsular process (although the vagus nerve was not enclosed), and a secondary tympanic membrane was almost certainly present. The otic capsule of *Sphenosuchus* is extremely crocodilian, but the strong subcapsular process ossification still does not enclose the vagus nerve

(WALKER, 1972, Figs. 1b, 6a; note that in Fig. 1b the cultriform process is too high up). Thus a gradual evolutionary change in otic structure in early archosaurs seems more likely.

Furthermore, it seems probable that wherever the vagus canal is diverted back externally by a strong bony ridge between it and the otic opening, as it is in many (but not all) dinosaurs, this ridge represents a subcapsular process. In the hypsilophodontid dinosaur *Zephyrosaurus* (SUES, 1980) the structure seems quite clear and unequivocal. The perilymphatic loop of the opisthotic has been broken off, apparently leaving only the base of the crista interfenestralis. The secondary tympanic membrane would have been stretched across the large aperture behind the crista, that is the lateral aperture of SUES's 'fissura metotica' (better called the m e t o t i c   f o r a m e n, since the metotic fissure is, strictly, an embryonic structure) or, more accurately, the lateral aperture of the recessus scalae tympani. The opening identified by SUES as the fenestra pseudorotunda is almost certainly the external vagus foramen, the canal for this nerve passing through (or behind) a subcapsular process ossification which forms a ridge on the exoccipital. It should be pointed out that the foramina for the two roots of nerve XII are exceptionally large in *Zephyrosaurus*. Hadrosaurs are essentially similar in otic structure (OSTROM, 1961). Parallel evolution in otic capsule structures is thus a distinct possibility.

The metotic cartilage of birds shows considerable variation in its development and cannot be dealt with here; it seems probable that it represents a further elaboration of the basic archosaurian subcapsular process.

(3) It seems to me now that the structure figured as a 'pterygoid process' of the quadrate in *Sphenosuchus* (WALKER, 1974) is, in reality, the lower portion of the pterygoid wing which has been split off below a ridge leading to the 'orbital process'. This ridge is evidently that named 'crest B' by IORDANSKY (1964) in the modern crocodialian skull. Recognition that there was no separate pterygoid process of the quadrate, and a reconsideration of bone contacts previously thought to have been moveable, leads me to conclude that the skull of *Sphenosuchus*, at least in the adult, was a k i n e t i c and m o n i m o s t y l i c.

(4) Another serious drawback to the hypothesis of a particularly close relationship between birds and crocodiles lies in the timing of events. Crocodiles are known to have attained a prootic contact of the quadrate head with presumed enclosure of the stapedial artery in a canal, typical otic capsule and subcapsular process by the Upper Triassic (WALKER, 1972; CROMPTON & SMITH, 1980). On the other hand, according to the interpretation presented here, *Archaeopteryx* in the Upper Jurassic still had the quadrate head in the 'normal' archosaurian position with no possiblity of stapedial artery enclosure and an otic capsule which, although primitive in certain details, is typically avian and somewhat different from that of *Sphenosuchus*, assuming that the latter represents the primitive crocodilian condition, or immediately pre-crocodilian in the strict sense.

The possibility that *Archaeopteryx* might be a 'pseudobird', that is, of separate descent from true birds, seems unlikely in the extreme. The structure and arrangement of the feathers, the furcula, the incipiently avian carpometacarpus and the build of the braincase, especially the otic capsule, make such a suggestion almost impossible to believe.

Space does not permit of a more detailed analysis here, but it does seem that the original concept of a particularly close relationship between birds and crocodiles has become so tenuous that it is very difficult to sustain.

Acknowledgements

Among the many who have helped me I particularly wish to thank the following: Mr. G. S. COWLES, Dr. A. J. CHARIG, Mr. C. A. WALKER, Dr. A. MILNER, Mr. P. WHYBROW (British Museum, Natural History); the British Antarctic Survey; Mr. E. MORTON (Hancock Museum, Newcastle upon Tyne); Dr. G. VIOHL, Herr J. BAUCH (Jura-Museum, Eichstätt); Dr. H. JAEGER (Humboldt Museum, Berlin).

References Cited

ALBRECHT, W. (1967): The cranial arteries and cranial arterial foramina of the turtle genera *Chrysemys*, *Sternotherus* and *Trionyx*, etc. – Tulane Studies Zool., **14**: 81–99.

BAIRD, I. L. (1960): A survey of the periotic labyrinth in some representative Recent reptiles. – Univ. Kansas Sci. Bull., **41**: 891–981.

BIGNON, F. (1889): Contribution à l'étude de la pneumaticité chez les oiseaux. – Mém. Soc. Zool. France, **2**: 260–320.

BRUNER, H. L. (1907): On the cephalic veins and sinuses of reptiles, etc. – Am. J. Anat., **7**: 1–117.

CHATTERJEE, S. (1978): A primitive parasuchid (phytosaur) reptile from the Upper Triassic Maleri Formation of India. – Paleontology, **21**, 1: 83–127.

CROMPTON, A. W. & SMITH, K. K. (1980): A new genus and species of crocodilian from the Kayenta Formation (Late Triassic?) of northern Arizona. – In: JACOBS, L. L. (Ed.): Aspects of vertebrate history: 193–217 (Museum of Northern Arizona Press).

Cruickshank, A.R.I. (1971): Early thecodont braincases. – Second Gondwana Symposium, internat. Union geol. Sci., South Africa 1970: 683–685.
De Beer, G.R. (1954): *Archaeopteryx lithographica*. – British Museum (Natural History), London: 1–68.
Hafferl, A. (1921): Zur Entwicklungsgeschichte der Kopfarterien beim Kiebitz (*Vanellus cristatus*). – Anat. Hefte, **59**: 521–576.
Hochstetter, F. (1906): Beiträge zur Anatomie und Entwicklungsgeschichte der Blutgefäßsystemes der Krokodile. – In: Voeltzkow, A.: Reise in Ostafrika, **4**: 1–139.
Hughes, A.F.W. (1934): On the development of the blood vessels in the head of the chick. – Phil. Trans. Roy. Soc. Lond., B, **224**: 75–129.
Iordansky, N.N. (1964): The jaw muscles of the crocodiles and some relating structures of the crocodilian skull. – Anat. Anz., **115**, 3: 256–280.
Jollie, M.T. (1957): The head skeleton of the chicken and remarks on the anatomy of this region in other birds. – J. Morph., **100**: 389–436.
Martin, L.D. (1983): The origin of birds and of avian flight. – Current Ornithology, **1**: 105–129.
O'Donoghue, C.H. (1921): The blood vascular system of the tuatara, *Sphenodon punctatus*. – Phil. Trans. Roy. Soc. Lond., B, **210**: 175–252.
Oelrich, T.M. (1956): Anatomy of the head of *Ctenosaura pectinata* (Iguanidae). – Misc. Publs. Mus. Zool. Univ. Mich., **94**: 1–122.
Ostrom, J.H. (1961): Cranial morphology of the hadrosaurian dinosaurs of North America. – Bull. Am. Mus. nat. Hist., **122**: 37–186.
Price, L.I. (1946): Sobre um novo pseudosuquio do Triassico superior do Rio Grande do Sul. – Bolm. Div. Geol. Miner. Bras., **120**: 7–38.
Pycraft, W.P. (1898): Contributions to the osteology of birds. Part I. Steganopodes. – Proc. Zool. Soc. Lond., 82–101.
Pycraft, W.P. (1902): Contributions to the osteology of birds. Part V. Falconiformes. – Proc. Zool. Soc. Lond., 277–320.
Romer, A.S. (1956): Osteology of the reptiles. – 772 pp. Chicago (Ill.) (University of Chicago Press).
Saiff, E.I. (1974): The middle ear of the skull of birds. The Procellariiformes. – Zool. J. Linn. Soc., **54**: 213–240.
Säve-Soderbergh, G. (1947): Notes on the brain-case in *Sphenodon* and certain Lacertilia. – Zool. Bidr. Uppsala, **25**: 489–516.
Shiino, K. (1914): Das Chondrocranium von *Crocodilus* mit Berücksichtigung der Gehirnnerven und der Kopfgefäße. – Anat. Hefte, **50**: 253–382.
Stellbogen, E. (1930): Über das äußere und mittlere Ohr des Waldkauzes (*Syrnium aluco* L.). – Z. Morph. Ökol. Tiere, **19**: 686–731.
Stresemann, E. (1927–34): Sauropsida: Aves. – In: Kükenthal, W. (Ed.) Handbuch der Zoologie, **7**, 2: 1–899, Berlin.
Sues, H.D. (1980): Anatomy and relationships of a new hypsilophodontid dinosaur from the Lower Cretaceous of North America. – Palaeontographica, (A), **169**: 51–72; Stuttgart.
Suschkin, P.P. (1899): Beiträge zur Morphologie des Vogelskeletts. I. Der Schädel von *Tinnunculus*. – Nouv. Mém. Soc. (imp.) Nat. Mosc., **16**, 2: 1–163.
Twining, G.H. (1906): The embryonic history of carotid arteries in the chick. – Anat. Anz., **29**: 650–663.
Walker, A.D. (1961): Triassic reptiles from the Elgin area: *Stagonolepis, Dasygnathus* and their allies. – Phil. Trans. Roy. Soc. Lond., B, **244**: 103–204.
Walker, A.D. (1972): New light on the origin of birds and crocodiles. – Nature, Lond., **237**: 257–263.
Walker, A.D. (1974): Evolution, organic. – McGraw-Hill Yearb. Sci. & Technol., **1974**: 177–179.
Walker, A.D. (1977): Evolution of the pelvis in birds and dinosaurs. – In: Problems in Vertebrate Evolution (Andrews, S.M., Miles, R.S. & Walker, A.D., Eds.), Linn. Soc. Symp. Ser., **4**: 319–358.
Walker, A.D. (1980): The pelvis of *Archaeopteryx*. – Geol. Mag., **177**: 595–600.
Wellnhofer, P. (1974): Das fünfte Skelettexemplar von *Archaeopteryx*. – Palaeontographica, A, **147**, 4–6: 169–216; Stuttgart.
Whetstone, K.N. (1983): Braincase of Mesozoic birds: I. New preparation of the "London" *Archaeopteryx*. – J. Vert. Paleont., **2**, 4: 439–452.
Whetstone, K.N. & Martin, L.D. (1979): New look at the origin of birds and crocodiles. – Nature, **279**: 234–236; London.
Whetstone, K.N. & Martin, L.D. (1981): Common ancestry of birds and crocodiles: a reply. – Nature, **289**: 98; London.

Author's address: Dr. R. Alick D. Walker, Department of Geology, University of Newcastle upon Tyne, Newcastle upon Tyne, NE 1 7RU, Great Britain.

Paul Bühler

On the Morphology of the Skull of *Archaeopteryx*

Abstract

The comparison between living reptiles and birds shows that in the anagenesis of birds (besides the emergence of the feather, the disappearance of the teeth, and the changes of the tail and limb systems) two feature complexes are transformed in a conspicuous mode: (1) The brain is enlarged in consequence of the increasing input and processing of information which is, in turn, a result of the complicated relationship to the environment of an animal which is able to fly, and (2) a highly movable upper jaw of the prokinetic type has been evolved which together with an elongated and flexible neck was able to balance the loss of the forelimbs in feeding and grooming activities.

The remains of the skulls of the specimens in Eichstätt, Berlin, and London together show that in *Archaeopteryx* the quadrate bone was movable against the skull and the upper jaw was movable in a prokinetic mode; furthermore the braincase was "inflated" and much more voluminous than former reconstructions of the skull showed. The forebrain and the cerebellum, in relation to the braincase, were enlarged and the optic lobes of the midbrain were separated by the cerebellum. All this together indicates that *Archaeopteryx* developed to a much more birdlike level than it has been assumed hitherto. This also includes a high probability that *Archaeopteryx* was able to fly. What is still lacking is the detailed knowledge of the form of the dorsal and the orbital processes of the quadrate and their articulations.

Zusammenfassung

Zur Schädelmorphologie von *Archaeopteryx*.
Der Vergleich rezenter Reptilien und Vögel zeigt, daß während der Anagenese der Vögel (neben der Ausbildung von Federn, der Reduktion der Zähne und der Umgestaltung des Schwanzes und der Gliedmaßen) zwei Merkmalskomplexe sich in auffälliger Weise verändert haben: (1) Das Gehirn hat sich vergrößert, weil der flugfähige Organismus durch seine komplizierte Beziehung zur Umwelt einen gesteigerten Informations-Input bewältigen mußte; und (2) eine wirkungsvolle prokinetische Oberkieferbeweglichkeit wurde entwickelt, die zusammen mit einem verlängerten und biegsamen Hals den Verlust der Vorderextremitäten für Nahrungsaufnahme und Körperpflege ausgleichen konnte.

Die Schädelreste der Exemplare in Eichstätt, Berlin und London lassen in einer vergleichenden Untersuchung erkennen, daß bei *Archaeopteryx* das Quadratum gegenüber der Schädelkapsel beweglich war, daß der Oberkiefer in prokinetischer Weise auf und ab bewegt werden konnte, und daß die Gehirnkapsel "blasig aufgetrieben" und wesentlich geräumiger war als die bisherigen Schädelrekonstruktionen vermuten ließen. Vorder- und Kleinhirn waren relativ groß, und die beiden Lobi optici des Mittelhirns waren durch das Kleinhirn getrennt. Aus diesen Befunden läßt sich ableiten, daß *Archaeopteryx* ein wesentlich höheres Evolutionsniveau erreicht hatte, als bisher angenommen worden war, was unter anderem auch wahrscheinlich macht, daß *Archaeopteryx* schon die Fähigkeit zu fliegen entwickelt hatte. Was noch immer fehlt, ist die genaue Kenntnis von Dorsalfortsatz und Orbitalfortsatz des Quadratums und damit der Gelenkungen zwischen Quadratum, Schädelkapsel und Pterygoid.

Introduction

Within the last 125 years the fossilized Jurassic birds of the Solnhofen area have evoked a more intensive scientific interest than perhaps any other group of fossilized animals at all. The skull and jaws – besides the features of the feathers, limbs, and tail – have always been in the center of the interest (DAMES 1884, HEILMANN 1912–1916, 1926, KLEINSCHMIDT 1951, DE BEER 1954, WELLNHOFER 1974, OSTROM 1976, WHYBROW 1982, WHETSTONE 1983, MARTIN 1983). On the other hand until now there have been found only three specimens of *Archaeopteryx* with cranial remains. So the question has to be raised if there actually is anything new to say about the skull of *Archaeopteryx*. When, in 1978, the author for the first time had the possibility to examine the Eichstätt specimen he was astonished – with the reconstructions of HEILMANN und WELLNHOFER in mind – to see how large the frontal and parietal parts of the braincase were and how birdlike their inflation was – inspite of its postmortal crushing (cf. e. g. Fig. 3 A in WHETSTONE 1983). When, in 1980, PETER WHYBROW did a preparation of the braincase of the London specimen the newly exposed details agreed nicely with these features of the Eichstätt specimen and confirmed very clearly that size and volume of the braincase in all previous lateral and dorsal reconstructions are too small. Furthermore some details of the frontal bone and the ventrolateral area of the braincase make it likely that *Archaeopteryx* moved the upper jaw in a prokinetic manner. Therefore within this paper a new restoration of the skull and the brain is represented.

Cranial kinesis and braincase in modern birds

If one carefully observes feeding birds it is striking that either the whole upper jaw can be moved (Fig. 2 in BÜHLER 1973) or a part of it (Fig. 8.7 and 8.11 in BÜHLER 1981).

To avoid confusion I want at first to point out how the author – following HOFER (1949) – is using the terms of the upper jaw mobility (BÜHLER 1981): If only the rostral part of the upper jaw is movable as in waders, ostriches, and hummingbirds the mode is called r h y n c h o k i n e s i s; if the whole upper jaw is moved against the braincase and the orbital area as in most modern birds we talk about p r o k i n e s i s; if a rostral part of the dermal braincase is moved together with the upper jaw as in geckos or in the coelacanth *Latimeria* we have to talk about m e s o k i n e s i s; and if the whole dermal braincase together with the upper jaw is moved as a unit against the "occipital segment" we call it m e t a k i n e s i s.

The morphological equipment of the upper jaw mobility in birds is composed of a pair of highly movable quadrate bones (Fig. 2 and 5 in BÜHLER 1970), a pair of palatopterygoid bridges, a pair of jugal bars, and a set of muscles (Fig. 8.1, 8.3 and 8.6 in BÜHLER 1981). The general basic type of the avian upper jaw mechanism is – as mentioned above – prokinesis. An important condition for this mobility is the craniofacial flexion zone transverse to and within the ridge of the upper jaw. This bendable section of true bone is morphologically characterized by being thin and flattened.

Another typical feature of modern birds is the inflation of the braincase. The outer form of the braincase – very much in contrast to the reptilian condition – is relatively similar to the rounded form of the enlarged brain inside (Fig. 10 and 11 in BÜHLER 1972).

The fossilized material

For the permission to examine the specimens and for supporting help I am very much obliged to G. VIOHL (Eichstätt), K. FISCHER, J. HELMS and B. STEPHAN (Berlin), and A. J. CHARIG and C. A. WALKER (London). Also I have to point out that I have discussed several aspects concerning this paper with L. D. MARTIN (Lawrence, Kansas) and M. A. RAATH (Johannesburg).

In the London specimen the parts of the skull are scattered to such a degree that for a time it has been thought that the skull is lacking at all (DE BEER 1954). In 1980 when PETER WHYBROW – after a suggestion of L. D. MARTIN and K. N. WHETSTONE – removed the area of the "endocranial cast" from the limestone slab and exposed a beautifully preserved median and left region of the cranium it became visible that the London specimen has a well preserved braincase (WHYBROW 1982) even if this braincase is deformed somewhat obliquely and the parietal and frontal bones are broken up near the dorsal midline.

In the Berlin specimen the outline of the skull in profile is nicely preserved but with exception of the orbit and the snout region there are only few and badly preserved details, and especially the area around the upper quadrate and the occipitalia is lacking altogether.

In the Eichstätt specimen we have the skull with the most details. But this skull is very much deformed obliquely, the braincase is depressed, and especially the area around the upper quadrate articulation is difficult to interpret.

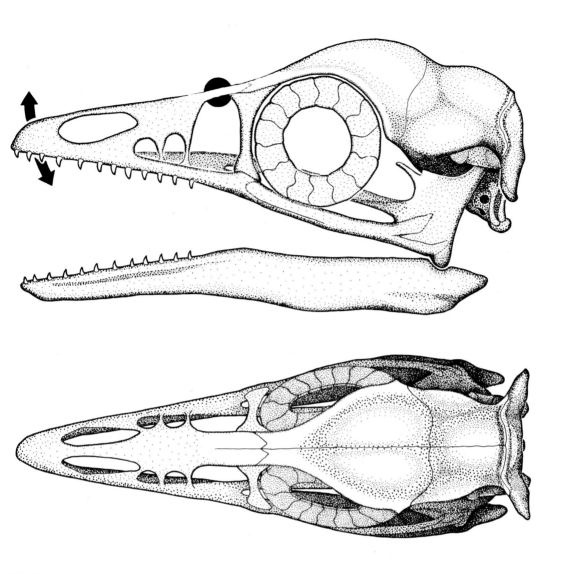

Fig. 1
The reconstruction of the *Archaeopteryx* skull showing prokinesis. The braincase is larger than it had been supposed hitherto. The round dot marks the hypothetic craniofacial flexion zone.

As none of the three skulls shows enough details to allow a reconstruction by itself, and as there are remarkable differences in size one may doubt that the three specimens are so nearly related that it makes sense to combine their features: (1) The Berlin and the Eichstätt specimens inspite of certain differences for instance in the dentition or the form of the nasal opening, are very similar in the form of the whole snout inclusively the lower jaw, the orbit, and the skull-angle. (2) The same is the case if one compares the right Berlin frontal bone seen from the right side with a mirror image of the left London frontal seen from the left. (3) The loose lacrimal in London is well comparable with the pieces of it in Eichstätt (inclusive what WELLNHOFER called prefrontal). And (4) the well known view from above of the

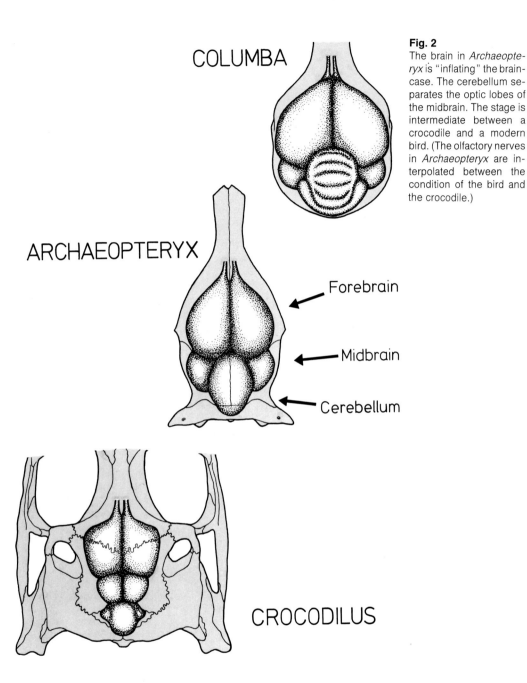

Fig. 2
The brain in *Archaeopteryx* is "inflating" the braincase. The cerebellum separates the optic lobes of the midbrain. The stage is intermediate between a crocodile and a modern bird. (The olfactory nerves in *Archaeopteryx* are interpolated between the condition of the bird and the crocodile.)

Eichstätt braincase is very much like a mirror image of an oblique dorsolateral view from the left side of the London braincase. Especially the similarity in the way the bones of the frontal and parietal region are broken is striking, and also the form of the paroccipital process (WHETSTONE 1983, called squamosum by WELLNHOFER 1974) is very similar.

The cranial kinesis in *Archaeopteryx*

The different theories on the kinesis of the *Archaeopteryx* skull which have been brought up in the past, for example by HEILMANN (1926), BOCK (1964), WELLNHOFER (1974), and WHETSTONE (1983) shall not be discussed here. At the moment the question is, which features of the available material can tell us something about whether the animal has had a certain kinetic or an akinetic condition. (1) Though we still don't know exactly what the dorsal process of the quadrate looked like, the London cranium shows that the quadrate was not fused with the braincase because in the caudolateroventral area of the braincase there is clearly no surface of a fracture visible. (2) This specimen also supports the theory of WELLNHOFER that neither a postorbital nor a preorbital rigid bony bridge had been formed between the jugal bar and the braincase. (3) Moreover, the London specimen shows that the rostral process of the frontal bone had been flattened and thin, and the Eichstätt specimen shows the same for the caudal end of the nasal bone. (4) In the Eichstätt specimen a bone – called laterosphenoid by WELLNHOFER and prootic by WHETSTONE – is located below the right parietal. By its position this bone fits in nicely with the hitherto lacking squamosal bone. It seems to have an intermediate position between the quadrate and the braincase, comparable for instance to the position of the squamosum of the theropod *Syntarsus* (cf. M. A. RAATH in this volume).

All these details agree with the theory that in *Archaeopteryx* the flattened ridge of the upper jaw was bendable in the zone between the braincase and the snout, that the ventral part of the movable quadrate bone was in kinematic connection with the ventrobasal part of the upper jaw and that the whole upper jaw was moved in a prokinetic way as in most modern birds (Fig. 1). But this prokinesis has been realized on a basic level without the highly developed articulations in modern birds, with poorly developed bending zones within bones, and the connections of connective tissue between the bones playing an important role. Nevertheless this stage can well be interpreted as being on the way to the more advanced prokinetic stage of modern birds.

The form of the braincase and the brain in *Archaeopteryx*

The theory of the enlarged and inflated braincase which was initiated by the Eichstätt specimen was very convincingly supported by the newly prepared details of the London braincase. The combination of the important details of all three specimens results in the reconstructions shown in Fig. 1. An additional fascinating feature of the newly exposed braincase is that its inflations show the location of the three main parts of the brain. It is of special interest that the form of the braincase shows clear similarities to the form of the brain itself and that the optic lobes of the midbrain are separated by a comparatively large cerebellum. The details are clearly intermediate between the features of a crocodile as a living archosaur and a modern bird (Fig. 2). These results are nicely in accordance with a widely ignored hypothesis of JERISON (1968). Finally it has to be stressed that the brainsize also supports the theory that *Archaeopteryx* had already evolved the ability to fly.

References Cited

BEER, G. DE (1954): Description of the British Museum specimen of *Archaeopteryx* and comparison with the Berlin specimen. – *Archaeopteryx lithographica*, Trustees of the British Museum; London.
BOCK, W. J. (1964): Kinetics of the Avian Skull. – J. Morph., **114**, 1: 1–42; Philadelphia.
BÜHLER, P. (1970): Schädelmorphologie und Kiefermechanik der Caprimulgidae (Aves). – Z. Morph. Tiere, **66**: 337–399; Berlin.
BÜHLER, P. (1972): Sandwich Structures in the Skull Capsules of various birds – The Principle of Lightweight Structures in Organisms. – Mitteilungen des Instituts für leichte Flächentragwerke **4**: 39–50; Stuttgart.
BÜHLER, P. (1973): Sandwichkonstruktionen – Leichtbau im Vogelschädel. – Der Deutsche Baumeister BDB **2**: 100–103; Stuttgart.
BÜHLER, P. (1981): Functional anatomy of the avian jaw apparatus. In: Form and Function in Birds **2**: 439–468; Academic Press London.

DAMES, W. (1884): Über *Archaeopteryx*. – Palaeont. Abh. **2**, 3: 119–196.
HEILMANN, G. (1912–1916): Vor nuvaerende Viden om Fuglenes Afstamning I – V. – Dansk Ornithologisk Forenings Tidsskrift **7**: 1–71, **8**: 1–92, **9**: 1–160, **10** 73–144; Kopenhagen.
HEILMANN, G. (1926): The origin of birds. – 208 pp. London (WITHERBY).
HOFER, H. (1949): Die Gaumenlücken der Vögel. – Acta Zoologica **30**: 209–248.
JERISON, H. J. (1968): Brain evolution and *Archaeopteryx*. – Nature **219**: 1381–1382; London.
KLEINSCHMIDT, A. (1951): Über eine Rekonstruktion des Schädels von *Archaeornis siemensi* DAMES 1884 im Naturhist. Museum, Braunschweig. – Proc. X. Int. Ornithol. Congr., 631–635; Uppsala.
MARTIN, L. D. (1983): The origin and early radiation of birds. – Perspectives in Ornithology: 291–338; Cambridge (University Press).
OSTROM, J. H. (1976): *Archaeopteryx* and the origin of birds. – Biol. J. Lin. Soc. **8**, 2: 91–182; London (Academic Press).
WELLNHOFER, P. (1974): Das fünfte Skelettexemplar von *Archaeopteryx*. – Palaeontographica, A, **147**, 4–6: 169–216; Stuttgart.
WHETSTONE, K. N. (1983): Braincase of Mesozoic birds: I. New preparation of the "London" *Archaeopteryx*. – J. Vertebr. Paleontol. **2**, 4: 439–452; Oklahoma.
WHYBROW, P. J. (1982): Preparation of the cranium of the holotype of *Archaeopteryx lithographica* from the collections of the British Museum (Natural History). – N. Jb. Geol. Palaeont. Mh. **3**: 184–192; Stuttgart.

Author's address: Dr. PAUL BÜHLER, Institut für Zoologie, Universität Hohenheim, 7000 Stuttgart 70, Federal Republic of Germany.

J.R. Hinchliffe

'One, two, three' or 'Two, three, four': An Embryologist's View of the Homologies of the Digits and Carpus of Modern Birds

Abstract

The embryological evidence concerning the identity of the digits and carpus of modern birds (chick) is surveyed. Isotope labelling using $^{35}SO_4$ of the chondroitin sulphate component of the matrix of precartilage elements is more sensitive than traditional staining methods used for examination of the developing chondrogenic pattern. Only five precartilaginous carpal elements can be identified: radiale, ulnare (which disappears after chondrifying), distal carpal 3, X and the pisiform. The position of the pisiform allows the three main digits to be identified as II-III-IV. This interpretation is supported by evidence that cell death removes a substantial area of mesenchyme from the preaxial part of the wing bud. The subsequent contribution of these elements to the adult wing skeleton is described.

The pattern of digital reduction in theropods is usually considered to be from the postaxial side inwards: surviving digits being identified as I-II-III. If this I-II-III interpretation is correct, the embryological evidence for II-III-IV does not support the theory of theropod ancestry of *Archaeopteryx* and modern birds.

Zusammenfassung

Der embryologische Befund bezüglich der Identität der Finger und des Carpus moderner Vögel (Huhn) wird geprüft. Die Isotopen-Markierung mit $^{35}SO_4$ des Chondroitinsulfat-Anteils der Matrix praecartilagener Elemente ist empfindlicher als die traditionellen Färbemethoden, die für die Untersuchung sich entwickelnder Knorpelmuster verwendet wurden. Es können nur 5 praecartilagene Carpalelemente identifiziert werden: Radiale, Ulnare (das nach der Verknorpelung verschwindet), distales Carpale 3, X und das Pisiforme. Die Lage des Pisiforme erlaubt es, die drei Hauptfinger als II-III-IV anzusprechen. Diese Interpretation wird durch den Nachweis untermauert, daß der Zelltod eine wesentliche Mesenchymsubstanz vom praeaxialen Teil der Flügelknospe wegnimmt. Die nachfolgende Beteiligung dieser Elemente am adulten Flügelskelett wird beschrieben.

Gewöhnlich wird angenommen, daß die Reduktion der Finger bei Theropoden von der postaxialen Seite nach innen erfolgte: Die bleibenden Finger werden somit als I-II-III identifiziert. Wenn diese I-II-III-Interpretation zutrifft, wird durch den embryologischen Nachweis für II-III-IV die Theorie der Theropodenabstammung von *Archaeopteryx* und der modernen Vögel nicht bestätigt.

Introduction

Embryologists and palaeontologists have long followed different conventions of identification of the three main digits of modern birds. While embryologists have followed a convention of II-III-IV, palaeontologists and anatomists have favoured I-II-III. The reasons for this difference lie in the different traditions of the two schools, as discussed in HINCHLIFFE & HECHT 1984. The embryological view derives particularly from HOLMGREN (1955) and MONTAGNA (1945). These authors claimed to find either a transient digit or its associated distal carpal preaxial to that developing digit which eventually forms the anterior digit of the adult wing. Thus they identified the four adult digits (including the rudimentary posterior one) as being II-III-IV-V. In addition these two authors also claimed to find all 13 archetypal embryonic precursors of the carpus.

However, as pointed out previously (HINCHLIFFE, 1977; HINCHLIFFE & GRIFFITHS, 1983), the histological evidence for these interpretations is equivocal, and the identification has been influenced by the recapitulationary assumption that the pentadactyl limb in its development passes through an archetypal stage common to all tetrapods. The present paper reviews the new embryological evidence emerging from more modern techniques as to the identity of carpal and digit elements in the development of the chick wing. It further considers the implications of these studies on the present debate on the evolution of birds and the reptilian origins of *Archaeopteryx*.

Identification of the developing skeletal elements of the wing

The wing bones are formed through a complex developmental process of cartilage replacement involving three successive stages: condensation, chondrogenesis and finally osteogenesis (HINCHLIFFE & JOHNSON, 1980). In the condensation phase, prechondrogenic areas start to show increased cell packing, probably involving changes in the adhesive properties of the cell surface of the mesenchyme, and contrasting with the loose packing of undifferentiated mesenchyme (THOROGOOD & HINCHLIFFE, 1975; EDE, 1983). Simultaneously with the first appearance of a condensation, cartilage matrix synthesis begins, involving two main components, collagen and the proteoglycan, chondroitin sulphate. The latter component may be labelled using $^{35}SO^4$, a technique which gives a particularly clear image of the emergence of chondrogenic pattern (SEARLS, 1965; HINCHLIFFE, 1977), far more precise than the older haematoxylin and eosin stained preparations which formed the basis of the classical interpretations.

It is at the condensation stage that the pattern described here diverges most radically from the classical accounts. Once matrix has accumulated during the chondrogenic phase, the chondrogenic pattern may be followed safely by such traditional techniques as the VAN WIJHE Methylene blue staining of the matrix in whole mounts (HAMBURGER, 1960). At this later stage, there is less disagreement with the pattern described in the classical accounts, but since individual elements may be followed through in their development from the condensation stage, the i d e n t i f i c a t i o n of the cartilage elements is now more securely based.

Interpretation of the digits

Autoradiography using $^{35}SO_4$ does not show the presence of the 'missing' preaxial digit claimed by HOLMGREN. Only the elements which become chondrogenic are found, identified here as follows: digit II metacarpal and phalange, digit III large metacarpal and two phalanges, digit IV metacarpal and phalange, and digit V rudimentary metacarpal (Figs 1–4).

Two features are helpful in identification of the digits. The first is the presence postaxially of the pisiform element of the carpus. The second is the area of preaxial mesenchymal cell death found in the early wing bud.

Topographical relationships enable identification with reasonable certainty of radiale, ulnare and pisiform in the carpus (Figs 1–4, 9) in relation to the radius, ulna, and postaxial border. Comparable homologous carpal bones are identified in the reptile archetype (ROMER & PARSONS, 1977). Recognition of the pisiform enables the adjacent rudimentary metacarpal to be identified as V. (The alternative, that this is IV, can be ruled out, since the missing V would have had to have been placed well posterior to the pisiform, a topographically incorrect position). The consequence of identifying the posterior rudimentary metacarpal as V is to make the missing digit the anterior one (I).

Consideration of the cell death patterns of the wing bud strengthens this argument. Beginning at 3½ days through to 5 days (HAMBURGER-HAMILTON stages 22–6) a substantial area of mesenchymal cell

Figs. 1–4 ▶
Autoradiographs of $^{35}SO_4$ uptake in the chick forelimb. 80μ Ci $^{35}SO_4$ applied to vitelline circulation for 4h (see HINCHLIFFE & EDE, 1973 for details). Fig 1, stage 28; Fig 2, stage 29; fig 3, stage 29/30; Fig 4, stage 30 (ventral aspect). Abbreviations (for all Figs): dc, distal carpal 3; mc, metacarpals (II–V); p, pisiform; R, radius; r, radiale; U, ulna; u, ulnare; X, element X.

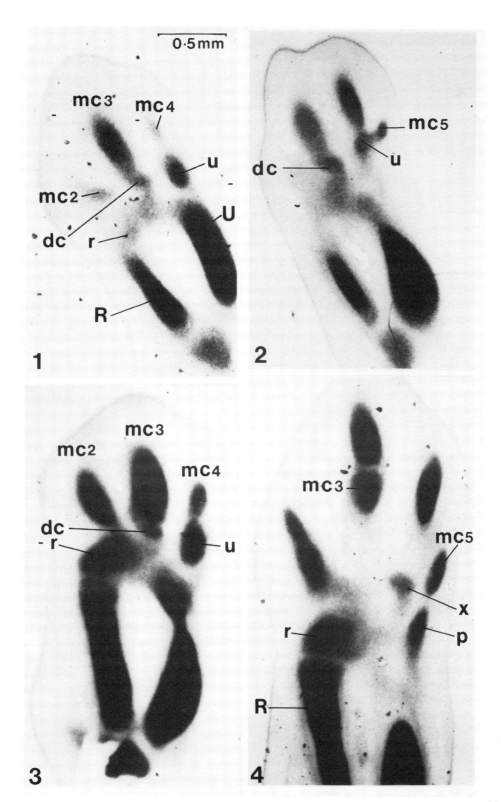

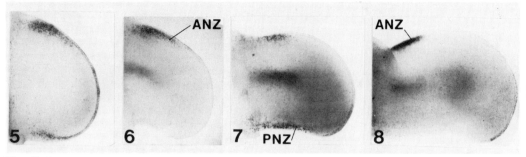

Figs. 5–8
Areas of mesenchymal cell death in the chick wingbud, shown by vital staining using Neutral Red (see HINCHLIFFE et al, 1981). ANZ, PNZ; anterior and posterior necrotic zones. Fig 5, stage 23 (4 days); Fig 6, stage 23/4; Fig 7, stage 24/5; Fig 8, stage 26 (5 days).

death removes anterior and distal tissue (HINCHLIFFE, 1982). A posterior necrotic zone makes a brief appearance later (Stage 24), but this area, unlike the anterior one, is absent in other avian species. There is now a good deal of evidence that the antero-posterior dimension of the distal mesenchyme of the wing bud is related to the number of digits formed (EDE, 1971; SUMMERBELL, 1981). The anterior cell death may well represent the mechanism for eliminating the tissue which in past evolutionary history formed digit I.

An interesting comparison is provided by the development of the fully pentadactyl mouse limb. The mouse limb bud lacks comparable anterior and posterior necrotic zones and has relatively a much greater antero-posterior dimension to its digital plate than the wing bud (MILAIRE, 1971; HINCHLIFFE, 1982).

In summary, therefore, the presence of i) the pisiform and ii) anterior cell death suggest that it is the most anterior of the pentadactyl digits which is missing in modern birds.

Identification of the carpal elements

The pattern of carpal elements revealed by $^{35}SO_4$ autoradiography has been described before (HINCHLIFFE, 1977; HINCHLIFFE & GRIFFITHS, 1983; HINCHLIFFE & HECHT, 1984) and only the main features will be outlined here. The carpal condensations make their first appearance at stage 27 (5½ days) and, by stage 28, four of them are clearly defined : radiale, ulnare, pisiform (ventral to the midline) and distal carpal 3 (at the base of metacarpal III) (Fig 1–49). All these elements become cartilaginous and can be seen using Methylene blue staining of whole mounts at stage 29 (6½ days). By stage 30 (7 days) a fifth element is formed at the base of metacarpal 4, but ventral to the midline. This element has been labelled 'X', following MONTAGNA (1945) and because its homology is not clear. At this time, the ulnare cartilage ceases to synthesize matrix and its cells die, so that it has completely disappeared by stage 32 (HOLMGREN, 1955; HINCHLIFFE & HECHT, 1984). Thus each of the five carpal condensations becomes cartilaginous, and the older interpretations (MONTAGNA, 1945; HOLMGREN, 1955), which describe a much larger initial number (13) of carpal condensations followed by a series of fusions and disappearances, must now be considered incorrect. The avian carpus must therefore be regarded as already specialized at the condensation pattern stage, and not as passing through a generalized tetrapod archetypal phase, only later modified by specialisation.

Subsequent development of the carpal cartilages

In spite of careful stage-by-stage studies of the development of the avian wing skeleton (e.g. PARKER, 1888), different anatomists have each given very different identifications of the adult carpal structure (e.g. BELLAIRS, 1960, identifies a radiale anteriorly and a piso-ulnare posteriorly). This is an inevitable

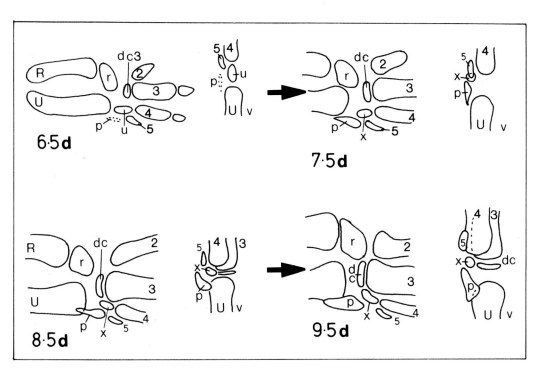
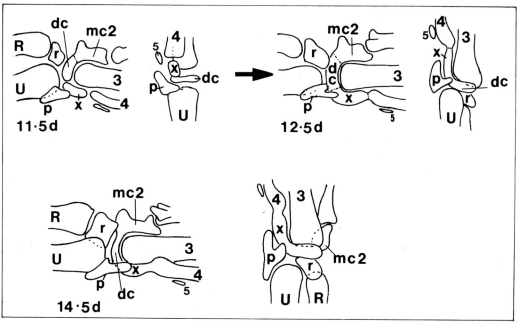

Figs. 9–10
The development of the carpal cartilages in the chick wing from 6.5 – 14.5 days. Drawn from Methylene blue stained preparations. Note the 'replacement' of ulnare by X between 6.5 and 7.5 days. The second drawing for each stage represents the posterior aspect (V, ventral surface).

consequence of the confusion over identification of the initial condensations. The present account, based on a clearer identification of the pattern of these (no doubt identification of individual condensations will continue to give rise to debate), attempts to give a precise account of the transition from the pattern of chondrogenic elements to the adult bony skeleton (Figs 9,10). In practice, this means following the development of the cartilages from 7 to 15 days of development, by which time the definitive adult skeleton has been established.

All five condensations become cartilaginous, and four of them (i.e. radiale, distal carpal 3, pisiform and X) become incorporated into the adult skeleton. As the ulnare disappears at 7 days (Fig 9 – HOLMGREN, 1955 p270, HINCHLIFFE & HECHT, 1984), at the same proximo-distal level, but ventral to the midline, cartilage X is forming simultaneously. It is the failure to recognise this 'replacement' process which has led many anatomists to identify an ulnare in the adult carpus.

Of the four elements which survive, the radiale and the pisiform remain as separate elements, respectively anterior and posterior. Distal carpal 3 fuses with metacarpal II anteriorly and posteriorly with X which moves distally, pushing the base of metacarpal IV distally, and fusing with it. Thus a half circle of fused elements is formed around the base of metacarpal III, and in a final later stage fuses with it, thus forming the carpometacarpus of the adult.

Having an accurate picture of the development of the carpus in modern birds does not enable us to identify clearly the three carpal elements in *Archaeopteryx*. The anterior element distal to the radius may be identified with reasonable certainty as the radiale. The large semilunate element may represent distal carpal 3, but in modern birds this does not extend to the base of metacarpal II, as does the semilunate. The posterior element is usually identified as an ulnare, but it could alternatively represent either the pisiform or X. Assuming *Archaeopteryx* to be ancestral to modern birds the most stringent hypothesis is that it is the pisiform, since otherwise this element would have to be reinvented by modern birds. The disappearance and 'replacement' of the ulnare in modern birds makes interpretation of the *Archaeopteryx* carpus particularly problematic since we do not know whether it too is the result of a similar process of developmental regression and replacement of carpus elements.

Phylogenetic implications

The embryological evidence on the modern bird wing skeleton enables two points to be made. The first is the need for caution in application of the principle of homology to adult structures. It is evident that there is no archetypal tetrapod pattern of carpal elements in development, and that the pattern of condensations in birds is already very specialized, emphasizing that developmental processes as much as adult morphology are altered by selection. The condensation pattern is thus not a sure guide to adult skeletal homologies. The problems in identification of adult structures are exemplified by the case of the ulnare, identified by many anatomists in the adult bird carpus. As has been already pointed out, the developmental disappearance of the ulnare makes this identification incorrect. If cartilages form and disappear in development, then identification of elements in fossil forms is problematic, and requires caution. Ideally, the full developmental history of the element and the area is needed.

More positively, it can be asserted that the embryological evidence, in particular that provided by i) new $^{35}SO_4$ autoradiographic evidence as to pisiform position and by ii) the anterior cell death in the wingbud, supports the traditional view of embryologists that the three main wing digits in modern birds are II-III-IV. This is similar to the pattern of reduction in other tetrapods (e.g. lizards) where digits are lost or reduced from both the anterior and posterior margins. The point is important in relation to current dicussion as to whether modern birds and *Archaeopteryx* are derived from theropod dinosaurs (OSTROM, 1976) or from thecodonts (TARSITANO & HECHT, 1980). In theropods it is argued that the reduction of digits is postaxial so that the surviving digits are I-II-III (ROMER, 1966; TAQUET, 1977). If this interpretation is correct, then the developmental evidence of digit reduction in birds suggests the pattern of digits is different from that in theropods.

References cited

BELLAIRS, A. D'A. & JENKIN, C.R. (1960): The skeleton of birds. – In: Biology and comparative physiology of birds (ed. Marshall, A.J.) vol **1**, pp. 241–300. Academic Press, New York.

EDE, D.A. (1971): Control of form and pattern in the vertebrate limb. – In: Control mechanisms of growth and differentiation (eds. DAVIES, D.D. & BALLS, M.) pp. 235–254. Symposium **25** of the Society for Experimental Biology. Cambridge University Press.

EDE, D.A. (1983): Cellular condensations and chondrogenesis. – In: Cartilage, **2** (ed. HALL, B.K.), pp. 143–185. Academic Press, New York.

HAMBURGER, V. (1960): A Manual of Experimental Embryology. – University of Chicago Press.

HINCHLIFFE, J.R. (1977): The chondrogenic pattern in chick limb morphogenesis: a problem of development and evolution. – In: Vertebrate limb and somite morphogenesis (eds. EDE, D.A., HINCHLIFFE, J.R. & BALLS, M.), pp. 293–309. Cambridge University Press.

HINCHLIFFE, J.R. (1982): Cell death in vertebrate limb morphogenesis. – Progress in Anatomy **2**: 1–17. Cambridge.

HINCHLIFFE, J.R. & EDE, D.A. (1973): Cell death and the development of limb form and skeletal pattern in normal and wingless (ws) chick embryos. – J. Embryol. exp. Morph. **30**: 753–772.

HINCHLIFFE, J.R., GARCIA-PORRERO, J.A. and GUMPEL-PINOT, M. (1981): The role of the zone of polarising activity (ZPA) in controlling the maintenance and antero-posterior differentiation of the apical mesenchyme of the chick wing bud: Histochemical techniques in the analysis of a developmental problem. – Histochem J. **13**: 643–658.

HINCHLIFFE, J.R. and GRIFFITHS, P.J. (1983): The prechondrogenic patterns in tetrapod limb development and their phylogenetic significance. – In: Development and Evolution (eds. GOODWIN, B.C., HOLDER, N. and WYLIE, C.C.) pp. 99–121. Cambridge University Press.

HINCHLIFFE, J.R. and HECHT, M. (1984): Homology of the bird wing skeleton: embryological versus paleontological evidence. – Evolutionary Biology **30**: 21–39. New York.

HINCHLIFFE, J.R. & JOHNSON, D.R. (1980): The Development of the Vertebrate Limb. – Oxford University Press.

HOLMGREN, N. (1955): Studies on the phylogeny of birds. – Acta. Zool. **36**: 243–328. Stockholm.

MILAIRE, J. (1970): Evolution et déterminisme des dégénerescences cellulaires au cours de la morphogenése des membres et leurs modifications dans diverse situations tératologiques. – In: Malformations congénitales des mammiféres (ed. TUCHMANN-DUPLESIS, H.). Colloque Pfizer, Amboise, pp. 131–149.

MONTAGNA, W. (1945): A re-investigation of the development of the wing of the fowl. – J. Morphol. **76**: 87–113. New York.

OSTROM, J.H. (1976): *Archaeopteryx* and the origin of birds. – Biol. J. Linn. Soc. **8**: 91–182. London.

PARKER, W.K. (1888): On the structure and development of the wing in the common fowl. – Phil. Trans. Roy. Soc. Lond. B, **179**: 385–398. London.

ROMER, A.S. (1966): Vertebrate Paleontology. – University of Chicago Press.

ROMER, A.S. & PARSONS, T. (1977): The Vertebrate Body. – Saunders, Philadelphia.

SEARLS, R.L. (1965): An autoradiographic study of the uptake of S^{35} sulphate during differentiation of limb bud cartilage. – Devl. Biol. **11**: 155–168. New York.

SUMMERBELL, D. (1981): The control of growth and the development of pattern across the antero-posterior axis of the chick limb bud. – J. Embryol. exp. Morph. **63**: 161–180. Cambridge.

TAQUET, P. (1977): Variation ou rudimentation du membre antérieure chez les Theropodes (Dinosauria)? – In: Mécanismes de la rudimentation des organes chez les embryons de vertébrés, pp. 333–339, Colloques Internationaux CNRS, **266**, Paris.

TARSITANO, S. & HECHT, M. (1980): A reconsideration of the reptilian relationships of *Archaeopteryx*. – Zool. J. Linn Soc. **69**: 149–182. London.

THOROGOOD, P.V. & HINCHLIFFE, J.R. (1975): An analysis of the condensation process during chondrogenesis in the embryonic chick hind limb. – J. Embryol. exp. Morph. **33**: 581–606. Cambridge.

Author's address: Dr. J.R. HINCHLIFFE, Dept. of Zoology, The University College of Wales, Aberystwyth, Dyfed, SY 23 3DA, Wales, United Kingdom.

Max K. Hecht

The Biological Significance of *Archaeopteryx*

Abstract

The significance of the discovery in 1861 of *Archaeopteryx* is presented as historical evidence for the Darwinian theory of evolution. The debate as to its phyletic position and its relation to the origin of birds is considered. The three major hypotheses as to the phyletic relationships of *Archaeopteryx* are the theropod, crocodilomorph and primitive thecodontian hypotheses. The evidence, strengths and weaknesses of all three are examined. They are found valid but dependent upon a given interpretation of the fossil data, weighting of the morphological evidence, methodology of character analysis, and systematic approach. Each hypothesis supports different models as to the origin of flight which also determines the mode of evolution. The two different modes of evolution, phyletic gradualism and punctuated equilibrium, are discussed in terms of different scenarios. The terminology of adaptation is evaluated in terms of the origin of birds and bird flight.

Zusammenfassung

Die Bedeutung der Entdeckung von *Archaeopteryx* im Jahre 1861 wird als historischer Beweis der DARWIN'SCHEN Evolutionstheorie dargestellt. Es wird auf den Streit über die phylogenetische Stellung und die Beziehung zum Ursprung der Vögel eingegangen. Die drei Haupthypothesen zur Stammesgeschichte von *Archaeopteryx* sind die Theropoden-, Krokodilomorphen- und Primitive-Thecodontier-Hypothese. Die Beweiskraft, die Stärke und Schwäche aller drei Hypothesen werden geprüft. Sie sind vertretbar, aber abhängig von einer gegebenen Interpretation der Fossildaten, der Gewichtung der morphologischen Merkmale, der Methode der Merkmalsanalyse und von der systematischen Betrachtungsweise. Jede Hypothese stützt verschiedene Modelle der Entstehung des Vogelfluges, was zugleich den Evolutionsablauf bestimmt. Die beiden verschiedenen Evolutionsarten, phyletischer Gradualismus und Punktualismus, werden in Form von verschiedenen Scenarios diskutiert. Die Terminologie der Anpassung wird im Sinne des Ursprungs der Vögel und des Vogelfluges kritisch beurteilt.

The Biological Significance of *Archaeopteryx*

The revolution in biological thought that was the result of the Darwinian interpretation of evolution produced profound changes in our view of the fossil record and the relationship of higher systematic categories. The impact of *Archaeopteryx* upon the latter half of the 19th and the first half of the 20th century was due to the expectation of many Darwinians for a connecting link as evidence for the continuous process of evolution. An early fulfillment of that expectation was the discovery in 1861 of *Archaeopteryx* (Dawn-bird or Urvogel). Since *Archaeopteryx* only the discovery of intermediate forms of fossil man has attracted such attention of the scientific and lay public.

It would be correct to say that "ever since DARWIN" the biological community and paleontologists have been searching for the connecting links between the large taxa or morphological groups that should characterize the diversity of life if evolution is a continuous process (RUSE, 1982). The discovery of *Archaeopteryx* was the unexpected early fulfillment of the verification of that process and the resulting continual improvement that one would expect in the progression of life. Furthermore in this simplified version of the evolutionary process *Archaeopteryx* dramatized a link between two distinct morphological types, reptiles and birds, that are easily recognized and exemplified in the public's mind.

An examination of general evolutionary biological papers and popular books on biology and paleontology demonstrates the grip that the Dawn-bird has on the imagination of the interested public. SIMPSON

(1983) in his new popular book on the significance of fossils uses *Archaeopteryx* as an outstanding example of the kind of information that can be found in the fossil record. Interest in *Archaeopteryx* even extends to commentaries in such international non-biological journals as the Economist (ANONYMOUS, 1984) which describes the importance of this fossil and calls attention to this conference as an intellectual endeavor.

As indicated in the Economist, the interested public is amazed that after so many years of research and debate that we have not developed a consensus of opinion on the phyletic position of the fossil. The public does not understand that science does not progress by consensus. Historical sciences have an even more difficult problem of interpreting the past because they can only compare conditions with the present or with other better known past events. It is to the surprise of non-comparative biologists that we are here debating the phyletic position of this form and still not in agreement on morphological or functional interpretations. The methodology of phylogenetic inference (still debated as to the proper approach and interpetation) or the study of any historical process requires a complex system of analysis not fully understood by our colleagues in other fields of biology. It is difficult for many observers to understand that the use of different techniques of analysis can result in different conclusions and interpetations from the same data base.

In order to evaluate the biological significance of *Archaeopteryx* as a fossil form, it is first necessary to place the form phyletically or to determine its closest relatives. There have been many phyletic positions suggested but at the present time there are essentially three major positions on the relationships of this form, they are: A) theropod relationships, B) crocodilomorph or crocodilian relationships, C) early archosaurian or thecodontian level relationships. The proposed relationship of *Archaeopteryx* to the Ornithischia (GALTON, 1970) has been abandoned by its own proponent.

Theropod relationships

Theropod relationships of *Archaeopteryx* have been expressed as three hypotheses of relationships, they are: 1) *Archaeopteryx* is a primitive bird and a direct descendent of the Theropoda; 2) *Archaeopteryx* is a sister group of the Theropoda; 3) *Archaeopteryx* is a member of the Theropoda and not a bird.

The foremost recent proponent of hypotheses 1 (or possibly 2) has been OSTROM (1976a, b; 1979). Following the general observations of OSBORN (1900, 1903), ABEL (1912), HEILMANN (1926), LOWE (1935, 1944) and WELLNHOFER (1974) OSTROM has called attention to the many similarities between *Archaeopteryx* and theropod dinosaurs. He has noted the similarity in body proportions, postulated locomotory patterns and described distinct morphological features, such as the presence of a semilunate bone in the carpus, phalangeal formula of the manus being identical to the theropod plan, and the similarity of the pelvic morphology of *Archaeopteryx* to that of coelurosaurs. OSTROM demonstrated specific similarities between *Archaeopteryx* and particular theropods such as coelurosaurs (*Ornitholestes, Deinonychus, Velociraptor, Compsognathus*, etc.).

He recognized that a more anteriorly directed pubis in the Eichstatt specimen would more closely approach his interpretation of the pubic condition of *Deinonychus* and *Velociraptor* (for further comments on this see WELLNHOFER, this volume). He furthermore elaborated his interpretation of the proximal bones of the tarsus and hallux of theropods in terms of their basic similarity to that of *Archaeopteryx*.

THULBORN and HAMLEY (1982) agreeing with OSTROM's general hypothesis evaluated the studies of TARSITANO and HECHT (1980) and MARTIN, STEWART and WHETSTONE (1982). The basic premise of THULBORN and HAMLEY (1980) is that the shared and derived characters OSTROM found among theropods and *Archaeopteryx* are essentially correct with the possible exception of the numbering of the digits of the manus. THULBORN and HAMLEY (1982) accepted the probability that the embryological digital numbering of the bird manus was correct and provided some morphological evidence in support of the 2-3-4 hypothesis. They only insisted that the digital identifications applied equally to the manus of theropods and *Archaeopteryx*. Furthermore THULBORN and HAMLEY (1982) agreed with OSTROM's identification of the ascending process of the astragalus.

A critique of THULBORN and HAMLEY'S evaluation reveals some important points of disagreement, which were pointed out in HECHT and TARSITANO (1982). THULBORN and HAMLEY (1982) ignored the interpretation of the reduction of the theropod manus as interpreted by ROMER (1956, 1966) and TAQUET (1977). If the transformation series of TAQUET (1977) is essentially correct then the reduction of the digits and metacarpals are from the postaxial to the preaxial side. HINCHLIFFE and HECHT (1984) have discussed this type of reduction as unusual in the Amniota.

THULBORN and HAMLEY (1982) criticized both TARSITANO and HECHT (1980) and MARTIN et al. (1980) on their interpretations of the tarsus of *Archaeopteryx*. TARSITANO and HECHT (1980) claimed that they found no evidence of an astragalar process in *Archaeopteryx*. It is possible that it was retained as a fused structure not visible in the adult. TARSITANO and HECHT (1980), HECHT and TARSITANO (1982), and MARTIN et al. (1980) were in agreement that if there was such a process, the homologies were wrong. McGOWAN (1984) presented new embryological evidence indicating that there are embryological vestiges (or neomorphs) in the ratites of both astragalar and calcaneal processes preformed as cartilages. McGOWAN'S data indicates a possibility of partial homology among birds and between birds and theropods concerning these processes. McGOWAN (1984) demonstrates that the carinate ankle probably has only one of these elements, the pretibial bone. It is difficult to homologize the tarsal elements in *Archaeopteryx* since TARSITANO and HECHT (1980) could find no evidence for a tarsal process. This does not mean that a fused state did or did not exist. Only new material or possible better preparation can accomplish this.

The question of the reflexed hallux has been used by THULBORN and HAMLEY (1982) as a point of similarity between *Archaeopteryx* and the primitive theropod *Compsognathus* (OSTROM, 1978). They also accept that at least a semireflexed hallux is present in *Archaeopteryx*. According to THULBORN and HAMLEY the similar condition of the reflexed hallux can be interpreted as a synapomorphy indicating relationships between the primitive coelurosaur and *Archaeopteryx*.

In summary THULBORN and HAMLEY (1982) argue that the theropod relationships of *Archaeopteryx* are supported by two synapomorphies, the presence of a semilunate carpal and the ascending process of the astragalus. Both of these characters are contested by TARSITANO and HECHT (1980) and one of these by MARTIN et al. (1980).

PADIAN (1982) presents another view of OSTROM'S phyletic hypothesis by proposing to answer his critics with a cladistic analysis of the relationships of *Archaeopteryx*. PADIAN proposes a cladogram defending this given phyletic position adopting a fundamentalist cladistic approach. PADIAN (1982) has listed 15 characters which supposedly represent synapomorphies that indicate the monophyletic relationship of *Archaeopteryx* and the Theropoda. An analysis of these characters shows that 13 of them are actually primitive states for the Archosauria or features of the Thecodontia (recognized as a primitive or paraphyletic group difficult to define). Such primitive features are not useful in the indication of relationships.

His first character, the semilunate carpal, was recognized by OSTROM as being an important similarity between the two groups. TARSITANO and HECHT (1980) and HECHT and TARSITANO (1982) proposed that the semilunate represented a fused state of unknown carpalia and therefore of unknown homology. Such fused structures can be made up of different elements in different lineages yet present the same phenotype (HINCHLIFFE and HECHT, 1984). It is difficult to determine the number of components that are found in a fused structure in a fossil. It appears that in the semilunate carpal of an apparently younger specimen of *Deinonychus* (M.C.Z. 4371), there are suture lines indicating at least two carpalia. In birds the best estimate at this time is only one distal carpal (HINCHLIFFE and HECHT, 1984). Therefore the utilization of this character as a synapomorphy, indicating relationship of birds and theropods is dubious.

The second character of PADIAN represents an almost unique character state in the reptilian amniota, that is the length of metacarpal I is less than that of phalanx 1 of digit I. PADIAN supports the hypothesis that the digits of the theropod, *Archaeopteryx* and modern birds are digits I–II–III. If the digits are identified as I–II–III, then this arrangement is a synapomorphy for a monophyletic origin for *Archaeopteryx* and theropods. Yet there is other evidence, particularly embryological evidence in birds, which precludes identification of the bird digits as I–II–III and requires an identification of II–III–IV (HINCHLIFFE and HECHT, 1984; HINCHLIFFE, this volume). Assuming the data from modern embryology is accepted, then the digits of birds cannot be homologous with those of theropods. If *Archaeopteryx* is a bird, then it must have digits II–III–IV. Therefore theropods have a totally different developmental program than birds. If PADIAN'S theropodan digital pattern is accepted, then this character and the digital reduction pattern cannot be utilized as a synapomorphy because they are not homologous but a convergence.

PADIAN further lists characters (characters 6 and 7 in his cladogram) which are essentially reduction characters in the manus (digital number) and are redundantly associated with the identification of the digits of manus. The other characters that PADIAN lists as synapomorphies are primitive states for reptilian amniotes or archosaurs, such as: metacarpal I shorter than metacarpals II and III; pubis longer than ischium; and ossified gastralia. The remaining seven characters are either erroneous, have a wider taxic distribution or are not present in *Archaeopteryx*.

In summary the hypothesis of relationship between Theropoda, *Archaeopteryx,* and birds are based on unique similarities which are interpreted by each of their proponents within their own frame of reference. Their major argument is based on similarity of obvious character states such as the manus, carpals, and pelvis. Further support for the position of OSTROM (1976) lies in the comparison of known fossil forms to which ancestral or primitive states can be referred. In fact a major argument in favor of the theropod hypothesis of relationships from the paleontological point of view is that a group of known reptiles can be identified as being either a sister or ancestral group. In conclusion the theropod relationships of *Archaeopteryx* remains a viable hypothesis.

Crocodilian Relationships

MARTIN (1983a, b) has summarized the crocodilian relationship hypothesis. The crocodilian hypotheses have two aspects: 1) What is meant by a crocodilian; 2) At which level of the Crocodilia are the comparisons being made? Many comparative morphologists referring to the Crocodilia mean the living eusuchians and make comparisons between birds and these highly derived crocodilians. The paleontological proponents of crocodilian relationships are discussing early true crocodilians (the Protosuchia) and the Paracrocodylia of WALKER (1970, 1972). The two subgroups are often placed in a higher and broader taxon also termed the Crocodylomorpha. The Protosuchia have been defined and discussed by CROMPTON and SMITH (1980) and HECHT and TARSITANO (1983) and are true crocodilians having the major features characterizing the Mesosuchia and Eusuchia. WALKER (1970) has placed in the Paracrocodylia many genera formerly placed in the Thecodontia by ROMER (1972) including such forms as *Sphenosuchus, Hesperosuchus, Pseudohesperosuchus, Terrestrasuchus, Saltoposuchus,* etc. The Paracrocodylia are considered more primitive in many features than the Protosuchia. This group is distinguished from the Protosuchia by having a vertical quadrate with a different pattern of fenestration, primitive pelvis (no acetabular opening), etc. Both groups are highly terrestrial forms with a well-developed crocodilomorph tarsus (HECHT and TARSITANO, 1982; WHETSTONE and WHYBROW, 1983). Many fossil forms, such as *Hallopus* are poorly preserved or unprepared and therefore almost irrelevant. WALKER (1972, 1974, 1977) has listed at least 18 characteristics which unite his assemblage of primitive crocodilians and paracrocodilians. It appears that the positions of WALKER and MARTIN are quite similar in that they are either referring to the Protosuchia or to the Paracrocodylia.

MARTIN (1983a, b) lists about 30 or more features (including the 18 characters of WALKER) which indicate the close relationship of birds and crocodilians. Some of these features are: 1. possession of laterosphenoids; 2. an external mandibular fossa; 3. quadrate head articulating with the squamosal and prootic; 4. crescentic shape of the occipital condyle; 5. similar carotid circulation; 6. pneumatic basisphenoid; 7. elongate cochlear duct; 8. similar pattern of digit reduction; 9. similar morphology of the palatines; 10. pneumatization of the skull; 11. reduction of the first metatarsal (general amniote pattern).

Such characters are dubious evidence for either sister group or ancestral relationships of crocodiles and birds because they are primitive for the Thecodontia.

Other characters listed by MARTIN (1983a) are not precluded as being primitive features at a lower hierarchial level, they are: 1. the tooth replacement pattern in crocodiles and toothed birds; 2. eustachian tube pattern; 3. presence of a fenestra pseudorotundum; 4. implantation of teeth in lower jaw and similarity of form in recent crocodilians and toothed fossil birds; 5. pneumatic quadrate and foramen aerosum in the articular. These features may be primitive but more likely indicate relationship at an advanced thecodontian or a paracrocodilian level.

MARTIN (1983a) utilizes the presence of a fenestra pseudorotundum as evidence for relationship of crocodiles and birds and excluding theropods. The basis of his observation is the lack of the structure in theropods. The variety of reptilian fenestra rotundum-like structures is great as described by WEVER (1976). Many of these structures would be difficult to determine in fossils.

Another character which is difficult to interpret is the quadrate cotylus of birds as being homologous to one of the sutured heads (quadrate cotylus of WHETSTONE and MARTIN, 1979) of the quadrate in the crocodilians. Their interpretation of homology is dependent on the partially published work of WALKER on *Sphenosuchus*. The condition in the paracrocodylians indicated by *Sphenosuchus* (WALKER, 1972) is a closer approximation to the bird condition with a true cotylar head.

The pneumatic cavities of crocodilians and birds (BREMER, 1940; MÜLLER, 1967) form in a similar embryological manner. The siphonium present in both is of a different plan because in the crocodilians the siphonium passes through the quadrate and into the articular whereas in birds it bypasses the quadrate and goes directly into the foramen aerosum of the articular. Furthermore in birds there are tympanic diverticula which invade the quadrate and therefore it is not certain that the two types of siphonia are strictly homologous. Whereas internal vacuities and an external foramen are known in some theropodan quadrates, it is not possible to determine in fossils whether they were true pneumatic spaces or cavities filled with tissue in the living organism. Finally, it is impossible to determine whether they are derived from the same embryological structures.

In summary the crocodilian and crocodylomorph hypotheses are similar enough to be considered a single hypothesis despite the fact they are not derived exactly from the same data base. Essentially the phyletic question is at what level of the crocodilian clade is the origin of birds to be placed? Both theories postulated an arboreal crocodilian ancestor but that is the one adaptive type that is presently not known in the early Crocodilia or Paracrocodylia. If the two positions are integrated then it seems clear that the ancestry of birds must be derived from this clade at the paracrocodilian level.

WHETSTONE (1983) and WHETSTONE & WHYBROW (1983) are other proponents of crocodilian relationships of *Archaeopteryx* and birds. This position is based on his interpretation of the protosuchian *Lesothosuchus* and the braincase of *Archaeopteryx*. An examination of both materials reveals that much of WHETSTONE'S morphological interpretations and their homologies are overinterpretations. HECHT and TARSITANO (1983) have commented on the structure of *Lesothosuchus* and demonstrated that it is a typical protosuchian. The limb morphology indicates a terrestrial form. Yet on the basis of these studies WHETSTONE proposed a cladogram which inverts the relationships of the possible sister groups of the crocodylomorphs and the birds. WHETSTONE proposes that the theropod condition is primitive and the development of the antrum pneumaticum dorsale and bipartite quadrate articulation are synapomorphies for the sphenosuchid, crocodilian, and bird lineage. If one accepts this arrangement, it would require that the tarsal transformation series reverse itself two times.

WHETSTONE (1983) has described the new preparation of the braincase of the London *Archaeopteryx* using the ratite skull as a model for the reconstruction of the braincase. The London specimen is crushed and distorted and despite this WHETSTONE has interpreted it in great detail with remarkable similarities to living birds. WALKER (this volume) disputes some of the identifications and our own examination of the same specimen reveals that WHETSTONE has again overinterpreted the fossil material.

Thecodontian level or early archosaurian level

TARSITANO and HECHT (1980) and HECHT and TARSITANO (1982) stated that *Archaeopteryx* was derived from a thecodontian ancestor in contrast to a theropodan or crocodilian ancestor. In consideration of the observation that most of the forms now placed in the Paracrocodylia were placed in the Pseudosuchia, the phyletic distances between the hypotheses of WALKER and MARTIN versus TARSITANO and HECHT is somewhat diminished.

The major arguments utilized by TARSITANO and HECHT (1980) and HECHT and TARSITANO (1982) against a crocodilian affinity for *Archaeopteryx* was the presence of the elongated proximal carpals, procumbent elongated quadrate and the structure of the pelvis (such as exclusion of the pubis from acetabulum) in early crocodilians. It is only the presence of the elongated carpals in the Paracrocodylia that exclude this group from sister group or ancestral relationship to the avian clade.

TARSITANO and HECHT'S (1980) criticism of the theropod relationships was based on their interpretation of the structure of the pectoral girdle (the reduction and curvature of the coracoid), the carpus (homologies of the carpal bones), manus (reduction pattern of the metacarpals and digits), pelvic girdle (public orientation), and tarsus (see above) in *Archaeopteryx*. One of the major pieces of evidence in support of the non-theropodan relationships is the pattern of carpus and manus development (HINCHLIFFE, this volume). This developmental pattern precludes theropodan relationships and implies that many of the post-axial resemblances are convergences related to bipedalism.

A major criticism of the early separation of the avian lineage and therefore of the TARSITANO and HECHT (1980) and HECHT and TARSITANO (1982) position is the lack of any known relative or sister group in the fossil record. HECHT and TARSITANO (1982) agreed with WILD (pers. com.) that the bird-like Triassic *Megalancosaurus* (Calzavara, Muscio, and WILD, 1981), is probably an arboreal thecodontian, indicating an early development of the avian clade. The fossil record is still inadequate for this group of thecodonts.

Further data for a more ancient separation of birds from the archosaurian stock can be found in morphology. Studies on the phallus and circulatory system seem to indicate an earlier branching of the avian clade than postulated by the theropod hypothesis (KING, 1981).

In summary the three major hypotheses of the phyletic position of *Archaeopteryx* can be presented as a series of cladograms (Fig. 1). Each interpretation has its own validity and its own critical points. It is not judicious to reject any of these hypotheses at this time.

Biological Conclusions to be derived from Phyletic Hypotheses

Each of the three hypotheses outlined above require different ancestral configurations and adaptations. These implied adaptations outline steps to the origin of flight in birds. The models that are developed must be consistent with the morphological data, physical laws governing flight, ecological and energetic limitations and should be within the biological known panaroma of analogues.

The theropod hypothesis is based on the general similarity of *Archaeopteryx* to coelurosaurs. Based on a bipedal model OSTROM supported the terrestrial flight transformation theory of NOPSCA (1907, 1923) which proposed a running – gliding – flight transition. It was evident to OSTROM that there were problems with NOPSCA'S model. OSTROM (1974, 1976a, b; 1979) therefore proposed the insect net hypothesis which represented *Archaeopteryx* at a preflight stage. In this model the flapping of the wings was to swat small insects and therefore the original adaptation of the wing was a food-getting mechanism. One of the difficulties with the proposed mechanism was that a swatting adaptation with an aerodynamic wing would not be adept at catching any insects. CAPLE, BALDA and WILLIS (1983) (in the following: CBW) have attempted to correct for this error by describing a model which, allowing for physical constraints, could develop a flight mechanism from a terrestrial bipedal ancestor.

This CBW hypothesis (as published in 1983) assumes only a theropod ancestry for birds and an insectivorous diet. Both of these assumptions are in question and of dubious validity. The CBW hypothesis suffers from oversimplification of both the arboreal hypothesis and the conditions for the feeding strategy of their hypothetical ancestor. Their hypothesis insists that the arboreal hypothesis actually means an approximation to a modern arboreal insect-eating bird which feeds and lives among the trees. These conditions set the immediate problem of landing controls and landing areas. Actually the utilization of the term (arboreal) merely means that an organism must attain height to glide, cover distance and land in some general area and not on a single twig or small branch. In this sense (BOCK, 1965; HECHT and TARSITANO, 1982) of the term "arboreal" there are no landing problems. In fact the model of HECHT and TARSITANO (1982) requires the beaches of the Solenhofen sea as a landing area and therefore no specific adaptation as for landing under difficult conditions. Furthermore the presence of free digits indicates the retention of climbing abilities by *Archaeopteryx*, a condition not required by CBW hypothesis.

The primary weakness of the CBW hypothesis and the model lie in the improbability of its ecological foundation, particularly its first three steps. It assumes that an organism can energetically function by jumping into the air, catching insects in its mouth. If this improbable model were feasible under some unique ancient environment, it is difficult to envisage how the organism can shift from the ballistic pattern to active flight. Ballistic patterns of locomotion have been attempted by vertebrates, particularly by frogs. It is important to note that this method of locomotion is not involved directly in the feeding process, or in gliding or powered flight. Yet the initial stages of the CBW ballistic hypothesis depends on adaptation to a special feeding mechanism and not simply a locomotory pattern for escape from predators, or to cover distances as was probable for frogs (HECHT, 1963). In no living analogue is the ballistic type of locomotion used as a vehicle for feeding and therefore an escape scenario would be preferable as the initial step in this hypothesis. The analogue with frogs has one flaw which is, of course, that frogs are not bipedal. The models postulated by J. M. V. RAYNER, U. M. NORBERG and R. A. NORBERG (this volume) presented an alternative and modified arboreal hypothesis (properly termed "from the heights-down hypothesis") in which the development of gliding and powered flight take advantage of the downward glide potential of a climbing animal (YALDEN, this volume). Furthermore RAYNER (this volume) indicates that the initial stages of flight must have evolved in a small animal about the size of a flying lizard (DRACO). Certainly no known theropod fits this description.

If it were accepted that the origin of flight in birds was a unique event as described by the CBW hypothesis (CAPLE et al., 1983) then one would have to accept a saltational mechanism as the evolutionary mode. A gradual transformation would not be possible because it would require a rapid radical overhaul of the morphology of the ancestral forms. Therefore CBW model requires punctuated equilbria conditions (GOULD and ELDREDGE, 1977). It is interesting that the one example that GOULD and ELDREDGE (1977, p. 147) cannot fit on morphological grounds into the punctuated equilibrium model is that of *Archaeopteryx*. Under the conditions set by the CBW hypothesis, punctuated equilibrium is the only feasible mechanism.

The Crocodylomorph hypothesis

This hypothesis requires no special mechanism. The proponent's insistence upon an arboreal crocodilomorph ancestor is incongruent with the known fauna. All the biomechanical adaptations of this group are related to quadrupedal locomotion of a highly terrestrial mode as indicated by the structure of the femur, tibia and tarsus (HECHT and TARSITANO, 1983). The evolution into a bipedal avian ancestor is hypothesized as a gradual transformation.

The early archosaurian hypothesis

The modern hypothesis of thecodontian or pre-thecodontian ancestry (TARSITANO and HECHT 1980; HECHT and TARSITANO, 1982) is based on the model of BOCK (1965). This hypothesis requires a terrestrial quadrupedal ancestor leading to a climbing adaptation and transformation to gliding and finally powered flight. The only disagreement with the original assumption is that the original thecodontian ancestor was

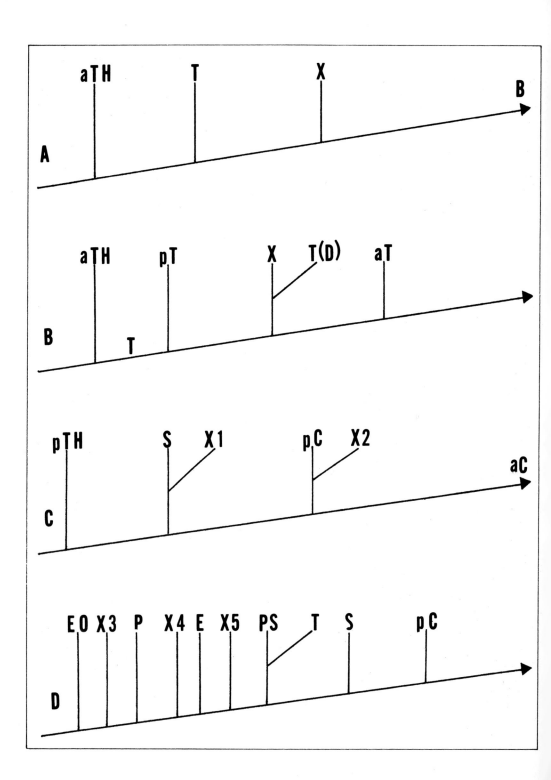

fully bipedal as originally assumed by most paleontologists (HEILMANN, 1926; ROMER, 1966). In this hypothesis HECHT and TARSITANO (1982) assume bipedality only in the transformation from climbing to gliding. BOCK'S hypothesis is basically gradualistic and does not require any special mechanisms. The HECHT and TARSITANO (1982) model (derived from BOCK, 1965 and FEDUCCIA, 1980) does not necessarily require a gradualistic model or a punctuated equilibria model, but it is more consistent with a gradualistic model.

The gradualistic model of HECHT and TARSITANO (1982) has six steps in the transformation. They are: I. Quadrupedal locomotion at a primitive thecodont or earlier level; II. Arboreal transformation at the more advanced thecodontian level probably similar to a *Megalancosaurus* level; III. Leaping between trees and other structures; IV. Parachuting; V. Gliding and bipedalism; VI. Gliding with assist from musculature for limited powered flight; VII. Powered flight. In this model it is possible to place morphologically equivalent fossils at only a few of the steps, for example, step I and II. Within this series depending on the morphological and functional interpretation of the fossil, *Archaeopteryx* could be placed at stage VI (OLSON and FEDUCCIA, 1979; HECHT and TARSITANO 1982). The fossil record is by no means complete or explicit to support this. OLSON and FEDUCCIA (1979), FEDUCCIA (1980), HECHT and TARSITANO (1982), MARTIN (1983a, b) are in agreement in that morphological characteristics of *Archaeopteryx* imply a glider but with limited powered flight as indicated by the furcula, asymmetric flight feathers and the form of the coracoid.

The problems addressed by the model can be summarized as follows: 1. the landing problem; 2. the retention of the digits; 3. the feeding strategy; 4. topographic environment. HECHT and TARSITANO (1982) envisaged *Archaeopteryx* as a scavenger and not the usual reconstruction as an insect eater. Gliding and partially powered flight afforded the ancient bird extended range capabilities to find food among the beaches surrounding the Solenhofen sea and its islands. Once having landed *Archaeopteryx*, utilizing its functional digits, could climb trees or high points in this habitat and commence its gliding and partially powered flight. This is correlated with the known morphological and paleontological data.

GOULD and ELDREDGE (1977) and SIMPSON (1983) have stated that *Archaeopteryx* represented a mosaic of reptilian and avian characters. It is for such reasons that most observers have accepted *Archaeopteryx* as an intermediate type. In a typological sense the genus is characterized by an assemblage of reptile characters indicating three possible reptilian grade origins. The theropod hypothesis describes the fossil as either a small coelurosaur with feathers or a bird with theropod affinities. In contrast the other hypotheses imply that it is further advanced on the avian lineage. On the basis of the crocodilomorph, thecodontian or pre-thecodontian relationships, *Archaeopteryx* must represent an early dichotomy probably originating in early Triassic or Permian times. In fact *Archaeopteryx* could be considered a relic living in the late Jurassic as *Sphenodon* is a relic living in the Recent.

The Adaptive Program

BOCK'S original hypothesis in developing a series of steps in the evolution of flight in birds was to demonstrate stages in adaptation. *Archaeopteryx* fulfills one of those stages and helps envisage the transformation of structures. REGAL (1975) suggested how feathers could be' modified from a

Fig. 1
Diagrammatic representations of the phylogenetic relationships of *Archaeopteryx*. A: Theropod hypothesis proposed by OSTROM. B: Modified theropod hypothesis of PADIAN. C: Crocodylomorph (X1; WALKER) and crocodilian hypotheses (X2; MARTIN). D: Primitive archosaurian relationships of *Archaeopteryx* at different nodes: X3 = pre-Proterosuchus; X4 = post-Proterosuchus; X5 = post-Euparkeria (of HEILMANN). Abbreviations: a = advanced; p = primitive; A = Archosauria: B = birds; C = Crocodylia; D = dromaeosaurs and relatives; E = Euparkeria; EO = Eosuchia; P = Proterosuchus; PS = Pseudosuchia; S = Crocodylomorpha or Sphenosuchia; T = Theropoda; TH = Thecodontia; X = *Archaeopteryx*.

thermoregulatory mechanism into a flight adaptation. BRUSH (1983) reinforces that interpretation with some biochemical evidence of the relationship of feathers and epidermal scutes. The combination of comparative morphology, physiology and paleontology provides an adaptive model for the transformation of a reptilian bauplan into an avian bauplan.

In the case of bird origins it seems that the modified reptile scale, now an expanded feather, could be used primitively as a thermoregulating mechanism (REGAL, 1975). This allows for the development of homeothermy and other functions as part of a dual function (BOCK, 1975). The preadaptive value of the keratinized structure as part of a behavioural and insulating mechanism allows for its expansion and development as flight feathers. The final transformation to gliding and partially powered flight is seen in *Archaeopteryx*.

The question of the adaptive program for the evolution of birds and flight have been utilized by morphologists for illustrating major trends and transformations. GOULD and LEWONTIN (1979) have criticized the extreme adaptational stands of some biologists. In the process they have apparently abandoned the idea of an adaptational program for morphological trends. There are factors of chance affecting the populational and species level characters (the raw material of major trends) but to ascribe major overhauls of morphology and biomechanical trends in higher taxa to chance phenomena seems difficult. In contradiction to this line of criticism it should be noted that all biomechanical systems examined have been shown to be competitively efficient, even if not optimally developed. Although the organism is trapped and limited by its past history and ontogenetic patterns, it still must remain well integrated at the business of survival. Therefore those characters which are widespread among groups of species must be under positive selective control because chance alone cannot account for their maintenance among many species. A case in point are the digits of *Archaeopteryx* which were not vestiges (YALDEN, this volume), but important aids to climbing. Those structures which are not maintained by selection are lost or reduced, as demonstrated by the so-called "regressive evolution" of eyes in cave animals.

The origin of the birds has often been used as an example of the process of adaptation (BOCK, 1965) using the terms preadaptation, dual function and key innovation. GOULD and VRBA (1982) have attempted to create a new set of terms to illustrate the process. Their use of the terms aptation, exaptation, nonaptation, etc. have "atomized" concepts and terminology. REIF (1984 pg. 92) has commented on the new terminology by stating "The usage of an exaption is called an effect. To give an example, feathers of birds have probably evolved as insulation mechanisms; hence they are 'adaptations' for insulation and 'exaptions' for flight." Note how complex the terminology would become if the atomized concepts and vocabulary of GOULD and VRBA (1982) were to be accepted. Based on our interpretations of morphology the feather plays a dual role as a thermoregulatory mechanism and display device, but later as a key innovation in the development of a wing surface. I am in agreement with REIF (1984), that these old terminologies are easier to conceptualize than the new terminology of a "taxonomy of fitness" (GOULD and VRBA, 1982).

Conclusions

The historical significance of the discovery of *Archaeopteryx* in 1861 is that it was presented as evidence for the Darwinian theory of evolution. It is clear that in the 19th and early 20th century morphologists and paleontologists accepted *Archaeopteryx* as evidence that their type of data could demonstrate that evolution was a continuous process.

Since that time there has been a vigorous debate as to the phyletic position of *Archaeopteryx*. The debate has resolved itself into three major hypotheses of relationships, they are: theropod, crocodilomorph and thecodontian relationships. The validity of these three positions rests upon differing interpretation of the fossil data, weighting of the morphological evidence, methodology of character analysis and systematic criteria. None of the theories can be rejected until more material and better morphological studies are made. Each hypothesis supports different models as to the evolution of flight which also determines the mode of evolution, phyletic gradualism, or punctuated equilibria.

Acknowledgments

The author thanks the National Science Foundation Grant no. BSR 8307345 and Research Foundation of the City University of New York Grand no. 6–63174 for their support of this project. The author also wishes to thank Dr. S. TARSITANO, Mrs. B. M. HECHT and Dr. W. BOCK for their suggestions and criticisms.

References Cited

ABEL, O. (1911): Die Vorfahren der Vögel und ihre Lebensweise. – Verh. Zool.-Bot. Ges. Wien **61**: 144–191.
ANONYMOUS (1984): Origin of flight. – The Economist 5–11 May, **1984**: 96–97.
BOCK, W. J. (1965): The role of adaptive mechanisms in the origin of the higher levels of organization. – Systematic Zoology, Washington, **14**: 272–287.
BREMER, J. L. (1940): The pneumatization of the common fowl. – J. Morph. **92**: 143–157.
BRUSH, A. H. and WYLD, J. A. 1980: Molecular Correlates of Morphological Differentiation: Avian Scutes and Sclaes. – Jour. Exp. Zool. **212**: 153–157.
CALZAVARA, M. MUSCIO, G. & WILD, R. (1981): *Megalancosaurus preonensis*, N. G., N. Sp., a new reptile from the Norian of Friuli, Italy. – Atti Mus. Friul. Storia. Nat., Fruili **2**: 49–64.
CAPLE, G. R., BALDA, R. P., and WILLIS, W. R. (1983): The physics of leaping reptiles and the evolution of pre-flight. – Amer. Nat. **121** (4): 455–476.
CROMPTON, A. W., and SMITH, K. K. (1980): A new genus and species of crocodilian from the Kayenta Formation (late Triassic) of northern Arizona. – In: L. JACOBS (ed.) Aspects of vertebrate history. pp. 193–217. Flagstaff Museum of Northern Arizona Press.
FEDUCCIA, A. (1980): The age of birds. – 196 pp., Cambridge, Mass. (Harvard University Press).
GALTON, P. M. (1970): Ornithischian dinosaurs and the origin of birds. – Evolution, **24**: 448–462.
GOULD, S. J. & N. ELDREDGE (1977): Punctuated Equilibria Reconsidered. – Paleobiology **3**: 115–149.
GOULD, S. J., and R. C. LEWONTIN (1979): The Spandrels of San Marcos or the Adaptive Programs. – Proc. Roy. Soc. London (B) **205**: 581–598.
GOULD, S. J. and VRBA, E. S. (1982): Exaptation – a missing term in the science of form. – Paleobiology **8**: 4–15.
HECHT, M. K. (1963): A reevaluation of the early history of the frogs. Part I. – System. Zool. **11**: 20–35.
HECHT, M. K. & TARSITANO, S. (1982): The paleobiology and phylogenetic position of *Archaeopteryx*. – Geobios Memoire speciale, **6**: 141–149.
HECHT, M. K. & TARSITANO, (1983): The tarsus and metatarsus of *Protosuchus* and its phyletic implications. – In: Advances in Herpetology and Evolutionary Biology; Essays in honor of ERNEST E. WILLIAMS. Museum of Comparative Zoology, Cambridge, Mass.: 332–349.
HEILMANN, G. (1926): Origin of the Birds. – London (Withherby).
HINCHLIFFE, J. R. & M. K. HECHT (1984): Homology of the Bird Wing Skeleton: Embryological versus Paleontological Evidence. – Evol. Biol., **18**: 21–39.
KING, A. S. (1981): Phallus. – In: A. S. KING and J. MCLELLAND (eds.) Form and Function in birds. New York, Academic Press: 1–469.
LOWE, P. R. (1935): On the relationships of the Struthiones to the dinosaurs. – Ibis, **13**: 398–432.
LOWE, P. R (1944): An analysis of the characters of *Archaeopteryx* and *Archaeornis*. Were they reptiles or birds? – Ibis, **86**: 517–543.
MARTIN, L. D., STEWART, J. D. & WHETSTONE, K. N. (1980): The origin of birds: structure of the tarsus and teeth. – The Auk, Lawrence, **97**: 86–93; Lawrence, Kansas.
MARTIN, L. D. (1983): The origin of birds and avian flight. – Curr. Ornithol., **1**: 105–129.
MARTIN, L. D. (1983): The origin and early radiation of birds. – In: Perspectives of ornithology (eds. A. H. BRUSH and S. A. CLARK Jr.), 291–338.
MCGOWAN, C. (1984): Evolutionary relationships of ratites and carinates. – Nature (**307**) 5953: 733–735.
MÜLLER, F. (1967): Zur embryonalen Kopfentwicklung von *Crocodylus cataphractus* Cuv. – Rev. suisse Zool. **74**: 1898–2294.
NOPSCA, F. VON (1907): Ideas on the origin of flight. – Proc. Zool. Soc. London, **1907**: 223–236.
NOPSCA, F. VON (1923): On the origin of birds. – Proc. Zool. Soc. London, **1923**: 463–477.
OLSON, S. L., and FEDDUCIA, A. (1979): Flight capability and the pectoral girdle of *Archaeopteryx*. – Nature (London) **278**: 247–248.
OSBORN, H. F. (1900): Reconsideration of the evidence for a common dinosaur-avian stem in the Permian. – Amer. Nat. **34**, 406: 777–799.
OSBORN, H. F. (1903): *Ornitholestes hermanni*, a new compsognathoid dinosaur from the upper Jurassic. – Bull. Amer. Mus. Nat. Hist., **19**: 459–464.

OSTROM, J. H. (1974): *Archaeopteryx* and the origin of flight. – Quart. Rev. Biology, **49**: 27–47.
OSTROM, J. H. (1976a): Some hypothetical anatomical stages in the evolution of avian flight. – Smithsonian Contr. Paleobiol. **27**: 1–21.
OSTROM, J. H. (1976b): *Archaeopteryx* and the origin of birds. – Biol. J. Linnean Soc., **8**: 91–182.
OSTROM, J. H. (1976c): On a new specimen of the Lower Cretaceous theropod dinosaur *Deinonychus antirrhopus*. – Brevoria, **439**: 1–21.
OSTROM, J. H. (1978): The osteology of *Compsognathus longipes* Wagner. – Zitteliana, **4**: 73–118.
OSTROM, J. H. (1979): Bird flight: how did id begin? – Am. Scient., **67**: 46–56.
PADIAN, K. (1982): Macroevolution and the Origin of Major Adaptions: Vertebrate Flight as a Paradigm for the analysis of Patterns. – Proc. Third North American Paleo Conv., **2**: 387–392.
REGAL, P. J. (1975): The evolutionary origin of feathers. – Quart. Rev. Biol., **50**: 35–65.
REIF, W. E. (1984): Preadaptation and the change of function – a discussion. – N. Jb. Geol. Paläont. Mh. **1984** (2): 90–94.
ROMER, A. S. (1956): Osteology of the Reptiles. – Univ. Chicago Press, Chicago (Univ. Chicago Press).
ROMER, A. S. (1966): Vertebrate Paleontology, 3rd ed. Chicago (University of Chicago Press).
ROMER, A. S. (1972): The Chanares (Argentina) Triassic reptile fauna, XVI. Thecodont classification. – Brevoria, **395**: 1–24.
RUSE, M. (1982): Darwinism Defended. – Reading, Mass. (Addison-Wesley Publishing Company).
SIMPSON, G. G. (1983): Fossils and the History of Life. – New York (Freeman).
TAQUET, P. (1977): Variation ou rudimentation du membre anterieur chez les Theropodes (Dinosauria)? – In: Mécanismes de la rudimentation des organes chez les embryons de vertébrés, pp. 333–339, Colloques Internationaux CNRS, **266**, Paris.
TARSITANO, S. & HECHT, M. K. (1980): A reconsideration of the reptilian relationships of *Archaeopteryx*. – Zool. J. Linnaean Soc., London, **69**: 149–182.
THULBORN, R. A. & HAMLEY, T. L. (1982): The reptilian relationships of *Archaeopteryx*: a reconsideration. – Australian J. Zool., **30**: 611–634.
WALKER, A. D. (1970): A revision of the Jurassic reptile *Hallopus victor* (MARSH), with remarks on the classification of crocodiles. – Phil. Transact. Royal Soc. London, (B), **257**: 323–372.
WALKER, A. D. (1972): New light on the origin of birds and crocodiles. – Nature, **237**: 257–263; London.
WALKER, A. D. (1977): Evolution of the pelvis in birds and dinosaurs. – In: S. M. ANDREWS, RS. S. MILES & A. D. WALKER (eds.), Problems in Vertebrate Evolution: 319–357. London (Academic Press).
WALKER, A. D. (1980): The pelvis of *Archaeopteryx*. – Geol. Mag., **117**: 595–600.
WELLNHOFER, P. (1974): Das fünfte Skelettexemplar von *Archaeopteryx*. – Paleontographica, (A), **147**: 169–216.
WEVER, E. G. (1978): The Reptile Ear. – New Jersey (Princeton University Press).
WHETSTONE, K. N. & MARTIN, L. D. (1979): New look at the origin of birds and crocodiles. – Nature, **279**: 234–236; London.
WHETSTONE, K. N. (1983): Braincase of Mesozoic birds: I. New preparation of the "London" *Archaeopteryx*. – Journ. Vert. Paleont., **2** (4): 439–452.
WHETSTONE, K. N. and WHYBROW, P. (1983): A "cursorial" crocodilian from the Triassic of Lesotho (Basutoland), South Africa. – Occasion. Papers Mus. Nat. Hist. Univ. Kansas, **106**: 1–37.

Author's address: Prof. Dr. MAX K. HECHT, Dept. Vertebrate Paleontology, American Museum of Natural History, New York, N. Y. 10024, and Queens College, Flushing, N. Y. 11367.

John H. Ostrom

The Meaning of *Archaeopteryx*

Abstract

The title of this presentation is precisely the reason the *Archaeopteryx* Conference was convened. From the first discovery in 1861, *Archaeopteryx* has been controversial; considered a true bird by some, but merely a feathered reptile by others. Even today, there are some who accept it as the Urvogel, others who consider it a feathered dinosaur and a few who think it unimportant.

The objectives of this conference were: First, to agree on the identity of this taxon; Second, to establish its relevance to the question of bird origins; and Third, to evaluate its significance concerning the development of avian flight.

My position on these three issues is well known, but I will repeat them. One: the evidence of feathers and an ossified furcula, two apparently uniquely avian characters, clearly establishes *Archaeopteryx* as a bird. Two: as the oldest certifiable bird, these specimens m u s t be the foundation of any hypothesis on bird origins – or else be p r o v e n to be irrelevant. Three: the co-existence of specialized bipedality in *Archaeopteryx* and the paradoxical combination of modern-like "wings" together with the absence of nearly all skeletal specializations of the flight apparatus provide the o n l y available clues as to how bird flight might have begun.

Since my 1976 and 1979 papers, several contrary interpretations have appeared, namely TARSITANO and HECHT (1980), HECHT and TARSITANO (1982, 1983), MARTIN ET AL (1980). This is a partial response to those papers.

Zusammenfassung

Der Titel dieses Beitrages ist genau der Grund, weshalb diese Konferenz organisiert wurde und warum wir hier sind: um die Bedeutung von *Archaeopteryx* zu verstehen. Seit der ersten Entdeckung im Jahre 1861 löste dieses Geschöpf Kontroversen aus: Von einigen für einen echten Vogel, von anderen für ein befiedertes Reptil gehalten. Sogar heute gibt es einige, die ihn für den "Urvogel" und andere, die ihn für einen befiederten Dinosaurier halten, und einige, die glauben, dies sei unwichtig.

Die Themen dieser Konferenz sollten sein: Erstens, Übereinstimmung im Hinblick auf die Identität dieses Taxons zu erzielen; zweitens, seine Relevanz zur Frage des Ursprungs der Vögel festzulegen; und drittens, seine Bedeutung im Hinblick auf die Entstehung des Vogelfluges zu bewerten.

Meine Position zu diesen drei Fragen ist wohl bekannt, aber ich will sie nochmals herausstellen:

1) Der Nachweis von Federn und einer verknöcherten Furcula, zweier ausschließlicher Vogelmerkmale, etabliert *Archaeopteryx* klar als Vogel.

2) Als ältester nachweisbarer Vogel müssen die Exemplare die Grundlage für jede Hypothese über den Ursprung der Vögel sein – es sei denn, es ist nachzuweisen, daß sie irrelevant sind.

3) Die Koexistenz von spezialisierter Bipedie bei *Archaeopteryx* und die paradoxe Kombination von modern anmutenden "Flügeln", zusammen mit der Abwesenheit fast aller Skelettspezialisierungen des Flugapparates, liefern die einzigen verfügbaren Anhaltspunkte dafür, wie der Vogelflug begann.

Diese Anhaltspunkte mögen unzureichend sein, um hier eine Übereinstimmung zu erzielen, aber wir haben bei diesem Treffen eine ungewöhnliche Gelegenheit, diese Fragen – produktiv – zusammen zu überdenken.

Seit meinen Arbeiten von 1976 und 1979 erschienen einige gegenteilige Deutungen, nämlich von TARSITANO & HECHT (1980), HECHT & TARSITANO (1982, 1983), MARTIN et al. (1980). Dies ist eine teilweise Erwiderung hierzu.

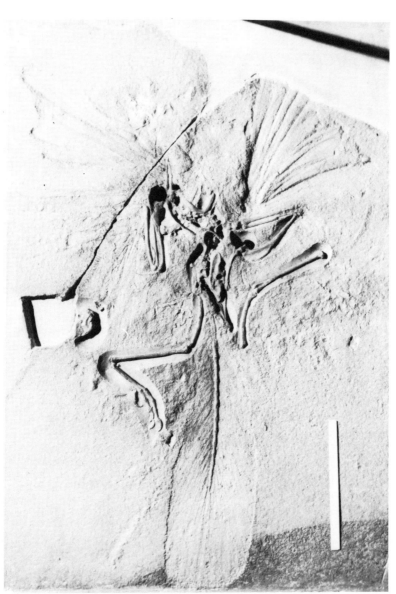

Fig. 1
The London specimen of *Archaeopteryx*, the holotype of *Archaeopteryx lithographica*.

The Systematic Position of *Archaeopteryx*

There should be no controversy over the identity of *Archaeopteryx* as avian. As we all know very well, the London, Berlin, and "Maxberg" specimens clearly preserve two uniquely avian characters: feathers and a furcula* (the latter only partially preserved in the Berlin and "Maxberg" specimens). It is not necessary to review here the ancient history of WAGNER's (1861) claim that the first specimen was a "feathered reptile" versus the counter assertions by OWEN (1862, 1863) and HUXLEY (1868) that it was a true bird. But it is necessary to consider some recent statements on this issue.

---

*BARSBOLD (1983) reported furcula-like structures in newly discovered theropod remains in Late Cretaceous strata of Mongolia.

BAKKER and GALTON (1974) declared "the avian radiation is an aerial exploitation of basic dinosaurian physiology and structure" and on that belief proposed that all birds be grouped in a subclass of their new class Dinosauria. (What do we know about dinosaurian physiology anyway?) Their proposal was answered by THULBORN (1975) who proposed that the ancestors of birds (the entire dinosaurian suborder Theropoda) be transferred to the class Aves. Recent studies by PADIAN (1982) and GAUTHIER (MS) conclude that *Archaeopteryx* should be placed in the saurischian suborder Theropoda. In my opinion, none of these suggestions is necessary or useful. BARSBOLD (1983) reported furcula-like structures in newly discovered theropod remains in Late Cretaceous strata of Mongolia.

The rationale behind all three of these proposals seems to be in the view of each author that their new alignments better express the evolutionary relationship of birds. Using that logic, then others who favor an avian-crocodilian affinity must classify birds as crocodiles, or vice versa, while others, who advocate a thecodontian-bird relationship must classify thecodontians as birds, or perhaps it would be birds as thecodontians! So far, no one has demonstrated the presence of any definitive character in any other archosaur, unless BARSBOLD's furcula-like structures prove to be fused clavicles instead of merely gastralia elements. Just as REED (1960) and VAN VALEN (1960) failed to establish that pelycosaurs and therapsids were mammals, these proposed re-alignments of birds and various archosaurs fail to meet the requirements of a utilitarian and stable systematic framework. I recommend that the class Aves be left where it is and include *Archaeopteryx* as its most archaic member.

Archaeopteryx and the Origin of Birds

It seems safe to conclude that no one who attended this Conference doubted the importance of the five specimens of *Archaeopteryx* concerning the question of bird origins. As the oldest certifiable bird remains they represent the only ancient evidence available to address that question. It is not important whether we believe that *Archaeopteryx* was on the main evolutionary lineage to later birds or was an aberrant side branch, because we can agree it was a bird. Rather, debate focuses on how different investigators compare the anatomy of *Archaeopteryx* with that of several proposed reptilian ancestral candidates – namely thecodontian (TARSITANO and HECHT, 1980, 1982), crocodilian (WALKER, 1972, 1974; MARTIN et al., 1980; MARTIN, 1983a, 1983b) and theropodan (OSTROM, 1973, 1975, 1976).

Obviously, it is highly improbable that *Archaeopteryx* actually represents a "mainline" antecedant of modern birds, but there is no evidence to rule out that possibility.

HECHT and TARSITANO (1983) stated that "it is possible to select the most probable relationship by evaluation of the characters and their states." I agree with that statement. HECHT and TARSITANO rejected both crocodilian and theropod affinities, favoring a thecodontian relationship. While I agree with most of their reasons for dismissing a crocodilian connection, not surprisingly, I disagree emphatically with their rationale for discarding theropods as closely related to *Archaeopteryx*. All five of their anatomical judgements are incorrect. To repeat their five 1983 reasons: 1) the fusion of the proximal tarsals to the tibia forming a tibiotarsus in *Archaeopteryx*; 2) the absence of a reflexed hallux in theropods; 3) the presence of a non-reduced and unique bent coracoid in *Archaeopteryx*; 4) the pubis in *Archaeopteryx* reflexed at the avian level of development; and 5) the presence of a semilunate-like element in the carpus of *Archaeopteryx* and theropods cannot be used as a synapomorphy because of the similar occurrence of this character state in thecodontians and some lepidosaurs.

My response is as follows:
Number one: the "fusion" of the proximal tarsals to the tibia in *Archaeopteryx* is no more so than that displayed in a variety of theropods (e.g. *Compsognathus, Coelophysis, Ornithomimus, Deinonychus*, etc.) and not at all comparable to the complete fusion of these elements in adult modern birds.

Number two: a reflexed hallux is n o t absent in theropods as the Munich specimen of *Compsognathus* clearly and unmistakably shows in Figure 2 (See also THULBORN and HAMLEY, 1982). TARSITANO (this volume) attempts to deny this by claiming that metatarsal I is preserved abutting against metatarsal IV, not metatarsal II. Figure 2 clearly shows that is not the case.

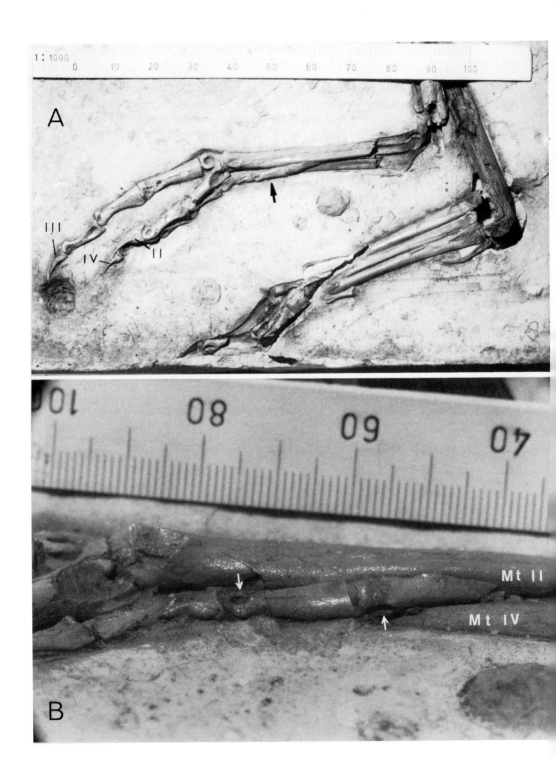

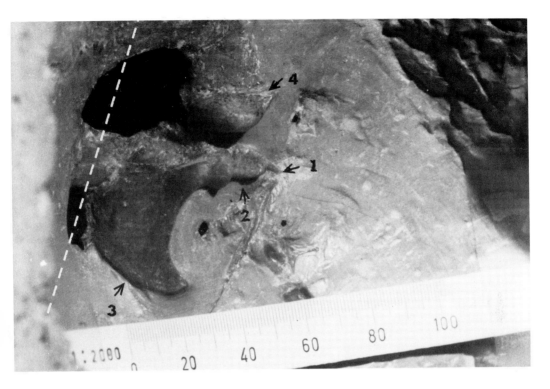

Fig. 3
Antero-lateral view of the left coracoid and scapula of the London specimen of *Archaeopteryx*. Arrow 1. indicates the external edge of the scapular blade; arrow 2. points to the glenoid; arrow 3. indicates the ventral articular margin of the coracoid that contacted a cartilaginous sternum. Arrow 4. marks the head of the left humerus. The transverse breadth of the coracoid indicates a broad-chested rib cage with the scapula lying nearly horizontally on the proximal segments of the laterally directed dorsal ribs. This clearly shows a laterally directed glenoid. The dashed line indicates the approximate orientation and position of the sagittal plane. Scale units equal 0.5 mm.

Number three: the presence (in their terms) of a "non-reduced and unique bent coracoid in *Archaeopteryx*" is irrelevant. That bone does not differ in any significant way from the coracoid morphology of typical theropods, and contrary to some interpretations, the glenoid of *Archaeopteryx* appears to have faced laterally. (see Figs. 3 and 4.)

Number four: their assertion that "the pubis of *Archaeopteryx* is reflexed at the avian level of development" is an overstatement at the very least. There is considerable disagreement over the pubis position in the Berlin specimen and important evidence exists that argues against their assertion. Both the "Maxberg" and the Eichstätt specimens clearly refute their supposed "avian" orientation of the pubis in the Berlin specimen. No one who has carefully examined the Berlin specimen closely can

◀
Fig. 2
The feet and metatarsi of *Compsognathus longipes* (B.S.P.A.S. 1563). A) Left (above) and right (below) feet. Arrow points to the hallux (digit I). Scale units equal 1.0 mm. B) Microphotograph of the left tarsus to show the hallux in which the proximal phalanx and ungual have been rotated 90° or more about their long axis relative to the metatarsal which is firmly articulated against metatarsal II, not metatarsal IV as Tarsitano claims (this volume). Arrows point to the collateral ligament fossae, the axes of flexion at the two joints of the hallux. These axes were parallel in life, clear evidence that the hallux was reflected and thus opposed the three main digits. Scale units equal 0.5 mm. Mt = Metatarsal.

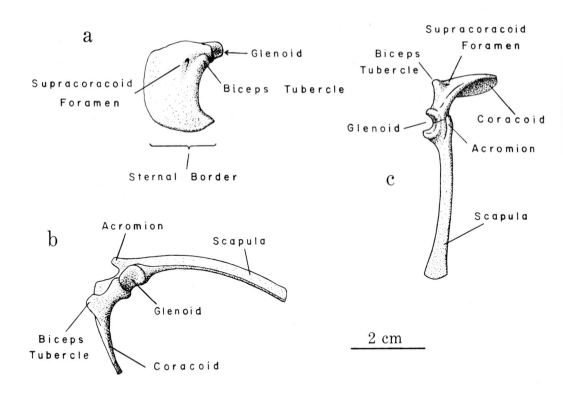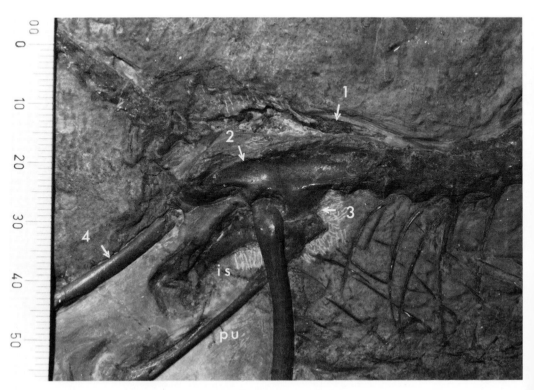

◀
Fig. 4
Three views of the restored shoulder skeleton of *Archaeopteryx* reconstructed by the author from the London, Berlin and "Maxberg" specimens. a) anterior view, b) lateral view, c) dorsal view.

possibly claim that the pubes there are preserved in their natural undisturbed position. I challenge anyone to prove that they are not distorted. Moreover, partially reflexed pubes are now known in several theropods: *Deinonychus* (OSTROM, 1976), *Velociraptor* (BARSBOLD, 1976, 1979), *Segnosaurus* and *Erlikosaurus* (BARSBOLD and PERLE, 1980).

Number five: HECHT and TARSITANO claim that the semilunate carpal cannot be used as a synapomorphy because "of similar occurrence of this character state in theocodontians" is not substantiated. No such element has been described or illustrated. To the best of my knowledge this feature is known only in *Archaeopteryx* and the theropods *Deinonychus*, *Velociraptor*, *Stenonychosaurus* and *Coelurus* (the latter not yet reported).

In their zeal to persuade the scientific community that a thecodontian ancestry of birds is the correct hypothesis, they have attempted to cast doubt on the avian uniqueness of feathers and furcula by remarking that "the presence of ectodermal feathers and a furcula is paralleled by the bizarre thecodontian *Longisquama*." First of all it is not clear that the only known specimen of *Longisquama* is in fact referable to the Order Thecodontia (whatever that assemblage represents). Second, it is not yet generally accepted that the sagittal dorsal projections in *Longisquama* are feather-like or have any connection with true feathers. Third, the object in *Longisquama* that they equate with the avian furcula is not at all comparable and most probably is a primitive feature – an interclavicle.

To justify their hypothesis, HECHT and TARSITANO state that "the fragmentary nature of the thecodontian record makes direct comparison between *Archaeopteryx* and thecodonts at this time difficult. However, the morphology of known thecodonts provides the necessary features from which the primitive avian level of organization can be derived." I agree with these statements but I enthusiastically endorse their statement that preceded them. "The thecodontian level of organization can easily be considered ancestral to the Crocodylia, t h e r o p o d s and birds." Now all that remains is to recognize the anatomical evidence and logic that *Archaeopteryx* arose from a thecodontian ancestry b y w a y of a coelurosaurian theropod. I am not convinced that dromaeosaurid theropods are more remote from *Archaeopteryx* than some hypothetical "thecodont".

In their papers of 1980, 1982 and 1983, HECHT and TARSITANO have stated or implied that my methodology was faulty when I arrived at the conclusion of a coelurosurian ancestry of *Archaeopteryx*. They criticized me for basing my hypothesis "only on general similarities and not on synapomorphies," as though that sin makes all the d e t a i l e d similarities that I cited irrelevant. Let me ask these authors where are the synapomorphies with their thecodontians? The assertions by TARSITANO and HECHT in no way invalidate the anatomical conditions found in certain theropods and *Archaeopteryx*.

Archaeopteryx and the Origin of Avian Flight

As with the question of bird origins, any hypothesis about the beginnings of bird flight must also be compatible with the evidence preserved in these specimens – no matter how well or how poorly

Fig. 5
The pelvis of the Berlin specimen of *Archaeopteryx* from the right side showing that the right and left sides are displaced, yet the pubes are still solidly fused at their distal extremities. Arrow 1. points to the upper margin of the left ilium; arrow 2. indicates the dorsal margin of the right ilium. As the left ilium was displaced up and backward relative to the right side, the fused pubes were distorted backward and upward. Arrow 3. points to the fracture that resulted between the right pubis and ilium. Notice that the left femur (arrow 4.) is also displaced back and upward relative to the right femur which is firmly articulated in the acetabulum. Scale units equal 1.0 mm. Abbreviations: is. = ischium; pu. = pubis.

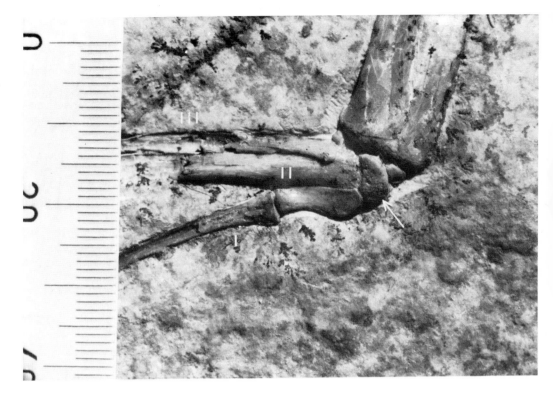

Fig. 6
Right wrist and metacarpus of the Eichstätt specimen of *Archaeopteryx*. Arrow points to the distinctive semi-lunate carpal element that articulates with the first and second metacarpals. So far, this element is known only in certain theropods and *Archaeopteryx*. Scale units equal 0.5 mm.

Archaeopteryx may have flown. As yet, there are no pre-*Archaeopteryx* or other Jurassic birds known. *Ichthyornis* of the Late Cretaceous shows that avian flight had been fully achieved by that time. What can we conclude about the flight capability of *Archaeopteryx*? It appears to have been a feeble flapper according to YALDEN (1970) and others, but what does the fossil evidence indicate about how it became airborne? I have maintained (OSTROM, 1974, 1979 and in press) that there is no evidence to support the argument that *Archaeopteryx* or its ancestors were arboreal. In fact, the evidence seems to me to be just the opposite. As a specialized obligate biped, the hind limb proportions indicate that *Archaeopteryx* was an accomplished cursor, despite MARTIN's (1983) disclaimer. In fact MARTIN's comment there that *Archaeopteryx* "has a considerable amount of closure of the inner wall of the pelvic acetabulum" is most misleading. Also misleading is his notation that the head of the femur is at 45° "to the sagittal plane" when in fact the femoral head is close to 90° to the femoral shaft. His implication is that the hind limbs could not asume a parasagittal orientation.

With regard to his first point, the acetabulum is visible unobstructed only in the London specimen where it obviously is fully and openly perforated as in theropods. Secondly, the right femur is firmly and parasagittally articulated in the Berlin specimen and the head of the femur is visible in place in the acetabulum as can be seen in the under side (medial aspect) of that specimen. Finally, femora preserved in the London and Berlin specimens, in addition to showing a nearly 90° offset of the head, together with the "Maxberg" specimen clearly preserve a longitudinal anterior-posterior curvature of the shaft perpendicular to the axis of flexion-extension of the distal condyle, rather than the sigmoidal curvature of the shaft typical of lizards and crocodilians. Moreover, the obviously digitigrade pes and the mesotarsal ankle of *Archaeopteryx* are consistent o n l y with parasagittal limb position and excursion. These anatomical conditions leave absolutely no question that *Archaeopteryx* was a highly adapted cursorial biped with limb excursion comparable to modern cursorial birds.

Not one of these special features – parasagittal bipedality, mesotarsal ankle, digitigrady – is an *a priori* adaptation for arboreality. In fact, except for modern birds, which I firmly believe became secondarily arboreal a f t e r the acquisition of powered flight, no other arboreal animal possesses these character states. This leads to the inescapable conclusion that bipedal stance and gait preceded avian flight (whatever that flight form may have been) in a g r o u n d - d w e l l i n g precursor of *Archaeopteryx* rather than in some unknown arboreal antecedent. That conclusion is reinforced by *Archaeopteryx* itself in which the pedal unguals are not at all like those of perching birds or birds of prey that have strong grasping powers. In those birds, the unguals are strongly curved with prominent flexor tubercles and the hallux ungual is larger than the others. In *Archaeopteryx*, the unguals are more like those of ground-dwelling birds, like pheasants, where these bones are not strongly curved, the flexor tubercles are weakly developed and the hallux ungual is shorter than those on the other toes. Also in *Archaeopteryx*, the hallux is short and positioned high on the metatarsus, quite unlike the long and lower-placed hallux of perching or predatory birds.

In summary, it is my conviction that *Archaeopteryx* was still learning to fly – from the ground up – and that avian flight began in a running, leaping, ground-dwelling biped. The studies of CAPLE, BALDA and WILLIS (1983, 1984) provide powerful support to this hypothesis, and the evidence preserved in the specimens of *Archaeopteryx* reinforce it further. By contrast, the arboreal hypothesis of bird flight origins is only a hypothetical scenario with no fossil, historical or physical evidence to support it. I emphatically do not claim that *Archaeopteryx* had no arboreal capability, but I do say that there is no compelling evidence in any of the specimens that points in that direction. (But see YALDEN's paper, this volume.)

Additional Comments

This occasion presents a convenient opportunity to comment on a few other points that have been raised by recent authors. In 1980 TARSITANO and HECHT offered their interpretation of the reptilian relationships of *Archaeopteryx*. Space does not permit me to comment on all of their many points that I disagree with but I must challenge a few of them here. Most important is their interpretation of the homologies of the hand digits. Because of the a p p a r e n t phalangeal formula (2-3-4) of the three remaining fingers of *Archaeopteryx*, these digits have traditionally been identified as the three inside digits (I, II, and III). Embryologic studies by MONTAGNA (1945), HOLMGREN (1955) and many recent workers, most notably by fellow participant J. R. HINCHLIFFE (1977) and associates (1980, 1983), have produced a consensus that the fingers of modern birds are the three middle digits (II, III and IV). TARSITANO and HECHT (1980) concluded from this that theropods can have nothing to do with bird ancestry. I say that is false logic because their digit homology interpretations are based on non-comparable data. There is no embryological evidence for digit identification in theropods, but by tradition, the hand digits have been labeled I, II and III for the same reasons they were so labeled in *Archaeopteryx*. However, they c o u l d be digits II, III and IV also, as suggested by THULBORN and HAMLEY (1982). Without the embryologic evidence, we cannot know what the digits in theropods represent, nor do we know the digital identities in *Archaeopteryx*! Neither can we dismiss their apparent homology with the manus digits of birds. It was improper of TARSITANO and HECHT to use embryologic data for one taxon (modern birds) to dismiss possible homology in another taxon (theropod dinosaurs) on the basis of non-embryologic reasoning. Their non-homology argument fails.

On another point in their 1980 paper, TARSITANO and HECHT imply that the phalangeal formulae of the fingers in *Archaeopteryx* is 2-3-3, rather than the long-believed 2-3-4. They believe that the proximal phalanx of the external digit is broken into two fragments wherever it is preserved or visible (both hands in the Berlin specimen and one hand in the "Maxberg" specimen). They concluded that "there are no indications of articulating surfaces and that these represent physical breaks rather than articulations." I

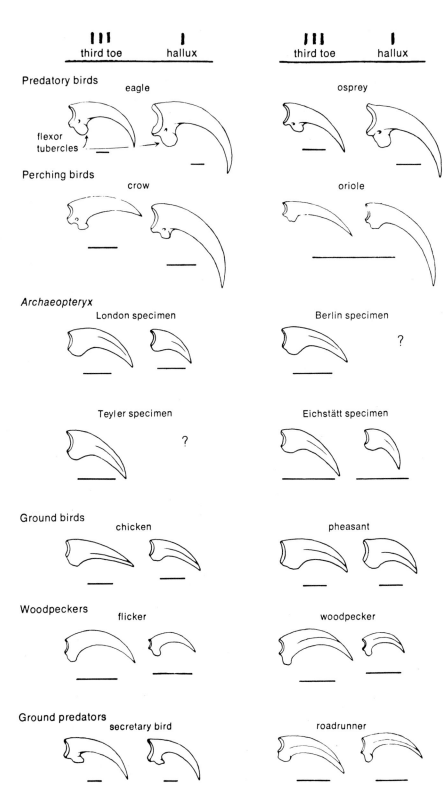

re-examined the Berlin specimen in 1982 to evaluate their interpretation and am convinced that the "broken finger" hypothesis has no basis in fact. In the left hand of the Berlin specimen the proximal "broken half" of the first phalanx clearly shows the dorsal lobes of the distal articular ginglymus. Their distal "broken half" is rotated about its long axis by nearly 90° so that the medial aspect is exposed and the internal collateral ligament fossa is quite evident. The "third" or external finger of the right hand shows almost the same condition. The presence of the articular ginglymus at the distal extremities of both of the proximal "broken halves" refutes the TARSITANO-HECHT "broken finger" hypothesis. HOWGATE (1983) came to the same conclusion.

MARTIN et al. (1980) raised several other points which they maintain argue for a crocodilian-bird affinity and against a theropod-bird relationship. They emphasize that the teeth of Cretaceous birds (*Hesperornis* and *Ichthyornis*) are very similar to crocodilian teeth (and they are) with simple unserrated conical crowns separated from an expanded root by a distinct neck or constriction. In their conclusion they extend this morphology to *Archaeopteryx* stating that it "has unserrated teeth with constricted bases and expanded roots." They note that by comparison theropods have serrated teeth with straight roots and no constriction. This leads them to conclude that the bird-crocodile relationship is supported by this and that the theropod-bird affinity is not.

First of all, they are correct that serrations have not been detected on any of the adequately exposed teeth of *Archaeopteryx*, but contrary to their implication, non-serrated teeth a r e known in some theropods (most of the teeth in *Compsognathus* and perhaps all of the teeth in *Ornitholestes*). Secondly, it is not true that the teeth of *Archaeopteryx* have constrictions at the base of the crown and the only complete root exposed in any of the specimens (the London specimen) does not have an expanded or inflated root. In fact, the teeth of the Eichstätt specimen are remarkably similar to those of the theropod *Compsognathus* and quite unlike those of known Cretaceous birds or crocodiles. Perhaps the absence or rarity of serrations in these taxa is related to their very small size or perhaps similar predatory (insectivorous?) habits. In any case, I conclude that dental evidence presented by MARTIN et al. (1980) does not support a crocodilian-bird relationship, but could well prove supportive of a theropod connection.

◀

Fig. 7
Comparison of terminal phalanges (without horny claws) from the feet of selected bird types with those of four specimens of *Archaeopteryx*. Each example includes the ungual of the median toe (III) on the left and the hallux ungual (I) on the right. All digit III unguals (the first and third columns) are drawn to unit length for convenient comparison and each companion sketch of I unguals (columns 2 and 4) is drawn to the same scale. Scales are indicated by the horizontal lines, all of which equal 5.0 mm. Orientations are standardized with the articular facets drawn in a vertical position. In nearly all respects, the unguals of *Archaeopteryx* resemble those of ground-dwelling birds more closely than all others, featuring slight curvature, robust construction and shallow, poorly defined flexor tubercles. Notice that in all (except the ground birds) the flexor tubercle is very prominent and sharply defined. Also, the hallux ungual is longer than the other unguals in most passerines (and many predatory birds). In addition to the four specimens of *Archaeopteryx* indicated, the examples sketched include: *Haliaeetus leucocephalus* and *Pandion haliaetus; Corvus brachyrhynchos* and *Icterus spurius; Gallus gallus* and *Phasianus colchicus; Colapter cafer* and *Picus viridis; Sagittarius serpentarius* and *Geococcyx californianus*.

Fig. 8 ▶
The left hand and metacarpus of the Berlin specimen of *Archaeopteryx* (A) and a close up photograph of the disputed region of the third or external finger (B). Arrow 1. points to the supposed fracture of the proximal phalanx claimed by TARSITANO and HECHT (1980), but note a) the bosses of the distal ginglymus (small arrows) and b) the twisted position of the adjacent phalanx (arrow 2.). Note also the expanded proximal end of that phalanx (or "broken fragment") – not an expected feature at mid-shaft of a broken phalanx. The evidence preserved here clearly indicates a phalangeal formula of 2-3-4 for the digits of the hand of *Archaeopteryx*, whether those digits are I, II and III, as labeled here, or digits II, III and IV as some embryologic evidence suggests for modern birds. Scale units equal 0.5 mm.

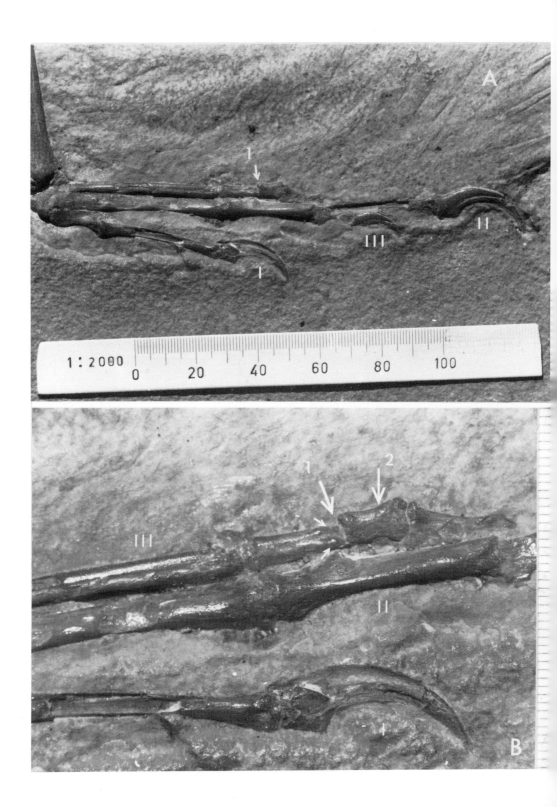

Fig. 9
Ankles of three of the specimens of *Archaeopteryx*. A) Left tarsus of the Berlin specimen; B) left tarsus of the London specimen; C) right tarsus of the Eichstätt specimen. Scale units in all equal 0.5 mm. Arrows point to the ascending process of the astragalus. In A and B, the tarsus lies on its lateral surface. In C, the lateral aspect is exposed and the calcaneum is not present here. The calcaneum is not preserved in contact with the fibula on the counterpart slab. Abbreviations: as. = astragalus; ti. = tibia.

In the same paper, MARTIN et al. (1980) challenged my interpretation of what I identified as the ascending process of the astragalus in *Archaeopteryx* – a characteristic element of the theropod tarsus. They pointed out that a similar feature in modern birds, termed the "pre-tibial" bone, fuses with the calcaneum late in development and only occasionally contacts the astragalus. Also, the pre-tibial bone is usually developed on the antero-lateral surface of the tibia in birds, whereas the ascending process of the astragalus in theropods is usually centered on the anterior distal surface of the tibia. I was not aware of the developmental history of the avian pre-tibial bone and out of ignorance made what seemed to me at the time to be a logical comparison. I further admit to confusing the issue by illustrating this process in *Archaeopteryx* in my 1976 paper in a much more medial position than it is actually preserved in the London and Berlin specimens, as Figure 9 clearly shows. Nevertheless, these photographs also show that this process is in direct contact with the antero-lateral surface of the tibia in both the London and Berlin specimens. However, the Eichstätt specimen may be even more important. On the main slab, it appears to show continous ossification between the dorsal process and the body of the astragalus in the right tarsus. On the counterpart slab, the fibula is complete (non bird-like) and extends almost to the end of the tibia. But there is no clear evidence of the calcaneum or of any contact with a "pre-tibial" bone. Damage to this region does not permit indisputable identification of the calcaneum – or even whether it is present, but there does not appear to be any lateral extension of the dorsal astragalar process to the calceneal region. Therefore, I conclude that *Archaeopteryx* probably did have an ascending process of the astragalus rather than a "pre-tibial" bone. That is consistent with MCGOWAN's (1984) embryologic findings of an astragalar process in ratites and a pre-tibial bone in carinates. Thus the ascending process of the astragalus may be considered primitive for *Archaeopteryx* and ratites, and as MARTIN et al. (1980) concluded, the pre-tibial bone is derived for carinates. Since both processes appear to have formed from ossification centers separate from the astragalus and the calcaneum respectively, they may in fact be the same bone which simply has fused with different tarsals in different taxa. But since the tarsal condition is equivocal in the Eichstätt specimen and has not yet been clearly resolved in the Berlin or London specimens, this feature cannot be used either for or against the theropod hypothesis of *Archaeopteryx* ancestry.

Summary and Conclusions

1) There can be no doubt that *Archaeopteryx* was a true bird and I urge that the class Aves be retained with *Archaeopteryx* as its most important taxon.

2) The five known specimens of *Archaeopteryx* preserve the only solid physical evidence of the earliest recognizable stage of bird evolution and thereby provide the most compelling evidence about bird origins – which all point to a coelurosaurian theropod ancestry – not crocodilian and not thecodontian.

3) Those same five specimens preserve no evidence to support the arboreal theory of the origin of avian flight. That *Archaeopteryx* was capable of feeble powered flight is open to question, but the skeletal anatomy of the hind quarters clearly demonstrates that this animal was a highly adapted bipedal and cursorial ground-dwelling predator. That lends powerful credence to the cursorial theory of the origin of bird flight.

4) The controversy over the homologies of the fingers of *Archaeopteryx* are addressed. No final conclusion is possible because of a lack of embryologic data for *Archaeopteryx* or theropods. But the phalangeal formula of the hand digits is resolved at 2-3-4. The "broken finger" hypothesis is disproved.

5) The teeth of *Archaeopteryx* do not resemble those of crocodilians or other Mesozoic birds and therefore do not support the theory of a crocodilian relationship.

6) The apparent conflict between the presence in *Archaeopteryx* of a "pre-tibial" bone versus an ascending process of the astragalus cannot be resolved with present material. But that does not preclude a theropod ancestry of *Archaeopteryx*; neither does it support a crocodilian or a thecodontian origin of birds.

Acknowledgements

I am deeply indebted to many colleagues in the pursuit of these and earlier studies, for all their assistance and hospitality. I especially acknowledge ALAN CHARIG and CYRIL WALKER of the British Museum (Natural History), London; HERMANN JÄGER, Humboldt Museum für Naturkunde, East Berlin; THEO KRESS, Solenhofen Aktien Verein, Maxberg; the late C.O. REGTEREN ALTENA, Teyler's Stichting, Haarlem, Netherlands; PETER WELLNHOFER, Bayerische Staatssammlung, Munich; and GÜNTER VIOHL of the Jura Museum, Eichstätt for their many favors and permission to study the *Archaeopteryx* specimens under their charge. Special thanks go to Dr. VIOHL and Dr. WELLNHOFER for organizing this conference.

References Cited

BAKKER, R.T. & GALTON, P.M. (1974): Dinosaur monophyly and a new class of vertebrates. – Nature **248**: 168–172; London.
BARSBOLD, R. (1976): On the evolution and systematics of Late Mesozoic dinosaurs. – Trudy Sovm. Sov. – Mong. Palaeont. Exped. **3**: 68–75; Moscow.
BARSBOLD, R. (1979): Opisthopubic pelvis in the carnivorous dinosaurs. – Nature **279**: 792–793; London.
BARSBOLD, R. (1983): Carnivorous dinosaurs from the Cretaceous of Mongolia. – Joint Soviet-Mongolian Paleontological Expedition: Vol. **19**: 1–119.
BARSBOLD, R. & PERLE, A. (1980): Segnosauria, a new infraorder of carnivorous dinosaurs. – Acta Palaeo. Polon. **25**: 187–195; Warsaw.
CAPLE, G., BALDA, R.P. & WILLIS, W.R. (1983): The physics of leaping animals and the evolution of pre-flight. – Amer. Nat. **121**: 455–476; Chicago.
CAPLE, G., BALDA, R.P. & WILLIS, W.R. (1984): Flap about flight. – Animal Kingdom **87** (4); 33–38; New York.
GAUTHIER, J. (1984): personal communication (manuscript in preparation); Berkeley.
HECHT, M.K. & TARSITANO, S. (1982): The paleobiology and phylogenetic position of *Archaeopteryx*. – Geobios Spéc. Mém. **6**: 141–149; Lyon.
HECHT, M.K. & TARSITANO, S. (1983): *Archaeopteryx* and its paleoecology. – Acta Palaeo. Polon. **28**: 133–136; Warsaw.
HINCHLIFFE, J.R. (1977): The chondrogenic pattern in chick limb morphogenesis. – In: Vertebrate Limb and Somite Morphogenesis; EDE, D.A., HINCHLIFFE, J.R. & BALLS, M. (Eds.): 293–308; Cambridge (England). Cambridge University Press.
HINCHLIFFE, J.R. & JOHNSON, D.R. (1980): The Development of the Vertebrate Limb. – 268 pp. Oxford (England), (Clarendon Press).
HINCHLIFFE, J.R. & GRIFFITHS, P.J. (1983): The prechondrogenic patterns in tetrapod limb development and their phylogenetic significance. – **In**: Development and Evolution; GOODWIN, B.C., HOLDER, N. & WYLIE, C.G. (Eds): 99–121. Cambridge (England), (Cambridge University Press).
HOLMGREN, N. (1955): Studies on the phylogeny of birds. – Acta Zool. **36**: 243–328; Stockholm.
HOWGATE, M. (1983): *Archaeopteryx* – no new finds after all. – Nature **306**: 644; London.
HUXLEY, T.H. (1868): Remarks upon *Archaeopteryx lithographica*. – Proc. Roy.. Soc. London, **16**: 243–248; London.
MARTIN, L.D. (1983a): The origin of birds and of avian flight. – **In**: Current Ornithology, JOHNSTON, R.F. (Ed.); **1**: 105–129; New York and London, (Plenum Press).
MARTIN, L.D. (1983b): The origin and early radiation of birds. – **In**: Perspectives in Ornithology, BRUSH, A.H. & CLARK, Jr., G.A. (Eds.): 291–353; Cambridge (England, London & New York), (Cambridge University Press).
MARTIN, L.D., STEWART, J.D. & WHETSTONE, K.N. (1980): The origin of birds: structure of the tarsus and teeth. – Auk **97**: 86–93; Washington, D.C.

McGowan, C. (1984): Evolutionary relationships of ratites and carinates: evidence from ontogeny of the tarsus. – Nature **307**: 733–735; London.

Montagna, W. (1945): A re-investigation of the development of the wing of the fowl. – Jour. Morph. **76**: 87–113: New York.

Ostrom, J.H. (1973): The ancestry of birds. – Nature **242**: 136; London.

Ostrom, J.H. (1974): *Archaeopteryx* and the origin of flight. – Quart. Rev. Biol. **49**: 27–47; Stony Brook (New York).

Ostrom, J.H. (1975): The origin of birds. In: Ann. Rev. Earth Planet. Sci., Donath, F.A. (Ed.), **3**: 55–77; Palo Alto, California.

Ostrom, J.H. (1976): *Archaeopteryx* and the origin of birds. – Biol. Jour. Linn. Soc., **8**: 91–182; London.

Ostrom, J.H. (1979): Bird flight: how did it begin? – Amer. Sci., **67**: 46–56; New Haven.

Ostrom, J.H. (in Press: 1985): The cursorial origin of avian flight. – Proc. Calif. Acad. Sci., San Francisco.

Owen, R. (1862): On the fossil remains of a long-tailed bird (*Archaeopteryx macrurus* Ow.) from the lithographic slate of Solenhofen. – Proc. Roy. Soc., **12**: 272–273; London.

Owen, R. (1863): On the *Archaeopteryx* of Von Meyer, with a description of a long-tailed species from the lithographic stone of Solenhofen. – Phil. Trans. Roy. Soc. **153**: 33–47; London.

Padian, K. (1982): Macroevolution and the origin of major adaptations: Vertebrate flight as a paradigm for analysis of patterns. – Proc. Third North Amer. Paleont. Conv. **2**: 387–322; Montreal.

Reed, C.A. (1960): Polyphyletic or monophyletic ancestry of mammals, or: what is a class? – Evolution **14**: 314–322; Lancaster, Pa.

Tarsitano, S. & Hecht, M.K. (1980): A reconsideration of the reptilian relationship of *Archaeopteryx*. – Zool. Jour. Linn. Soc., **69**: 149–182; London.

Thulborn, R.A. (1975): Dinosaur polyphyly and the classification of archosaurs and birds. – Australian Jour. Zool. **23**: 249–270; Melbourne.

Thulborn, R.A. & Hamley, T.L. (1982): The reptilian relationships of *Archaeopteryx*. – Australian Jour. Zool., **30**: 611–634; Melbourne.

Van Valen, L. (1960): Therapsids as mammals. – Evolution, **14**: 304–313; Lancaster, Pa.

Wagner, J.A. (1861): Ueber ein neues, augenblich mit Vogelfedern versehenes Reptil aus dem Solenhofer lithographischen Schiefer. – Sitz. bayer. Akad. Wiss., **2**: 146–154; Munich.

Walker, A.D. (1972): New light on the origin of birds and crocodiles. – Nature **237**: 257–263; London.

Walker, A.D. (1974): Evolution, organic. – McGraw-Hill Yearbook Sci. & Techn., **1974**: 177–179; New York.

Yalden, D.K. (1970): The flying ability of *Archaeopteryx*. – Ibis **113**: 349–356; London.

Yalden, D.K. (1985): Foelimb function in *Archaeopteryx*. – This volume.

Author's address: Prof. Dr. John H. Ostrom, Peabody Museum of Natural History, Yale University, 170 Whitney Avenue, P.O. Box 6666, New Haven, Conn. 06511, U.S.A.

Larry D. Martin

The Relationship of *Archaeopteryx* to other Birds

Abstract

Many characters of the skull, postcranial skeleton, and feathers show that *Archaeopteryx* is more closely related to other birds than to any known group of reptiles. *Archaeopteryx* is not ancestral to any modern birds and is instead a member of an extinct subclass of birds, the Sauriurae. The Sauriurae includes all of the known terrestrial birds of the Mesozoic.

Zusammenfassung

Zahlreiche Merkmale an Schädel, postcranialem Skelett und Federn zeigen, daß *Archaeopteryx* näher mit den anderen Vögeln verwandt ist als mit irgendeiner Gruppe der Reptilia. *Archaeopteryx* ist kein Vorläufer der modernen Vögel, sondern ein Vertreter der Sauriurae, einer erloschenen Unterklasse der Vögel. Die Sauriurae umfassen alle bekannten terrestrischen Vögel des Mesozoikums.

Introduction

The origin and early radiation of the class Aves is one of the most interesting problems in vertebrate evolution. The earliest fossil evidence concerning this topic comes from five specimens of the late Jurassic bird *Archaeopteryx*, which have been studied in considerable detail. Although *Archaeopteryx* is known from very complete remains, crushing and preparation difficulties have hindered its study. Until recently the most commonly accepted hypothesis of avian origins was that birds have a common ancestor with both dinosaurs and crocodilians no later than the Triassic Pseudosuchia. This viewpoint has been largely replaced by a hypothesis that birds were derived directly from theropod dinosaurs. Finally, some workers still support a pseudosuchian ancestry, but consider birds closer to crocodilians than to dinosaurs.

The relationships of *Archaeopteryx*, have inspired controversy since its discovery in 1861. At that time, WAGNER (1861) declared that *Archaeopteryx* could not be a true bird and must be regarded as a feathered reptile. This view was quickly set aside by other workers (OWEN, 1863; HUXLEY, 1868) who *Archaeopteryx* avian credentials requires a re-examination. Another aspect of *Archaeopteryx* also has held until quite recently. During the last few years LOWE's position that *Archaeopteryx* is a feathered theropod has begun to appear again (for instance see THULBORN, 1985), and the question of *Archaeopteryx*'s avian credentials requires a re-examination. Another aspect of *Archaeopteryx* also needs a new look. This is its relationships with later birds. Because it is the oldest known bird and remarkably primitive in much of its anatomy, most workers have treated it as though it were ancestral to the entire subsequent avian radiation, even though it has usually been recognized that this is unlikely to be literally true (OSTROM, 1976). The question of whether *Archaeopteryx* is a bird or not is in part semantic (dependent on how we define a bird). JOLLIE (1977) gives a very lengthy diagnosis for the Class Aves including: the loss of all teeth; the presence of a keeled sternum, and air sacs which extend throughout the body. These features are absent from *Archaeopteryx*, which JOLLIE (1977, p. 107) thinks "should be viewed as too similar to the reptile and not modified enough to be a bird". A strict application of his criteria would also exclude such other Mesozoic toothed birds as *Hesperornis* and *Ichthyornis*. "Solutions" of this sort may be as numerous as authors willing to propose new definitions, and they do

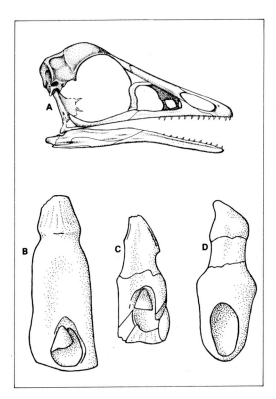

Fig. 1
A restoration of the skull of *Archaeopteryx* based on the London cranium and Wellnhofer's (1974) restoration of the Eichstätt skull. B.-D. Lingual views of teeth showing oval resorption pits: B. Alligator.
C. *Parahesperornis alexi*. D. *Archaeopteryx lithographica*. Not to scale.

not in any way change our opinions of how taxa are related. For most ornithologists the "key" character for the class Aves has been the presence of feathers, and in this feature *Archaeopteryx* is clearly a bird. Another approach to the problem is to attempt to more closely ally *Archaeopteryx* with some other group (usually some subset of the theropod dinosaurs) than to other known birds. The dinosaur examples demand that feathers be fairly widely distributed and the absence of evidence for dinosaurian feathers ascribed to accidents of preservation. For instance, THULBORN's (1985) phylogeny suggests that we might reasonably expect some feathers on *Tyrannosaurus rex*. Unfortunately very little theropod integument has been described. OSTROM was unsuccessful in a very determined effort to identify feathers on *Compsognathus* (OSTROM, 1978) although the preservation of the *Compsognathus* specimens is the same as that of the *Archaeopteryx* specimens where feathers are preserved.

The answer to the question of how close *Archaeopteryx* is to other birds, as compared to some hypothetical reptilian stem-group, can be addressed using features more likely to be preserved than feathers. We can at the same time evaluate the relationship of *Archaeopteryx* to other birds and try to ascertain if it is on the ancestral line to modern birds. If we find that it is not, then we must be cautious in our interpretations of what is primitive in its anatomy.

Primitive Features not found in other Birds

Archaeopteryx has a number of features that are not known from any other birds and on the basis of outgroup comparisons with reptiles, can be considered primitive character states. In the skull the premaxillary bones are small for a bird. The premaxillaries, the maxillaries, and the dentaries are toothed. The nasals meet each other on the midline. The maxillaries are large. The jugal and the quadratojugals have distinct ascending processes, and the quadratojugal joint with the quadrate does not appear to have the peg and socket arrangement found in later birds. The post-cranial skeleton lacks many of the fusions that characterize modern birds. Sutures can be found in adult *Archaeopteryx* throughout the carpometacarpus, and at the junctions of the ilium, ischium, and pubis in the pelvis. A few structures that are lost in later birds such as: some of the manus phalanges, the gastralia, and the teeth are still present and functional. Some features that are ossified in later birds may have been present in *Archaeopteryx*, but were cartilaginous, including: the sternum and the uncinate processes on the ribs. The latter two structures are vital to the characteristic breathing of modern birds and this suggests that breathing using the air-sac system and the sternum probably was not yet developed in *Archaeopteryx*.

Features Characteristic of Birds in General

The skull of *Archaeopteryx* is fundamentally bird-like and does not closely resemble the skull of any known reptile (Fig. 1A). The brain is already somewhat enlarged and has reached avian proportions. The postorbital seems to have been lost and the skull is no longer diapsid. There is no contact between the squamosal and the quadratojugal. The teeth are in sockets within a groove bounded on both sides by dense bone, and tooth replacement occurs with most of the development of the replacing tooth within the root of its predecessor, as in crocodilians, but very different from the tooth implantation and replacement of dinosaurs. The teeth themselves are essentially identical to those known from other toothed birds (*Hesperornis, Parahesperornis,* and *Ichthyornis*, Fig. B–D), and once again resemble crocodilian teeth. They are not similar to the teeth of any known dinosaur. Bird teeth have flattened triangular crowns that have a distinct waist at their base. They are not serrated, and the root is expanded and cement-covered. It usually contains an oval resorption pit and a developing tooth crown. This is because that early in the ontogeny of an avian or crocodilian tooth the tooth bud inclines labially, enters the root of its predecessor, straightens out, and completes its development. In dinosaurs most of the ontogeny of a developing tooth is lingual to its predecessor.

Archaeopteryx has feathers of an essentially modern type composed of a quill supporting a system of vanes, rachi, and barbs. Barbules and hamuli also seem to be present although the best evidence for the latter is the even arangement of the barbs on the specimens. The next oldest fossil feathers have been described by SCHLEE (1973) from the Lower Cretaceous (Neocomian) amber of the Lebanon Mountains. These feathers clearly show the presence of barbules and hamuli. Taken as a whole the fossil record shows that the microstructure of the feathers is one of the oldest diagnostic avian features. The feathers of *Archaeopteryx* are also divided into rectrices and remiges as in modern birds and the remiges have the anterior vanes narrower than the posterior vanes. This relationship is characteristic of modern flying birds where it helps to form the airfoil. FEDUCCIA and TORDOFF (1979) have argued persuasively that asymmetrical feathers indicate that *Archaeopteryx* had some aerodynamic capabilities. The flight feathers are divided into primaries and secondaries. Coverts are also present and the whole plumage has a very modern aspect (DE BEER, 1954). An important aspect of many of these features is that they are necessary for the production of a flight-adapted "avian" wing, and are probably of no real use for direct food capture or thermoregulation. The feathers of *Archaeopteryx* commit it to flight in anatomically the same way as do the feathers of modern birds. Even if we choose to propose hypothetical feathers for reptiles, we must place *Archaeopteryx* closer to birds or argue flight-adapted feathers for the competing group. I do not think that anyone has argued for flight adaptation in any known dinosaurs, crocodilians or pseudosuchians, and I think that the anatomy of the feathers does strongly ally *Archaeopteryx* with the rest of the Aves.

The scapula and coracoid in birds have a very unusual relationship to each other and are rotated about 90° so that the scapula lies nearly parallel to the vertebral column. In mammals this positioning occurs in primates and bats and might have something to do with climbing. In reptiles, I have only observed it in a few small Triassic forms like *Scleromochlus*. *Archaeopteryx* is fully avian in this respect. Although it has been reported that the scapula and coracoid are fused in *Archaeopteryx* (DE BEER, 1954) this does not seem to be the case. In all examples except the right side of the London specimen, the coracoids and scapulae show some degree of disarticulation.

The forelimbs of *Archaeopteryx* have undergone a series of positional rotations that enhance their ability to function as wings. These are exactly the same rotations that occur in the forelimbs of modern birds and

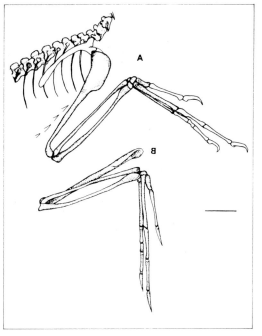

Fig. 2
Different restorations of the posture of the forelimb in *Archaeopteryx*: A. As commonly restored B. In the avian position. The latter arrangement conforms better to the position of these elements as found on the fossils. Scale 2 cm.

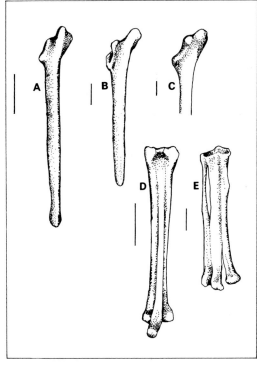

Fig. 3
Diagnostic skeletal elements of birds belonging to the subclass Sauriurae. A–C dorsal views of scapulae: A. *Archaeopteryx lithographica*. B. Enantiornithine bird. C. *Gobipteryx* (after ELZANOWSKI, 1981). D–E anterior views of tarsometatarsi: D. *Archaeopteryx lithographica*. E. Enantiornithine bird. All scales are 1 cm. except that C = 1 mm.

have to do with the ability to fold the wing without the feathers dragging on the ground. The humerus, radius and ulna have rotated 90° from their usual orientation so that when the wing is folded the palmar surface of the humerus faces laterally and that of the ulna faces medially. A semilunate bone (composed of fused distal carpals) articulates in a joint formed by the radiale and ulnare so that the manus can rotate posteriorly on what in other animals would be its lateral edge. This moves the tips of the primaries up and tucks them into the rest of the wing. This is a very complicated system that is fully developed in *Archaeopteryx*, and the positioning of the wings in the London, Berlin, and Eichstatt specimens is a direct result of this kind of a folding apparatus. It does not seem to be present in any known reptile and indeed we would not expect it unless that reptile was thought to have an avian wing. It should be pointed out that most restorations of the skeleton of *Archaeopteryx* have not been done by ornithologists and the wing has generally been restored in the posture it would have in a normal quadruped (Fig. 2).

Archaeopteryx is like other birds in having a fully opisthopubic pelvis with both the ischium and the pubis strongly rotated posteriorly. The ischium is essentially parallel with the postacetabular ilium as it is in other birds but not as it is in most other reptiles including dinosaurs. The acetabulum is partially closed (MARTIN, 1983a) and bipedalism may not have been as fully developed as in later birds.

In the evolution of the avian ankle the distal tibia expanded laterally until it overlaid and captured the calcaneum, which still retained its important articular function. In fact, it forms the entire outer condyle of the distal tibiotarsus. This expansion of the tibia separated the fibula from its normal contact with the calcaneum. The distal contact of the fibula in all birds (including *Archaeopteryx*) is entirely with the lateral side of the tibia. The pretibial bone in birds (see MARTIN, STEWART and WHETSTONE, 1980) is a separate ossification which begins on the anterolateral face of the tibiotarsus usually above the junction of the calcaneum and astragalus. It does not chondrify until after the chondrification of the astragalus and the calcaneum and is often not figured in early stages of embryological development. It generally fuses with the anterodorsal surface of the calcaneum but it may not contact the astragalus, although it usually expands medially until it contacts the lateral side of the astragalus. In some birds (ostrich) the base becomes very broad and almost the whole dorsal surface of the astragalus is contacted by the pretibial bone. The association of the pretibial bone with the lateral rather than the medial side of the tarsus can be observed in various Mesozoic birds including the tarsi of *Archaeopteryx*, the hesperornithiforms, *Ichthyornis*, and virtually all modern birds. The loss of contact between the fibula and the calcaneum, and the fusion of the pretibial bone with the lateral condyle (calcaneum) are important avian features which I do not think have been reported in reptiles.

The function of the pretibial bone is to stabilize the mesotarsal joint. Without it this joint would have both a distal and a proximal kinetic surface and might act as a roller bearing (CHARIG, 1972). This is especially true for birds because the avian leg is strongly flexed. Many animals which may be at least occasionally bipedal have faced this problem and developed a lock for the upper joint. The basilisk lizard uses a process of the calcaneum for this purpose and dinosaurs have developed a variety of solutions. The ornithischian dinosaur *Hypsilophodon* has an ascending process of the astragalus (GALTON, 1974) fully comparable to that in many theropods, and the theropods themselves show a variety of ascending processes. In some forms *(Coelophysis)* the process extends into the distal end of the tibia rather than in front of it. In other taxa the process lies in front of the tibia as it does in birds but in no known dinosaur does it overlap the calcaneum, and even in those forms (struthiomimids) where the astragalus completely forms the functional joint, there is no fusion of the ascending process with the calcaneum. WELLES (1983) has recently described a specimen of *Dilophosaurus* that seems to have a suture between the ascending process and the astragalus. However, this form is not bird-like because the ascending process lies within rather than in front of the tibia. The ankle joint of *Archaeopteryx* is bird-like and differs from known reptiles in the same ways as do the ankle joints of other birds. It adds additional support to an avian classification for *Archaeopteryx*.

Relationship to other Birds

Archaeopteryx solves some morphological problems in a very different way from that found in modern birds. The quadrate articulation on the basicranium seems to be especially unusual. WHETSTONE (1983) argues that the primitive reptilian articulation of the squamosal with the quadrate (found in all modern birds) has been lost. The squamosal itself was either extremely small, lost, or if present is not preserved on any of the known specimens, and had a unique morphology. My own observations lead me to conclude that the squamosal could not have had the normal morphology of either a modern bird or a reptile, and that it is probably very reduced or lost. This would be a derived character state separating *Archaeopteryx* from modern birds and most other Mesozoic birds, but possibly not from *Gobipteryx* which also has a very peculiar quadrate morphology. The scapula of *Archaeopteryx* (Fig. 3) has a broad medially directed process anterior to the glenoid that braces against the axial skeleton. A similar scapular morphology is presently known only for *Gobipteryx* (ELZANOWSKI, 1981), *Alexornis* (MARTIN, 1983b), and the Enantiornithes (WALKER, 1981; MARTIN, 1983b). Because this scapular morphology is not known for any of the suggested reptilian out-groups, I consider it a derived character state. The pelvis of both *Archaeopteryx* and the Enantiornithes share a large dorsally directed ischial process that seems to be a derived character with no direct analog (MARTIN, 1983b) in modern birds. The tarsometatarsus of *Archaeopteryx* is interesting, as it lacks the large tarsal cap that is characteristic of the modern avian tarsometatarsus. The metatarsals are also not fused distally although this is the first region to fuse in modern bird ontogeny, the metatarsals are in a transverse row proximally (in modern birds the proximal end of the middle metatarsal is slightly posterior to the other two). All of these features suggest that the common ancestor of *Archaeopteryx* and modern birds (including the Mesozoic Hesperornithiformes and Ichthyornithiformes) did not have a fused tarsometatarsus. A tarsometatarsus of exactly the same basic form as that found in *Archaeopteryx* does occur throughout the Enantiornithes (see WALKER, 1981).

Conclusions

Archaeopteryx has some primitive traits that separate it from all other birds and give us a unique insight into a very early stage of avian evolution. It has many other features that are typical of birds and I think conclusively show that *Archaeopteryx* is a genuine bird in the sense that it is much closer to other birds than it is to any of the proposed reptilian out-groups. Many of the features that most strongly unite it with the Class Aves are related to flight.

While most workers have assumed that *Archaeopteryx* had some limited capabilities of powered flight, the extent of these capabilities has been hotly contested. OSTROM (1976) has presented a convincing argument that *Archaeopteryx* was a cursorial, terrestrial predator based upon similarities to cursorial dinosaurs, and upon the absence of avian specializations in the shoulder girdle. In modern birds, the presence of a large, keeled sternum appears to be essential for effective powered flight. *Archaeopteryx* appears to lack an ossified sternum since the one reported by DE BEER (1954) is actually some disassociated vertebrae. On this basis OSTROM questions its ability to fly and argues, instead, that the enlargement of the wing feathers might have been initially evolved as a net to capture insects. There are some obvious difficulties with this interpretation. For the feathers to be an effective "net", air would have to pass through them on the downstroke so that the prey would not be forced away form the predator's grasp. This would make any utilization of the wings for flight impossible. If the wings were used for gliding flight, the feathers would still result in a loss of ground speed and maneuverability, hindering both pursuit and escape from pursuit. I believe that *Archaeopteryx* is still best interpreted as an arboreal, flying animal and its anatomy still supports an arboreal origin for flight.

Many of the arguments against flight capabilities for *Archaeopteryx* have been dispelled by OLSON and FEDUCCIA (1979) who have shown that its large furcula or "wish-bone" (relatively the largest known in any bird) may have served most of the function of an ossified sternum. FEDUCCIA and TORDOFF (1979) have also provided the strongest argument for flight in *Archaeopteryx*, pointing out that the flight feathers were asymmetrical in the same way as in modern flying birds and that this configuration provides lift.

Archaeopteryx is not ancestral of any group of modern birds. It has specializations in its tarsometatarsus and skull which show conclusively that it is on a side branch of avian evolution. These characters include reduction or loss of the squamosal in the skull and the fusion of the metatarsals without the characteristic tarsal cap of modern birds. This means that it can no longer act as a primitive model for all birds, and that we must consider not only it, but other well-known Mesozoic birds such as *Hesperornis* and *Ichthyornis*, and Recent birds, in order to get a full picture of what the hypothetical "proavis" was like. *Archaeopteryx* is the earliest known member of a totally extinct group of birds including *Gobipteryx* (ELZANOWSKI, 1981) from the Upper Cretaceous of Mongolia; *Alexornis* (BRODKORB, 1976) from the Upper Cretaceous of Baja, California, and the Enantiornithes (WALKER, 1981) from the Upper Cretaceous of Argentina. These birds may be united by the presence of most of the following derived characters: 1) reduction of the proximal quadrate; 2) reduction or loss of the squamosal; 3) fusion of the anterior thoracic vertebrae with a special process on the scapula abutting against the axial skeleton; 4) tarsometatarsus fusing proximally but not distally; 5) metatarsal bones fused in a straight line; 6) distal tarsal bones either absent or fused as small individual bones (not forming a large tarsal cap). All of the known Mesozoic terrestrial birds belong to this group which may be characterized as the subclass Sauriurae (HAECKEL, 1866) which has formerly contained only *Archaeopteryx*. All other known Mesozoic and later birds fall into the subclass Ornithurae HAECKEL, 1866. I have divided the Ornithurae into two infraclasses: the Odontoholcae STEJNEGER, 1884 (new rank) for the hesperornithiform birds and the Neornithes GADOW, 1893 (new rank) for *Ichthyornis* and all later birds.

Acknowledgements

For allowing me to examine specimens, I thank J. P. Lehman, D. Goujet, F. Poplin, and D. E. Russell (Muséum National d'Histoire Naturelle, Paris); A. J. Charig, A. Milner, and C. A. Walker [British Museum (Natural History) London]; G. S. Cowles and C. J. O. Harrison (British Museum, Ornithological Department, Tring); A. D. Walker (University of Newcastle Upon Tyne); G. Viohl (Jura Museum, Eichstätt); H. Jaeger and H. Fischer (Humbold Museum für Naturkunde Berlin); P. Wellnhofer (Bayerische Staatssammlung, Munich); Z. Kielan-Jaworowska, A. Elzanowski, and H. Osmolska (Polaska Akademia Nauk, Warsaw); P. Ellenberger (Laboratoire de Paléontologie des Vertébrés, Montpellier); and J. H. Ostrom and M. Turner (Yale Peabody Museum, New Haven).

I have benefited from many stimulating conversations with A. J. Charig, J. Cracraft, M. Jenkinson, A. Fedducia, R. Mengel, P. Humphrey, C. Harrison, C. A. Walker, A. Milner, J. D. Stewart, L. M. Witmer and K. N. Whetstone. M. A. Klotz prepared the figures. L. M. Witmer and J. D. Stewart critically read the manuscript. Funding was provided by the University of Kansas (sabbatical leave) and University General Research Grant 3251–5038, NSF DEB 7821432 and National Geographic Grant 2228–80.

References Cited

de Beer, G. (1954): *Archaeopteryx lithographica*, a study based upon the British Museum specimen. – 68 pp. British Museum (Natural History) London.
Brodkorb, P (1976): Discovery of a Creataceous bird, apparently ancestral to the orders Coraciiformes and Piciformes (Aves: Carinatae). – Smithson. Contrib. Paleobiol., **27**: 67–73.
Charig A. J. (1972) The evolution of the archosaur pelvis and hindlimb: An explanation in functional terms. – Studies in Vertebrate Evolution (K. A. Joysey and T. S. Kemp, eds.), pp. 121–155; Oliver and Boyd, Edinburgh.
Elzanowski, A. (1981): Embryonic bird skeletons from the Late Cretaceous of Mongolia. – Palaeontol. Polon. **42**: 147–179.
Feduccia, A., and Tordoff, H. G. (1979): Feathers of *Archaeopteryx*: asymmetric vanes indicate aerodynamic function. – Science **203**: 1021.
Galton, P. M. (1974): The Ornithischian dinosaur *Hypsilophodon* from the Wealden of the Isle of Wight. – Bull. Br. Mus. Nat. Hist. **251**: 1–152.
Huxley, T. H. (1868): On the animals which are most nearly intermediate between the birds and reptiles. – Ann. Mag. Nat. Hist. Lond. **24**: 66–75.
Jollie, M. (1977): A contribution to the morphology and phylogeny of the Falconiformes – Part 4. – Evol. Theory, **2**: 1–141.
Lowe, P. R. (1933): On the relationships of the Struthiones to the dinosaurs and to the rest of the avian class, with special reference to the position of *Archaeopteryx*. – Ibis, **13**: 398–432.
Martin, L. D. (1983a): The origin of birds and of avian flight. – Current Ornithology, **1**: 105–129.
Martin, L. D. (1983b): The origin and early radiation of birds. – Perspectives in ornithology A. H. Brush and G. A. Clark, Jr. eds.) 291–338. New York (Cambridge University Press).
Martin, L. D., Stewart, J. D. and Whetstone, K. N. (1980): The origin of birds: structure of the tarsus and teeth. – Auk, **97**: 86–93.
Olson, S. L. and Feduccia, A. (1979): Flight capability and the pectoral girdle of *Archaeopteryx*. – Nature, **278**: 247–248.
Ostrom, J. H. (1976): *Archaeopteryx* and the origin of birds. – Biol. J. Linn. Soc. **8**: 91–182.
Ostrom, J. H. (1978): The osteology of *Compsognathus longipes* Wagner. – Zitteliana, **4**: 73–118.
Owen, R. (1863): On the *Archaeopteryx* of von Meyer, with a description of the fossil remains of a long-tailed species from the lithographic stone of Solenhofen. – Phil. Trans. R. Soc. **153**: 33–47.
Schlee, D. (1973): Harzkonservierte fossile Vogelfedern aus der untersten Kreide. – J. Ornithol. **114**: 207–219.
Thulborn, R. A. (1985): The avian relationships of *Archaeopteryx*, and the origin of birds. – Zool. J. Linn. Soc. (In Press).
Wagner, J. A. (1861): Über ein neues, angeblich mit Vogelfedern versehenes Reptil aus dem Solenhofener lithographischen Schiefer. – Sitz. Bayer. Akad. Wiss. **2**: 146–154.
Walker, C. A. (1981): New subclass of birds from the Cretaceous of South America. – Nature **292**: 51–53.
Welles, S. P. (1983): Two centers of ossification in a theropod astragalus. – Journ. Paleo. **57** 2: 401.
Whetstone, K. N. (1983): Braincase of Mesozoic birds: I. New preparation of the "London" *Archaeopteryx*. – Journ. Vert. Paleo. **2** 4: 439–452.

Author's address: Prof. Dr. Larry D. Martin, Department of Systematics and Ecology and Museum of Natural History, University of Kansas, Lawrence, Kansas 66045, USA.

Jacques Gauthier and Kevin Padian

Phylogenetic, Functional, and Aerodynamic Analyses of the Origin of Birds and their Flight

Abstract

The origin of birds must be examined in light of evidence concerning their phylogenetic relationships within Archosauria. Functional analyses can only be applied to evolutionary questions if they are hierarchically constructed according to inferred phylogenetic patterns; otherwise, there is no independent means to choose among possible evolutionary scenarios. Choice among alternative scenarios of the origin of flight should be based as far as possible on the concordance of independent patterns derived from phylogenetic, functional, and aerodynamic evidence. We begin with phylogeny.

Archosauria is a monophyletic taxon composed of the living crocodiles and birds, and of fossil taxa that share their most recent common ancestor. "Thecodonts" is a meaningless term in this context. Any extinct monophyletic taxon must either (1) have arisen before the origin of the common ancestor of crocodiles and birds; (2) be more closely related to crocodiles than to birds; or (3) be closer to birds. Several archosaur groups, including Parasuchia, Aetosauria, and Rauisuchia, are successively closer to crocodiles. Others, including *Euparkeria*, Ornithosuchidae, *Lagosuchus*, Pterosauria, and non-avian dinosaurs, are successively closer to birds. Within Dinosauria, the ornithischians, sauropodomorphs, and non-avian theropods are successively closer to birds. Theropoda includes a number of early Mesozoic forms and the sister-taxa Carnosauria and Coelurosauria. As redefined, Coelurosauria includes birds and those theropods that are closest to birds, such as Ornithomimidae and Deinonychosauria, and is no longer paraphyletic.

Thus, birds are members of the coelurosaur, theropod, saurischian dinosaurs. We reject alternate hypotheses of relationship: crocodiles and birds share few putative synapomorphies, which are deduced to be convergences; and "thecodont" ancestry is a red herring because it is based on unknown members of a paraphyletic collection of archosaurs that merely retain plesiomorphic features. The long-standing controversy surrounding the origin of birds is artificial: it has much less to do with evidence than with philosophy and methodology, particularly in treating paraphyletic groups as if they were real and in asserting convergence instead of demonstrating it.

Phylogenetic analysis shows that most characters traditionally considered "avian" were already present in non-avian coelurosaurs; therefore they did not evolve in the context of flight, but were later co-opted for flight. We compare the "arboreal" and "cursorial" theories of flight to the phylogenetic pattern and find that the former hypothesis explains very little of this pattern: most characters presumed to have evolved to advantage in an arboreal setting actually evolved in terrestrial, non-avian coelurosaurs. The "cursorial" hypothesis is consistent with all known evidence and accounts for the presence of features perfected in later birds.

Central problems in the origin of flight include the evolution of the flight stroke, the role of arboreality, and the role of gliding. The evidence suggests that (1) *Archaeopteryx*, like all coelurosaurs, was a cursorial biped, but unlike the others, its forelimbs were probably used very little for non-flight functions; (2) the skeletal modifications prerequisite to powered flight, including characteristic changes in the forelimbs that allow the functional equivalent of the down-and-forward flight stroke, arose within coelurosaurs before the origin of birds; (3) the principal structural and functional modifications enabling avian flight, including small size and flight feathers, were in place in *Archaeopteryx*, though not to the degree seen in modern birds; (4) scansorial and arboreal adaptations are not plesiomorphic characters of birds; and (5) although the possibility of arboreal or gliding stages in the origin of avian flight cannot logically be ruled out, such stages (and hypotheses about them) are less parsimonious in view of the available historical, functional, and aerodynamic evidence.

Zusammenfassung

Der Ursprung der Vögel muß im Lichte ihrer phylogenetischen Beziehungen innerhalb der Archosaurier untersucht werden. Funktionelle Analysen können auf Evolutionsfragen nur angewendet werden, wenn sie entsprechend der gefolgerten phylogenetischen Modelle hierarchisch aufgebaut sind. Andernfalls gibt es keine Wahl zwischen möglichen stammesgeschichtlichen Scenarios.

Die Wahl unter alternativen Scenarios der Entstehung des Fluges sollte so weit wie möglich auf das Zusammenwirken von unabhängigen Modellen begründet sein, die von phylogenetischen, funktionellen und aerodynamsichen Befunden abgeleitet sind. Wir beginnen mit der Phylogenie.

Die Archosauria sind ein monophyletisches Taxon, zusammengesetzt aus den lebenden Krokodilen und Vögeln und fossilen Taxa, die ihren jüngsten gemeinsamen Vorfahren teilen. "Thecodontier" ist in diesem Zusammenhang ein bedeutungsloser Begriff. Jedes ausgestorbene monophyletische Taxon muß entweder (1) vor dem Ursprung des gemeinsamen Ahnen von Krokodilen und Vögeln entstanden sein, (2) mit Krokodilen näher verwandt sein als mit Vögeln, oder (3) näher mit Vögeln verwandt sein. Mehrere Archosaurier-Gruppen, einschließlich der Parasuchia, Aetosauria und Rauisuchia, sind in dieser Reihenfolge näher den Krokodilen. Andere, einschließlich *Euparkeria*, Ornithosuchidae, *Lagosuchus*, Pterosauria und nicht-vogelartige Dinosaurier stehen zunehmend den Vögeln näher. Innerhalb der Dinosaurier sind die Ornithischier, Sauropodomorphen und die nicht-vogelartigen Theropoden zunehmend den Vögeln näher verwandt. Die Theropoda umfassen eine Reihe von frühmesozoischen Formen sowie die Schwester-Taxa Carnosauria und Coelurosauria. Nach unserer Neudefinition beinhalten die Coelurosauria die Vögel und jene Theropoden, die den Vögeln am nächsten stehen, wie die Ornithomimidae und die Deinonychosauria. Sie sind nicht mehr paraphyletisch.

Somit gehören die Vögel zu den coeluriden, theropoden Saurischiern. Andere Hypothesen zur Verwandtschaft weisen wir zurück: Krokodile und Vögel teilen wenige vermeintliche Synapomorphien, die auf Konvergenzen zurückgeführt werden. "Thecodonten"Abstammung ist ein gegenstandsloser Begriff, da sie auf unbekannte Vertreter einer paraphyletischen, nicht definierten Sammlung von Archosauriern begründet ist, die alle plesiomorphe Merkmale enthalten. Die lang dauernde Kontroverse um den Ursprung der Vöge ist künstlich. Sie hat viel weniger mit Beweisen zu tun als mit Philosophie und Methodologie, insbesondere in der Behandlung vermeintlich realer paraphyletischer Gruppen, und mehr im Behaupten als im Beweisen von Konvergenz.

Die phylogenetische Analyse zeigt, daß die meisten traditionell als "vogelartig" angesehenen Merkmale schon bei den nicht-vogelartigen Coelurosauriern vorhanden waren. Deshalb entstanden sie nicht im Zusammenhang mit dem Flug, sondern wurden später für den Flug verwendet. Wir vergleichen die "Arboreal"- und die "Cursorial"-Theorien des Fluges mit dem phylogenetischen Modell und finden, daß die erstere Hypothese sehr wenig dieses Modells erklärt: Die meisten Merkmale, von denen angenommen wurde, daß sie zum Vorteil in einer arboricolen Nische entstanden seien, entwickelten sich tatsächlich bei terrestrischen, nicht-vogelartigen Coelurosauriern. Die "Cursorial"-Hypothese steht in Einklang mit allen bekannten Tatsachen und erklärt das Vorhandensein von Merkmalen, die bei den späteren Vögeln vervollkommnet sind.

Zentrale Probleme der Entstehung des Fluges umfassen die Evolution des Flügelschlages, die Rolle des Baumlebens und die Rolle des Gleitfluges.

Die Befunde deuten folgendes an: 1.) *Archaeopteryx* war wie alle Coelurosaurier ein laufender Bipede, der aber im Gegensatz zu den anderen seine Vordergliedmaßen kaum zu anderen als Flugfunktionen eingesetzt hat. 2.) Die Skelettmodifikationen entstanden als Prärequisit des Kraftfluges innerhalb der Coelurosaurier vor dem Ursprung der Vögel (charakteristische Veränderungen in der Vorderextremität, die das funktionelle Äquivalent des Ab- und Vorwärts-Flügelschlages ermöglichen). 3.) Die prinzipiell strukturellen und funktionellen Modifikationen (einschließlich geringe Größe und Flugfedern) waren bei *Archaeopteryx* schon ausgebildet, obwohl noch nicht in dem Maße wie bei modernen Vögeln. 4.) Kletter- und Baumanpassungen sind keine plesiomorphen Merkmale der Vögel. 5.) Obwohl die Möglichkeit arboricoler oder Gleitflug-Stadien bei der Entstehung des Vogelfluges logischerweise nicht ausgeschlossen werden kann, entsprechen solche Stadien (und Hypothesen über sie) im Hinblick auf die verfügbaren historischen, funktionellen und aerodynamischen Befunde nicht dem Ökonomieprinzip.

I. The Origin of Birds

1. The Phylogenetic Method

Probably the greatest obstacle to the understanding of phylogenetic relationships has been treating paraphyletic groups (e.g., "Thecodontia") as if they were monophyletic. Evolution proceeds by the splitting of lineages, which are monophyletic. New monophyletic taxa are subdivisions of the older

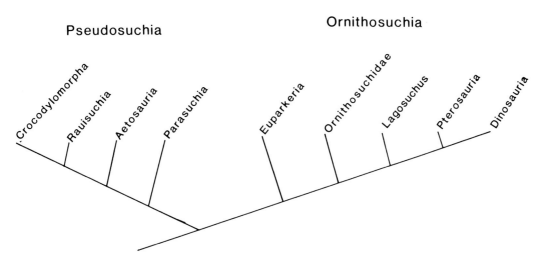

Fig. 1
Cladogram of the Archosauria, with its major divisions Pseudosuchia (**n. comb.**: crocodiles and archosaurs closer to crocodiles than to birds) and Ornithosuchia (**n. comb.**: birds and archosaurs closer to birds than to crocodiles). From GAUTHIER (1984).

monophyletic taxa, and the latter remain monophyletic only if the new taxa are included in them. Paraphyletic taxa have no evolutionary identity, they cannot be diagnosed, and they remain only as an admission of ignorance about actual evolutionary relationships. Therefore, we insist on the use of monophyletic groups in assessing the question of the origin of birds and their phylogenetic relationships.

In the question of bird origins, the use of paraphyletic groups is not only counterproductive: it is unnecessary. Archosauria, including crocodiles, birds, and all fossil taxa that share their most recent common ancestor, is monophyletic. It is united by at least 16 soft-part synapomorphies and 10 discernible from skeletal anatomy alone (GAUTHIER 1984). These include an antorbital fenestra set in a prominent fossa, ossification of the laterosphenoid, deeply arched otic notch, fully thecodont dentition, and so on. It is necessary to note that these features are basal to the Archosauria: they do not preclude the transmutation of these features in later Archosauria. Transformation states of these and other characters become synapomorphies at less inclusive levels within Archosauria, as evidenced in the following discussions. Archosauria as so constituted includes forms such as Parasuchia, Aetosauria, Rauisuchia, and Crocodylomorpha; *Euparkeria*, Ornithosuchidae, *Lagosuchus*, Pterosauria, and Dinosauria. (We will argue presently that the last is monophyletic only if it includes birds.) Other forms, including Proterosuchia and Erythrosuchia, are usually included in Archosauria (CHARIG 1976), but GAUTHIER (1984) preferred to use HUENE's Archosauromorpha for this more inclusive taxon, because the Proterosuchia and Erythrosuchia are outside the common ancestor of crocodiles and birds (Fig. 1).

For the reasons given above, we stress that only monophyletic taxa have been used in this analysis. "Thecodontia" has been a convenient term for archosaurs that are not crocodiles, pterosaurs, dinosaurs, or birds, but its use has obscured relationships more than clarified them. In phylogenetic analysis, this admittedly paraphyletic grouping is useless. Claims that birds descended from an "unknown thecodont", as favored by TARSITANO and HECHT (1980), MARTIN (1983), and several others, have no meaning apart from the statement that birds are archosaurs, which no one denies. But what are their interrelationships? Only phylogenetic analysis can address this question. No alternate method has been proposed that explicitly analyzes the sequences of evolutionary character acquisitions and branching events. We suggest that other authors use phylogenetic analysis to evaluate our results here, or to organize their own. If another kind of analysis is used, its methodology and testability should be made explicit.

The logical structure of cladograms must also be stressed. Previously, the only published cladistic analysis of the ancestry of birds was a preliminary one by PADIAN (1982). MARTIN (1983), in reviewing this and other phylogenetic hypotheses, merely listed some of the 50 synapomorphies proposed by PADIAN, ignoring the fact that cladograms and lists have very different logical structures. Collapsing a cladogram into a list destroys its logical hierarchy, its sequence of acquisition of synapomorphies, and its evolutionary implications. MARTIN'S (1983) analysis of character distributions among archosaurs is explicitly non-phylogenetic because (1) he asserts convergence without deducing it and (2) he rejects as synapomorphies characters that appear in any third group, however remote from the two at hand.

As Archosauria is diagnosed here, only three possibilities can express the membership of any monophyletic taxon within Archosauria. The taxon can be closer to birds than it is to crocodiles, it can be closer to crocodiles than to birds, or it can be outside the clade that includes the most recent common ancestor of crocodiles and birds (in which case it is not an archosaur as defined here).

2. Archosaurs closer to crocodiles than to birds

Phylogenetic analysis indicates that the monophyletic groups Parasuchia, Aetosauria, and Rauisuchia are successively closer to crocodiles. Members of the Crocodylomorpha are recognized by their unique ear region, elongate radiale and ulnare, and so on. Rauisuchia is joined to it by possession of parapophysis and apophysis of the axis on the odontoid process, anterolateral process on parasagittal osteoderms, enlarged pneumatic basipterygoid processes, and several other characters. The Aetosauria is joined to Crocodylomorpha + Rauisuchia by the "screw-joint" tibio-astragalar articulation, osteoderms on the ventral surface of the tail, a "fully developed" crocodile-normal tarsus (sensu BRINKMAN 1981 and CHATTERJEE 1982a, b), and several other synapomorphies. The Parasuchia joins these three taxa on the basis of the "crocodile-normal" crurotarsal joint with enlarged calcanear tubercle; the "peg" is on the astragalus and the "socket" on the calcaneum. This character and several others apply to the common ancestor of all the taxa mentioned above (GAUTHIER 1984).

3. Archosaurs closer to birds than to crocodiles

Phylogenetic analysis indicates that *Euparkeria*, Ornithosuchidae, *Lagosuchus*, and Pterosauria are successively closer to Dinosauria, which will be shown presently to include birds. Dinosauria, diagnosed below, shares 14 synapomorphies with Pterosauria. The scapula is elongated and inclined posterodorsally; the femur has a medially rotated head, a distinct neck, and a bowed shaft, and the distal end has a pronounced intercondylar groove and fibular condyle and a deep lateral condyle; the antitrochanter faces mostly ventrally, and the calcanear tubercle is lost. *Lagosuchus* joins this group on the basis of the "three-regionalized" vertebral column (BONAPARTE, 1975) with a more S-shaped neck, a posteroventrally facing glenoid, a moderately pronounced lateral condyle of the femur, reduced fibula and calcaneum, an ascending process of the astragalus, a double condyle formed by astragalus and calcaneum that articulates with three compressed distal tarsals (mesotarsal joint), and several other characters. Ornithosuchidae is united with this group because, like the other taxa listed above (except *Lagosuchus*, for which some characters are not certain), it has at least three sacral vertebrae, distal condyles of the first metacarpal conspicuously offset (so that the pollex is directed more medially), enlarged unguals, a markedly asymmetrical manus in which digits IV and V are reduced, and at least 9 other synapomorphies. *Euparkeria*, though it has no synapomorphies of its own as far as we can tell, is united with the four taxa just mentioned on the basis of its "crocodile-reversed" ankle, elongate ischium, at least six inclined cervical centra, and four other synapomorphies. (For details see GAUTHIER, 1984.)

4. "Pseudosuchia", "Flying Dutchmen", and a reorganization of Archosauria

Traditionally, "thecodonts" that were not parasuchians, proterosuchians, or erythrosuchians have been relegated to a group known as "Pseudosuchia". By its very definition, this group is explicitly paraphyletic because it is supposed to include the "primitive" archosaurs and the "unknown" ancestors of the "higher" archosaurs, such as dinosaurs and birds, but not the latter taxa themselves (HUENE 1956; ROMER 1956, 1966, 1972; CHARIG 1976; TARSITANO and HECHT 1980; MARTIN 1983; TARSITANO 1984). The paraphyly is ironic because ZITTEL (1890) originally coined the term for reception of what we now call Aetosauria, a monophyletic taxon. The corruption of the name in this century seems to be due mainly to HUENE (KREBS 1976), who expanded it to include some South American Triassic forms. In current usage, "Pseudosuchia" is a refuge for archosaurs of uncertain phylogenetic relationships that do not belong to any well-defined monophyletic taxon; and as its ranks swell, its identity becomes progressively murkier. Authors such as MARTIN, TARSITANO, and HECHT have treated these "Flying Dutchmen" as if they formed a natural taxon, and have even claimed "that the avian ancestor would not belong to the same group of pseudosuchians which were ancestral to any dinosaur taxon" (TARSITANO 1984); yet these authors have not been able to diagnose the "pseudosuchians" or document their phylogenetic relationships to each other or to other archosaurs.

We suggest instead, following GAUTHIER (1984), that the crocodiles and all archosaurs closer to crocodiles than to birds be called Pseudosuchia. As defined, Pseudosuchia is monophyletic and includes the taxon for which it was originally named (Aetosauria); and, inasmuch as the included extinct taxa are closer to crocodiles than to birds, they are truly "false crocodiles" or "would-be crocodiles".

Likewise, birds and all archosaurs that are closer to birds than to crocodiles may be named Ornithosuchia, because this monophyletic taxon includes not only birds, but the eponymous Ornithosuchidae. *Euparkeria*, Ornithosuchidae, *Lagosuchus*, and Pterosauria are successively closer to dinosaurs, of which some members are closer to birds. Birds, as HUXLEY (1868) first realized, and as OSTROM (1973, 1976a, etc.) fully documented, are not as closely related to crocodiles and other archosaurs as they are to dinosaurs, particularly theropods.

5. Dinosauria

The phylogeny of the Dinosauria is expressed in Fig. 2, based on GAUTHIER (1984). Inasmuch as non-theropod dinosaurs are not currently considered potential ancestors of birds (MARTIN 1983), and the monophyly of the Dinosauria is not currently in question, we will not detail all of the supporting synapomorphies here. Synapomorphies present in the common ancestor of Dinosauria include a reduced number of phalanges in the outer two digits of the manus (formula = 2–3–4–3–2); a semiperforate acetabulum and prominent supra-acetabular buttress (note that the pelvis is not fully perforate, contrary to MARTIN'S [1983] assertion); a fossa on the ventral margin of the postacetabular ilium for origin of the m. caudofemoralis brevis; a prominent anterior (= "lesser") trochanter on the femur; a prominent cnemial crest, projecting beyond the femoral condyles, and curving anterolaterally; a tibia in which the proximal end is broadened anteroposteriorly and the distal end is broadened transversely ("twisted" tibia), with a fossa in the distal end for reception of the ascending process; and several other synapomorphies.

The major monophyletic taxa in Dinosauria are Ornithischia and Saurischia (n. comb.); a few dinosaurs, such as *Herrarasaurus*, are outside the common ancestor of these two taxa. Saurischia includes Sauropodomorpha and Theropoda, of which the latter may be divided into Carnosauria, Coelurosauria, and several early Mesozoic forms that belong outside the common ancestor of Carnosauria + Coelurosauria (e. g. Ceratosauria, n. comb). As here defined (GAUTHIER 1984), Carnosauria includes

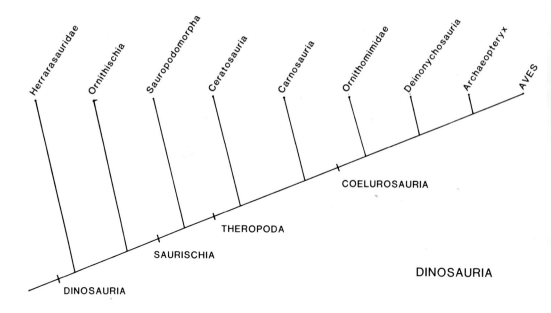

Fig. 2
Cladogram of the major groups of Dinosauria, with divisions of the Theropoda stressed. After GAUTHIER (1984).

mainly *Allosaurus* and Tyrannosauridae; these forms are very large, with large heads and short necks, keyhole-shaped orbits, enlarged teeth, shortened forelimbs with the third digit either shorter than the first or lost, and so on. Carnosauria + Coelurosauria have a maxillary fenestra (= 2nd antorbital fenestra), an expanded ectopterygoid with ventral fossa, entirely antorbital tooth rows, a hand at least half the length of the humerus plus radius, an obturator process on the ischium, an expanded pubic foot, a winglike anterior trochanter, a tall, broad ascending process of the astragalus, and the losses of an enlarged dentary tooth and digit IV of the hand, and a reduced metatarsal I, among other characters. This diagnosis removes theropods such as *Coelophysis* and *Syntarsus* from Coelurosauria, because they lack synapomorphies of Carnosauria + Coelurosauria as defined here. Likewise, *Dilophosaurus* and *Ceratosaurus* are excluded from Carnosauria. These four species are here included in Ceratosauria (n. comb. GAUTHIER, 1984).

Interrelationships of the Coelurosauria (n. comb.) require detail because this group includes the taxa most often implicated in avian ancestry. As here defined, Coelurosauria includes birds, Deinonychosauria, Ornithomimidae and related groups, and some less well known forms, such as *Coelurus, Ornitholestes*, and *Compsognathus*. Regardless of later modifications, such as the loss of teeth in some birds, the most recent common ancestor of these taxa possessed an extensive series of synapomorphies. Coelurosaurs have triangular frontals, a subsidiary fenestra between the pterygoid and palatine, a deeply excavated pocket on the ventral surface of the ectopterygoid flange, a fenestra pseudorotundum, cervical ribs fused to the centra, a straplike scapula, a ventrally elongated coracoid, furcula, fused sternal plates (at least in adults), a forelimb at least half the length of the hindlimb (and of the presacrum), a long hand with metacarpal I less than one-third the length of metacarpal II, the third phalanx of digit III equal to or longer than the first two phalanges combined, a weak or absent fourth trochanter, an enlarged ascending process that covers most of the anterodistal tibia, and a posterior location of metatarsal I on metatarsal II.

Within Coelurosauria, Deinonychosauria (dromaeosaurs + saurornithoidids) is the sister-group of *Archaeopteryx* and birds. Synapomorphies of this group include a prefrontal that is reduced or absent;

modified haemal arches and long prezygapophyses on the anterior caudals that extend nearly to the base of the tail; a subrectangular, elongate coracoid; a forelimb nearly 75% of the presacral length, and a hand equal to or longer than the foot; a semilunate carpal (distal carpals 1 + 2); a thin, bowed metacarpal III; posteroventrally curved dorsal margin of ilium; ilium elongated anteriorly so that the acetabulum is well in the posterior half; a pubic peduncle longer than the ischiadic, and posteroventrally directed, with the pubis directed posteroventrally; a pubic foot reduced anteriorly and curved posteriorly; an ischium half the length of the pubis (or less); an enlarged, distally placed obturator process; an anterior trochanter nearly confluent with the head of the femur; no fourth trochanter; and pedal digit IV longer than II and nearly as long as III. (For details see GAUTHIER, 1984.)

6. The ancestry of birds

The above review is prefatory to the central question: where do birds fit? *Archaeopteryx* has all the synapomorphies of dinosaurs listed above. It also has all those of Coelurosauria, though none of Carnosauria. Among coelurosaurs, the deinonychosaurs, by virtue of the suite of synapomorphies listed above, share a common ancestor with birds that excludes ornithomimids and all other forms. Birds, therefore, arose well within the Coelurosauria; OSTROM'S conclusions about the ancestry of birds are fully sustained by over 120 synapomorphies (GAUTHIER 1984; PADIAN 1982). "Thecodont" ancestry is inviable because "thecodonts" have no phylogenetic reality. The proposal of close crocodile-bird common ancestry has been based on only a few synapomorphies (MARTIN 1983), and these appear to be shared primitive features of archosaurs or homoplastic resemblances (PADIAN 1982; MARTIN 1983; GAUTHIER 1984).

Archaeopteryx is the earliest, most primitive taxon considered a bird, and it is generally regarded as the sister-group to all other birds. We can find no apomorphies of *Archaeopteryx*, based on known evidence, and so there is no obstacle to considering it "ancestral" to other birds, were that the objective. We do not find MARTIN'S (1983) putative apomorphies of *Archaeopteryx* convincing: it is primitive for theropods to have separate (unfused) distal tarsals, which fuse to form the metatarsal caps of birds (RAATH 1969; PADIAN, MS.), so this is not an apomorphy of *Archaeopteryx*; and despite MARTIN'S claim that *Archaeopteryx* has reduced or lost the squamosal, it appears to be present on the counterslab of the London specimen, on which MARTIN based his conclusion. We think this presumed synapomorphy is a misinterpretation, but even if MARTIN is right, it makes little difference to the relationships of *Archaeopteryx* to other birds and dinosaurs.

II. The Origin of Avian Flight

1. Methods of macroevolutionary analysis

Macroevolution is the study of events above the species level; it is in no way inconsistent with microevolution (population processes), but it has its own patterns and processes. These are discerned by comparing evolutionary patterns that reveal the historical sequence of events, and using other lines of evidence (functional morphology, biogeography, physicochemical laws, etc.) to arrive at the most robust working scenario that explains all the lines of evidence. The origin of flight is a major feature of adaptive evolution. As such, a theory of the evolution of flight should be concordant with the known facts derived independently from phylogeny, and with functional-morphological interpretations of the biology of the taxa involved. In approaching this problem, we accept only this line of reasoning and that the usual goal of science is to explain all the known evidence, while minimizing *ad hoc* hypotheses.

We will now proceed to build on the phylogenetic analysis that reveals the evolutionary sequence of characters usually associated with birds or with avian flight. Examination of the lists of synapomorphies given above for dinosaurs, theropods, coelurosaurs, and deinonychosaurs + birds shows that most "avian" characters traditionally cited (e. g. in textbooks) were already present in theropods before birds evolved. In many cases the avian form of these characters is only slightly modified over the non-avian form, and the evolution of many "avian" or "flight-related" features (such as hollow bones, a fused sternum, elongate coracoid, fused clavicles, long arms and hands, etc.) occurred in a completely different functional context. Therefore, many "avian" characteristics apply to a more inclusive functional context than flight alone. This leaves less to explain about the evolution of features traditionally presumed to have been advantageous only for birds, or for flying animals.

2. The central problem: evolution of the flight stroke

Most attention given to the origin of flight has focused on the ecological milieu in which flight evolved. But so far little attention has been given to the central problem of flight: the evolution of the flight stroke. A gliding membrane provides lift, but only a flapping stroke can provide thrust; furthermore, the animal cannot simply move its limbs up and down: it must have an airfoil designed to create a pressure differential over the upper and lower surfaces, and a flight stroke that actually propels the animal forward. This stroke, as observed in modern birds and bats, is down-and-forward, up-and-backward in medium speed flight (see RAYNER 1981 and references therein). No theory that does not account for the origin of this stroke can account satisfactorily for the origin of flight.

The phylogenetic sequence indicates that virtually all the apparatus necessary for this stroke was already present in non-avian relatives of *Archaeopteryx*. Deinonychosaurs and birds share a semilunate carpal, noted above (OSTROM 1969, 1974, 1976b), that allowed them to keep their long hands flexed against the lateral side of the forearm. When the deinonychosaurs extended the forelimb to seize prey, the humerus would be protracted outward and forward, the forelimb extended and protracted, and the hand swung forward (Fig. 3). These motions are only trivially different from the flight stroke of birds, and therefore the latter only requires explanation beyond the predatory motion already present in all coelurosaurs whose common ancestor had such long arms, long hands, and a semi-lunate carpal. We do not claim that the two functions are identical, only that most of the flight stroke motion was already possible for coelurosaurs. (This has yet to be demonstrated for any other Mesozoic archosaurs.) The flight stroke is characteristic of a phylogenetic subset of a group with raptorial forelimbs – not, as far as any evidence suggests, of climbing forelimbs. A "proto-bird" leaping into the air, following the model of CAPLE et al. (1983: see below), could have perfected this stroke little by little: it would be advantageous from the start, and more so when repeated. Even a small airfoil could provide the crucial difference in catching a fleeing insect with the jaws, or, more importantly, in escaping a larger predator. And the forelimbs, even when manipulated to no aerodynamic advantage, were still useful for other functions until powered flight took over completely.

3. "Arboreal" vs. "cursorial" theories: predictive differences

The two major competing theories of the origin of avian flight have been reviewed several times recently (e.g., OSTROM 1979, MARTIN 1983), which obviates basic discussion here. These theories ought to be amenable to "testing" against independent lines of evidence. They ought to have discernible, predictable differences that might be resolved by some empirical evidence. We believe that these differences exist, that they predict different consequences in design, and that at least some of these differences can be resolved by comparative biology, with the help of functional morphology and aerodynamics.

The "arboreal" theory. BOCK (1965, 1983), MARTIN (1983), and other proponents of the idea that avian flight began in trees face a difficult challenge: they must explain why so many features of modern birds, usually deemed to have evolved in the context of flight and arboreality, are present in non-flying coelurosaurian sister groups of birds; and they must grapple with the absence of any obvious arboreal adaptations in *Archaeopteryx*. Many such adaptations are recognized in modern birds (BOCK and MILLER 1959; FEDUCCIA 1973), yet not even the "reversed" hallux of *Archaeopteryx* was long enough to have

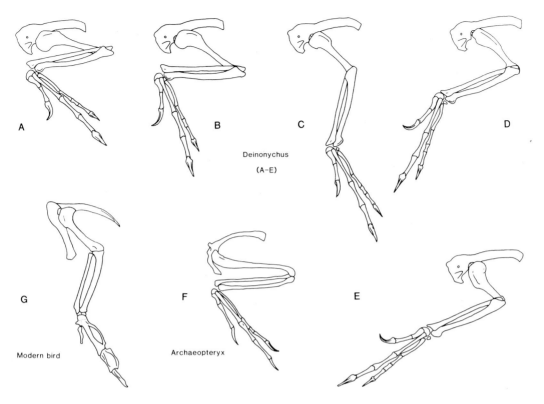

Fig. 3
Pectoral girdles and forelimbs of (A–E) the coelurosaurs *Deinonychus* (after OSTROM, in part), (F) *Archaeopteryx* (after HEILMANN), and (G) a modern bird. The sequence (A–E) reconstructs the predatory motion of *Deinonychus* as it may have reached out to grasp prey: the humerus is protracted, rotated, and abducted as the forearm and hand extend. Note the similarities of position A to F and of C to G. The flight stroke of birds requires only slight modifications from the (A–E) sequence illustrated.

been an effective perching device (OSTROM 1979). In fact, the earliest fossil birds generally lack arboreal specializations of the skeleton. Part of this lack certainly reflects the usual biases in the fossil record toward aquaphiles and away from silviphiles, and would merely be negative evidence were it not for the fact that non-arboreality is also a plesiomorphic characteristic of living birds. Arboreality is variably present or facultative in neognaths; neither ratites nor tinamous live in trees, and arboreality can hardly be said to be either common or primitive in shorebirds and seabirds, any more than in *Hesperornis* or crocodiles. Therefore, arboreal habits are not apparently plesiomorphic for birds (GAUTHIER et al., MS.).

Proponents of an arboreal stage in the evolution of flight also have the burden of explaining the benefit of the "proto-wings" in an arboreal setting. The usual explanation is that they slowed the fall when jumping between branches or from tree to tree (BOCK 1983). No one has calculated how large such feathers would have to have been before a significant effect was realized. Moreover, there are the forces and mechanisms involved in landing to consider, and a fully arboreal protobird would not have had the luxury of a running landing in a tree. Being bipedal, as all theropods are, would have complicated the issue: CAPLE et al. (1983) have shown the technical difficulty of controlling torque during arboreal landings. So far, proponents of the arboreal theory (1) have not addressed these problems and (2) have failed to show that their theory explains a n y of the observed evidence of morphology and phylogeny.

The "cursorial" theory. Phylogenetic analysis indicates that bird ancestors were terrestrial bipeds that had long arms and legs, prehensile hands, and hindlimb proportions and features adapted for quick progression. *Archaeopteryx* is recognized as a bird by its feathers which, according to most authors, initially evolved for use in thermoregulation (REGAL 1975; OSTROM 1979; BOCK 1983). OSTROM (1974) used the predatory context of avian ancestors to suggest that long feathers originally may have been predatory "insect nets" that later were co-opted for flight. Although this idea was not widely accepted by other authors, it stimulated investigation into the aerodynamic consequences of long feathers when used in running and leaping. CAPLE et al. (1983) showed that even a surface capable of lifting 1% of the animal's weight would have significant stabilizing effects, if positioned far enough away from the body wall (as they are in coelurosaurs). As the feather area increased, so would lift and stability, leading to longer time in the air, longer leaps, and improved flight stroke while in the air. Eventually the flight stroke would be perfected, and feathers would have filled in the inner arm as well. Flight would have evolved from the ground up, without a gliding stage, and without unbalancing the animal during predatory lunges. This scenario is consistent with what is known of the phylogenetic sequence and the associated functional morphology, as well as with ontogeny (see below).

The South American hoatzin *(Opistocomus cristatus)* has often been invoked by authors who argue for an arboreal origin of birds, because its juveniles clamber among the branches using unusually developed first and second digits of the hand. (They also dive and swim under water [HEILMANN 1927], but arboreally inclined authors generally do not choose to make much of this.) To do so, however, requires retarding development of the primaries on both sides of the first and second phalanges, so as not to interfere with grasping. The outermost primaries, which attach to the phalanges, are especially shortened for climbing, and some authors have argued that this was the primitive habit for birds. However, this evidence actually argues better against the arboreal theory than for it. In the juvenile hoatzin, the outer two primaries do not extend beyond the second digit, the next two are scarcely longer, and the remaining primaries gradually increase in length. In *Archaeopteryx* even the outermost primary is longer than digit II, the next is longer still, and those bracketing the distal joint extend far beyond the end of the digit, comparable to those of a pigeon (HEILMANN, 1927, Fig. 21). This is especially striking in view of the fact that *Archaeopteryx* had two more outermost primaries than the hoatzin has; therefore, even the shortest primaries of *Archaeopteryx*, which are longer than the shortest of modern birds, are not homologous to the latter. If feathers attached to the phalanges grew as quickly as do other flight feathers, the fingers could not have continued to grasp branches and trunks without damaging the feathers (HEILMANN 1927: 102). The pattern of its feathers and its long, gracile fingers indicate that *Archaeopteryx* was not an arboreal climber, and that the hoatzin is secondarily specialized, not primitive. It is worth noting that in ontogeny the downy feathers appear first, and the outer wing feathers (which generate thrust) form before the inner wing feathers (which contribute mostly to lift) (HEILMANN, 1927). If this pattern is taken at face value, it supports the idea that insulation was the first function of feathers, and is also consistent with the sequence of evolution of feathering envisioned by CAPLE et al. – though not at all with the arboreal theory, which still has to explain why, in a presumably arboreal ancestor, the hand did not flex mainly dorsoventrally, but mediolaterally; why the feathers develop from the outside in during ontogeny; and why there is no parallel arrangement of the phalangeal rows, such as one finds in the hands of primates that climb trees, and in the feet of bats.

Scenarios for the evolution of flight are complex arguments that consist of series of statements predicated on initial assumptions or observations. Some statements are based on independent lines of evidence; others rely on assumptions about the evolutionary process. These scenarios are too complex to be falsified, but their consequences can be clarified.

(1) If one accepts that a fully gliding stage (comparable to that seen in modern gliders) was necessary to avian flight, then arboreality (or an environment unusually replete with canyons and cliffs) must be postulated as well: it is not possible to sustain gliding flight from the ground up without special circumstances (gliding squirrels can do so in high winds).

(2) Conversely, if one accepts that an arboreal state was necessary, then a full gliding stage must be postulated as well, because experimentation with flight from arboreal heights is very risky.

(3) If one accepts that a fully gliding stage is unnecessary, then flight must begin from the ground up: an arboreal animal could hardly evolve flight directly in trees without the "safety net" of a gliding membrane (point 2).

(4) Conversely, if a terrestrial origin is accepted, gliding does not play a large role in the evolution of flight (point 1). However, some gliding is necessary even to CAPLE et al.'s model – at least in the earliest stages, when only lift and stability are provided by the proto-wings.

In the arboreal theory, then, some sort of gliding membrane must be postulated. Inasmuch as feathers are not attached to the hindlimbs, one can only conclude that they must have initially been capable of supporting the animal's glide. In that case they would have had to become as large as they are in *Archaeopteryx* before they could even be commensurate with the wing loading of birds (YALDEN 1971). What selective pressure caused flight feathers to grow so large in an arboreal setting?

A final problem with the arboreal theory: if birds passed through a gliding stage, why does *Archaeopteryx* (to say nothing of modern birds) show none of the typical characteristics of vertebrate gliders – whereas bats, which probably went through an arboreal, gliding stage, and which even today move poorly on the ground, show all the morphological correlates of gliders (PADIAN 1983)?

The terrestrial theory has a simpler and more conservative set of hypotheses. It builds on OSTROM'S predictive functional analysis of the forelimb (without the insect net function) and on the calculations of CAPLE et al. During a running takeoff, the bird builds up speed in order to reach that part on the U-shaped curve of speed vs. power required at which the power required is lowest: medium-speed flight. The wings begin to beat, especially the outer portion, which provides thrust, some of which is converted into lift. As speed increases, the thrust generated by the wings and legs is taken over by the wings; the bird is now moving too fast for its legs to keep up with it, and it is airborne. To land, the flight speed is slowed by a combination of increased flapping amplitude and increased angle of attack: the bird descends and slows enough to reach maximum running speed, and the legs touch the ground. Once on the ground, the bird uses both legs and wings to brake incrementally. This is the most economical way for a beginning flapper to operate because it puts the least power burden on the forelimb apparatus. It also coordinates the problems of landing with the ability to run, and makes experimentation with flight far less dangerous (as hang-gliding humans have discovered).

Taking off from a standing position and landing on one spot are both energetically and aerodynamically much more difficult. We agree that it is possible for an animal to launch itself from a branch, glide, and begin to flap. This argument may be plausible, but what facts evident in *Archaeopteryx* and its ancestry does it explain? The biology of *Archaeopteryx* indicates that (1) it was not especially arboreally adapted and (2) landing in trees would have been much more difficult for it than landing on the ground. By contrast, taking off and landing from the ground would have been a simple and standard part of its behavioral repertoire already. The gradual development of sustained flight – thrust – would merely have enhanced locomotory capabilities already present. From then on, as the cliché goes, the sky was the limit – and so were the trees.

4. Conclusion

Phylogenetic analysis is the basis for studying more complex macroevolutionary problems, and for erecting paradigms that account for adaptive evolution. We feel that the inclusion of phylogeny only makes paradigms stronger – first, by using the best evidence of what actually happened, and second, by restricting paradigmatic explanations only to changes that require explanation. Scenarios, such as the reconstruction of the evolution of flight in birds, are very complex ideas of which only their components (i.e., sub-hypotheses) may be explicitly testable. Many scenarios can be proposed for each evolutionary event, but only one can be correct. We cannot prove which is correct, but we can choose which is most robust: that which explains the most evidence, requires the fewest *ad hoc* assumptions, and is supported by logically independent lines of evidence. The closest known monophyletic sister group to *Archaeopteryx* and modern birds is the Deinonychosauria; all three of these taxa are members of the

Coelurosauria and share a unique suite of evolutionary novelties. Study of the order of acquisition of these synapomorphies helps to detail the origin of avian flight. Our analysis finds that the classic arboreal hypothesis, though consistent with evolutionary theory, is supported by no evidence; parts of it are even extremely unlikely. The cursorial hypothesis, as we have restated it here, is also consistent with evolutionary theory, but is far more conservative with respect to known facts, and explains far more of the observed evolutionary, functional, and aerodynamic patterns.

Acknowledgments

We thank KEN WARHEIT and TIM ROWE for many useful suggestions in the organization and preparation of this manuscript. To the organizers and hosts of the First International *Archaeopteryx* Conference, our deepest appreciation for the stimulating and hospitable environment of the conference; and to each participant, our gratitude for providing so many new ideas and for helping us to develop our own. Travel to the conference and preparation of the manuscript was facilitated by grants from the National Science Foundation, the Museum of Paleontology of the University of California, and the Committee on Research of the University of California at Berkeley. GAUTHIER extends special thanks to the California Academy of Sciences and NSF grant BSR-8304581 for their partial support of this work.

References cited

BELLAIRS, A. (1969): The Life of Reptiles. – 590 pp. London (Weidenfeld and Nicolson).

BOCK, W. J. (1965): The role of adaptive mechanisms in the origin of higher levels of organization. – Syst. Zool., **14**: 272–287; Lawrence (Kans.).

BOCK, W. J. (1983): On extended wings. – The Sciences, **23**, 2: 16–20; New York.

BOCK, W. J., and MILLER A. H. (1959): The scansorial foot of the woodpeckers, with comments on the evolution of perching and climbing feet in birds. – Amer. Mus. Novit., **1931**: 1–45; New York.

BONAPARTE, J. F. (1975): Nuevos materiales de *Lagosuchus talampeyensis* Romer (Thecodontia – Pseudosuchia) y su significado en el origen de los Saurischia. Chañarense inferior, Triasico medio de Argentina. – Acta Geol. Lilloana, **13**, 1: 5–90; Tucuman.

BRAMWELL, C. D. (1971): Flying ability of *Archaeopteryx*. – Nature, **231**: 128; London.

BRINKMAN, D. (1981): The origin of the crocodiloid tarsi and the interrelationships of thecodontian reptiles. – Breviora, **464**: 1–23; Cambridge (Mass.).

CAPLE, G. R., BALDA, R. T., and WILLIS, W. R. (1983): The physics of leaping animals and the evolution of pre-flight. – Amer. Nat., **121**: 455–467; Chicago.

CHARIG, A. (1976): Archosauria. Handbuch der Palaeoherpetologie, Teil 13: Thecodontia, pp. 1–6. Stuttgart (Gustav Fischer Verlag).

CHATTERJEE, S. (1982a): Phylogeny and classification of thecodontian reptiles. – Nature, **295**: 317–320; London.

CHATTERJEE, S. (1982b): Significance of ankle structure in archosaur phylogeny: Reply to Thulborn. (Nature, **299**: 657–658.

FEDUCCIA, A. (1973): Evolutionary trends in the Neotropical ovenbirds and woodhewers. – Orn. Monogr., **13**: 1–69.

FEDUCCIA, A., and TORDOFF, H. B. (1979): Feathers of *Archaeopteryx*: asymmetric vanes indicate aerodynamic function. – Science, **203**: 1021–1022; Washington (D.C.).

GAUTHIER, J. (1984): A Cladistic Analysis of the Higher Systematic Categories of the Diapsida. – Ph. D. Thesis, University of California, Berkeley. 565 pp. (# 85-12825, University Microfilms, Ann Arbor, MI, U.S.A.)

GAUTHIER, J., WARHEIT, K., and DEQUIEROZ, K. (MS.): The ontogeny of the avian tarsus and its implications for avian phylogeny: comments on McGowan's analysis.

HEILMANN, G. (1927): The Origin of Birds. – v + 210 pp. London (Witherby) and New York (D. Appleton Co.).

HUENE, F. von (1956): Palaeontologie und Phylogenie der niederen Tetrapoden. – 716 pp. Jena and Berlin (Gustav Fischer Verlag).

HUXLEY, T. H. (1868): On the animals which are most nearly intermediate between the birds and reptiles. – Ann. Mag. Nat. Hist., **2**, 4: 66–75; London.

KREBS, B. (1976): Pseudosuchia. Handbuch der Palaeoherpetologie, Teil 13: Thecodontia, pp. 40–98. Stuttgart (Gustav Fischer Verlag).

MARTIN, L. D. (1983): The origin of birds and of avian flight. – Current Ornithology, : 105–129; Lawrence, Kansas.

OSTROM, J. H. (1969): Osteology of *Deinonychus antirrhopus*, an unusual theropod from the Lower Cretaceous of Montana. – Bull. Yale Peabody Mus. Nat. Hist., **30**: 1–165; New Haven (Conn.).

OSTROM, J. H. (1973): The ancestry of birds. – Nature, **242**: 136; London.
OSTROM, J. H. (1974): *Archaeopteryx* and the origin of flight. – Quart. Rev. Biol., **49**: 27–47; Stony Brook (N.Y.).
OSTROM, J. H. (1976a): *Archaeopteryx* and the origin of birds. – Biol. J. Linn. Soc., **8**: 91–182; London.
OSTROM, J. H. (1976b): Some hypothetical anatomical stages in the evolution of avian flight. – Smiths. Contrib. Paleob., **27**: 1–21; Washington (D.C.).
OSTROM, J. H. (1979): Bird flight: how did it begin? – Amer. Scientist, **67**, 1: 46–56.
PADIAN, K. (1982): Macroevolution and the origin of major adaptations: vertebrate flight as a paradigm for the analysis of patterns. – Proc. Third N. Amer. Paleont. Conv., **2**: 387–392; Montreal.
PADIAN, K. (1983): A functional analysis of flying and walking in pterosaurs. – Paleobiology, **9**, 3: 218–239.
PADIAN, K. (MS.): On the type material of *Coelophysis* (Saurischia: Theropoda), and a new specimen from the Petrified Forest of Arizona (Upper Triassic: Chinle Formation).
RAATH, M. (1969): A new coelurosaurian dinosaur from the Forest Sandstone of Rhodesia. – Arnoldia, **4**: 1–25; Rhodesia.
REGAL, P. J. (1975): The evolutionary origin of feathers. – Quart. Rev. Biol., **50**: 35–66; Stony Brook (N.Y.).
RAYNER, J. M. V. (1981): Flight adaptations in vertebrates. pp. 137–182 In Day, M. H. (ed.): Vertebrate Locomotion; New York (Academic Press).
ROMER, A. S. (1956): Osteology of the Reptiles. – xxi + 772 pp. Chicago (University of Chicago Press).
ROMER, A. S. (1966): Vertebrate Paleontology (3rd ed.). – 468 pp. Chicago (University of Chicago Press).
ROMER, A. S. (1972): The Chañares (Argentina) Triassic reptile fauna XVI. Thecodont classification. – Breviora, **395**: 1–24; Cambridge (Mass.).
TARSITANO, S. (1984): Stance and gait in theropod dinosaurs. – Acta Palaeontologica Polonica, **28**, 1–2: 251–264; Warsaw.
TARSITANO, S. & HECHT, M. K. (1980): A reconsideration of the reptilian relationships of *Archaeopteryx*. – Zool. J. Linn. Soc., **69**, 2: 149–182; London and New York.
YALDEN, D. W. (1971): The flying ability of *Archaeopteryx*. – Ibis, **113**: 349–356.
ZITTEL, K. A. VON (1890): Handbuch der Palaeontologie. 1. Abt.: Palaeozoologie, 3. – xii + 900 pp., 719 figs. München and Leipzig (R. Oldenbourg).

Authors' addresses:
Dr. JACQUES GAUTHIER, Department of Herpetology, California Academy of Sciences, San Francisco, CA 94106. Present address: Museum of Zoology, University of Michigan, Ann Arbor, MI 48104, U.S.A.
DR. KEVIN PADIAN, Department of Paleontology, University of California, Berkeley, CA 94720, U.S.A. Correspondence should be addressed to Padian.

Walter J. Bock

The Arboreal Theory for the Origin of Birds

Abstract

Theories for the origin of birds, including the evolution of avian flight, fall under the heading of historical-narrative (H-NE) explanations in evolutionary biology. These must be based on pertinent nomological-deductive explanations (N-DE) if they are to be scientific. Most important is that H-NE must be presented in proper chronological order because past events affect all subsequent events. H-NE in evolutionary biology must be based on the known nomological causal explanations of evolutionary change as well as pertinent biological, physical and other N-DE. An application of these ideas to the analysis of avian features that must have preceded and which were essential for the evolution of flight in birds provides strong support for the arboreal theory of avian flight. According to this theory, flight originated from the heights down to the ground. The pull of gravity provides all, or at least part, of the force required for forward airborne movement in the early gliding stages preceding powered flight. Although it is not possible to provide conclusive disproof of the terrestrial theory of avian flight – from the ground up against the downward pull of gravity – the shortcomings of this theory appear to be fatal for it at the present time.

Zusammenfassung

Theorien über den Ursprung der Vögel unter Einschluß der Evolution des Vogelfluges fallen unter die Kategorie historisch-narrativer Erklärungen im Rahmen der Evolutionsbiologie. Historisch-narrative Erklärungen müssen auf zutreffende nomologisch-deduktive Erklärungen begründet werden, wenn sie wissenschaftlich von Belang sein sollen. Wichtig ist, daß historisch-narrative Erklärungen in der richtigen zeitlichen Ordnung angegeben werden, weil vorangegangene Gegebenheiten alle folgenden Abläufe mitbestimmen. Von historisch-narrativen Erklärungen der Evolutionsbiologie ist zu fordern, daß sie auf den bekannten nomologischen Kausalerklärungen des Evolutionswandels aufbauen und sich auf einschlägige biologische, physikalische und andere nomologisch-deduktive Erklärungen stützen. Die Anwendung dieses Prinzips auf die Analyse der Vogelmerkmale, die der Evolution des Fluges vorausgegangen sein müssen und die für dessen Entwicklung wesentlich waren, liefert starke Argumente für die Baumtheorie des Vogelfluges. Dieser Theorie zufolge, entstand das aktive Fliegen der Vögel aus einem Gleitfliegen von erhöhten Lagen. Dabei lieferte die Schwerkraft einen Teil der Kräfte, die für die vorwärtsgerichtete Bewegung der frühen Stadien des Gleitfluges, die dem aktiven Flug vorangingen, verantwortlich sind. Obgleich es nicht möglich ist, terrestrische Theorien des Vogelfluges absolut schlüssig zurückzuweisen, erscheinen die offensichtlichen Unzulänglichkeiten dieser Theorien, die eine Entwicklung des Fluges vom Boden aus, gegen den Zug der Schwerkraft, postulieren müssen, so groß, daß die ganze Vorstellung nicht haltbar ist.

Introduction

Analyses of the ancestry of *Archaeopteryx*, of its relationships to modern birds, and of the evolutionary history of the origin of avian features, including avian flight, are all historical-narrative explanations. Although systematists, paleontologists, and evolutionists have been dealing with historical-narrative explanations (hereafter H-NE) for over 100 years, little attention has been given to the nature of such explanations in science in general and in evolutionary biology in particular. H-NE are a perfectly valid form of explanation in science, probably constituting over 90% of all scientific work. And contrary to those biologists who label such attempts as "just-so stories", H-NE are respectable, and indeed represent the most difficult form of explanation in science. H-NE for the evolutionary origin of birds are dependent on a complex set of underlying nomological-deductive explanations (hereafter N-DE), including deductive causal agents for evolutionary change and deductive explanations in morphology, physiology, ecology, physics, etc. Moreover, H-NE for the evolution of birds must be as fully holistic as possible. It is not possible to discuss the origin of avian flight in the absence of discussion of the evolution of other avian features.

I would like to inquire into the theoretical bases of H-NE to show how they should be applied to the origin of birds and of avian flight and to argue that consideration of these ideas supports the arboreal theory of avian flight. By arboreal theory, I mean the origin of flight from an elevated position in trees down to the ground using the acceleration of gravity. By the terrestrial theory, I mean the origin of flight from the ground up into the air against the downward pull of gravity.

Explanations in Science

Explanations in science cover a wide range of theoretical statements characterized by the requirement of being testable against empirical observations. Explanations can be divided into several categories (e.g. NAGEL, 1961: 20—26, 1965). I will use a simple classification and group them into two classes – nomological-deductive (N-DE) and historical-narrative explanations (H-NE) – both of which are theoretical statements regardless of whether they are called theories, hypotheses, concepts, laws, etc., and must be tested against empirical observations. These two types of explanations differ sharply from each other and must be discussed separately.

Nomological-deductive explanations are in the form, given a set of facts (e.g., initial and boundary conditions) and a set of laws, both of which form the e x p l a n a s or explanatory sentence, a particular conclusion or the e x p l a n a n d u m follows (HEMPEL, 1965: 335—338). A N-DE answers the question: Why has a particular explanandum phenomenon occurred? N-DE apply to universals (indefinite sets of phenomena), do not depend on the past history of the objects or phenomena being explained, and their premises (the nomological statements) are always true. If the explanandum did not always result from the conjunction of the set of facts invoked (initial and boundary conditions) and the set of general laws, then the N-DE is not valid (it has been falsified) and one must search for the reason for the falsification. Perhaps one of the general laws on which the explanation depends is not correct, possibly the set of initial and boundary conditions has not been stated properly, or perhaps one of the empirical observations used for the test was in error. Examples of N-DE include NEWTON'S three laws of motion and the HARDY-WEINBERG law of genetic equilibrium.

Historical-narrative explanations provide an account of the existing attributes of a particular set of objects at any point in time; these explanations depend on the past history of these objects and must use pertinent N-DE. The objects being explained by a H-NE are not universals, but are particulars – a finite set having definite spatial-temporal positions. H-NE are given on an nondeductive basis with the hope of obtaining the most reasonable and probable explanation for the objects under study. Four aspects of H-NE must be stressed. First is that the explanation is always given on a probability basis of being correct (NAGEL, 1961:26) which results from the number and often conflicting N-DE that must be used and from uncertainty over the initial and boundary conditions. Second, is that any H-NE must be based on pertinent N-DE, and hence these N-DE must be part of the chain of arguments used in testing the H-NE. Third, is that N-DE are not general. A successful H-NE for one phenomenon, e.g., the origin of flight in bats, need not hold for a similar phenomenon, e.g., the origin of avian flight. And lastly, H-NE must be stated clearly and unambiguously because of their complexity, possible confusion between conflicting theories, and difficulty in identifying valid confirming and falsifying tests. It makes a great deal of difference whether arboreal means life only in trees or a combination of trees and on the ground, or whether terrestrial means upward movement from the ground against the pull of gravity or leaping downward along a slope. Unless H-NE are stated completely and unequivocally, it is not possible to test them and to appraise rival H-NE.

Although both N-DE and H-NE are scientific under the criterion of demarcation advocated by POPPER, they differ in many ways of how they are expressed, how they are tested and how they can be used to test other theoretical statements. Because most philosophical analyses apply to N-DE, I would like to restrict my remarks to H-NE (see BOCK, 1981; BOCK and CAPLAN, in prep.).

Being theoretical scientific statements, H-NE are available to tests by falsification, but such tests are often extremely difficult and frequently inconclusive. It is also possible to test H-NE by confirmation with the addition of more and more positive support; thus testing of H-NE is closely akin, if not identical, to induction in the strict sense. Objections to this should not exist because H-NE are theoretical statements about a finite number of objects in contrast to N-DE which cover universals. Testing of H-NE depend on argument chains involving pertinent N-DE and a large number of background assumptions (many about the initial and boundary conditions) and finally tested against empirical observations. One must proceed to the empirical observations as quickly as possible although the argument chain may often be complicated. The empirical observations and their roles as tests must be designated clearly.

H-NE cannot be tested using parsimony as a criterion although parsimony may be used to sort out H-NE for serious empirical testing. Frequently, two or more reasonable H-NE may be available for the same set of objects being explained. These conflicting explanations may not differ greatly in their degree of probability or plausability. Distinction between them and acceptance of one in preference to others cannot be made on simple grounds of parsimony. The only basis for choice of one over the others is by additional testing against empirical evidence, with the final decision generally made on the basis of the number and conviction of the positive (inductive) tests. Only rarely can the decision between opposing H-NE be made by falsifying one of the explanations. And, of course, H-NE can never be proven, no matter how convinced one may be of its correctness.

Statements about the history of life, about the evolution or phylogeny of any group of organisms, about the classification of any group, about the historical pattern of drifting continents, and the origin of the solar system are all H-NE (BOCK and VON WAHLERT, 1963; BOCK, 1978). The arguments by many philosophers and scientists that H-NE are not proper scientific explanations, but just story-telling, are based on an extremely narrow view that science consists only of N-DE, a position that I do not accept.

Historical-narrative explanations in evolutionary biology

Evolutionary biology abounds with all types of H-NE; discussion of the whole range of these explanations lies well beyond the scope of this paper. I will restrict my comments only to H-NE for the origin of new features and the origin of new groups – the area of macroevolutionary explanation. The synthetic model of macroevolutionary change (see BOCK, 1979) is completely reducible to the causal agents of microevolutionary change – that is the mechanisms of small scale phyletic evolution as studied by population geneticists and animal and plant breeders, and the mechanisms of speciation as discussed by MAYR (1963). No distinct causal agents of macroevolution, i.e., any N-D causal agents separate from those acting at the microevolutionary level, have been shown to exist. Macroevolutionary change is adaptive throughout in that selective agents act during the whole period of the major evolutionary event and drive the change. Most important is that the pattern of selective agents changes throughout the macroevolutionary episode and that preadaptation is extremely common and important. It is difficult to think of a major change in any organic feature that did not pass through one and usually several preadaptive stages. It is not valid to examine a major adaptive change using only the set of selective agents acting at the final stage as the basis for explaining the whole evolutionary event.

If major evolutionary changes are adaptive throughout, then one must be concerned with the nature of the selective agents arising from the external environment. If the selective agents change during a macroevolutionary episode, this means that the external environment is changing or at least that the relationships between the organisms and their environment are changing. Thus it is necessary to inquire closely into the possible environments with which the evolving organisms are interacting. Perhaps the most important external environmental source of selective agents are species interactions (BOCK, 1972). Identification of the selective agents and their environmental sources is most difficult when dealing with fossil organisms. Nevertheless, H-NE of evolutionary changes must be based on detailed analyses of functional and ecological properties, e.g., studies of functional and ecological morphology which are major ingredients of evolutionary morphology.

Major evolutionary changes probably occur rarely as a single, isolated phyletic lineage. Rather they almost always take place as series of radiations which involve, of course, speciations. Most likely rates of major evolutionary changes are correlated with rates of speciation in the evolving group. The importance of speciation is the production of species and hence the number and diversity of species interactions (BOCK, 1979: 36). The latter provide the major source of selective agents that drive the evolutionary change.

A critical part of any macroevolutionary explanation is that the changes must be arranged and discussed in the correct chronological sequence. Events and changes at one time set the stage for the next series of modifications. Individual steps in this series should be as small as possible, preferably at the magnitude of differences between congeneric species or less. The greater the gaps between successive steps in an evolutionary H-NE, the more difficult it is to provide a convincing argument.

Ever since DARWIN, the classic and still best approach in developing an evolutionary H-NE, such as for the origin of flight in birds, is to establish a pseudophylogeny using known analogous organisms and hypothetical forms. The organisms at each step must be viable and must be able to interact successfully with their presumed environments. This requirement is especially critical for hypothetical forms. Given the pseudophylogeny, one must be able to show that change can progress from one stage to the next using the known causal agents of evolutionary change, that is, one must be able to show the basis of adaptive change. It is desirable, but not essential to have known organisms as analogues for the stages in an evolutionary H-NE, i.e., in the pseudophylogeny. If a H-NE is proposed for which no known organisms exist at the critical stages, that deficiency would be a serious, but not fatal, argument against the explanation. This particular point becomes important when evaluating rival H-NE of which one has a number of analogous known organisms at critical stages and the other has none. However, this lack can never provide absolute reasons for discarding a particular H-NE explanation.

Evolutionary H-NE, as all H-NE in science, are never general. What may be a strongly substantiated H-NE for the origin of avian flight may be completely wrong for the origin of flight in other tetrapods. I must emphasize that the lack of generality applies only to H-NE and not to the underlying N-DE, e.g., the constraints imposed by physical laws. One of the reasons why no general H-NE exist in evolutionary biology is because of the accidental component of phyletic evolution (MAYR, 1962).

Lastly, the certainty of any H-NE depends very largely on the nature of the empirical evidence available to test the explanation. If the available evidence is scanty, it may be very difficult to distinguish between alternative H-NE, even those that differ very widely. For evolutionary H-NE, such as the evolution of birds and of avian flight, the tests depend very strongly on the completeness of the fossil record. In the case at hand, the only available empirical evidence comes from the single fossil taxon *Archaeopteryx* which is only one stage in a complex evolutionary history between the reptilian ancestors of birds and modern birds. The skeleton of *Archaeopteryx* is reasonably well known as is its possession of wing and tail feathers. But nothing is known about the rest of its anatomy, its physiology and other biological attributes (as is usual for fossils) and absolutely nothing is known about its ecology. Therefore it can be difficult or impossible to distinguish between alternative, reasonable H-NE. And because *Archaeopteryx* represents one stage in a long sequence of steps between reptiles and birds, it has only a minor role in the testing and evaluation of alternative theories.

Critical Features

A full discussion of the evolutionary origin of birds would include analysis of a large number of features (BOCK, 1965, in press); consideration of all of them would lie outside the scope of this paper. I would like to comment only on those which I believe to be most useful in an evaluation of the arboreal theory for the origin of avian flight. I must stress that proper analyses and evaluation of H-NE for the origin of flight in birds must be holistic and therefore include a number of characteristics considered in their proper chronological order.

Birds, including *Archaeopteryx*, possess a flight surface composed largely of elongated stiffened feathers. Disagreement exists as to whether feathers evolved as temperature regulating structures first and then as flight structures (REGAL, 1975, 1985) or the reverse (PARKES, 1976). I accept the REGAL analysis because it correlates better with the evolution of other features such as homoiothermy.

The evolution of homoiothermy is closely associated with the evolution of an insulating layer of the skin, which is the plumage of birds. Homoiothermy has many advantages in that the organism is constantly operating at high activity and is independent of external sources of heat. But it also has serious disadvantages which include a much higher energy budget and hence the need to obtain greater amounts of food at a relatively constant rate. Moreover, the evolution of homoiothermy required a major reconstruction of the heart, lungs and pulmonary circulation, and of a portion of the somatic circulation. Evolution of homoiothermy in birds must not be treated as a minor change, associated with trivial environmental shifts.

Homoiothermic birds and mammals face two conflicting problems. The first is staying warm during periods of inactivity. The second is losing heat sufficiently rapidly during times of high activity. Both problems are exasperated by the narrow range between normal operating temperature and upper lethal levels. Staying warm is solved, and especially in small species by the evolution of an insulating layer in the skin. This insulation becomes difficult to bypass when the animal is overheating and wishes to lose excess heat. Therefore, homoiothermy more likely evolved to maintain a uniform level of body temperature when the animal was inactive than when it was active.

Birds are basically daytime creatures in contrast to the funamentally nocturnal mammals. It is difficult to think of biological roles and selective agents associated with homoiothermy in a terrestrial, diurnal protobird, especially in view of the disadvantages of homoiothermy. On the other hand, homoiothermy can be of advantage to such animals if they were in trees. A comparison of microclimates of trees compared with the ground will show that the climate in trees is cooler because of increased wind, greater shade, and lack of heat reflection from the ground. It is easy to think of a number of advantages for protobirds in trees. They could climb trees to escape predators, to sleep in trees during the day or overnight, and to use trees as a reproductive site, e.g., a place to place the eggs in a nest away from predators. In the last case, successful development of the embryos would most likely depend upon the adults providing heat to the eggs. If these suggestions are reasonable, than the evolution of homoiothermy and of feathers in early birds would be associated with maintenance of a constant high body temperature when they were arboreal and inactive.

Life in trees would favor smaller size of the animal. Small size has several obvious advantages such as being supported by smaller branches, increasing the muscle/body weight ratio, reducing the energy expended for vertical climbing, and hitting the ground with a smaller impact force if the animal falls. Small size could also be associated with the need for evolution of homoiothermy and an insulating layer of feathers. And it would be important for the evolution of flight.

The environment of protobirds does not have to be a simple one. Certainly it does not have to be e i t h e r arboreal or terrestrial. The assumption by many workers interested in the origin of birds has been that arboreal life precludes activity on the ground, and vice versa, which is too simplistic. Several groups of modern birds feed on the ground, and roost and nest in trees as, for example, many gallinaceous birds. Or they feed on the water and nest on land. No objections exist to postulate a complex environment for protobirds which feed on the ground and roost and nest in trees.

If *Archaeopteryx* was arboreal for some part of its daily life, than the degree of specialization of its arboreal adaptations would depend on the exact nature and extent of its activity in trees. If trees were used only for a refuge and a nesting site, and not for feeding, then *Archaeopteryx* need not possess adaptations as specialized as those found in modern climbing birds such as woodpeckers. Moreover, if no other arboreal groups existed, then the adaptations in *Archaeopteryx* need not be as specialized because of the absence of selective agents arising from competition. It is invalid to insist that climbing adaptations, e.g., the structure of the claws, in *Archaeopteryx* must be as specialized as those found in modern climbing birds before considering it to be arboreal. Protobirds living partly in trees and partly on the ground in the absence of competition from other arboreal groups, need not possess climbing and other arboreal adaptations any more specialized than those found in modern birds such as the galliforms.

The ability for three-dimensional orientation and associated features, such as a well developed inner ear sense, would evolve in arboreal forms, but not in terrestrial species which live in a two-dimensional world. This is the contrast between dogs and cats, the latter having a good three-dimensional sense and can right themselves when falling.

Bipedal locomotion clearly evolved early in protobirds and in association with locomotion on the ground. It is most difficult to imagine the evolution of bipedal locomotion in an arboreal tetrapod. Moreover, it is difficult to envision the evolution of the avian wing and the feathered tail if protobirds were quadrapedal, strictly arboreal tetrapods. (See GUTMANN and PETERS, 1985, for a discussion of the relationship of the center of mass and of the center of lift in airborne tetrapods).

The reversed hallux characteristic of birds and present in *Archaeopteryx* cannot be explained with bipedal terrestrial locomotion. The reversed hallux permits the foot to function as a grasping structure. Grasping can have a number of biological roles. One is to hold onto branches firmly in an arboreal animal. Another is to hold food which is possible but would be difficult to argue.

Archaeopteryx still possesses three free digits in its hand, each of which is provided with a strong claw. Free digits terminating in claws are not new features, but characteristic of tetrapods at least as far back as the beginning of reptiles. However, since *Archaeopteryx* possesses a well developed wing and was certainly a good glider if not a weak active flier, the possession of claws on its manus is interesting. Because of the large flight feathers a sizeable drag would result if the hands were moved rapidly to catch prey; hence, it is difficult to envision a prey catching function of the grasping hand. Possibly *Archaeopteryx* held prey with its hands after capture, but this would be as difficult to show as would be a prey holding role of the foot. The free, clawed digits of the hand could have a role in grasping the trunk and branches when the animal is climbing trees (see YALDEN, 1985, for an analysis of the structure and possible roles of claws in *Archaeopteryx*). The flight feathers, being attached to the posterior edge of the forelimb, would not interfere with a climbing role of the grasping hand.

The arboreal theory for avian flight

The important difference between arboreal and terrestrial theories for the origin of avian flight is whether the protobird is able to make use of the force provided by the acceleration of gravity in the early stages of the evolution of flight or has to provide muscular force to combat the pull of gravity as would be necessary in the terrestrial theories. I would like to provide a reasonable sequence of chronological stages that could be expected in an arboreal H-NE for avian flight, and to show how the above features support the arboreal theory. Some of the proposed changes in the arboreal theory cannot be placed in a precise order because they could have taken place over a wider range of the macroevolutionary event.

The beginning stage would be a reasonably small reptile living on the ground, with either quadrupedal or bipedal locomotion. It would be facultatively warm-blooded, controlling its body temperature with the help of behavioral mechanisms using solar radiation as the external source of heat. It is not possible to judge whether the evolution of bipedal locomotion occurred before or after the start of arboreal life, nor does it matter.

The first important step was the invasion of trees which could be used for hiding, sleeping overnight or for nesting. Presumably the first roles for arboreal dwelling were for hiding and/or sleeping, with nesting being a more specialized role. The start of arboreal life would be associated with the evolution of homoiothermy and of feathers as insulation. This would permit the animal to maintain higher body temperatures when inactive in a cooler microclimate. All of the modifications in the circulatory and respiratory systems associated with homoiothermy as well as increase in the rate of feeding would evolve.

Trees need not be used for the whole life of protobirds which could and probably still searched for food on the ground. Increased food requirements would require more intense food searching behavior. Moreover, evolution of homoiothermy would facilitate active locomotion. Possibly bipedal locomotion evolved at this time with lengthening of the limb and modification of the pelvic girdle. Regardless of whether bipedalism evolved prior to or after the origin of arboreal life, it had to evolve prior to the beginnings of flight.

With the evolution of bipedal locomotion, the reversed hallux developed as a grasping foot. Protobirds could have climbed trees using all four limbs or using only the hind limbs. Many modern groups of birds are excellent bipedal climbers. Other arboreal specializations evolved early, such as three-dimensional orientation and specialization of the inner ear. Presumably the size of the animals became small, possibly in the range of 20 to 200 grams, if they had not been small already. Various locomotory adaptations such as greater ability to climb, to leap from the tree to the ground, to leap from branch to branch, to stabilize the body during arboreal locomotion and during leaps, to aim for the branch and grab it, etc., would develop as the animals became more specialized to arboreal life.

Arboreal animals sometimes fall out of trees, and therefore are faced with the problem of impact forces when hitting the ground. Features associated with falling, whether accidentally or purposefully, to the ground would evolve to decrease the rate of descent and/or lessen the impact. These would include proper orientation of the body, spreading out and flattening of the body. Increase in length and stiffening of feathers would increase body surface and improve parachuting ability, and could have been the first stage in the evolution of wing and tail feathers. Parachuting need not imply a strictly vertical drop, but could be at a steep angle. Indeed no sharp separation exists between parachuting behavior and gliding.

With evolution of adaptations for parachuting and steep glides, protobirds could depend more and more on these forms of locomotion rather than climbing to leave trees. Parachuting and/or gliding would be faster, require less energy and permit the animal to reach the ground further away from the tree, all of which could have selective advantages. Specialization of the glide would require a series of modifications, including increase in the length and stiffness of wing and tail feathers, need for stronger muscles to hold the forelimbs horizontally and development of glide control. These modifications would lead to the beginning of the enlargement of flight muscles, strengthening of the pectoral girdle and specializations within the forelimb. With specialization of the glide, including the ability for longer glides, would come increased development of control to change direction, to slow and to terminate the glide. Also important would be the development of stalling behavior to check forward speed when the animal wished to land. The analysis of many of these control mechanisms has been well presented by CAPLE et al. (1983). It is incorrect to think of a glide as an automatic action determined only by the initial conditions. All known tetrapod gliders exercise considerable control over the glide path (LULL, 1906). Many tetrapod gliders can execute complicated maneuvers during the flight, control (e.g., shorten) the length of the flight at will, and stall at landing. These actions in protobirds, presumably, would involve modifications in the shape of the wing (partly folding it), change in the curvature of the flight surfaces, and actual movement of the wing relative to the body (flapping). All of these control movements would lead to further development in the strength and specialization of the forelimb and flight muscles and in the pectoral bony apparatus.

RAYNER (1981, 1982, 1985a, 1985b) and NORBERG (1985) have shown that gliding tetrapods the size of *Archaeopteryx* could start to flap and thereby acquire forward thrust without any flight disadvantages as claimed by advocates of terrestrial theories (e.g. CAPLE et al., 1983, 1984; BALDA et al., 1985). Further they have shown that the origin of flapping flight would be energetically easiest if the animal was already moving with a reasonable flight speed as in a glide, and would be most costly, if not impossible, for a hovering flight. Although suitable flight speed may be theoretically reachable if the animal ran at a very high speed before leaping into the air, the needed running speed and the distance that would be gained in the resulting flight path do not make sense ecologically, energetically and selectively.

It is easily possible to envision a whole set of selective agents acting on the evolving flight apparatus and a whole series of preadaptive levels bridging the gap between the stage of gliding and that of active flapping flight. It is not necessary, and probably wrong, to analyze the evolution of active flapping avian flight from the gliding stage using selective agents associated only with fully developed active flight. Advantages for improving gliding and for the origin of flapping flight include flattening the slope and hence increasing the length of the flight path, and increasing maneuverability during flight, including ending the flight at will. PENNYCUICK (in press) presented a detailed analysis of the evolution of gliding and flapping flight, including pointing out the basis of selective advantages. Suffice it to say that flight, either gliding or powered, is energetically, the cheapest form of tetrapod locomotion. It is doubtful that flight, either gliding or early stages of powered, flapping avian flight, were associated directly with feeding. Most likely it was used to reach feeding areas, escape from predators, and otherwise cover long distances rapidly and efficiently. Most likely powered flight was used for feeding only after it had become well developed. Most modern birds do not hunt for food on the wing, and use flight only for a restricted set of daily activities.

In this H-NE of avian flight, the gaps between successive stages are relatively small, selective advantages can be proposed for all changes, and known organisms can be pointed to as analogues for critical stages. A number of features appear to be prerequisites for powered, flapping avian flight. These include feathers, homoiothermy, three-dimensional orientation, smaller body size, bipedal locomotion, and some development of the muscle-bone system of the whole pectoral apparatus. Most of these features can be shown to have with reasonable probability evolved in connection with arboreal life. Moreover, there are a number of additional avian features which point to life in trees as being basic in the evolution of birds, but it is not possible to show with any certainty that these must have evolved prior to the origin of powered flight.

References Cited

BALDA, R. P., CAPLE, G. & WILLIS, W. R. (1985): Comparison of the gliding to flapping sequence with the flapping to gliding sequence. This volume.
BOCK, W. J. (1965): The role of adaptive mechanisms in the origin of higher levels of organization. – Syst. Zool. **14**: 272–287.
BOCK, W. J. (1972): Species interactions and macroevolution. In: Evolutionary Biology (T. DOBSZHANDKY, M. K. HECHT, and W. C. STEERE, eds.), **5**: 1–27; New York (Appleton-Century-Crofts).
BOCK, W. J. (1978): Comments on classifications as historical narratives. – Syst. Zool., **27**: 362–364.
BOCK, W. J. (1979): The synthetic explanation of macroevolutionary change – a reductionistic approach. – Bull. Carnegie Mus. Natl. Hist., **13**: 20–69.
BOCK, W. J. (1981): Functional-adaptive analysis in evolutionary classification. – Amer. Zool. **21**: 5–20.
BOCK, W. J. (in press): The arboreal theory of the origin of flight in birds. – San Francisco (Calif. Acad. Sci.).
BOCK, W. J. & CAPLAN, A. (in prep): An analysis of explanations in evolutionary biology.
BOCK, W. J. & VON WAHLERT, G. (1963): Two evolutionary theories – a discussion. – Brit. Journ. Phil. Sci., **14**: 140–146.
CAPLE, G., BALDA, R. P. & WILLIS, W. R. (1983): The physics of leaping animals and the evolution of flight. – Amer. Nat. **121**: 455–476.
CAPLE, G, BALDA, R. P. & WILLIS, W. R. (1984): Flap about flight., Anim. Kingdom, **87**: 33–38.
HEMPLE, C. G. (1965): Aspects of scientific explanation. – New York (Free Press).
LULL, R. S. (1906): Volant adaptation in vertebrates. – Amer. Nat., **40**: 537–566.
NAGEL, E. (1961): The structure of science. – New York (Harcourt, Brace and World).
NAGEL, E. (1965): Types of causal explanation in science. – In: Cause and Effect (D. LERNER, ed), pp. 11–26. New York (The Free Press).
NORBERG, U. M. (1985): Evolution of flight in birds: aerodynamic, mechanical and ecological aspects. – This volume
PARKES, K. C. (1966): Speculations on the origin of feathers. – The Living Bird, **5**: 77–86.
PENNYCUICK, C. (in press): Mechanical constraints on the evolution of flight. – San Francisco (Calif. Acad. Sci.).
PETERS D. S. & GUTMANN, W. FR. (1985): Constructional and functional preconditions for the transition to powered flight in vertebrates. – This volume.
RAYNER, J. M. V. (1981): Flight adaptations in vertebrates. – Symp. Zool. Soc. London, **48**: 137–172.
RAYNER, J. M. V. (1982): Avian flight energetics. – Ann. Rev. Physiol., **44**: 109–119.
RAYNER, J. M. V. (1985a): Vertebrate flapping flight mechanics and aerodynamics, and the evolution of flight in bats. – In: Biona Report **5**, Fledermausflug, ed. W. NACHTIGALL (Gustav Fischer).

Rayner, J. M. V. (1985b): Mechanical and ecological constraints on flight evolution. – This volume.
Regal, P. J. (1975): The evolutionary origin of feathers. – Quart. Rev. Biol., **50**: 35–66.
Regal, P. J. (1985): Common sense and reconstructions of the biology of fossils: *Archaeopteryx* and feathers. – This volume.
Yalden, D. W. (1985): Forelimb function in *Archaeopteryx*. – This volume.

Author's address: Prof. Dr. Walter J. Bock, Department of Biological Sciences, Columbia University, New York, NY 10027 U.S.A.

R. E. Molnar

Alternatives to *Archaeopteryx*: A Survey of Proposed Early or Ancestral Birds

Abstract

Alternative or additional ancestral birds to *Archaeopteryx* have been proposed but never widely accepted, or surveyed. Few are clearly related to birds, although several bear on the origin of birds, suggesting that "avian" character states arose before birds themselves. Several states considered characteristic of birds are also found in theropods, including some previously believed to be common only to birds and crocodilians.

Zusammenfassung

Neben *Archaeopteryx* hat man zwar weitere oder alternative Vogelvorfahren vorgeschlagen; sie wurden aber niemals in weiterem Umfang akzeptiert oder untersucht. Obwohl einige im Zusammenhang mit der Entstehung der Vögel stehen, sind nur wenige eindeutig mit Vögeln verwandt, was darauf hindeutet, daß "Vogel"-Merkmale vor den Vögeln selbst entstanden sind.

Einige Merkmale, die für vogelartig gehalten werden, trifft man auch bei Theropoden an, einschließlich solcher, von denen man früher glaubte, sie seien nur Vögeln und Krokodilen gemeinsam.

Introduction

The problem of the origin and ancestors of birds is one that has attracted attention from both the scientific and the lay communities. Most of this attention has centred on *Archaeopteryx* because of its hitherto unique combination of reptilian and avian character states. The origin of birds, unlike that of mammals or reptiles, is linked to one specific genus. However, even in the late nineteenth century, other early or ancestral birds have been proposed none receiving as much study as *Archaeopteryx*. A recent survey (MOLNAR & ARCHER 1984), based on acceptance of published descriptions at face value, suggested that some of these forms deserved more attention. This contribution has resulted from a critical appraisal of the literature.

References cited after taxonomic names herein do not necessarily refer to the original description, but rather to that paper which documents the feature under consideration.

Proposed ancestral birds and ancestors of birds

Considered here are those forms suggested as very primitive birds, other than *Archaeopteryx*. Many of these forms are incompletely described, and all could use both more thorough study and description.

Laopteryx was described by MARSH (1881), and restudied briefly by SIMPSON (reported in LAMBRECHT 1933). The specimen consists of an incomplete cranium, together with an associated tooth that has apparently been lost (LAMBRECHT 1933), found in the Morrison Fm. at Como Bluff, Wyoming. Nothing significant can be said of *Laopteryx* save that it needs further study.

Palaeopteryx has been named, but not yet formally described, by JENSEN (1981a, 1981b). It is based on a tibiotarsus from the Morrison Fm. of southwestern Colorado. While it is certainly premature to conclude anything about it prior to complete description, it should be remembered that tibiotarsi are known in at least two theropods: *Avimimus* (KURZANOV 1981) and *Syntarsus* (RAATH 1969). Furthermore the proximal head of this tibiotarsus seems to bear a detailed resemblance to those of *Avimimus* and *Tugulusaurus* (DONG 1973).

Praeornis is based on a set of feathers or feather-like structures found in the Upper Jurassic lake deposits of the Karatau Range in Soviet Central Asia (RAUTIAN 1978). These structures are simpler than the feathers of *Archaeopteryx*, for while they possess both central rhachis and barbs, they lack barbules. They thus exhibit a possible stage in the development of feathers, and if feathers define birds, *Praeornis* may be a bird more primitive than *Archaeopteryx*.

Dry Mesa femur 1 is the designation used here for a femur mentioned by JENSEN (1981b) and attributed to *Archaeopteryx*. This femur derives from the Morrison Fm. of southwestern Colorado. From the published information, this femur would seem to be either avian or theropod.

Dry Mesa femur 2 is the designation here used for a second femur from the same locality attributed (presumably) to *Palaeopteryx*. This cannot be discussed on the basis of currently published information alone.

The Dry Mesa sacrum is again from the same locality and horizon, and has been mentioned by JENSEN (1981a, 1981b). It consists of at least 5 vertebrae including a possible dorsosacral. It is not a synsacrum which consists of sacrals, together with caudals, "lumbars" and some dorsals. Fused sacrals are found in theropods: *Saurornithoides junior* has five sacrals fused together with one caudal (BARSBOLD 1971). Thus there is no compelling reason to consider the Dry Mesa sacrum avian, but neither does it resemble any known theropod sacrum. Presumably it is one or the other, and thus indicates the existence of virtually unknown birds or theropods.

The carnavians are a group established by ELLENBERGER (1974) on small footprints from the Lower Jurassic Stormberg of Lesotho. Some of these tracks *(Masitisisauropus palmipes* and *Ralikhomopus aviator)* are described by ELLENBERGER (1974) as bearing the impressions of plumes or plume-like structures. Only the first of these taxa is illustrated, and although multiple impressions of *M. palmipes* are reported, only one is illustrated. The plume-like impressions are not designated on the photographs, but are presumably the triple lineations. There is no suggestion from the depth of the impressions that these lineations were made by structures actually attached to the digits, as the manual and pedal impressions are substantially impressed into the substrate, but the lineations seem surface features. Thus they are not impressed into the substrate as would be expected if they were in fact attached to the digits (ELLENBERGER 1974 Pl. 28). There seems no compelling evidence for plume-like structures in carnavians: the lineations may simply be invertebrate trails trod over by the *M.* track-maker.

Cosesaurus is based on a virtually complete skeleton from the Muschelkalk of Spain (ELLENBERGER & DE VILLALTA 1974). It was described as an avian ancestor, but only a preliminary description has yet appeared. It was later referred to the prolacertoid eosuchians by Olsen (1979). ELLENBERGER & DE VILLALTA described several avian characteristics: general cranial form, with relatively large orbits and beak-like rostrum, hollow limb bones and relatively large forelimb. As they themselves note the skull seems similar to that of a passeriform, not to be expected in an ancestor of *Archaeopteryx*. The relatively large orbits and beak-like rostrum may be juvenile features. However in relative size of orbits, form of "beak" and size and form of teeth, *Cosesaurus* resembles *Megalancosaurus* (CALZAVARA et al. 1980). *Cosesaurus* has eight cervicals, as opposed to fourteen or fifteen usual for modern birds. It has a pentadactyl (plesiomorphic) manus and pes. The forelimb is relatively long as might be expected for an avian ancestor, although the metapodials are not enlarged as expected from the considerations set forth by CAPLE et al. 1983. The pelvis is not clearly illustrated. In the absence of an elongate manus, or other avian-like character states, there seems no good reason to suspect this form of a position in the ancestry of the birds.

Cosesaurus does apparently possess an antorbital fenestra, an archosaur autapomorphy. OLSEN (1979) suggests that it is an eosuchian, ignoring the antorbital fenestra, and stating it shows the pedal characters, vertebral formula and pectoral girdle characters of tanystropheids, even implying that the large heterotopic bones near the proximal caudals may be present (OLSEN 1979: 6). These large heterotopic bones have escaped the notice not only of ELLENBERGER & DE VILLALTA but also of the illustrator's camera. OLSEN himself admits that the short, hooked fifth metatarsal evolved at least four times independently among reptiles. None of the tanystropheid pectoral character states, nor the elongate proximal phalanx of pedal digit V, are described by ELLENBERGER & DE VILLALTA nor are apparent in the photographs. The given vertebral count for *Cosesaurus* is 8 cervicals, 15 to 16 dorsals, 2 to 3 sacrals, and 28 to 30 caudals, while OLSEN gives that for tanystropheids as 12, 13, 2, to 46 respectively. There are almost the same number of presacrals in both, but OLSEN has yet to show that the cervical and dorsal counts for *Cosesaurus* are incorrect. For these reasons the assignment of *Cosesaurus* to the eosuchians must be rejected until further evidence is presented.

Proposed groups ancestral to birds

Crocodilians

For much of the twentieth century it was accepted that birds had evolved, like dinosaurs and crocodilians, from thecodonts. The work of OSTROM during the 1970's can fairly be said to have shifted the majority from this view to that that birds evolved from theropod dinosaurs. Another idea however, was also proposed: that birds evolved from, or were the sister-group of, the crocodilians.

Evidence in favour of a special relationship between birds and crocodilians has been gathered in WALKER (1972), WHETSTONE & MARTIN (1979, 1981) and WHETSTONE & WHYBROW (1983) among others. Here only those features determinable in fossils (unlike heart structure or formation and early growth of infrapolar cartilages) and rigorously stated (unlike "characteristic build of otic capsule") are given.

The major features given as derived states shared by both birds and crocodiles are the following: 1 – a squamosal shelf (WALKER 1972, WHETSTONE & MARTIN 1981); 2 – palatine elongate (WALKER 1972); 3 – a choanal trough on the palatine (WALKER 1972); 4 – a "bipartite" quadrate articulation with the braincase, the quadrate contacting the laterosphenoid, squamosal, prootic, and otoccipital (WALKER 1972, WHETSTONE & MARTIN 1979 & 1981, WHETSTONE & WHYBROW 1983); 5 – an elongate, tubular cochlear recess (WALKER 1972); 6 – paroccital processes projecting well posterior to the quadrates to form the posterior walls of the tympanic cavities (WALKER 1972); 7 – a complex series of air spaces connecting the middle ear and oral cavities (WALKER 1972); 8 – the course of the internal carotids through two of these cavities (WALKER 1972); 9 – an antrum pneumaticum dorsale located between the superoccipital, otoccipital, squamosal and parietal (WHETSTONE & MARTIN 1979, WHETSTONE & WHYBROW 1983); 10 – an antrum pneumaticum rostrale located just anteriad to the foramen for cranial nerve VII and just posteriad to that for V (WHETSTONE & WHYBROW 1983); 11 – quadrate pneumatic with medial foramen (WHETSTONE & MARTIN 1981, WHETSTONE & WHYBROW 1983); 12 – a sinus chamber in the base of the paroccipital process (WHETSTONE & MARTIN 1979); 13 – a foramen aerosum in the lower jaw (WHETSTONE & MARTIN 1981); 14 – bony tooth roots with an enclosed resorption pit (WHETSTONE & MARTIN 1981); 15 – a functional linkage of the elbow and wrist joints (WALKER 1972).

Many of these features also appear in theropods. This of course renders them unsuitable to demonstrate any special relationship of birds to crocodiles. Taking these in order: 1 – Not reported in theropods; 2 – Generally present in theropods (compare for example those of *Sphenosuchus* WALKER 1972 Fig. 3a with those of *Allosaurus* MADSEN 1976 Pl. 2B); 3 – Not reported in thereopods. 4 – Not present in theropods in this form. The quadrate of *Tyrannosaurus rex* (LACM 23844), the only *T.* quadrate in which the proximal articular surface may be examined does not show a simple convex condyle. It has instead a roughly saddle-shaped surface concave from anteromedial to posterolateral and convex from anterolateral to posteromedial. The posterolateral portion is highest. This does not show, but does suggest, an approach to the bipartite head of the quadrate. The significance of this is not clear in this context, however, as WALKER (pers. comm. 1984) has retracted his claim of a bipartite quadrate head in *Archaeopteryx* and in birds in general; 5 – Not reported in theropods, although present in ornithopods (LANGSTON 1960). This suggests that this state is not clearly a synapomorphy of birds and crocodiles, but needs more work; 6 – Not reported in theropods; 7 – Present in some theropods, and described in the next section; 8 – The canal for the internal carotids broadly communicates with at least one sinus chamber in Tyrannosaurs. Also discussed in the next section; 9 and 10 – Not reported in theropods; 11 – Not reported in theropods. Nevertheless it is found in *Tyrannosaurus* and *Labocania*, as discussed in the next section; 12 – Not reported in theropods, but again present in *Tyrannosaurus*, as reported in the next section; 13 – Not reported in theropods. *Tyrannosaurus rex* does possess an internal sinus chamber and medial pneumatic foramen in the articular; 14 – Not reported in theropods; 15 – Present in *Avimimus* (KURZANOV 1983).

Of the fifteen character states alleged to not occur among theropods six either have been reported in the literature (2, 8, 15), or do occur although they have not yet been reported in the literature (7, 11, 12). A seventh (5) has not been reported among theropods but does occur among ornithopods. Of the remaining eight, one (13) is obscure and may occur among theropods and one (4) is approached among theropods, but may not occur among all birds. This leaves less than half of the states. Since an unfortunate number of these states is found among some theropods but is not reported in the literature, we must conclude that the other states cannot be presumed absent among theropods without a thorough study of available material. We must also conclude that theropods need more work.

Sinus chambers in theropods

The system of sinus chambers in the skulls of crocodilans has played a large role in relating crocodilans to birds. Therefore it is necessary to set forth the anatomy of sinus chambers in theropods in some detail to demonstrate that this is not a set of features linking birds to crocodilians alone.

Although sinus chambers occur extensively in the theropod skull, little has entered the literature. For example, WHETSONE & WHYBROW (1983) were able to write "To the best of my (sic) knowledge a hollow quadrate bone has never been described in a dinosaur...", (p. 31). Indeed, it has not – nonetheless two theropods, *Tyrannosaurus rex* and *Labocania anomala* either have a demonstrably hollow quadrate or exhibit a similar pneumatic foramen and thus may be inferred to have had a hollow quadrate. The chambers of the ectopterygoid have been the most often described, mentioned by several authors. These are known to occur in *Albertosaurus lancensis* (GILMORE 1946), *Allosaurus fragilis* (MADSEN 1976), *Daspletosaurus torosus* (RUSSELL 1970), *Deinonychus antirrhopus* (OSTROM 1969), *Dromaeosaurus albertensis* (COLBERT & RUSSELL 1969) as well as in *Tyrannosaurus rex*. OSBORN (1912), MOODIE (1915), RUSSELL (1970) and MADSEN (1976) also describe other cranial sinuses in theropods, but with the exception of the first, all have been treated in the course of other topics.

The following description of the skull of *Tyrannosaurus rex* relies heavily on LACM 23844. Elements that enclose sinus chambers are the maxilla, lachrymal, jugal, epipterygoid and quadrate, and chambers also occur in some bones of the braincase. Only those of the quadrate and braincase are here described, as these bear most on the question of avian-crocodilian relationships.

The ventral portion of the body of the quadrate contains two chambers, a smaller lying lateral to a larger. This larger chamber communicates to the outside through a foramen at the medial edge of the anterior face of the quadrate, just above the condyle*. A third, flattened chamber of undetermined extent projects dorsally into the pterygoid process. A break at the top of the quadrate of LACM 23844 reveals that a chamber (presumably this one) extends almost to the dorsal articulation. Two channels project into the lateral process.

The quadrate of the type and only specimen of *Labocania* (MOLNAR 1974) is incomplete. The medial portion is missing, revealing a central cavity in the body of uncertain extent. Such of the medial portion as remains indicates that this chamber opened broadly medially.

The basal portion of the braincase of *T. rex* contains a set of sinus chambers (the recessus basisphenoideus of OSBORN 1912), these are figured in detail for *Albertosaurus* by RUSSELL (1970). In both the gorgosaurs (RUSSELL 1970) and *Tyrannosaurus* the canal for the internal carotid runs in close

---

* This foramen also occurs in *Labocania* (MOLNAR 1974), allowing the inference that that form also had a hollow quadrate. (Editors' insertion)

proximity to the central chamber of the basisphenoid. In the sectioned braincase, AMNH 5029, the carotid canal actually opens into the chamber, continuing as a groove in the anterior portion of the lateral wall (OSBORN 1912, Pl. 3). This seems similar to the situation described by WALKER (1972) in *Sphenosuchus*.

At least one other sinus, of unknown extent, occupies the lateral portion of the exoccipital. This is indicated by the braincase AMNH 5117 in which a portion of the posterior surface of the exoccipital has collapsed forward into this chamber. A large sinus chamber also occurs in the articular, and opens to the exterior via a pneumatic foramen.

Theropods

During the nineteenth century the hypothesis of the theropod ancestry of birds was quite acceptable. This hypothesis has regained acceptance through the work of OSTROM, who demonstrated the distinct resemblance between *Archaeopteryx* and small theropods. Accepting *Archaeopteryx* as an ancestral bird, this demonstrates a relationship between birds and theropods.

Without doubt the most significant discovery since *Archaeopteryx*, regarding the origin of birds, not to mention the relation between birds and theropods, is that of *Avimimus* (KURZANOV 1981, 1983).

Avimimus

Avimimus is based on a skeleton found in the Nemegt Fm. of Mongolia, and exhibits a suite of avian character states. KURZANOV (1983) gives eighteen: 1 – long, slender hind limbs; 2 – hind limbs with finely formed articular surfaces; 3 – femur with lateral condyle having a trochlea fibularis; 4 – femur with large popliteal surface indicating the existence of cruciate ligaments; 5 – a true tibiotarsus; 6 – tibia with large, prominent, anteriorly directed cnemial crest; 7 – intertarsal joint with enlarged articular surface; 8 – hypotarsus; 9 – ilium with antitrochanter (of the avian rather than the hadrosaurian form); 10 – pelvic bones fused; 11 – radial and ulnar articular surfaces of humerus shifted onto ventral face; 12 – fused scapulocoracoid; 13 – 160 degree angle between scapula and coracoid; 14 – elongate coracoid; 15 – less than eleven cervicals; 16 – dorsals with hypapophyses; 17 – cranial roofing bones fused; 18 – large cerebellum indicated by form of endocranial cavity.

All of these characters except 15 are found in modern birds, but many are also found among other theropods and, most interestingly, not all are found in *Archaeopteryx*. Thus to elucidate the relation between modern birds, *Archaeopteryx,* and theropods it is useful to determine the taxonomic range of these character states. Those for *Archaeopteryx* are taken from OSTROM (1976) unless otherwise attributed: 1 – This is stated vaguely, but comparing the ratio of overall length of tibia to diameter near midshaft two theropods approach closely the proportions of *Avimimus* – *Kakuru* (MOLNAR & PLEDGE 1980) and *Tugulusaurus* (DONG 1973). *Archaeopteryx* appears to be very close to *Avimimus*; 2 – This cannot be checked with sufficient confidence from the literature; 3 – A trochlea fibularis has been reported neither for *Archaeopteryx* nor for any theropod. (But see the femur of *Bahariasaurus* – STROMER 1934, Taf. 3, Fig. 5a); 4 – This cannot be checked from the literature; 5 – A tibiotarsus is found in *Archaeopteryx* and also in *Heptasteornis* (HARRISON & WALKER 1975) and *Syntarsus* (RAATH 1969) among theropods. It has also been reported in the theropods *Stenonychosaurus* by STERNBERG (1932) and *Struthiomimus* by LAMBE (1917); 6 – While *Archaeopteryx* has no external cnemial crest (OSTROM 1976), large, prominent crests are common among theropods, e.g. in ornithomimes (OSBORN 1916, OSMOLSKA, RONIEWICZ & BARSBOLD 1972), *Chilantaisaurus* (HU 1964), *Ceratosaurus* (GILMORE 1920), and the massive Lameta theropod (*Indosaurus?* – VON HUENE & MATLEY 1933); 7 – An enlarged intertarsal (mesotarsal) joint is found in *Deinonychus* (OSTROM 1969), *Bradycneme* and *Heptasteornis* (HARRISON & WALKER 1975) and *Stenonychosaurus* (RUSSELL 1969) among theropods, but seems not to be present (or at least is not well-developed) in *Archaeopteryx*; 8 – The hypotarsus is poorly developed at best in *Archaeopteryx* (OSTROM 1976), but is reported among other theropods: *Elmisaurus* (OSMOLSKA 1981) and *Syntarsus* (RAATH 1969); 9 – The avian antitrochanter is not found in *Archaeopteryx* nor has it been reported among other theropods. However a marked, projecting rim is found dorsally bordering the acetabulum in *Allosaurus* (MADSEN 1976, Pl. 46) and *Dilophosaurus* (this is most clearly illustrated by COLBERT 1983: 26). No articular surface for this on the femur has been reported in theropods, but the proximal face of the femur in theropods extends from the head to the greater trochanter unbroken by a sulcus as is found among ornithopods, or a slope as in ankylosaurs, paleopods, sauropods and stegosaurs; 10 – Fused pelvic elements are found in *Ceratosaurus* (GILMORE 1920), *Ornithomimus* (MARSH 1892), *Sarcosaurus* (Von HUENE 1932), *Syntarsus* (RAATH 1969) and *Tyrannosaurus* (STEEL 1970: this is seemingly a variable state for *T.*). Whether it also occurred in *Archaeopteryx* is not clear; 11 – A slight shift of the radial and ulnar articular surfaces onto the ventral face of the humerus is apparent in *Archaeopteryx* (OSTROM 1976), it can also be seen in *Allosaurus* (GALTON & JENSEN 1979 Fig. 3); 12 – The scapulocoracoid is apparently fused in *Archaeopteryx* and, among theropods, in *Syntarsus* (RAATH 1969) at least; 13 – This character state is not determinable from the literature; 14 – Neither any other theropod nor *Archaeopteryx* has an elongate coracoid. But KURZANOV is uncertain about this feature in *Avimimus* (THULBORN 1984); 15 – Theropod cervical numbers range from nine in, e. g., *Allosaurus* (MADSEN 1976) to twelve in *Struthiomimus* (OSBORN 1916). There are eight to nine in *Archaeopteryx* and usually fourteen to fifteen in modern birds; 16 – Dorsal hypapophyses have been reported in no other theropods, nor in *Archaeopteryx*; 17 – Fusion of the cranial roofing bones has been reported in no other theropods, and those of *Archaeopteryx* are not fused (DE BEER 1954); 18 – An enlarged cerebellum is reported in no other theropods, but was presumably present in *Archaeopteryx*.

Of these eighteen features three (2, 4, 13) cannot be determined from the literature and one (14) is doubtful. Nine of these (1, 5, 6, 7, 8, 10, 11, 12, 15) are also found in other theropods, although not together in any taxon so far as is known. *Archaeopteryx* has eight (1, 5, 8, 11, 12, 15, 18) of these character states – significantly fewer than *Avimimus* itself, and also fewer than are found piecemeal among theropods. This suggests a reassessment of the position of *Archaeopteryx* in the evolution of birds a topic not pursued here, but see THULBORN (1984).

It also substantiates a relationship between birds and theropods.

Other features known in birds, some not found in *Avimimus*, have also recently been found among theropods, particularly those of the Mongolian late Cretaceous. Unexpectedly theropods with a retroverted pubis, or opisthopubic pelvis, were discovered in the 1970's (BARSBOLD 1983a). These include not only the aberrant segnosaurs, but also the dromaeosaurid *Adasaurus*. BARSBOLD has reported a furcula in both *Oviraptor* and *Ingenia* (BARSBOLD 1983a, 1983b), and ossified sterna in those two genera and in *Velociraptor* (BARSBOLD 1983a, 1983b). Recently (McGOWAN 1984) an astragalar ascending process like those found in theropods has been confirmed in ratites. Thus the list of derived character states shared by theropods and birds is greater than those shown by *Avimimus*, or even *Archaeopteryx*.

In addition to those listed for *Avimimus* there are: opisthopubic pelvis; ossified sterna; ascending process of astragalus; furcula. The furcula is usually homologized with the clavicles, although OSTROM (1976) has pointed out that there are problems with this interpretation. Nonetheless the elements identified by BARSBOLD as furculae do not depend for their identification upon homology with clavicles.

In addition the following character states are long known to be found widely among theropods: tridactyl manus; tridactyl pes; hollow limb bones. Taken together these suggest that the hypothesis of theropod ancestry of birds is the strongest contender. They also suggest that more and detailed studies of theropod morphology are needed.

Thecodonts

Due to the work of HEILMANN the thecodont hypothesis of avian ancestry was almost universally accepted before the work of OSTROM. Since then there has been little serious attempt to support this hypothesis. However some new evidence has been discovered that bears on this question.

The whole group of thecodonts seems perpetually in need of thorough taxonomic revision. In its absence the relationships of the various groups of thecodonts both to each other and to the other archosaurs remain dubious.

Cosesaurus has been discussed previously. It seems to be a unique and interesting thecodont, but is not clearly related to birds.*

Longisquama is an extremely unusual beast from the early Triassic of Fergana, central Asia, with two character states regarded as similar to those found in birds. These are the furcula and the elongate, in some cases even plume-like, scales. The plume-like scales reportedly show a central rhachis-like thickening (SHAROV 1970). If accurately described, and there is no good reason to doubt the description, these suggest that both character states may be primitive archosaur states. And hence they may not be usable in identifying birds.

Megalancosaurus is a small form based on a partial skeleton from the Upper Triassic of northern Italy (CALZAVARA et al. 1980). This is the only form yet described that shows a significant feature – considerable elongation of the manus – to be expected in an ancestor to birds (or least flying vertebrates) according to the work of CAPLE et al. (1983). Hence this is a promising form for further research.

The problem with assessing the thecodont hypothesis is that neither of the other hypotheses denies that thecodonts are ancestral to birds, however they both propose more proximate non-avian ancestors than thecodonts. Thus this hypothesis must produce synapomorphic character states in thecodonts and birds. So far it has not done so to a credible extent: the furcula has also been reported in theropods (BARSBOLD 1983a), and while the plume-like scales of *Longisquama* bear some resemblance to feathers they are hardly close enough to qualify as a synapomorphy.

Conclusions

Some material suggested to be avian indicates the existence of virtually unknown theropods or birds (e.g. the Dry Mesa sacrum). A significant proportion of character states described in the literature as linking birds and crocodilians occur in theropods as well. Of these, a significant number have not been described in the literature, and point up the desirability of much more thorough study of the Theropoda, both large and small. Thus character states cannot be presumed to be absent in a group from a study of the literature alone, but all available material must be checked. Many character states generally regarded as being avian occur among theropods, including the furcula, hypotarsus, opisthopubic pelvis, etc. Thus birds need a rigorous osteological definition to separate them from theropods. The number of shared states leaves theropods as the best contender as the group ancestral to birds. Some character states are shared by birds and theropods but are not present in *Archaeopteryx*.

References Cited

BARSBOLD, R. (1974): Saurornithoididae, a new family of small theropod dinosaurs from central Asia and North America. – Paleont. Polonica, **30**: 5–22; Warsaw and Cracow.

BARSBOLD, R. (1981): Bessubie khishchnie dinozavri Mongolii. – Sovmest. Sovet.-Mongol. Paleont. Yeksped., Trudi, **15**: 28–39; Moscow.

BARSBOLD, R. (1983a): Khishchnie dinozavri Mela Mongolii. – Sovmest. Sovet.-Mongol. Paleont. Yeksped., Trudi, **19**: 1–20; Moscow.

BARSBOLD, R (1983b): O "ptitsikh" chertakh v stroenii khishchnikh dinosavrov. – Sovmest. Sovet.-Mong. Paleont. Yeksped., Trudi, **24**: 96–103; Moscow.

BEER, G. DE (1954): *Archaeopteryx lithographica*. – 68 pp. London.

CALZAVARA, M., MUSCIO, G. & WILD, R. (1980): *Megalancosaurus preonensis* n. g., n. sp., a new reptile from the Norian of Friuli, Italy. – Gortiana, **2**: 49–64; Udine.

---

* After completion of this paper, a paper by SANZ & LOPEZ-MARTINEZ, (1984) has come to my attention, in which *Cosesaurus* is re-examined. They concluded that *Cosesaurus* was probably a prolacertid lepidosaur.

CAPLE G., BALDA, R. P. & WILLIS, W. R. (1983): The physics of leaping animals and the evolution of preflight. – Amer. Nat., **121**: 455–476; Chicago.

COLBERT, E. H. & RUSSELL, D. A. (1969): The small Cretaceous dinosaur *Dromaeosaurus*. – Amer. Mus. Novitates, **2380**: 1–49; New York.

COLBERT, E. H. (1983): Dinosaurs of the Colorado Plateau. – Plateau, **54**: 2/3: 1–48; Flagstaff.

DONG, Z. (1973): Dinosaurs from Wuerho. – Mem. Inst. Vert. Paleont. Paleoanthro., Acad. Sin., **11**: 45–52; Beijing. (In Chinese.)

ELLENBERGER, P. (1974): Contribution à la classification des Pistes de Vertébrés du Trias: les types du Stormberg d'Afrique du Sud (IIeme Partie: le Stormberg supérieur I. Le biome de la zone B/1 ou niveau de Moyeni: ses biocenoses). – Paléovert. Mem. Extraord., **1974**: 1–141; Montpellier.

ELLENBERGER, P. & VILLALTA, J. F. DE (1974): Sur la présence d'un ancetre probable des Oiseaux dans le Muschelkalk supérieur de Catalogne (Espagne). Note préliminaire. – Acta Geol. Hispanica, **9**: 162–168; Barcelona.

GALTON, P. M. & JENSEN, J. A. (1979): A new large theropod from the Upper Jurassic of Colorado. – Brigham Young Univ. Geol. Studies, **26**: 1: 1–12; Provo.

GILMORE, C. W. (1920): Osteology of the carnivorous Dinosauria in the United States National Museum with special reference to the genera *Antrodemus (Allosaurus)* and *Ceratosaurus*. – Bull. U. S. Nat. Mus., **110**: 1–159; Washington.

GILMORE, C. W. (1946): A new carnivorous dinosaur from the Lance Formation of Montana. – Smithsonian Misc. Coll., **106**: 1–19; Washington.

HARRISON, C. J. O. & WALKER, C. A. (1975): The Bradycnemidae, a new family of owls from the Upper Cretaceous of Romania. – Palaeontology, **18**: 563–570; London.

HU, S. (1964): Carnosaurian remains from Alashan, Inner Mongolia. – Vert. PalAs., **8**: 42–43; Beijing.

HUENE, F. VON (1932): Die fossile Reptil-Ordnung Saurischia, ihre Entwicklung und Geschichte. – Monog. Geol. Paläont., **1**: 1–361; Leipzig.

HUENE, F. VON & MATLEY, C. A. (1933): The Cretaceous Saurischia and Ornithischia of the Central Provinces of India. – Geol. Soc. India Mem., Palaeont. Indica, ns. **21**: 1–74; Delhi.

JENSEN, J. A. (1981a): A new oldest bird? – Anima, **1981**: 33–39; Tokyo. (In Japanese.)

JENSEN, J. A. (1981b): Another look at *Archaeopteryx* as the world's oldest bird. – Encyclia, **58**: 109–128; Salt Lake City.

KURZANOV, S. M. (1981): O neobichnikh teropodakh iz verkhnego Mela MNR. – Sovmest. Sovet.-Mongol. Paleont. Yeksped. Trudi, **15**: 39–50; Moscow.

KURZANOV, S. M. (1983): *Avimimus* i problema proiskhozhdenia ptits. – Sovmest. Sovet.-Mongol. Paleont. Yeksped. Trudi, **24**: 104–109; Moscow.

LAMBE, L. M. (1917): The Cretaceous theropodous dinosaur *Gorgosaurus*. – Canada Geol. Surv. Mem., **100**: 1–84: Ottawa.

LAMBRECHT, K. (1933): Handbuch der Palaeornithologie. – Berlin (Gebrüder Borntraeger).

LANGSTON, W., Jr. (1960): The vertebrate fauna of the Selma Formation of Alabama. Part VI. The dinosaurs. – Fieldiana: Geol. Mem., **3**, 6: 313–361; Chicago.

MADSEN, J., Jr. (1976): *Allosaurus fragilis*: a revised osteology. – Utah Geol. Mineral Surv., Bull., **1091**: 1–163; Salt Lake City.

MARSH, O. C. (1881): Discovery of a fossil bird in the Jurassic of Wyoming. – Amer. J. Sci., (3) **21**: 341–342; New Haven.

MARSH, O. C. (1892): Notice of new reptiles from the Laramie Formation. – Amer. J. Sci., (3), **43**: 449–453; New Haven.

MARTIN, L. D., STEWART, J. D. & WHETSTONE, K. N. (1980): The origin of birds: structure of the tarsus and teeth. – The Auk, **97**: 86–93; Lawrence.

MCGOWAN, C. (1984): Evolutionary relationships of ratites and carinates: evidence from ontogeny of the tarsus. – Nature, **307**: 733–735; London and New York.

MCGOWAN, C. & BAKER, A. J. (1981): Common ancestry for birds and crocodiles? – Nature, **289**: 97–98; London and New York.

MOLNAR, R. E. (1974): A distinctive theropod dinosaur from the Upper Cretaceous of Baja California (Mexico). – J. Paleo., **48**: 1009–1017; Lawrence.

MOLNAR, R. E. & ARCHER, M. (1984): Feeble and not so feeble flapping fliers: a consideration of early birds and bird-like reptiles. – pp. 331–336. In: M. ARCHER & G. CLAYTON, eds. Vertebrate Zoogeography and Evolution in Australasia. Carlisle (Hesperian Press).

MOLNAR, R. E. & PLEDGE, N. S. (1980): A new theropod dinosaur from South Australia. – Alcheringa, **4**: 281–287; Sydney.

MOODIE, R. L. (1915): A sphenoidal sinus in the dinosaurs. – Science, **41**: 288–289; Washington.

OLSEN, P. E. (1979): A new aquatic eosuchian from the Newark Supergroup (late Triassic-early Jurassic) of North Carolina and Virginia. – Postilla, **176**: 1–14; New Haven.

OSBORN, H. F. (1912): Crania of *Tyrannosaurus* and *Allosaurus*. – Mem. Amer. Mus. Nat. Hist., ns., **1**: 1–30; New York.

OSBORN, H. F. (1916): Skeletal adaptations of *Ornitholestes, Struthiomimus, Tyrannosaurus*. – Bull. Amer. Mus. Nat. Hist., **35**: 733–771; New York.

OSMOLSKA, H. (1981): Coossified tarsometatarsi in theropod dinosaurs and their bearing on the problem of bird origins. – Palaeont. Polonica, **42**: 79–95; Warsaw and Cracow.

OSMOLSKA, H., RONIEWICZ, E. & BARSBOLD, R. (1972): A new dinosaur, *Gallimimus bullatus* n. gen., n. sp. (Ornithomimidae) from the Upper Cretaceous of Mongolia. – Paleont. Polonica, **27**: 103–143; Warsaw and Cracow.

OSTROM, J. H. (1969): Osteology of *Deinonychus antirrhopus*, an unusual theropod from the Lower Cretaceous of Montana. – Bull. Peabody Mus. Nat. Hist., **30**: 1–165; New Haven.

OSTROM, J. H. (1976): *Archaeopteryx* and the origin of birds. – Biol. J. Linn. Soc., **8**: 91–182; London.

RAATH, M. (1969): A new coelurosaurian dinosaur from the Forest Sandstone of Rhodesia. – Arnoldia, **28**: 1–5; Salisbury.

RAUTIAN, A. S. (1978): Unikalnoe pero ptitsi iz otlozhewnii urskogo ozera v khrebte Karatau. – Paleont. Zhurnal, **1978**, 4: 106–114; Moscow.

RUSSELL, D. A. (1969): A new specimen of *Stenonychosaurus* from the Oldman Formation (Cretaceous) of Alberta. – Can. J. Earth Sci., **6**: 595–612; Ottawa.

RUSSELL, D. A. (1970): Tyrannosaurs from the late Cretaceous of western Canada. – Natl. Mus. Nat. Sci., Publ. Paleo., **1**: 1–34; Ottawa.

SANZ, J. L & LOPEZ-MARTINEZ, N. (1984): The prolacertid lepidosaurian *Cosesaurus aviceps* Ellenberger & Villalta, a claimed "protoavian" from the middle Triassic of Spain. – Geobiosm 17, (6): 741–753.

SHAROV, A. G. (1970): Svoeobraznaia reptiliia iz nizhnego Triasa Fergani. – Paleont. Zhurnal, **1970**, 1: 127–130; Moscow.

STEEL, R. (1970): Saurischia. In: O. KUHN, ed., Handbuch der Palaeoherpetologie, T. 14. Stuttgart (Gustav Fischer).

STERNBERG, C. M. (1932): Two new theropod dinosaurs from the Belly River Formation of Alberta. – Canad. Field-Nat., **46**: 99–105; Ottawa.

STROMER, E. (1934): Ergebnisse der Forschungsreisen Prof. E. STROMERS in den Wüsten Aegyptens. II. Wirbeltierreste der Baharîje-Stufe (unterstes Cenoman). 13. Dinosauria. – Abhand. bayer. Akad. Wissen., Math.-naturwiss. Abt., nf. **22**: 1–79; München.

THULBORN, R. A. (1984): The avian relationships of *Archaeopteryx*, and the origin of birds. – Zool. J. Linn. Soc., **82**: 119–158; London.

WALKER, A. D. (1972): New light on the origin of birds and crocodiles. – Nature, **237**: 257–263; London and New York.

WHETSTONE, K. N. & MARTIN, L. D. (1979): New look at the origin of birds and crocodiles. – Nature, **279**: 234–236; London and New York.

WHETSTONE, K. N. & MARTIN, L. D. (1981): Common ancestry for birds and crocodiles? (reply) – Nature, **289**: 98; London and New York.

WHETSTONE, K. N. & WHYBROW, P. J. (1983): A "cursorial" crocodilian from the Triassic of Lesotho (Basutoland), southern Africa. – Occas. Papers Mus. Nat. Hist. Univ. Kansas, **106**: 1–7; Lawrence.

Author's address: Dr. R. E. MOLNAR, Queensland Museum, Fortitude Valley, Queensland 4006, Australia.

Michael A. Raath

The Theropod *Syntarsus* and Its Bearing on the Origin of Birds

Abstract

The late Triassic/early Jurassic theropod genus *Syntarsus* is known from relatively abundant and well preserved cranial and post-cranial material from terminal Karoo deposits in southern Africa. Recent discoveries indicate that it might also have occurred in the northern hemisphere. It possesses skeletal features in the skull (especially the braincase), forelimb (especially carpus and manus) and hindlimb (especially pubis, tarsus and perhaps the femur) that are noteworthy in their similarity to homologous features of *Archaeopteryx*. These features necessitate reconsideration of the conclusions of TARSITANO and HECHT (1980) that theropods were probably not implicated in the ancestry of *Archaeopteryx*. It is suggested that *Archaeopteryx* is more closely related to procompsognathid theropods, exemplified by *Syntarsus*, than to the later non-procompsognathid theropods that were its contemporaries.

Zusammenfassung

Die Theropoden-Gattung *Syntarsus* ist durch zahlreiches, gut erhaltenes Schädel- und Postcranialmaterial von obersten Karoo-Ablagerungen (Obertrias/Unterjura) im südlichen Afrika bekannt. Neuere Entdeckungen deuten darauf hin, daß *Syntarsus* auch in der nördlichen Hemisphäre vorgekommen sein könnte. Er besitzt Skelettmerkmale am Schädel (besonders der Hirnkapsel), an der Vorderextremität (besonders an Carpus und Manus) und an der Hinterextremität (besonders an Pubis, Tarsus und vielleicht am Femur), die in ihrer Ähnlichkeit mit homologen Merkmalen von *Archaeopteryx* bemerkenswert sind. Diese Merkmale erfordern ein Überdenken der Schlußfolgerungen von TARSITANO und HECHT (1980), wonach Theropoden wahrscheinlich nicht an der Ahnenreihe von *Archaeopteryx* beteiligt waren. Es wird die Ansicht vertreten, daß *Archaeopteryx* mit procompsognathiden Theropoden, vertreten durch *Syntarsus*, näher verwandt ist, als mit den zeitgleichen, nicht-procompsognathiden Theropoden.

Introduction

Syntarsus is a small (2 m long), lightly built, bipedal theropod from terminal Karoo deposits in southern Africa (RAATH, 1969; 1980) and there are indications that it might occur also in the northern hemisphere (TIMOTHY ROWE, pers. comm. 1980; DIANE WARRENER, pers. comm. 1980; OLSEN and GALTON, 1984). Its limb anatomy and muscle scars, together with the evidence of footprints attributed to it (RAATH, 1972), led me to suggest that it was a bipedal saltator (RAATH, 1977).

The excellent preservation of the recovered sample (more than 30 individuals represented), including many isolated cranial elements, offers unprecedented opportunities for studying osteological details of an early theropod.

The current debate on the affinities of *Archaeopteryx* and the origin of birds (WALKER, 1972; OSTROM, 1973, 1976; TARSITANO and HECHT, 1980; WHETSTONE, 1983) is the principal reason behind the organisation of this conference and it seems desirable to see what contribution *Syntarsus* can make towards it.

OSTROM'S (1973, 1976) detailed analysis of the evidence for close relationship between *Archaeopteryx* and theropod dinosaurs was based mainly on postcranial characters because reliable information on the skull of *Archaeopteryx* has been largely lacking hitherto. This contribution will consider the postcranial skeleton only to the extent that it adds something considered significant. It will instead concentrate on a comparison of the braincase in *Syntarsus* and *Archaeopteryx* because of new information provided by WHETSTONE'S (1983) detailed account of the *Archaeopteryx* braincase following Peter WHYBROW'S recent superb preparation of the skull of the London specimen (WHYBROW, 1982).

Comparison of Skeletal Features of *Syntarsus* and *Archaeopteryx*

OSTROM (1973) listed what he regarded as "the coelurosaurian features of *Archaeopteryx*", choosing to consider only "post-Triassic" theropods in his analysis because he was seeking clarity on the immediate (rather than the remote) ancestry of *Archaeopteryx*. In this sense *Syntarsus* would qualify as a "Triassic" form (but see discussion on age in OLSEN and GALTON, 1984); nevertheless it, too, shares with *Archaeopteryx* many of the features listed by OSTROM.

This study ignores the vertebral column except to note that the sacral vertebrae of *Syntarsus* undergo fusion in mature individuals in a manner similar to that seen in the synsacrum of birds – the neural spines coalesce, and so do the sacral ribs, producing continuous longitudinal sheets of bone that may obliterate any trace of the junctions between the processes of adjacent sacral vertebrae.

Postcranial features

1. Forelimb and girdle:
Archaeopteryx differs in forelimb structure from *Syntarsus* (and theropods in general) in the following ways:
Possession of furcula (? fused clavicles) (but some theropods do possess ossified clavicles – e.g. *Segisaurus*; CAMP, 1936);
Forelimbs markedly longer than hindlimbs (but many theropods show a tendency towards elongation of the forelimb – e.g. *Deinonychus*; OSTROM, 1976);
Coracoid large and of unique shape (but considered by OSTROM to be a variant of the theropod pattern); retention of three metacarpals (cf. four in *Syntarsus*, but many other theropods have only three).
Their similarities in forelimb structure include:
Possession of an enlarged semilunate carpal associated with metacarpals I and II (also shared with several other theropods – OSTROM, 1976) (but *Syntarsus* has a total of six carpals compared with three in *Archaeopteryx* – see GALTON, 1971);
Metacarpal I markedly shorter than the metacarpals of the other two functional digits, and the articular condyles inclined, deflecting the axis of articulation of digit I laterally;
Metacarpals II and III subequal in length;
Penultimate phalanges the longest in each digit of the manus (but this applies to pterosaurs as well as to theropods – TARSITANO and HECHT, 1980);
Although *Syntarsus* retains a reduced metcarpal IV and vestigial phalanx, its phalangeal formula for the functional digits I–III is otherwise very similar to that of *Archaeopteryx* (*Syntarsus* 2–3–4–1; *Archaeopteryx* 2–3–4).

Controversy continues over the homology of the digits in the manus of *Archaeopteryx* (e.g. OSTROM, 1976; TARSITANO and HECHT, 1980; HOWGATE 1983, 1984). HOWGATE has confirmed the conclusion of many other workers that digit III has two small proximal joints and a longer one between the metacarpal and the ungual; the two smaller phalanges do not represent a broken longer one as asserted by TARSITANO and HECHT (1980). This fact, together with the similarities in the proximal carpal row, metacarpal configuration, number of digits and relative size of the phalanges all suggest that the digits of *Archaeopteryx* and theropods are indeed homologous and are therefore digits I, II and III, not II, III and IV.

2. Hindlimb and girdle:

Archaeopteryx and *Syntarsus* are surprisingly similar in some features of the pubis – a bone often cited as a strong reason for linking *Archaeopteryx* with birds because of its conceivably birdlike opisthopubic attitude. Whether *Archaeopteryx* was propubic or opisthopubic, it nonetheless retains apparently primitive features such as a transverse pubic apron and a relatively long pubic symphysis along the thin medial blade of the apron; furthermore, the distal end of the pubis is not greatly expanded. These are considered archosaurian plesiomorphies, and among theropods they are confined to *Procompsognathus, Coelophysis, Syntarsus,* and possibly *Segisaurus* (all but *Segisaurus* belonging to the theropod family Procompsognathidae; OSTROM, 1981).

The ilia of both *Syntarsus* and *Archaeopteryx* are expanded anteriorly into relatively deep, long, thin blades of bone, while posteriorly they are relatively wide and shallow. *Syntarsus* bears a well-marked ventral "brevis shelf" posteriorly for the origin of part of the caudifemoral musculature (RAATH, 1977), but details in this postacetabular region are not clear in *Archaeopteryx*.

The ischium of *Archaeopteryx* is unique in shape.

OSTROM'S (1976) account of the *Archaeopteryx* femur indicates that it has proximal femoral trochanters similar in several respects to those of theropods. *Syntarsus* shows marked sexual dimorphism in these trochanters (RAATH, 1977). The fourth trochanter is indistinct in *Archaeopteryx* (OSTROM, 1976), and the same is true of other theropods. *Syntarsus* has only a very low ridge in the corresponding position on the posteromedial surface of the femur. It evidently defines the medial limit of the femorotibialis posterior origin scar (RAATH, 1977), and has no direct association with the caudifemoral muscles. However, just forward of this low ridge on the medial femoral surface are two contiguous well-defined pits which are also dimorphic, being sharply-defined and semicircular in "robust" femora but less clearly demarcated and more elongate in "gracile" femora; I interpreted them as insertion scars for the tendons of the caudifemoral muscles (RAATH, 1977). I have been able to examine this region only in the London *Archaeopteryx* specimen, whose left femur seems to bear two small depressions on the medial surface in this area; I therefore suggest that *Archaeopteryx* probably had caudifemoral tendon scars similar to those of *Syntarsus*.

Both WELLNHOFER (1974) and OSTROM (1976) state that *Archaeopteryx* has an ascending process on the astragalus that covers part of the anterior tibial shaft distally as in theropods. TARSITANO and HECHT (1980) ascribe the supposed process to post-mortem damage of the tibial shaft. Contrary to OSTROM (1976: 121), *Syntarsus* does have an ascending process on the astragalus, but it articulates with the distal end of the tibia in a shallow groove mainly on the anterolateral side (RAATH, 1977: Plate 26 a, c); it does not rise up the anterior tibia surface as a covering tongue of bone as it does in some theropods (e.g. *Deinonychus*; OSTROM, 1976: Fig. 29 d, e). Mature specimens of *Syntarsus* have the astragalus and calcaneum fused to each other and to the tibia, and the distal tarsals also fuse to the proximal ends of the metatarsals (RAATH, 1977: Fig. 20, Plate 24), except for distal tarsal 4 which remains free regardless of age or sex. CRACRAFT (1977) has remarked on the apparent similarity in tarsal condition in *Archaeopteryx* and *Syntarsus*.

The first metatarsal of *Syntarsus* is reduced to a small tapering shaft applied to the medial surface of metatarsal II in its distal half, and a small digit of two phalanges. It is not reflexed, but the distal articular surface of the metatarsal is almost hemispherical and the proximal phalanx has a strongly developed abductor tubercle, indicating that the digit was capable of a considerale amount of movement laterally. Overall, the pes of *Syntarsus* has the typical theropod structure – elongate metapodium, functional tridactyly, and symmetry about the long third toe. The fifth metatarsal is reduced to a slender, slightly curved splint lying beneath metatarsal IV proximally.

Cranial features

Skulls are preserved in three specimens of *Archaeopteryx*, mostly too poorly preserved to permit unequivocal statements on its cranial morphology. WHETSTONE'S recent (1983) detailed study of the braincase of the London specimen has contributed important new information about skull structure in this taxon.

The skull of *Syntarsus* is known from five well preserved braincases, two complete with the cranial roof and the entire occiput, and three consisting of basicranium and sidewalls plus a varying amount of

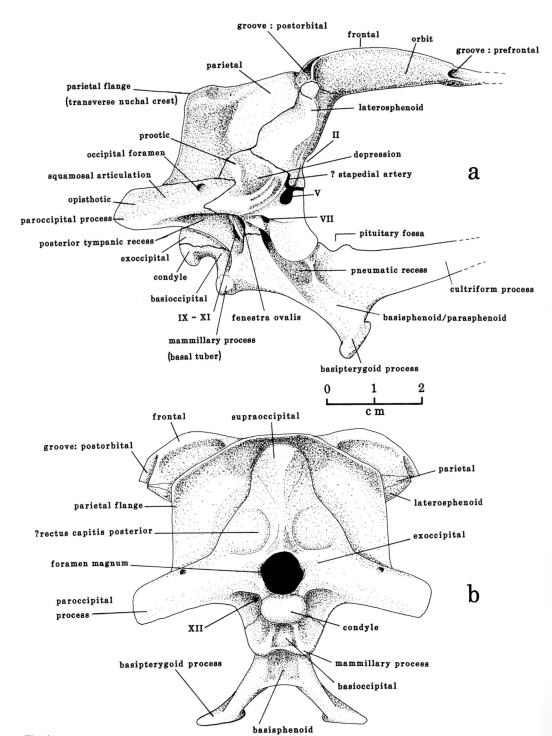

Fig. 1
Braincase of *Syntarsus rhodesiensis* (composite reconstruction based mainly on QG 193, 194 and 196): a) right side view; b) occipital view.

occiput; there is also a large collection of isolated cranial bones – mainly facial elements, temporal arch components and parts of jaws (RAATH, 1977).

1. Sidewall of the braincase (Fig. 1a):

WHYBROW'S new preparation of the London *Archaeopteryx* has revealed the well preserved left side of the braincase, together with the laterosphenoid and roofing bones. In *Syntarsus* the laterosphenoid has much the same shape, being relatively shallow and elongate, slightly swollen laterally and sharply inturned ventrally toward the midline following the contour of the optic lobe below the cerebral hemisphere. In *Syntarsus* it rises up to the skull roof more steeply than it seems to in *Archaeopteryx*. WHETSTONE (1983) notes that in the London specimen the anterior end of the laterosphenoid is precluded from contact with the parietal at the posterolateral corner of the orbit by a small projection from the frontal; WELLNHOFER (1974) reported a projection in a comparable position on the frontal of the Eichstätt specimen. *Syntarsus* has a complex structure for the articulation of the postorbital in this region of the frontal (Fig. 1), consisting of a very delicately constructed groove that accommodates an equally thin and fragile tongue from the anterior process of the postorbital. Because this structure is extremely fragile it is preserved intact in very few specimens; most show it only as a narrow shelf-like remnant. The same may possibly be true of the London *Archaeopteryx*, the projection noted above representing the remnant of a postorbital groove.

Both *Syntarsus* and *Archaeopteryx* appear to have a plain, edge-to-edge contact between the laterosphenoid and the overlying parietal. In *Syntarsus* this arrangement is interpreted as part of the cranial kinetic mechanism; the simple loose contact between the laterosphenoid (which is functionally united to the braincase) and the skull roof would permit a limited amount of movement to take place (RAATH, 1977).

WHETSTONE (1983) draws attention to a depression on the prootic of *Archaeopteryx*, just ahead of the parietal flange which he aptly terms the transverse nuchal crest. He suggests that it might be a posterior remnant of a once more extensive dorsal periotic sinus. *Syntarsus* has a similar depression in a comparable position, although it is much shallower than that of *Archaeopteryx*; it does not seem to have any obvious function other than as part of the attachment scar of the jaw musculature, but two distinct curving grooves run along its lower border. This depression is associated with a foramen that penetrates the paroccipital process in the same way as in *Archaeopteryx*, but there is no pneumatic sinus in the paroccipital process of *Syntarsus* such as reported by WHETSTONE (1983) in *Archaeopteryx*.

The foramina and structures of the braincase sidewall are very similarly arranged in the two taxa – indeed the two are more similar to each other than *Syntarsus* is to another theropod, namely *Dromaeosaurus* (cf. Fig. 1a and COLBERT and RUSSELL, 1969: Fig. 7). Both have a notch-like trigeminal foramen with the facial foramen (VII) situated behind it on the same level. Below these two foramina the wall of the braincase in both is deeply excavated. WHETSTONE (1983) terms the excavation in *Archaeopteryx* a "rostral periotic sinus" and notes that it houses three foramina (presumably pneumatic). I termed the comparable trough in *Syntarsus* the "parabasal canal", suggesting that it formed a duct for the palatine artery, lateral head vein, and palatine branch of the facial nerve (RAATH, 1977), but Dr. A. D. WALKER has corrected me, pointing out that archosaurs do not have a parabasal canal; he suggests that this recess is partly pneumatic (pers. comm.).

2. Otic region:

There is an apparent major conflict between WHETSTONE'S interpretation of the otic region of *Archaeopteryx* and mine of *Syntarsus*. In both taxa the tympanic cavity consists of a deep recess roofed over by a rim of the prootic and the lower edge of the paroccipital process, and walled behind by a vertical wing of the opisthotic below the paroccipital process; this cavity is divided by a steep, oblique septum of very thin bone (evidently an outgrowth of the opisthotic) into two subequal chambers – one anterodorsal and one posteroventral. WHETSTONE states that in *Archaeopteryx* the anterodorsal chamber is further subdivided into an upper and lower compartment, the upper one housing the fenestra ovalis and the lower the fenestra pseudorotunda. The posteroventral chamber, he states, bears at its upper rear end a "prominent, oval cotyle for articulation with the quadrate".

The equivalent of the anterodorsal chamber in *Syntarsus* is not subdivided; it is interpreted as housing only the fenestra ovalis ("fenestra ovale"; RAATH, 1977). The posteroventral chamber in *Syntarsus* houses the external opening of the slit-like metotic (vagal) fissure; I interpreted it as having transmitted the glossopharyngeal (IX), vagus (X), and accessory (XI) cranial nerves through a lower anterior foramen with the fenestra pseudorotunda ("fenestra rotunda" in RAATH, 1977) opening separately slightly above and behind the nerve exit. Further preparation of this region in each of the skulls in which it is preserved (particularly well shown in QG196) shows that this interpretation is mistaken; there is no separate foramen for the fenestra pseudorotunda distinct from the metotic slit. I can identify no structure as a clearly demarcated subcapsular process supporting the fenestra pseudorotunda, but Dr. A. D. WALKER (pers. comm.) regards the rear wall of the otic cavity as sufficiently developed to justify being termed a subcapsular process, which would have supported a secondary tympanic membrane. WHETSTONE remarked that the "interfenestral process" dividing the fenestra ovalis from the fenestra pseudorotunda in *Archaeopteryx* was indicated by "a remnant", and in his drawings it is dotted in (WHETSTONE, 1983: Fig. 7), raising doubts about his confidence in his interpretation. My examination of the London specimen revealed no such interfenestral "remnant".

The two taxa are in nearly every respect other than size so similar in otic morphology that it seems to me inconceivable that they could differ so fundamentally in the possession or position of the fenestra pseudorotunda.

Another apparent conflict in interpretation concerns the cotylus for the quadrate reported by WHETSTONE (1983) in the otic region of *Archaeopteryx*. The bone here is surely too thin and fragile to have housed an articulation for such a bone as the quadrate, subject to such stresses from jaw muscle action. WALKER (pers. comm.) suggests that the area designated by WHETSTONE as the quadrate articular area is the smooth floor of what he terms the posterior tympanic recess. *Syntarsus* too has a recess in this area, doming up slightly into the lower edge of the opisthotic posteriorly.

The question of the possible site of quadrate articulation in *Archaeopteryx* will be returned to below.

3. Occiput (Fig. 1b):
Neither taxon retains post-temporal fenestrae. WHETSTONE (1983) reports a small foramen on the posterodorsal surface of the paroccipital process of *Archaeopteryx* which he equates with the occipital foramen of Recent birds; a foramen exists in the same position in *Syntarsus*, penetrating through the upper edge of the paroccipital process immediately below its contact with the parietal flange ("transverse nuchal crest"). Both forms have clearly marked circular pits on either side below the median supraoccipital knob: WHETSTONE (1983) interprets them as scars of the external occipital vein in *Archaeopteryx*; I interpret them as insertion scars for the rectus capitis posterior muscle in *Syntarsus* (Fig. 1b; RAATH, 1977).

Both taxa have ventrolateral cavities or blind-ending pockets on either side of the occipital surface lateral to the condyle, bounded above by the lower edge of the paroccipital process and the exoccipital. Comparing the London specimen's skull as preserved with WHETSTONE'S reconstruction drawing (WHETSTONE, 1983: Fig. 9), it seems clear that he has under-emphasised the depth of the pockets in *Archaeopteryx*. In both taxa a group of three closely spaced small foramina emerge from each pocket. WHETSTONE interprets the largest of them in *Archaeopteryx* as the vagal (X) foramen – similar to that of Recent birds – and the other two as transmitting the hypoglossal nerve (XII); I interpreted them all as transmitting the hypoglossal in *Syntarsus* (RAATH, 1977).

WHETSTONE (1983) notes the pronounced arching of the occipital dorsal profile in *Archaeopteryx*, which probably is a reflection of its real cerebral enlargement, but it may also be somewhat exaggerated by the distortion that has affected the skull. The corresponding area in *Syntarsus* is not arched; the skull roof is in fact rather flat on top and the margin of the parietal crest on the occipital profile is slightly downturned laterally. This arching in *Archaeopteryx* constitutes about the only unequivocal major difference between the two taxa in occipital morphology. WHETSTONE (1983: 449) claims that *Archaeopteryx* lacks "the extensive plate which, in theropods, extends ventrally from the basitemporal region and below the occipital condyle", but this is debatable. I suggest that WHETSTONE has over-estimated the width of the basicranium in his restoration of the occiput (WHETSTONE, 1983: Fig. 9), and in so doing he has also overestimated the width of the supraoccipital and foramen magnum. The occiput of the London specimen is, in my opinion, exposed nearly to the midline, and a more correct restoration would be obtained by adding little more than a mirror-image of the part now exposed. This would restore the "extensive plate" and remove or diminish two other alleged points of difference between *Archaeopteryx* and theropods in occipital morphology: the much greater size of the supraoccipital and the greater relative size of the foramen magnum (WHETSTONE, 1983: 449). If my suggestions are correct, then far from being "typically avian" in braincase construction (WHETSTONE, 1983), *Archaeopteryx* would be much more typically theropodan, being practically indistinguishable from an early theropod in all verifiable features.

4. Structure of the upper temporal arch and jaw suspension:
A lingering problem in the cranial osteology of *Archaeopteryx* is the true nature of its temporal arches and jaw suspensorium. WHETSTONE'S (1983) suggestion about the quadrate articulation site has already been mentioned. He further suggests (p. 451) a pattern of upper temporal arch construction in *Archaeopteryx* that would be quite unique, involving a squamosal that has lost contact with the quadrate and a ligamentous equivalent of the postorbital to complete the arch.

Attention was drawn earlier to the delicate and easily lost groove on the frontal of *Syntarsus* which receives the corresponding tongue from the postorbital. A tapering posterior process of the triradiate postorbital articulates loosely in a reciprocally tapered groove on the outer anterior surface of the squamosal. The squamosal itself articulates with the braincase via a very loose squamous contact with the thin parietal flange and the anterior face of the paroccipital process. The lower surface of the domed, curved squamosal bears an unmistakable cotylus behind its ventral spur to house the single, rounded head of the streptostylic quadrate. The result of this arrangement in *Syntarsus* is that these elements (postorbital, squamosal, and quadrate) are almost invariably found detached from the skull, although they may be variably associated with each other. Apart from the easily obliteratet postorbital groove on the frontal, there is no unequivocal scar for attachment of any of these bones on the braincase.

It seems to me probable that *Archaeopteryx* had an upper arch of similar construction, and that no quadrate articulation scar is to be found anywhere on the braincase. The suggestion that *Archaeopteryx* might have had a theropod-like temporal arch and jaw suspension implies that it had a theropod-like single-headed quadrate. Until cranial material is found with all of these bones preserved in articulation, this fundamental question will probably remain unresolved.

Phylogenetic relationship of *Syntarsus* and *Archaeopteryx* (Fig. 2)

OSTROM (1981) doubtfully includes *Syntarsus* with *Procompsognathus, Halticosaurus* and *Coelophysis* in the family Procompsognathidae. Procompsognathids are among the most primitive coelurosaurian theropods: they retain a primitive transverse pubic apron, lack the distal pubic foot, retain a vestigial fourth metacarpal and digit, and they occur earliest in time. On these critieria there is no doubt that *Syntarsus* is a procompsognathid. Only the inadequately known *Segisaurus* seems to have characters that are more primitive in theropod terms than procompsognathids – it is reported to have solid limb bones and vertebral centra, and it possesses an ossified clavicle (CAMP, 1936). *Segisaurus* is also approximately coeval with procompsognathids.

Two hypotheses of relationship between *Archaeopteryx* and theropods are presented in Figure 2, one (Fig. 2a) representing the view advanced here that *Archaeopteryx* and procompsognathids (like *Syntarsus*) share a common ancestry closer than that of either with the non-procompsognathid theropods, and the other (Fig. 2b) representing a simplified version of the "A3" section of the cladogram in TARSITANO and HECHT (1980: Fig. 9) (because it purports to reflect OSTROM'S view, and he confined his analysis to "Post-Triassic" – i.e. non-procompsognathid – theropods). The comparison of braincase, pubic, carpal, tarsal, and pes structure presented obove seems to eliminate Figure 2b, leaving the

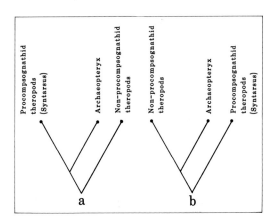

Fig. 2
Cladograms of relationship: *Archaeopteryx* and theropods (see text).

hypothesis summarised in Figure 2a as an apparently viable alternative hypothesis of relationship between the theropod *Syntarsus* and *Archaeopteryx*. This suggests that *Syntarsus* or one of its allies in the Procompsognathidae played an important, if not crucial, role in the origin of birds from theropods via *Archaeopteryx*.

In most respects the procompsognathids are ideal candidates as ancestors not only of the later theropods, but also of *Archaeopteryx* and, by implication, of birds; they do not seem to possess apomorphies that would obviously disqualify them. The similarities between *Syntarsus* and *Archaeopteryx* in the characters considered above are so great in my opinion that I find it difficult to accept their separation at infraordinal level, let alone at Class level – especially since in those character states they seem to be closer to each other than *Syntarsus* is to other undoubted (non-procompsognathid) theropods.

Acknowledgements

I thank Dr. GÜNTER VIOHL for accepting my late application to attend this symposium and for allowing my contribution onto the already full programme. Dr. ALAN CHARIG of the British Museum (Natural History) kindly permitted me to examine the London specimen on my way to Eichstätt, and CYRIL WALKER made the necessary facilities available to me to allow the work to proceed; I am grateful to them both. I have benefitted from an exchange of correspondence over several years with Dr. ALICK WALKER, in which some of the ideas and observations discussed here have been aired and debated. I thank the Council for Scientific and Industrial Research, South Africa, and the University of the Witwatersrand Council Research Committee for assistance to attend this symposium.

References Cited

CAMP, C. L. (1936): A new type of small bipedal dinosaur from the Navajo Sandstone of Arizona. – Univ. Calif. Publs Geol. Sci., **24** (2): 39–65.
COLBERT, E. H. and RUSSELL, D. A. (1969): The small Cretaceous dinosaur *Dromaeosaurus*. – Am. Mus. Novit., **2380**: 1–49.
CRACRAFT, J. (1977): Special review: John Ostrom's studies of *Archaeopteryx*, the origin of birds, and the evolution of avian flight. – Wilson Bulletin, **89** (3): 488–492.
GALTON, P. M. (1971): Manus movements of the coelurosaurian dinosaur *Syntarsus* and opposability of the theropod hallux (sic). – Arnoldia Rhod., **5** (15): 1–8.
HOWGATE, M. E. (1983): *Archaeopteryx* – no new finds after all. – Nature, **306**: 644–645.
HOWGATE, M. E. (1984): *Archaeopteryx's* morphology. – Nature, **310**: 104.
OLSEN, P. E. & GALTON, P. M. (1984): A review of the reptile and amphibian assemblages from the Stormberg Group of southern Africa, with special emphasis on the footprints and the age of the Stormberg. – Palaeont. afr., **25**: 87–110.
OSTROM, J. H. (1973): The ancestry of birds. – Nature, **242**: 136.

OSTROM, J. H. (1976) *Archaeopteryx* and the origin of birds. – Biol. J. Linn. Soc., **8** (2): 91–182.

OSTROM, J. H. (1981): *Procompsognathus* – theropod or thecodont? – Palaeontographica, Abt. A, **175** (4–6): 179–195.

RAATH, M. A. (1969): A new coelurosaurian dinosaur from the Forest Sandstone of Rhodesia. – Arnoldia (Rhod.), **4** (28): 1–25.

RAATH, M. A. (1972): First record of dinosaur footprints from Rhodesia. – Arnoldia (Rhod.), **5** (27): 1–5.

RAATH, M. A. (1977): The anatomy of the Triassic theropod *Syntarsus rhodesiensis* (Saurischia: Podokesauridae) and a consideration of its biology. – Unpubl. Ph. D. thesis, Rhodes University, Grahamstown, S. Africa.

RAATH, M. A. (1980): The theropod dinosaur *Syntarsus* (Saurischia: Podokesauridae) discovered in South Africa. – S. Afr. J. Sci., **76**: 375–376.

TARSITANO, S. & HECHT, M. K. (1980): A reconsideration of the reptilian relationships of *Archaeopteryx*. – Zool. J. Linn. Soc., **69**: 149–182.

WALKER, A. D. (1972): New light on the origin of birds and crocodiles. – Nature, **237** (5353): 257–263.

WELLNHOFER, P. (1974): Das fünfte Skelettexemplar von *Archaeopteryx*. – Palaeontographica, Abt. A, **147**: 169–216.

WHETSTONE, K. N. (1983): Braincase of Mesozoic birds: I. New preparation of the "London" *Archaeopteryx*. – J. Vert. Paleont., **2** (4): 439–452.

WHETSTONE, K. N. & MARTIN, L. D. (1979): New look at the origin of birds and crocodiles. – Nature, **279**: 234–236.

WHYBROW, P. J. (1982): Preparation of the cranium of the holotype of *Archaeopteryx lithographica* from the collections of the British Museum (Natural History). – N. Jb. Geol. Paläont., Mh. **3**: 184–192.

Author's address: Prof. Dr. MICHAEL A. RAATH, Bernard Price Institute for Palaeontological Research, University of the Witwatersrand, 1 Jan Smuts Avenue, Johannesburg 2001, South Africa.

Philippe Taquet

Two new Jurassic Specimens of Coelurosaurs (Dinosauria)

Fig. 1
Liassic coelurosaur from Morocco.
ACF. ; – 1A: the left hind foot partially prepared, in anterior view, with the astragalus and a tarsal bone in distal view.
ACF. 1 – 1B: the same in posterior view with the fingers I and V: with the astragalus and a tarsal bone in proximal view.
x : 2/3, Acfarcid. High Atlas Mountains, Morocco, Toarcian-Upper-Lias Jurassic.

Abstract

The detailed studies of an Upper Liassic coelurosaur from Morocco and of *Compsognathus corallestris* from France should contribute to a better understanding of the relationships between Jurassic coelurosaurs and *Archaeopteryx*.

Résumé

Les études détaillées d'un coelurosaure du Lias supérieur du Maroc et de *Compsognathus corallestris* de France devraient permettre une meilleure compréhension des relations phylogénétiques entre les coelurosaures jurassiques et *Archaeopteryx*.

Fig. 2
Compsognathus corallestris. MNHN CNJ 79.
The skeleton as described by BIDAR et al. (1972) with the two distinct blocks (scale on the photograph), Canjuers, Var, France, Upper Jurassic.

Zusammenfassung

Detaillierte Untersuchungen an einem oberliassischen Coelurosaurier von Marokko und an *Compsognathus corallestris* von Frankreich dürften zu einem besseren Verständnis der Verwandtschaftsverhältnisse zwischen jurassischen Coelurosauriern und *Archaeopteryx* beitragen.

Understanding the problems related to the origin of birds depends in part on future discoveries of new fossils from Lower and Middle Jurassic levels. In fact, the different specimens of *Archaeopteryx* are actually compared "faute de mieux" with Triassic or Upper Jurassic Reptiles, due to the scarcity of the continental faunas from the Lower and Middle Jurassic.
So the discoveries in Morocco of well preserved skeletal parts of coelurosaurs in a Lower Jurassic (Toarcian) locality located in Middle Atlas mountains are interesting.

Fig. 3 ▶
Compsognathus corallestris. MNHN CNJ 79.
The skeleton now in the Muséum National d'Histoire Naturelle (Paris), with the blocks in connection. Canjuers, Var, France, Upper Jurassic.

We gave elsewhere (JENNY et al. 1980) details on the circumstances of the discovery of this dinosaur and the data concerning the geological context in which it was found. We display here an anterior and a posterior view of the posterior left foot of this animal. The posterior view enables us to observe the elements of the first digit in articulation with practically no displacement.

On the other hand, the acquisition in 1983, by the Muséum National d'Histoire Naturelle (Paris) of the second *Compsognathus* specimen, which had been collected in the Upper Jurassic fossil locality of Canjuers (South Eastern France), enables us to redescribe in detail a coelurosaur whose phylogenetic relationships with *Archaeopteryx* are the focal point of many present-day discussions.

The *Compsognathus* from Canjuers was described by BIDAR et al. (1972a, 1972b); it is longer (1,20 meter in length) than the *Compsognathus longipes* specimen from Bavaria. It was named by these authors *Compsognathus corallestris*. Later, OSTROM (1978) expressed the view that it was a larger representative of only one species, *Compsognathus longipes*. Finally, GINSBURG (1973) and FABRE et al. (1982) considered that the specimen did belong to a distinct species.

The specimen from Canjuers was the property of a private collector who also kept in his collection other fossils from the same locality: many fishes, some rhynchocephalian reptiles, two chelonians, and one 3,60 meters long complete crocodile.

We plan to provide a detailed description of this specimen in the near future. A first remark can be made on this splendid coelurosaur: the animal is preserved in two blocks, which were not correctly fit together when they were first studied by BIDAR et al. (one block contains the major part of the skeleton, a smaller one contains most of the tail). In the first descriptions, the two blocks were placed together in such a way that the tip of the tail was in line with the curvature preserved by the first seven caudal vertebrae. In fact, this reconstruction is erroneous. We discovered that the block with the major part of the tail fits perfectly with the main block, with a down-turning of 45°. The contacts are clearly indicated and the small cracks visible on the main block extended perfectly on the second element.

After this new reconstruction, we noticed, to our surprise, that the tail of *Compsognathus corallestris* had been in fact broken just after death at the level of the seventh caudal vertebra. It also appears to have happened in the Bavarian specimen, *Compsognathus longipes*. The only two representatives of *Compsognathus* lying on the same side (right) in the sediment, with the same post-mortem posture: neck curved backwards above the back, base of the tail lifted upward, tail broken at the level of the seventh vertebra.

In conclusion, the detailed studies of this new Moroccan coelurosaur and of the *Compsognathus corallestris* should contribute to a better knowledge of the history of the group and should contribute also to a better understanding of the relationships between ante-Upper Jurassic coelurosaurs and *Archaeopteryx*.

References Cited

BIDAR, A., DEMAY, L. & THOMEL, G. (1972a): Sur la présence du Dinosaurien *Compsognathus* dans le Portlandien de Canjuers (Var) – C. R. Acad. Sc., **275**: 2327–2329, Paris.

BIDAR, A., DEMAY, L. & THOMEL, G. (1972b): *Compsognathus corallestris*, nouvelle espèce de Dinosaurien théropode du Portlandien de Canjuers (Sud-Est de la France) – Annales Mus. Hist. Nat. de Nice, **1**: 9–40, Nice.

FABRE, J., BROIN, F. DE, GINSBURG, L. & WENZ, S. (1982): Les Vertébrés du Berriasien de Canjuers (Var, France) et leur environnement – Geobios, **15–6** 891–923, Lyon.

GINSBURG, L. (1973): Paléoécologie des calcaires lithographiques portlandiens du petit plan de Canjuers (Var) – C. R. Acad. Sc. **276**: 933–934, Paris.

JENNY, J., JENNY-DESHUSSES, C., LE MARREC, A. & TAQUET, P. (1980): Découverte d'ossements de Dinosauriens dans le Jurassique inférieur (Toarcien) du Haut-Atlans central (Maroc) – C. R. Acad. Sc., **290**: 839–842, Paris.

OSTROM, J. (1978): The osteology of *Compsognathus longipes* Wagner – Zitteliana, **4**: 73–118, München.

Author's address: Prof. Dr. PHILIPPE TAQUET, Institut de Paléontologie, Muséum National d'Histoire Naturelle, 8 rue Buffon, 75005 Paris, France.

D. Stefan Peters & Wolfgang Fr. Gutmann

Constructional and Functional Preconditions for the Transition to Powered Flight in Vertebrates

Abstract

Active flight is a mode of locomotion that is restricted to well advanced coelomates. Therefore the problem arises which constructional properties precluded the evolution of flight in primitive forms and which were the preconditions for the transition to active flight in higher tetrapods.

Primitive coelomates, fishes included, obtain their propulsive force from sideways movements and possess small horizontal planes. Therefore the preconditions for lift generation which would be required in the necessary intermediate stage leading to active flight were lacking. In addition to that the body is too unstabile to provide the preconditions for effective movements. In fish constructions the body is kept constant in length, stability of the crossection of the body is however ensured by the densely packed but heavy musculature.

In primitive tetrapods the skeletal systems of the body trunk developed into a frame that is controlled by a decreasing mass of muscles. Consequently the body weight is reduced and the unrestrained deformability of the body trunk is effectively suppressed.

Furthermore the emergence of the appendages with their skeletal bars and deformation restricting joints relieve the body trunk of its older locomotory function and offer the functional elements for the emergence of lift generating planes and for effective steering movements that could be transformed into thrust generation of effective flight.

In organisms performing quadrupedal locomotion, transition to gliding requires an extended lift generating surface that is provided by lateral folds supported by the appendages. As prebirds had already turned to bipedal locomotion the hind limbs could not become part of the lift generating apparatus. So the planes were provided by the feathers of the wings and the tail with the laterally positioned feathers.

Active flapping was derived from steering movements of distal parts of the fore-limbs. Intensification of steering with smaller planes replaced steering movements of the body. The increased influence of wing movements on the gliding course resulted in the addition of propulsion to gliding.

Zusammenfassung

Aktiver Flug als eine Lokomotions- und Antriebsweise ist auf höherentwickelte Coelomaten beschränkt. Daraus erwächst das komplementäre Problem, welche konstruktiven Vorbedingungen in niederen Vertebraten oder Coelomaten den Übergang zum Flug verhindert haben.

Primitive Coelomaten unter Einschluß der Fische, entwickelten ihre Antriebskraft aus seitlichen Biegungen des Körpers und besitzen nur kleine horizontale Flächen. Daher fehlten die Vorbedingungen für die Antriebserzeugung im notwendigen Zwischenstadium des Gleitens. Zudem ist der Körper zu instabil, um die effektiven Mechanismen der Flugbewegung sicherzustellen. In Fischen, die in wenigen Fällen zum Gleiten nach Schlängelantrieb im Wasser übergehen, wird der Körper nur längenkonstant gehalten, die begrenzte Stabilität des Querschnittes wird durch dicht gepackte und daher schwere Muskeln gesichert.

Von den niederen Tetrapoden an entwickelt sich das Skelett-Muskel-System des Körperstammes zu einem skelettalen Rahmen weiter, der in seiner Form durch eine sich vermindernde Menge von Muskeln kontrolliert wird. Folglich wird das Körpergewicht reduziert und die nicht-restringierte Verformbarkeit des Körpers wirksam unterdrückt.

Weiterhin führt die Ausbildung der voll beweglichen Extremitäten zu einer Entlastung des Körperstammes, der in geringerem Maße durch Biegungen die Lokomotion unterstützen muß. Die paarigen Extremitäten liefern auch die Vorbedingungen für die Entstehung horizontaler Flächen, die ein Gleiten erlauben.

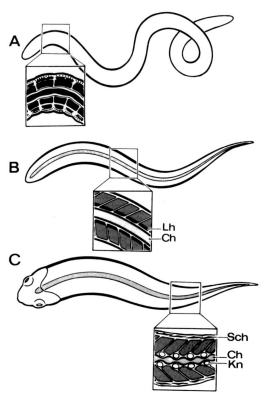

Fig. 1
Basic organization of coelomates.
A. Early worm-like stage; lateral position of coelomic cavities points to the primitiveness of lateral undulatory movements.
B. Chordate construction with a notochord ensuring constancy of length. Deformability is restricted but not totally suppressed (Ch-notochord, LH-coelomic cavity).
C. Craniote construction with developing skeletal structures (Sch – scales, Ch – notochord, Kn – bony vertebral elements). Weight of densely packed muscles and prevailing lateral propulsive movements preclude transition to active flight.

Von quadruped laufenden Organismen aus wird das Gleiten als Übergang zum Flug durch Ausbildung von seitlichen Körperfalten gesichert, die vor allem durch die Extremitäten gespannt werden. Im Falle der Vögel war vor dem Übergang zum Gleiten schon bipedes Laufen erreicht; die Stellung der Hinterextremitäten unter dem Körper ließ ihre Nutzung als Hilfseinrichtungen bei der Auftriebserzeugung nicht zu. Im Gefolge davon mußte der durch Federn verbreiterte Schwanz zusammen mit den Flügeln den Auftrieb besorgen.

Aktiver Antrieb durch Flügelschlag wurde bei allen fliegenden Wirbeltieren dadurch erreicht, daß distale Abschnitte der vorderen Extremitäten für das Steuern und Bremsen genutzt wurden. Auf diese Weise wurde der Körper von aufwendigeren Steuerbewegungen entlastet. Intensive Steueraktionen und Beeinflussung der Gleitbahn leiteten in Flügelschlag über, der dem Gleiten zugeschaltet wurde und den Gleitapparat zum Teil weiternutzt.

1. Introduction

The only organisms that show active flight as a mode of locomotion are of advanced coelomate constructions. This poses the question as to the biomechanical preconditions of the organismic construction that would open this phylogenetic pathway. The answer to the complementary question would give the reasons for the inability of primitive coelomates to develop actively flying representatives. In the attempt to elucidate the biomechanical aspects of the organismic construction, other physiological mechanisms, especially the ability to breath in the free air, are disregarded.

2. The skeleton-muscle-system in flying vertebrates

Comparison of flying vertebrates will reveal that the body construction of all of them is constituted of a well advanced skeleton-muscle system. The shape of the trunk and the appendages are effectively controlled by the muscle bracing of the skeletal frame. All movements are performed in an effective

Fig. 2
Flying Fish. The basic lateral undulating mechanism is used for acceleration in the water to break through the surface and to recover thrust by using the tail fin. (After KLAUSEWITZ 1960)

manner because well developed joints help to suppress unwanted deformations resulting in a loss of energy without contributing to propulsion. It is highly suggestive that the high level of efficiency was an essential precondition for the transition to gliding and flight. In the vertebrate series this state was not attained before the reptilian level.

As compared to the higher tetrapods primitive coelomates are highly deformable worm-like constructions that constitute fluid filled hydraulic systems (Fig. 1). Because of the lack of skeletal structures and joints every purposeful movement has to be accompanied by the suppression of a great number of unwanted deformations by active muscle control resulting in a high rate of energy expenditure that could not be directly converted into thrust (GUTMANN 1972, 1977).

The position of the coelomic cavities in the flanks of the body of coelomates indicates that the primitive mode of locomotion consisted of lateral bending movements that travelled down the body and generated thrust in the water. This is a reliable inference because fluid filled organs could reduce the stiffness of the moving construction in a way most favourable for lateral bending. Both aspects, the low level of efficiency of the motor system and the sideways bending, would have precluded the transition to active flight.

3. The horizontal planes required for flight

None of the flying vertebrates obtains its power for propulsion from sideways movements. This indicates that active flight could only develop from horizontal planes that already existed.

While the lateral bending movements persisted in the fish and even in the amphibian levels of vertebrate evolution the efficiency of the motor apparatus was considerably advanced. This is the result of the formation of the flexible body axis that guarantees constancy of length and suppresses a multitude of deformations still possible in the worm stage. However, the price for this advantage can be seen in the densely packed body muscles of fishes. Their weight would by itself preclude the transition to active flight. Nevertheless some fishes can leave the water and glide through the air (Fig. 2). But this is made possible by a powerful propulsion in the water that allows the flying fishes to break through the surface and to follow a ballistic trajectory through the air. In some flying fishes the paired fins underwent transformations into airfoils that allow gliding over longer distances. Propulsive power must however be

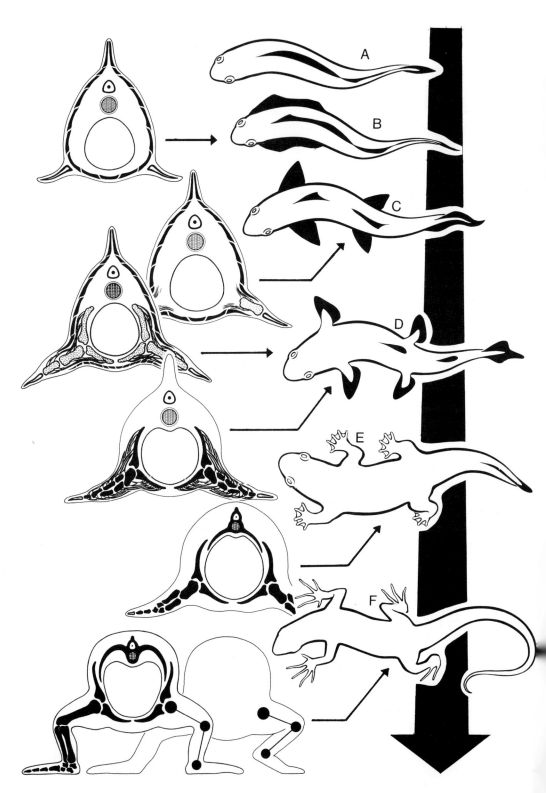

236

generated in the water by sideways movements of the tail fin. But there can be little doubt that this thrust producing mechanism can not be transformed in a way to allow active flight in the air (KLAUSEWITZ 1960).

From the contrast displayed by the body constructions of fishes and flying vertebrates respectively the further essential preconditions of flight become apparent. Only after the formation of limbs that could carry the body was the body trunk relieved of its older function of generating the main propulsive force for locomotion. Simultaneously the body trunk was transformed into a more perfected skeletal frame. The thickly packed muscles typical of fishes could become reduced because the muscles gained direct insertion on the skeletal frame and could generate movements in a much more effective way. Much weight of muscles was thereby saved (Fig. 3). The limbs themselves are constituted of discrete skeletal elements that are coupled together in well developed joints and can also be moved in an effectively controlled manner. This level of organization was well established in the reptiles and persisted in the mammalian level.

In all transformation sequences leading into gliding and flight the canonical system of the limbs is used. Their formation and organization could be explained by the requirements of a primitve fish with paired appendages which were used to walk on land. The combination of bending movements of the body trunk with the effective gait of the limbs and the avoidance of disadvantageous frictions on the ground was only possible if the basic pattern of bones and major joints were formed (PETERS & GUTMANN 1978; PETERS, in press). In the contribution of D. S. PETERS (this volume) the altered use of this system in birds will be discussed.

4. Preconditions for gliding and volplaning

Generation of lift in gliding must have preceded active flight because the intermediate stage of gliding had to bridge the energy gap between walking and powered flight. Direct transition to powered flight over the intermediate stage of hovering can be convincingly rejected (CLARK 1976). Such a change, starting from ambulatory movements, would have been a saltatory modification requiring an abrupt rise of energy output in flapping (Fig. 4).

But there is a second difficulty with such a highly improbable change. As the centre of mass of an elongate reptilian tetrapod lies between the fore and hind limbs, hovering movements executed by the arms would not support the body in a useful way.

However, the formation of the lift generating planes of tetrapods could not occur and become more effective in an arbitrary way. For the transition to gliding the elongate construction required the emergence of lift generating planes over a longer portion of the body. The centre of mass that lay between the two pairs of appendages had to be positioned within the lift producing plane. An extended lift-generating plane would have given each portion of the body its own lift.

Fig. 3
Formation of horizontal planes utilizeable for gliding and flight in the phylogeny of vertebrates.
A. Primitive craniote with typical lateral undulations.
B. Formation of ventral flattenings and lateral folds which prevent the immobile head region from sinking downwards.
C. Subdivision of lateral folds into two pairs of fins that are independently moveable.
D. Fish-like precursor of tetrapods.
E. Amphibian construction. Locomotion is performed by a combination of lateral movements and gaits of the short limbs.
F. Reptilian stage: Elongation of limbs, continuous relief of body trunk of movements aiding ambulation.
  The cross sections to the left show the organization of the developing limbs and the necessity of the observable subdivision of the bony bars and joints. Gliding and flight could only begin after the formation of the limbs and the considerable reduction of lateral bending mechanisms.

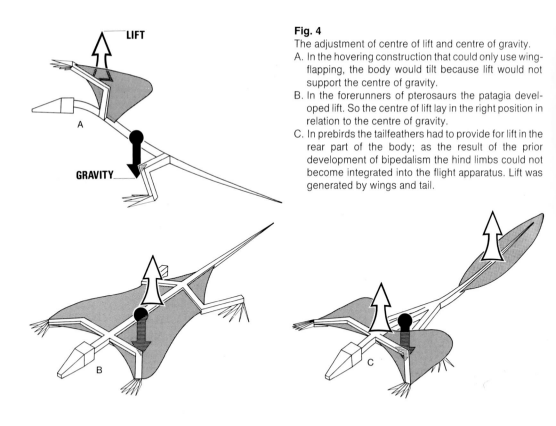

Fig. 4
The adjustment of centre of lift and centre of gravity.
A. In the hovering construction that could only use wing-flapping, the body would tilt because lift would not support the centre of gravity.
B. In the forerunners of pterosaurs the patagia developed lift. So the centre of lift lay in the right position in relation to the centre of gravity.
C. In prebirds the tailfeathers had to provide for lift in the rear part of the body; as the result of the prior development of bipedalism the hind limbs could not become integrated into the flight apparatus. Lift was generated by wings and tail.

Upward thrust produced by flapping arms would lift the anterior part of the body, but the whole body would be tilted because the centre of lift could not support the centre of mass. From this we can infer that lift generating planes had to extend over a longer part of the body. The centre of lift and the centre of mass had to be adjusted from the very beginning of gliding. The centre of mass had to lie in a forward position as compared to the centre of lift. This was only possible when there existed enough lift generating surface further back in the organism. If the centre of lift lay too close to the centre of mass gliding would be disturbed by a great instability, the body could not be held on a stabile course. The location of the centre of mass behind the centre of lift would not allow gliding at all.

The preconditions for effective gliding would be given if the body has a sufficiently long plane. In this case a relatively stable gliding course could be followed by the animal. It would also minimize the steering requirements in the early stages of gliding. However, postulate that gliding had to develop in a way that guaranteed stability of the gliding course would not contradict the necessity of steering movements from the very beginning. Only an in-built stability of the course would have allowed for the gradual perfection of steering movements on the basis of adequate changes in the nervous system. All flying tetrapods must have passed through stages that fulfilled these demands.

In the forerunners of pterosaurs it seems most likely that a lateral fold of the body wall, a patagium, developed between the fore and hind limbs (Fig. 5). When it gained width, it was spanned tight by the extremities. Further skinfolds may have appeared between the anterior part of the tail and the hind limbs and in front of the fore-limbs. The transition to gliding in the ancestors of bats was certainly based on similar changes. Both organismic groups had ancestors that walked in a quadrupedal manner. So the four limbs provided the support of the membranes that allowed volplaning.

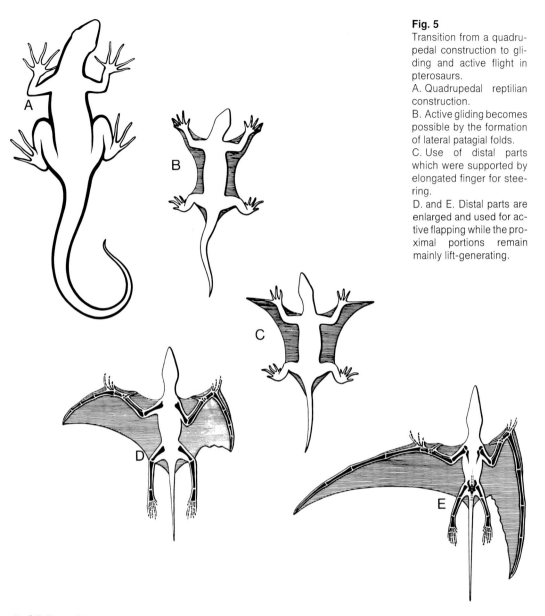

Fig. 5
Transition from a quadrupedal construction to gliding and active flight in pterosaurs.
A. Quadrupedal reptilian construction.
B. Active gliding becomes possible by the formation of lateral patagial folds.
C. Use of distal parts which were supported by elongated finger for steering.
D. and E. Distal parts are enlarged and used for active flapping while the proximal portions remain mainly lift-generating.

5. Gliding of bipedal prebirds

Volplaning of pterosaurs marks a significant difference with the ancestors of birds. Before prebirds took to gliding they had developed a bipedal mode of locomotion. In the formation of the lift generating planes only the forelimbs could offer support for the lateral extension of membranes. Therefore the birds show only very rudimentary membranes between the stylopodium and the zeugopodium and between the stylopodium and the body.

But most of the plane that allowed gliding was generated by feathers. These enlarged the plane supported by the fore-limbs. The continuation of the lift producing plane was interrupted by the hind-limbs that had gained a new position precluding their integration into the gliding and prospective flying apparatus. This was compensated by the transformation of the feathers of the tail. As they assumed a

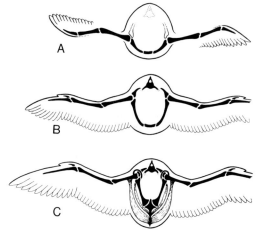

Fig. 6
Development of wings and active flight.
A. In a gliding pre-bird the distal parts could be used as steering devices and as brakes.
B. Enlargement of the distal portions improved steering and made flapping for the prolongation of gliding possible.
C. In the fully developed wing the proximal portion is mainly lift-generating while the distal portion effects propulsion.

lateral position the tail formed a posterior part of the lift generating apparatus. The low aspect ratio of the flattened out tail did not provide much lift but nevertheless it was an indispensable condition for the transition to gliding. As a result the requirements for a stabile gliding path with the centre of mass lying in front of the centre of lift, are met in a way that differs from the situation in prepterosauria and prebats. These differences were determined by the modes of walking that were developed before the onset of gliding.

While gliding must be considered a necessary intermediate step it does not automatically lead into active and powered flight. Gliding cannot become improved beyond gliding (JEPSEN 1970). So the question of this essential transition is posed. As far as birds and their flying machine is concerned, this problem will be treated in the article of PETERS (this volume). But in the following paragraphs we will try to give a very general explanation for the modifications that resulted in the powering of gliding.

6. Transition to active flight

From the beginning of gliding the requirements of steering must have been met. Perfection of steering certainly went in pace with the improvement of gliding. In the incipient steps control over the gliding path was effected by movements of the appendages that could vary the size of parts of the lift-generating planes. In addition to that, bending or twisting of the whole body proved to be useful mechanism to effect more abrupt alterations of the gliding course. More differentiated and improved steering was possible by changes of the angle of attack of small parts of the lift generating planes. For an optimal efficiency these had to be situated far from the body so that they could work over greatest leverage.

When these distal parts acted more and more effectively, energy consuming bending and twisting movements of the body itself for steering purposes could be avoided. This must have had a considerable economizing effect. Enlargement of the steering planes took over the control of the gliding path.

Fig. 7
Transition of a bipedal reptilian construction to flight.
A. Quadrupedal reptilian construction.
B. Sidebranch leading into gliders and flyers with four legs integrated into flight apparatus (Pterosaurs).
C. Bipedal reptile with feather-like structures that started gliding. Emergence of patagial planes.
D. Arrangement of "stiff" feathers that allowed improved gliding and steering movements with the distal parts of the forelimbs. Effective gliding required a longitudinally extended lateral fold that consisted of the wings and the tail with the laterally positioned feathers.
E. Stage for which *Archaeopteryx* might have been a representative. Steering movements had gradually changed into active flapping.
F. Completed bird construction. The body has acquired the form of least resistance (Laminarspindel), the wings are fully developed with proximal mainly lift-generating portions and distal propulsive parts.

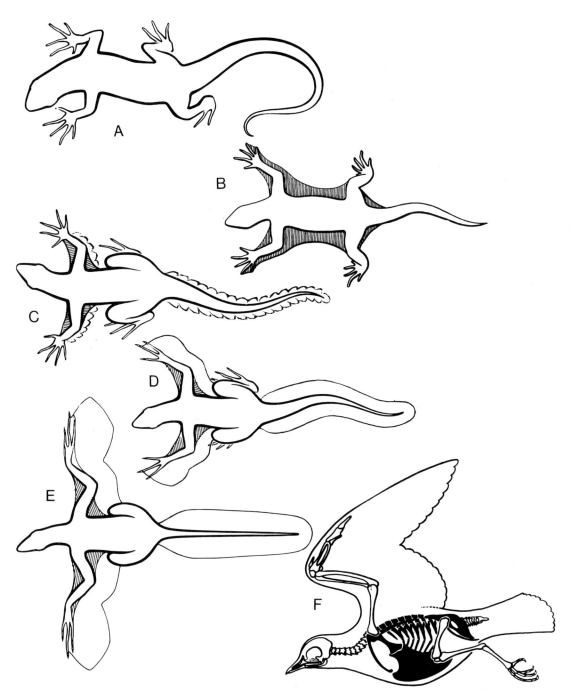

In addition to improvement of the steering capabilities the changes of the geometry of the lift-generating and steering planes by the increase of the aspect ratio allowed more effective gliding and extended the gliding path. This means that the several aspects of aerodynamics are not in conflict with each other, in reality they would drive the transformation process according to several superimposed selective benefits.

In some instances steering control could mean that the gliding path had to be cut short or that efforts would be necessary to prolong it beyond a landing site that was aimed at in an earlier stage of the descent. For this the enlarged steering planes could be used as brakes or by a controlled increase of the angle of attack for a swoop up. Such powerful movements required an effective action of muscles that held the developing wings against the air currents and moved them in a strictly controlled way (Fig. 6). Under the influence of the resulting selection constraints the whole apparatus underwent appropriate transformations. These resulted in a transcendence of effective steering movements to active flight by means of flapping wings. The intermediate stage was certainly a kind of well controlled mode of gliding which, when necessary, was prolonged by some powerful flapping movements. Intermittent flapping of wings was primarily a mechanism that prolonged the gliding course. Steering and increasingly effective propulsion were added to the gliding and volplaning mechanisms. This is in good accordance with the fact that not only in birds but also in bats and pterosaurs the more distal parts of the wings are used for propulsion while the proximal parts continue to function more as lift-generating planes. We can therefore conclude that the development of active flight followed the same principle in all flying tetrapods (Fig. 7).

7. Final remarks

In concluding, the methodology utilized here for this reconstruction should be mentioned. It is at least partly responsible for the results given and their divergence from earlier concepts of bird evolution summarized in the book of STEPHAN (1974). The methodology was derived from the organism-centred concept of evolution (GUTMANN & PETERS 1973, GUTMANN & BONIK 1981). According to this, gradualism of all transformations is dictated by the complexity of organisms. Therefore, in all stages the occurrence of saltatory changes is precluded because they would automatically be deleterious. Consequently our reconstructions were designed by starting from the organisms as energy converting systems that could only develop in an orderly way.

Since every stage can behave by utilizing degrees of freedom of the preexisting construction and exploit a broad spectrum of environmental conditions, the phylogenetic pathway was traced along constructional and not environmental constraints. The organismic construction decides which environmental conditions are essential and which properties of the environment would allow the next alterations to take place. So the scenario for the evolution of flight is designed by projecting functional requirements of the construction into the environment and by assuming that the organisms could have actively penetrated the environmental conditions conducive for the next transformation steps. As a result old alternative hypotheses for the evolution of flight are bypassed.

To formulate this in a more positive way we would like to point out that all stages of the highly versatile constructions on the phylogenetic path to birds and their flying capabilities could have lived in a wide spectrum of environmental conditions. Consequently there is no reliable methodology to find narrow ecological limitations; the resolving power of the theory is also restricted to more general and dominant organismic-environmental interactions.

Neither the epithets cursorial nor arboreal for the evolution of the bird construction could reasonably convey the decisive aspects of the reconstructed transformation processes that are based on evolutionary explanations. One should not rule out the possibility that all or several intermediate stages could climb in trees. But the main lineage must have been tied to the cursorial mode of locomotion and the use of any kind of elevation to gather speed for gliding.

References Cited

See Peters (this volume, page 248).

Authors' address: Priv.-Doz. Dr. D. STEFAN PETERS and Prof. Dr. WOLFGANG FR.GUTMANN, Forschungsinstitut Senckenberg, Senckenberg-Anlage 25, 6000 Frankfurt, Federal Republic of Germany.

Dieter Stefan Peters

Functional and Constructive Limitations in the Early Evolution of Birds

Abstract

Functional interpretations of the constructions of birds, considering also the physical limitations that confront every flying vertebrate, make it highly probable that:

1. The prebird was a bipedal endothermic animal covered with feathers;
2. The speed needed to begin successful flapping flight could be achieved by the prebird only with the additional contribution of some external force.

Since an arboreal origin of birds is not consistent with the development of avian construction, prebirds very likely lived on mountain slopes where they could leap downwards, gaining in this way the speed needed for active flight. Steering movements with the forelimbs could have evolved in connection with running and jumping in riven terrain. Later on such movements could be executed during gliding; they became more and more effective with higher speed and gave rise to the lengthening of the distal part of the forelimbs and to active wing flapping.

Zusammenfassung

Die funktionsmorphologische Interpretation anatomischer Befunde und die Beachtung der für alle fliegenden Vertebraten geltenden physikalischen Limitationen zwingen zu dem Schluß, daß:

1. der "Vorvogel" ein biped laufendes, endothermes, befiedertes Tier war;
2. für die Entwicklung des aktiven Fluges eine Geschwindigkeit erreicht werden mußte, die der Vorvogel aus eigener Kraft wohl nicht erzielen konnte.

Da die Entstehung der Vogelkonstruktion mit einer arboricolen Lebensweise kaum vereinbar ist, gewinnt die Annahme an Wahrscheinlichkeit, daß die Vorvögel auf Gebirgshängen lebten, wo sie bei Abwärtssprüngen die für den (zunächst nur mit wenig effektiven Mitteln betreibbaren) aktiven Flug nötige Geschwindigkeit erreichten. Steuerbewegungen mit den Vorderextremitäten, die schon beim Rennen in zerklüftetem Gelände eine Rolle gespielt haben mochten und zur Verlängerung der distalen Teile der Vorderextremitäten führten, konnten bei Gleitsprüngen eingesetzt werden und gewannen mit wachsender Geschwindigkeit immer größere Wirkung. Aus solchen Bewegungen konnte der aktive Flügelschlag entstehen.

I. Introduction

The goal of this paper is to analyse some anagenetical aspects of the evolutionary pathway terminating in the construction of a bird. I shall focus my attention on constraints which must have canalized this pathway regardless of the exact genealogy of the organisms involved; although I assume, of course, that the ancestors of birds are to be found among the reptiles or, more precisely, among the archosaurs.

What is a bird? Considering modern species, a long list of features characterizing these animals can be compiled (HENNIG 1983). If *Archaeopteryx* is included, however, only amazingly few characters remain which can be interpreted as structures peculiar to birds. Among these structures the most important ones seem to be:

1. Feathers.
2. The lateral flexion and the suppression of other flexions in the wrist.

3. The reversed hallux.
4. The strong and specific constructional and functional divergence between pectoral and pelvic limbs.

If we find plausible and noncontradictory explanations for the development of these four features we may gain from these explanations an understanding of the anagenetical evolution of early birds.

II. The feather

Since ancient times, feathers were regarded as the key feature of birds. At present there seems to be no doubt that feathers are derived from reptilian scales. Circumstantial evidence supporting this view was furnished by anatomical, embryological, and biochemical studies (BLASZYK 1935, RAWLES 1963, THOMSON 1964, MADERSON 1971). A featherlike structure found in Jurassic deposits of Kazakhstan and known as *Praeornis sharovi* RAUTIAN 1978 fits well as an intermediate stage between a scale and a feather assuming that it is really a feather, which seems to me not quite certain. Its general appearance is that of a feather, but the rami are provided with laminae instead of radii.

In modern birds, feathers fulfill a variety of functions. In any case feathers make a very efficient insulating covering, they protect the body against external influences, and they are part of an excellent flying apparatus. In addition to these main functions, feathers can be modified in color and form for use in courtship and threatening display, as a concealing pattern, and so forth. But, which was the first function? What kind of selection forces could have promoted the transformation of a scale into a feather?

Whatever the answer, it should be in concordance with the following prerequisites.

a) In ectothermic organisms an insulating covering would be unnessecary or even disastrous.
b) Organisms of small or moderate size are very probably able to afford endothermy only if the energetic costs can be minimized by an insulating covering.
c) A covering consisting of minutely split elements provides better insulating properties where there are no aerodynamic demands for a flying apparatus to consist of split elements.
d) Since active flight is an expensive form of locomotion it seems more likely that animals evolving active flight were endothermic.

If these four assumptions are realistic they suggest that feathers were primarily part of an insulating covering which preceded the evolution of active flight. Nevertheless the question remains: How can endothermy evolve without insulating structures, while insulating structures cannot precede the evolution of endothermy? This seems to be a difficult question which cannot be solved by the pseudoexplanation that there must have been some kind of coevolution. As far as I know the only model for the evolution of feathers yielding sufficient plausibility to solve this riddle was published by Regal (1975). It is a model which gives attention to continuity of different functions and adaptive advantages, starting with elongated scales acting as shields to solar radiation. To my knowledge, a similarly unbroken model for the origin of feathers in connection with flight is still lacking.

III. The avian wrist

The mobility of the avian wrist is unique among the tetrapods. Birds can flex their hand against the ulnar side of the forearm, other flexions being almost impossible. At the same time the mechanical arrangement of the skeletal elements and joints of the forearm and the hand results in automatically synchronized flexions or extensions at the elbow and wrist joints. This technical refinement is very useful e. g. when external forces acting on the hand and the forearm in the same direction tend to change the configuration of the wing. Since the elbow and the wrist flex (or extend) in opposite directions, the effect of these external forces is at least partly reduced by means of the linkage mentioned above. Thus the muscular effort required by the bird to stabilize its wings can be markedly reduced (PETERS 1984).

When analyzing other parts of the wing, this mechanism has to be kept in mind. We shall understand then for instance that the Musculus extensor metacarpalis radialis in spite of its name cannot be the chief unfolder of the wing, as OSTROM (1976) has suggested. On the contrary, this muscle flexes the elbow, and thus flexes indirectly the hand too. It can extend the hand only if the elbow is locked by other muscles. But even in this case the extension is suppressed largely by the linkage of the two joints. The main functions of M. ext. metacarp. rad. are (1) to strengthen the linkage of the elbow and the wrist by pressing together tightly the joints between the carpometacarpus and the radiale, between the radiale and the radius, and between the radius and the humerus and (2) to prevent undesired extensions of the wing (PETERS, 1985). This is especially important during the power stroke when centrifugal forces act to pull apart the skeletal elements of the wing. In accordance with these demands, the M. extens. metacarp. rad. and its skeletal sites of origin and insertion (Processus supracondylaris dorsalis humeri, Processus extensorius carpometacarpi) are most strongly developed in birds with sustained powered flight (e.g. Apodiformes, Charadriiformes).

In *Archaeopteryx* lateral flexion of the hand has developed already to a certain degree, whereas the linkage of the elbow and the wrist may be still lacking. Inferring from the anatomy of the wrist we may assume that lateral flexion could be reinforced beyond the *Archaeopteryx*-stage only with an increasing ability to shift the relative positions of ulna and radius, parallel to their long axis. Otherwise the lateral flexion would produce a gap between the radius (or the radiale) and the hand. It means that the further evolution of lateral flexion proceeded in conjunction with the evolution of the kinetic linkage of the elbow and the wrist.

Selecting forces acting in favour of such constructions are incompatible with the function of a climbing extremity. It is highly improbable that lateral flexion, and suppression of other flexions should have had any advantage in a hand used for climbing on stems or branches. An adaptive advantage of the linkage system of the elbow and the wrist is even less probable in a climbing prebird. However both constructions are very advantageous in a wing (PETERS, 1984).

IV. The reversed hallux

As was shown by HECHT (1980) the reversed hallux is a feature found in birds but not in other archosaurs. This feature very often was interpreted as an evidence of the arboreal origin of birds. However a grasping foot can have different functions as can be seen in modern birds. Thus we are not forced by the reversed hallux to believe in the "arboreal theory".

V. The divergence between pectoral and pelvic limbs

Considering the significance of the differently constructed limbs of birds we have to occupy ourselves with two items.

1. As stated by PETERS & GUTMANN (1978), and PETERS (in press) the "bauplan" of the tetrapod limb, its canonical pattern of elements, can be explained as the result of mechanical constraints. This pattern can be simplified or changed if and only if the limitations given by the constraints are changed too.

The type of construction realized in the legs of birds, including *Archaeopteryx*, corresponds to a limb which is not extended laterally but moving in a parasagittal plane while walking. When moving in this way such a limb is not forced to twist internally as for example the legs of a newt must do. Thus the zeugopodium and the autopodium can be simplified by reducing and fusing elements which were necessary in the primitive stage of tetrapod limbs to fulfil the twisting movements, but which have now lost their function. This kind of a limb with a reduced skeleton can evolve in cursorial animals (see for instance the limbs of Ungulata) whereas it is very improbable in arboreal creatures, which rather should retain and improve the structure capable of rotating movements (see the limbs of Primates).

In contrast to the hind limbs, the forelimbs of birds remain much closer to the canonical pattern of tetrapod-limbs in spite of losing two fingers. Apparently the forelimbs were not confronted with the same selection forces as the hind limbs. It means that they were released from the function of walking before or soon after they achieved a position in a parasagittal plane. In other words, the ancestors of birds became bipeds before their forelimbs could be adapted to walking in a parasagittal position. Maybe the ability of the forelimbs to rotate in pronation and supination continued to be of some use as is indicated by the unfused condition of ulna and radius.

When the distal parts lengthened as an adaptation for steering and gliding a new problem arose. Long feathers on the hand are a bulky structure when not used. The best way to place them in unoperative position is as done in modern birds, i.e. by lateral flexion of the hand. This kind of flexion could be improved because ulna, radius, and some carpalia were still available as free elements. In the course of the evolution of this highly specialized mechanism those bones changed their function, losing some of the functions they had in the primeval tetrapod limb, but aquiring new qualities very advantageous for a wing.

2. Doubtlessly, the evolutionary pathway of any flying organisms had to start with some kind of airfoil. The larger this initial lift-generating plane was, the better. In default of perfect wings, these flying "beginners" had to use what they had, which means that their whole body had to become an airfoil, as can be seen still in gliding tetrapods (e.g. flying frogs, flying squirrels etc). The anatomy of pterosaurs and bats indicates very strongly that their hind limbs were part of the initial flight apparatus. However no such indication can be found in birds. Apparently early birds were not able to spread their legs laterally in the same way they spread their forelimbs.

Combining these two points I would conclude that the ancestors of birds were bipedal before starting to fly. Their legs adapted to bipedalism could not be straddled in the same plane as the forelimbs and thus never became part of the lift generating plane. Based on this construction, in the course of the further evolution of birds, the wings and the legs remain independent from each other with respect to different selection forces. I suppose that this is the reason for the fact that there are swimming, diving or even flightless birds whereas no analogy can be found in pterosaurs or bats.

VI. The start to flight

Summarizing the foregoing interpretation of anatomical evidences I would assume that "proavis" was an endothermic, bipedal, cursorial animal when it entered the new pathway leading to active flight. Now the question remains how early birds became flying animals. We can try only to reconstruct in a model the events which might have happened. This model should be consistent with the results of the foregoing considerations and with the physical frame conditions which confronts every flying organism as explained by Gutmann & Peters elsewhere in this volume. In powered flight the power required is high both at high and low speed, and reaches its minimum at some intermediate speed (Pennycuick 1975). For this reason it seems highly improbable that birds entered their aerial career by active flapping flight. The energetic costs would have been too high. There must have been some initial speed generating behaviour which enabled the early birds to enter the stage of active flight at low energetic costs. The cursorial hypothesis of the origin of flight in its classical form (Nopcsa 1907) thus has to be abandoned. This hypothesis was modified in many details by Ostrom (1974, 1979) and differently by Caple, Balda & Willis (1983). As far as I can see, their modified theories explain, why the forelimbs and their feathers were selected for greater length and strength and how the ability to control and manoeuvre the body axes could have improved, but they do not explain the origin of flight. Lift simply does not increase running speed. But running combined with lift production is the only source of speed offered by these theories.

In the same way the hypothesis that the feathers on the forelimbs and on the tail were used for display (Stephan 1974, Cowen & Lipps 1982) explains only the lengthening of these parts of the body; the transition into active flight is not explained.

It seems inevitable that some external forces must have generated the initial speed, namely that gliding preceded active flight. In the foregoing paper (Peters & Gutmann, this volume) the general constructive constraints acting on an early birdlike glider were described. Since the hindlimbs could not be integrated into the gliding surface, the main parts of the lift generating plane were the forelimbs and the tail. The better the aerodynamical properties became, the longer could be the gliding distances, and the longer the distances, the greater the necessity for steering abilities. Very probably some steering mechanisms preceded flight. Since, concuring with STEPHAN (1974), we mentioned briefly in a previous paper (PETERS & GUTMANN 1976) that balancing could have been a prefunction of the wing, I favour in this respect (but not with regard to the beginning of flight) the model of CAPLE, BALDA & WILLIS (1983). Such steering abilities could be used during glides. Their efficiency even increased proportionally with the gliding speed. Relatively weak movements of the forelimbs could then produce strong influences. This could have been the beginning of active powered flight. The lenthening of the flight feathers and of the forelimbs, especially in the distal parts must have been positively selected by the aerodynamical demands of both steering and flight.

It seems tempting now to accept the theory of the arboreal origin of flight, because it offers a hypothesis concerning initial speed by the scenario of jumping and gliding from high branches. Nevertheless this theory is inconsistent with the evidences of the bird construction mentioned above. Another contradiction can be added. Climbing certainly has to be regarded as an essential ability of an arboreal animal. Even a glider has to return somehow to its arboreal environment. In this respect it is hardly imaginable how flight-feathers should evolve on a forelimb used for climbing. This fact does not make the arboreal theory very convincing. The auxiliary hypothesis that prebirds were bipedal arboreal animals does not sound plausible either.

This being the case the only other reasonable alternative available, is a modified cursorial theory. Since leaping horizontally is not more efficient than running, I suppose that ancestors of birds were living in a mountainous or hilly habitat with a rough structure of the surface where they could jump downwards along the slopes, gaining in this way the speed needed for the evolution of active flight. Possibly leaping was an essential element of their activity as described by CAPLE, BALDA & WILLIS, and long downward jumps happened at first only in escape. The details, I am afraid, are unattainable for reconstruction. Nevertheless some features seem to be indicative of a xerothermic habitat, for instance the high temperature of birds and the maintainance of the excretion of uric acid, contrary to mammals. REGAL's model for the origin of feathers concurs with this hypothesis too. Moreover this hypothesis is in accordance with the regrettable fact that no fossils of those earliest birds are known. In a habitat like that sketched above it is very difficult to become a fossil. It was a fissured habitat with boulders and scattered bushes where jumping was an adequate manner of locomotion besides running, and where a reversed hallux provided a firm foothold on edges of stones or on branches if the latter could be reached by jumping or by running on slanting stems and so forth.

Of course these are not strict proofs for my hypothesis but at least there are no contradictions.

My considerations can be summarized as follows. The ancestors of birds were endothermic bipedal animals covered with feathers. They acquired steering abilities in connection with running on cleft and more or less steep slopes or/and with leaps which they used while hunting insects and other animals or during escape. When jumping downwards along the slope speed was added by gravity. With the development of lift generating planes, such jumps could turn gradually to gliding flight. With higher speed steering movements became more efficient. They influenced the glides more and more and turned finally to active powered flight. In the course of this development the wings became longer and stronger so that the caudal part of the primeval lift generating plane could be reduced.

Acknowledgements

I wish to express my appreciation and thanks to Prof. Dr. WALTER BOCK and Prof. Dr. WOLFGANG F. GUTMANN who critically read the manuscript and offered suggestions.

Addendum

As far as I understood, most contributions and the discussion during the conference did not contradict the theses put forward in my paper. The only exception appeared to be Dr. YALDEN'S case analyzing the claws of *Archaeopteryx*. Dr. YALDEN argued very convincingly that these claws were adapted for climbing on tree-trunks. In my opinion his arguments have the following consequences: Since the selection forces acting on limbs used for climbing and the selection forces acting on limbs used for flight are so different and incompatible, we must assume that *Archaeopteryx* (or her ancestor) began to climb only after having transformed the forelimbs into primitive wings and the hind limbs into cursorial legs. Although at a first glance it seems improbable that a bipedal animal turns out to become a quadrupedal climber, such cases may happen indeed as can be seen in arboreal kangaroos *(Dendrolagus)*. This case was already emphasized by BOCK (1969). Of course such secondary climbers do not improve their former specialization of bipedality. And this is the point also in the case of *Archaeopteryx*. If *Archaeopteryx* was a climbing animal, she abandoned the evolutionary pathway towards real birds. The specialization of a recent bird's wing could not be achieved in a forelimb used for climbing. This means that, provided *Archaeopteryx* was indeed a climber, she could not be in the direct line of ancestry of modern birds.

References Cited

BLASZYK, P. (1935): Untersuchungen über die Stammesgeschichte der Vogelschuppen und Federn und über die Abhängigkeit ihrer Ausbildung am Vogelfuß von der Funktion. – Morph. Jb., **75**: 483–567; Leipzig.

BOCK, W. J. (1969): The origin and radiation of birds. – Ann. N. Y. Acad. Sci. **167**: 147–155.

CAPLE, G., BALDA, R. P. & WILLIS, W. R. (1983): The physics of leaping animals and the evolution of preflight. – Amer. Natur. **121** (4): 455–476; Salem.

CLARK, B. D. (1976): Energetics of hovering flight and the origin of Bats. – In: HECHT, M. K., COODY, P. C. & HECHT, B. M. (eds.): Major patterns in vertebrate evolution. New York, London. (Plenum Press).

COWEN, R. & LIPPS, J. H. (1982): An adaptive scenario for the origin of birds and of flight in birds. – Third N. – Am. paleont. Conv., Proc. **1**: 109–112.

GUTMANN, W. F. (1972): Die Hydroskelett-Theorie. – Aufs. u. Rd. senck. naturf. Ges. **21**: 1–91, Frankfurt am Main (W. Kramer).

GUTMANN, W. F. (1977): Phylogenetic reconstruction: Theory, methodology, and application to chordate evolution. – In: HECHT, M. K. et al. (eds.): Major patterns in vertebrate evolution: 645–669. New York, London (Plenum Press).

GUTMANN, W. F. & BONIK, K. (1981): Kritische Evolutionstheorie. – Hildesheim (Gerstenberg).

GUTMANN, W. F. & PETERS, D. S. (1973): Konstruktion und Selektion: Argumente gegen einen morphologisch verkürzten Selektionismus. – Acta biotheor. **22** (4): 151–180; Leiden.

HECHT, M. K. (1976): Phylogenetic inference and methodology as applied to the vertebrate record. – Evolutionary Biology. **9**: 335–363; New York, London.

HENNIG, W. (1983): Stammesgeschichte der Chordaten. – Hamburg, Berlin (Paul Parey).

JEPSEN, G. L. (1970): Bat origins and evolution. – In: WIMSATT, W. A. (ed.): Biology of Bats. **1**: 1–64; New York (Academic Press).

KLAUSEWITZ, W. (1960): Fliegende Tiere des Wassers. – In: SCHMIDT, H. (Hrsg.): Der Flug der Tiere. – Senckenberg – Buch **39;** Frankfurt am Main (Waldemar Kramer).

MADERSON, R. F. A. (1972): On how an archosaurian scale might have given rise to an avian feather. – Amer. Natur., **106**: 424–428; Salem.

MERTENS, R. (1960): Fallschirmspringer und Gleitflieger unter den Amphibien und Reptilien. – In: SCHMIDT, H. (Hrsg.): Der Flug der Tiere. – Senckenberg-Buch **39**; Frankfurt am Main (W. Kramer).

NOPCSA, F. (1907): Ideas on the origin of flight. – Proc. zool. Soc. London, **1907**: 223–236.

OSTROM, J. H. (1974): *Archaeopteryx* and the origin of flight. – Quart. Rev. Biol.; **49**; 27–27; Baltimore.

OSTROM, J. H. (1976): Some hypothetical anatomical stages in the evolution of avian flight. – Smiths. Contr. Palaeobiol., **27**: 1–21; Washington.

OSTROM, J. H. (1979): Bird flight: how did it begin? – Amer. Scientist, **67**: 46–56; Burlington.

PENNYCUICK, C. J. (1975): Mechanics of flight. – In: FARNER, D. S. et al. (eds.): Avian biology: 51–75; New York, San Francisco, London (Academic Press).

PETERS, D. S. (1984): Konstruktionsmorphologische Gesichtspunkte zur Entstehung der Vögel. – Nat. u. Museum, **114** (7): 199–210; Frankfurt am Main.

PETERS, D. S. (1985): Ein neuer Segler aus der Grube Messel und seine Bedeutung für den Status der Aegialornithidae (Aves: Apodiformes). – Senckenbergiana leth., **66**(1–2): 143–164; Frankfurt am Main.

PETERS, D. S. (in press): Mechanical constraints canalizing the development of tetrapod limbs. – Acta biotheor., Leiden.

PETERS, D. S. & GUTMANN, W. F. (1976): Die Stellung des "Urvogels" *Archaeopteryx* im Ableitungsmodell der Vögel. – Nat. u. Museum, **106** (9): 265–275; Frankfurt am Main.

PETERS, D. S. & GUTMANN, W. F. (1978): Ausgangsform und Entwicklungszwänge der Gliedmaßen landlebiger Wirbeltiere. – Nat. u. Museum, **108** (1): 16–21; Frankfurt am Main.

RAUTIAN, A. D. (1978): Unikalnoje pero ptitsy iz otlozheniy jurskogo ozera v khrebte Karatau. – Paleont. Zhur., **1978** (4): 106–114.

RAWLES, M. E. (1963): Tissue interactions in scale and feather development as studied in dermal-epidermal recombinations. – J. Embryol. exp. Morph., **11** (4): 765–789.

REGAL, P. J. (1975): The evolutionary origin of feathers. – Quart. Rev. Biol., **50**: 35–66; Baltimore.

STEPHAN, B (1974): Urvögel. – Die neue Brehm Bücherei **465**, 167 S.; Wittenberg Lutherstadt (A. Ziemsen).

THOMSON, J. L. (1964): Morphogenesis and histochemistry of scales in the chick. – J. Morph., **115**: 207–224; Philadelphia.

Author's address: Priv.-Doz. Dr. D. STEPHAN PETERS, Forschungsinstitut Senckenberg, Senckenberg-Anlage 25, 6000 Frankfurt am Main, Federal Republic of Germany.

Siegfried Rietschel

Feathers and Wings of *Archaeopteryx*, and the Question of her Flight Ability

Dedicated to my cherished colleague and former Professor Dr. ADOLF SEILACHER, Tübingen, on occasion of his 60th birthday on February 24, 1985.

Abstract

Preservation of feathers, wing and tail, of the Berlin specimen of *Archaeopteryx* is described and explained by a "precipitation model". The "impression hypothesis" is disapproved. On the basis of the highly developed feather and wing anatomy good flight ability is concluded for *Archaeopteryx*. According to the lack of a crested bony breastbone in combination with a long, bilaterally feathered tail, it is suggested, that this was a flapping flight by simultaneously beating wings, steered by the tail. This type of active flight is regarded as not derivable from a gliding nor from a "running Proavis", but from a "hopping" getaway on the ground. The claws of the fingers are explained as grooming organs for the plumage, and it is assumed, that they could not have been essential for use in climbing or grasping.

Zusammenfassung

Die Erhaltungsweise von Federn, Flügeln und Schwanz des Berliner Exemplares von *Archaeopteryx* wird beschrieben und durch ein "Niederschlag-Modell" erläutert. Mit diesem lassen sich, besser als mit der abzulehnenden "Abdruck-Hypothese" die verschiedenen überlieferten Strukturen erklären. Aufgrund der hochentwickelten Feder- und Flügelanatomie wird auf ein gutes Flugvermögen von *Archaeopteryx* geschlossen. In Bezug auf das Fehlen eines knöchernen, gekielten Brustbeines wird ein Zusammenhang mit dem langen, zweizeilig befiederten Schwanz gesehen. Danach müßte der Flug von *Archaeopteryx* durch ein gleichsinniges, synchrones Schlagen mit den Flügeln bewerkstelligt worden sein, wobei der Schwanz die Steuerung übernahm. Dieser Flug wird nicht von einem Gleitflug hergeleitet bzw. von einem "rennenden Proavis", sondern aus einem aktiven, hüpfenden Start vom Boden aus. Die Krallen der Finger werden als Organe zur Gefiederpflege gedeutet, die kaum zum Klettern oder zum Greifen eingesetzt werden konnten.

1. Introduction

In the literature not much attention has been paid to the conditions under which the preservation of the feathers of *Archaeopteryx* took place. Most authors followed two unreflected theories: (1) the *Archaeopteryx* individuals died on land and were transported as mummies to the burial place, (2) they died where they are found, on a more or less wet lagoonal mud. Consequently feather preservation was explained as "impressions" on a soft sediment surface, represented by the better preserved so-called "main-slab". Additional theories were necessary to explain certain structures, especially the "double-struck impressions". As a result, very contradictory conclusions on the number and character of the feathers in the wing of *Archaeopteryx* have been published. The number of primaries ranges in publications from 6 to 12, and many theories on flight ability of *Archaeopteryx* gave wings to fantasy more than to investigations of the plumage.

In 1976 I published a model of post-mortem and embedding history of *Archaeopteryx*; the foregoing ideas on feather preservation were the subject of a talk (Meeting, Paläontologische Gesellschaft, Karlsruhe 1976). All my investigations had to be based on casts of the originals. The original Berlin specimen, on which this paper is based, was studied in 1978; a gipsum cast of the counter-slab was at my disposal for essential detail studies by courtesy of Dr. J. HELMS and Dr. H. JAEGER, Museum Humboldt-Universität, Berlin.

The mentioned model (RIETSCHEL 1976) comes to the conclusion, that the known *Archaeopteryx* fossils are the result of drowned animals, floating for some time on the water surface until they sank to the bottom – a quite normal event. This is in accord with a subaquatic sedimentation of the Solnhofen limestone (e.g. BARTHEL 1970). Under these conditions it is most probable, that the feathers of wings and tail were not able to make an "impression" even in a very soft sediment. So the only way feather preservation could have resulted, is that the surface of the plumage was (1) covered by a very fine-grained sediment or (2) conserved by an overgrowth of bacteria and algae. The result of both mechanisms, here called precipitation, should have been, that the overlying slab has on its lower side a negative picture of this surface: a cast. If, according to the model, a carcass was lying with its back on the bottom, the cast would show the ventral side of feathers and wings. In accordance with the model, regarding the so-called "main-slab" as the roof, the cast has to be found on this slab and not on the counter-slab.

2. What does the Berlin specimen show?

The main-slab shows the solid skeleton together with a negative mold of the ventral side of feathers and wings; the counter-slab shows the cast of the skeleton as a negative – and the feathers as a positive cast from the negative mold of the main-slab. So we see the same feather structures as negative and positive on both slabs, and on the counter-slab not the expected "impression" of the dorsal side of feathers and wings (HEINROTH 1923, RAU 1969, HELMS 1982)*. There are in the Berlin specimen only very few details of plumage, which are not identical on main- and counter-slab.

2.1 The feathers

So, the counter-slab seems to give the exact topography of the wing's and tail's surface, as seen from the ventral side. We can identify minute structures of many feathers. Well preserved are the shafts (with ventral furrow) and the barbs. The latter show the very thin rim, which characterises as well the ventral side of a modern bird's feather. The medium distance of barbs ranges between 0.35 and 0.45 mm, which is less than the average distance of similar-sized feathers of modern birds (e.g. dove and magpie with 0.60 mm). The interlocking barbules must have been relatively short and are not visible. The original counterslab shows a very thin "rusty" network, which might be a relic of organic matter of barbs and barbules. The perfect arrangement of the barbs and the occurrence of splitted, isolated barbs, especially in coverts, show that the contour feathers of *Archaeopteryx* had the same anatomy and supposedly a comparable stability and flexibility as contour feathers of modern birds.

All feathers are partly overlapped by others, so none is visible completely. This overlapping system leads to the first of several statements:

(1) In general only the uppermost surface of the wing, represented by the ventral sides of the feathers, is preserved and visible. – Where a feather was covered by a neighbour, the neighbour consequently formed the surface and its surface was preserved. So, in all places where primaries, secondaries, or coverts formed the uppermost layer, they were fossilised (Plate 1).

(2) The arrangement of the wing feathers of a bird's carcass doesn't follow exactly a textbook order. Usually some feathers are a little bit dislocated. Shifted feathers sometimes do not overlap, and set the peripheries of the vanes more or less upright, so that the vanes adjoin each other with their edges. Under a certain pressure this common edge gets a zigzag structure and the barbs tend to form an undulating surface (Plate 2).

Statements (1) and (2) are sufficient for an explanation of many visible feather structures in *Archaeopteryx*. We can see, how always the uppermost feather layer is pictured by the very fine grained

---

* For the London specimen DE BEER (1954: 34) stated, that the main slab shows the underside but the counter-slab the dorsal side of the wing.

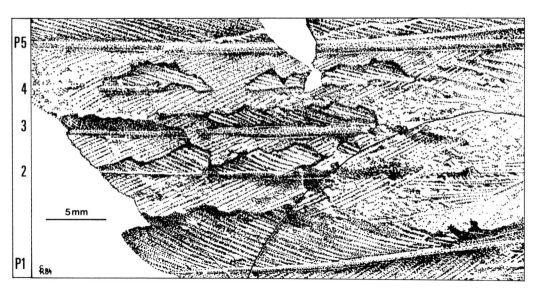

Fig. 1
Detail of right wing of *Archaeopteryx*, Berlin specimen, counter-slab, showing distal region between P1 and P5, where two layers of feathers are visible.

Solnhofen lime. A nice example on both wings gives a covert with splitted barbs at the end, which are crossing the barbs of the underlying P1, whilst under the coherent inner part of the covert, where the barbs are linked together, no structures of the underlying feathers can be seen. The vaulting of adjoining vanes with zigzag-structures and the undulating barbs can be seen e. g. in tossed parts of the left wing between P1/P2 and P3/P4.

(3) Fortunately there is no general rule without exception. So we find – in contradiction to (1) – at least one place on the counter-slab, where not only the ventral side of the uppermost feathers is preserved. P2 to P5 of the right wing show an area with a mosaic of barbs, preserved in two levels. The main-slab presents in this area only the higher level in poor preservation, but with small pieces of attached sediment. On the counter-slab, that means originally under this very thin sheet of sediment which has partially broken off, the structure of the ventral side of another underlying feather layer appears (fig. 1)! The barbs of the two levels form a criss-cross structure. So, a small window to a deeper layer of the wing is open, where underlying vanes can be identified. We conclude, that – at least in some places – not only the ventral features of the uppermost feathers have been preserved by a cover of sediment, but also below them the ventral sides of hidden feathers. That should be caused by fine sediment particles which were deposited between the feathers "in" the wing. Here we get a key to explain the so-called "double struck feather impressions" of DE BEER (1954: 34), and already observed by HEILMANN (1926).

(4) There are mysterious "shadows of feather shafts" on several vanes, which caused much trouble in the interpretation and for reconstruction of the number of primaries. These "shaft-shadows" have an inverted morphology, compared with the normal shafts. That is, they are positive on the main-slab (where the normal shafts are negative), and they are negative on the counter-slab (where the normal shafts are positive); they cannot be attributed to a vane, but are everywhere crossed by the barbs of the visible feathers. Many theories evolved for their explanation. Some authors thought, they were moulting feathers; according to the "impression-hypothesis" they could be explained as "double-struck" (e. g. DE BEER 1954, RAU 1969). The best descriptions and photographs of feather structures have been published recently by HELMS (1982), who also followed an "impression-hypothesis" and regarded the "shaft-shadows" as the dorsal sides of hidden feathers, printed through during diagenesis. There is no

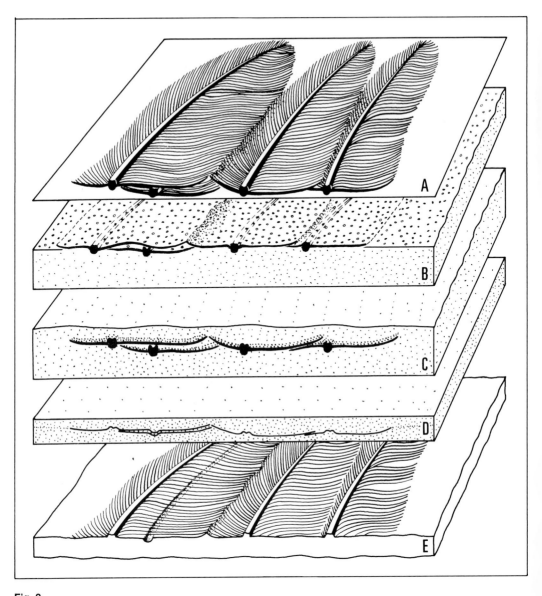

Fig. 2
Diagrammatic drawing to explain preservation of feathers according to the "precipitation model".
A: Section of distal parts of 4 primaries in different positions.
B: Section under sedimentation.
C: Sediment-covered section with early cementation upon the organic structures.
D: Settled sediment with discontinuities in place of former organic structures.
E: Section with uncovered fossil.

space to debate the "impression" arguments, but it is necessary to attempt a convincing explanation according to the proposed "precipitation model" of feather preservation (Fig. 2):

As stated above, precipitation of lime – as a very fine-grained sediment or as an algal mat – took place not only upon the wing's surface, but also between the feathers. There could accumulate more lime on the

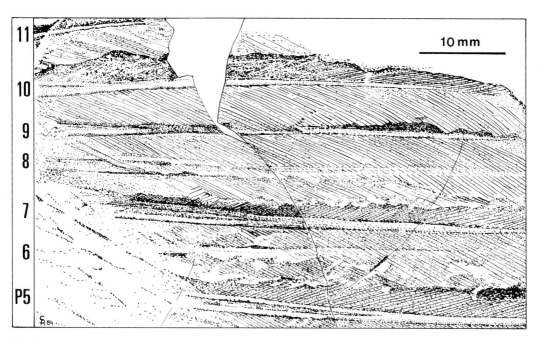

Fig. 3
Detail of left wing of *Archaeopteryx*, Berlin specimen, counter-slab, showing P5–P11; P6 completely as "shaft shadow", P8 with "diving shaft", that is: proximally (left side) as partially covered positive shaft, distally under the vane of P9 as negative "shaft-shadow".

hidden vane surfaces than on the hidden shafts, especially when the latter were in touch with an overlying vane. The sediment is supposed to have cemented first where it was in closest contact with the organic material – therefore the upper (ventral) sides and not the lower (dorsal) sides of the feathers were conserved (fig. 2C). By this mechanism, shafts of underneath feathers caused a debit balance of sediment, and when the shaft disappeared (fig. 2D), the overlying area of a vane settled more than in the surrounding area (stabilized by earlier cemented sediment underneath). Consequently a shallow furrow appears now on the upper vane (fig. 1, 2E). Concerning the overlying feathers, shafts and vanes were covered by sediment together, and the shafts are only slightly compressed. That means that the "shaft-shadows" of underlying shafts are not a structure of the original feather, nor a sedimentary structure. They have to be explained as early diagenetic. This explanation is supported by the fact that the barbs don't turn to the side in the "shaft-shadows"; according to the vertical settling, they only bend slightly downwards. All these considerations refer to a sedimentation plane, as represented in the so-called counter-slab. The main-slab, as the roof, still shows the original negative, which worked as a mold for the understratum during early diagenesis.

As an inevitable conclusion follows, that the "shaft-shadows" (I prefer this neutral term instead of "double-struck impressions") represent original feathers and have to be counted as such. A striking proof is P8 of the left wing: the proximal part of this feather's shaft is nearly uncovered by the neighbour's vane, and so preserved as a positive, whilst the distal part of the shaft lies under the inner vane of P9 and appears morphologically as a negative "shaft-shadow" (Fig. 3).

2.2 The wings

The wings shall be regarded in this study as mainly referring to the plumage, and not the skeleton. At first sight an *Archaeopteryx* wing shows two well marked areas: one, in which the feathers can be identified

Fig. 4
Reconstructed right wing of *Archaeopteryx*, Berlin specimen, seen from below. The wrist angle, in the original about 100°, is opened to 145°.

easily and another one, with more dubious structures. The first represents, roughly speaking, the outer parts of the hand, the second the inner part of the hand and most of the arm plumage.

The hand has 11 primaries, a 12th is suspected (HEILMANN 1926: fig. 20, 21; HELMS 1982: 188). As I was not able to identify shaft or barb structures but only a vague outline of this P12, I hesitate to confirm its presence or absence. P1–P11 of this study are identical with P1–P11 of HEILMANNS and HELMS' figures. P1–P8 have shafts which are bent slightly foreward, shafts of P9–P11 are bent slightly backward. In addition, there is a remarkable crest between P5 and P7. It's quite possible, that these structures, present equally on both wings, reflect the fixation of the P-quills to different bones: P1–P6 to the metacarpal, P7–P8 to the 1st, and P9–P11 to the 2nd phalanx of digit 2 – the shadow of P6 here still attributed to the metacarpal, for it appears shifted over. The primaries with even numbers except P10 – are covered by those of odd numbers, but in the right wing P2 and P4 are partially visible near the outer margin, as described above (fig. 3). This is not the same picture, as it appears in a folded wing of a recent bird, where the primaries are arranged in a simple sequence, and where the posterior vane covers the shaft of the following feather. But the arrangement in *Archaeopteryx* may have been effected not only by the attachment of the quills; we have to take into strong consideration, that the carcass was in a state of decay when it reached the sediment and furthermore, the finger bones were not fused. Therefore I don't dare to try too much interpretation of the post-mortem position of the primaries.

The minimum length of the primaries measures:
P1: 130 mm; P3: 135 mm; P5: 145 mm; P7: 140 mm; P9: 125 mm; P10: 95 mm; P11: 55mm.

The width of the vanes can be measured and amounts to between 1 mm and 8 mm, but the measurements themselves show no more than the first impression: the vanes are broken off at their anterior and posterior borders and are incomplete. 40% (P1) to 70% (P11) of the total length of the primaries is free distally, the proximal part is hidden by the under-coverts. The ventral furrow of the shafts reaches out from ½ (P10) to ¾ (P1) of the total shaft length. If we follow the shafts lengthwise, we can reconstruct the outer margin of the under-coverts. The under-coverts hide (like a screen) the basal parts of the primaries, of the secondaries, and of themselves, row by row. There is not a single feather totally preserved up to the quill. The overlapping by coverts along the feather axis interferes with an overlapping of feathers within the rows. A feather of the wing's underside generally is covered by the inner vane of it's outer neighbour (at the – not preserved – upper side of the wing by the outer vane of it's inner neighbour). The degree of coverage depends on the angle, under which the wing is folded. As the hand in birds is folded under the arm, the inner primaries in a ventral view cover the outer secondaries.

It is very difficult, to attribute the shaft and barb structures of the inner area of the wing to distinct coverts, as well as some fantasy is needed to locate the quills of primaries and to number exactly the quills of secondaries. A reconstruction of the area of the coverts is possible and makes sense as soon as a undistorted reconstruction of the wing is tried (Fig. 4: see although YALDEN 1970). Only the distal ends of single coverts are free and extraordinarily well preserved. The main area of the arm coverts presents the barbs as a "combed" structure, "floating" backward and down to the bundle of elbow feathers. One gets the impression, that the filigree structures of the vanes of coverts are coated by a thin, amorphous film of sediment, which might originate in an algal mat. The traceable structures are too long and irregular for barbs. Astonishing it is, how these structures, through which occasionally parts of the underlying secondaries can be seen, are nearly symmetrical between both wings. The secondaries, of which only the distal ends can be identified, can't be counted exactly, and there is no doubt, that some of the outer ones are covered by the inner primaries. The skiffleboard structure of their quills shows that there have been 12 or more of them.

2.3 The tail

The feathers of the tail show the same preservation as those of the wings. There is only the ventral side is exposed, in perfect detail. Structures like shafts, partly covered by vanes and the "shaft-shadows" are as nicely present as in the wings. But I couldn't find any traces of coverts. The vertebral column is accompanied by an uneven but smooth surface, which stretches out to both sides of the axis from 10 mm near the 12th to 5 mm at the last caudal. An analysis has to be given elsewhere.

3. Theoretical considerations

3.1 Function of the claws

Here one question, which at first sight seems to be more marginal to feathers and wings, has to be asked: what was the function of the finger claws? Could they have been important tools and, if yes, for what? Some authors and most who tried reconstructions of the life of *Archaeopteryx*, believe that the finger claws still were in use either for locomotion, especially for climbing, or for grasping. I doubt that. In any case, the locomotion function has been given up by early forerunners of *Archaeopteryx* amongst the Theropoda. The position of the claws is very curious. The fossil specimens show the claws lying in line, along the front margin of the wing. This position is caused by a spreading apart of the 1st finger and a flexion of the 3rd finger between the 2nd and 3rd phalanx. This is the case not only in the Berlin, but also in other specimens (Eichstätt, Maxberg, Teyler). Therefore I suppose this to be an original feature which signifies, that the fingers were not free, and only the movable claws were inserted free at the wing's margin. There their tips were directed in a ventral direction.

Let me now draw attention upon the fact, that modern birds dedicate much time of the day to plumage grooming. Only well-groomed feathers guarantee the best function in insolation and flight. Modern birds comb disarranged or dirty and fuzzy barbed feathers with the bill and (especially those of the wings and tail) by the claws of the feet. Grooming is essential for a bird's survival as soon as it leaves it's nest. There should be no doubt, that *Archaeopteryx* had to keep her plumage clean and well-groomed. But grooming activities of *Archaeopteryx* must have been different from those of modern birds: (1) for her neck was not so flexible, the bill could only reach a restricted area of the plumage, and (2) her long tail was difficult to reach, because of the axial vetebral column. So, the horny finger claws could have been of great value for keeping the feathers of body and tail clean and functional. The morphology of the claws, with a thin and double-shaped sharp blade of semicircular curvature at the inner margin, doesn't speak for a stressable climbing organ; the morphology of the claws is quite different from that e.g. of recent bats; the size of the claws, with an inner diameter of ⅓ to ⅔ of the vane width of coverts, could have made them useful grooming organs for *Archaeopteryx*. It can be expected, that during bird evolution such grooming finger claws, which interfered to a certain extent with better flight ability, became more and more superfluous when the saddle articulation of vertebrae gave the neck a higher flexibility and the pygostyle-type of the bird tail evolved.

3.2 Flight ability of *Archaeopteryx*

As we have seen, the feather detail and the wing plumage of *Archaeopteryx*, as far as is known, are very close to those of modern birds. The most evolved and complicated feather type (contour feather like primary etc.) is already present. Feather arrangement in the wing of *Archaeopteryx* strongly resembles that of a modern bird. Even the number of primaries corresponds directly to that of the majority of modern birds, as well as the secondaries are of a comparable number to the average in modern birds. Minor differences to modern birds probably existed in distance between barbs (which means density of vanes) and in size of ventral coverts, which apparently covered a more extensive area.

So the conclusion is warranted, that – by evidence of feathers and wings – the *Archaeopteryx* wing was capable of good flight. If flight capability of the bird itself was restricted, as several authors tried to establish, this restriction might only have had it's cause in the the inner flight mechanisms, by muscles, sinews and skeletal elements. Only the latter are present, and there is evidence in the lack of a well developed breast bone in all known specimens of *Archaeopteryx*.

Therefore we may ask, if a bony, keeled breastbone is an essential prerequisite for good flight? Undoubtedly this is the fact in modern birds, where steering during flight occurs by the wings, and where by steering movements the muscles of the wings are stressed with varying effort on both sides. Therefore they have to be attached to a solid structure of the body skeleton. As a matter of fact, the only conclusion we can draw from the lack of a bony carina in *Archaeopteryx* is: during flight she could not steer with her wings – but she could steer with her prominent long and flexible tail! In this connection I want to direct attention to a certain parallel evolution of pterosaurs, where the early Rhamphorhynchoids combined an unkeeled breastbone with a long tail, and the younger, short-tailed Pterodactyls have a distinct carina on their breastbone.

Archaeopteryx' flight was without doubt different from flight of a modern bird, but that does not mean that it was not a very successful flight. The pectoral muscles supposedly were attached at the thorax only by cartilaginous, may be only by connective tissue structures. But even connective tissue would have been sufficient for muscles which worked as bilateral counterparts – as long as they were stressed more antagonistically and synchronously than bilaterally independant of each other. *Archaeopteryx* has been obviously well outfitted for flapping flight by simultaneously beating wings, with the tail as a control surface and rudder. I don't see any reason to doubt her ability for a good or even perfect flight under favourable habitat conditions. We have to take into consideration, that this type of flight also demanded a smaller capacity of motion control. This leads us to the general aspects of early flight by reptiles, then respectively flight of early birds.

3.3 A "hopping Proavis"?

The hypothetical forerunners of *Archaeopteryx* – the so-called "Proavis" – were small, bipedal theropod reptiles. It is strongly suggested, that endothermy and, in connection with it, body insolation by feathers evolved, both early conditions for bird flight. Still these forerunners walked bipedally, steered quick motion by their tail, didn't use their forelimbs for locomotion and were predators. Some of them might have

Fig. 5
The "running *Compsognathus*" (from Heilmann 1926; reconstruction of pelvis region anatomically wrong!).

Fig. 6
The "hopping *Compsognathus*" (from Abel 1911; the kangaroo habit of this reconstruction is not transferable to Proavis or *Archaeopteryx*).

had a tendency for a hopping mode of locomotion, as e.g. ABEL (1911) suggested for *Compsognathus* (Fig. 6). Such a tendency could have been very favorable for flight. I think, it was a precondition for flight evolution. A bipedal ground animal goes a thorny way, if it tries to become a gliding bird via a climbing stage. A "running Proavis" (like HEILMANN's *Compsognathus* in Fig. 5) would have had not only great difficulties with balance especially by his getaway for flight, but further difficulties in coordination of alternating leg movement and synchronous wing movement. Hopping with synchronous flapping of the wings is to be observed during the start of flight in many recent birds, as well as in learning youngsters as in adults, especially if they have problems getting off the ground (e.g. great vultures, which hop in place until they get enough air under their wings). Many recent birds with a hopping locomotion possess a backward directed first toe (hallux). This seems to be useful for keeping balance and is, as an attribute (like in *Archaeopteryx*) not an argument against a "hopping Proavis".

Consequently I assume, a "hopping Proavis" could have had, even in very different paleoenvironments, the best chances to evolve in to a flying *Archaeopteryx*!

4. References Cited

ABEL, O (1911): Die Vorfahren der Vögel und ihre Lebensweise. – Verh. k. u. k. zool.-bot. Ges., **61**: 144–191; Wien.
BARTHEL, K. W. (1970): On the deposition of the Solnhofen lithographic limestone. – N. Jb. Geol. Paläont. Abh., **135**: 1–18; Stuttgart.
DE BEER, G. (1954): *Archaeopteryx lithographica*. – 11 + 68 pp., 15 pl.; London (Brit. Mus. Natur. Hist.).
HEILMANN, G. (1926): The origin of birds. – 2 + 208 pp.; London (Witherby).
HEINROTH, O. (1923): Die Flügel von *Archaeopteryx*. – J. Ornithol., **71**: 277–283; Berlin.
HELMS, J. (1982): Zur Fossilisation der Federn des Urvogels (Berliner Exemplar). – Wiss. Z. Humboldt-Univ., math.-naturwiss. R., **31**: 185–199; Berlin.
KEUPP, H. (1977): Ultrafazies und Genese der Solnhofener Plattenkalke. – Abh. naturhist. Ges., **37**: 128 pp.; Nürnberg.
RAU, R. (1969): Über den Flügel von *Archaeopteryx*. – Natur & Mus., **99**: 1–8; Frankfurt a. M.
RIETSCHEL, S. (1976): *Archaeopteryx* – Tod und Einbettung. – Natur & Mus., **106**: 280–286; Frankfurt a. M.
YALDEN, D. W. (1971): The flying ability of *Archaeopteryx*. – Ibis, **113**: 349–356; London.

Plate 1 (see appendix)
Archaeopteryx litographica H. v. MEYER. Berlin specimen, counterslab, gypsum cast. Left wing, plumage of the underside with distal parts of the primaries and covering secondaries (center and left side), showing in the primaries' region shaft-shadows, diving shaft (see Fig. 3) and the zigzag structures where primaries are tossed. X 2,8.

Plate 2 (see appendix)
Archaeopteryx litographica H. v. MEYER. Berlin specimen, counterslab, gypsum cast. Right wing, plumage of the underside with shaft-shadows, partial overlapping of feathers and a region with two layers of feathers (see Fig. 1); left side shows the primaries, lower right side the covering secondaries. X 4,1.

Author's address: Prof. Dr. SIEGFRIED RIETSCHEL, Direktor der Landessammlungen für Naturkunde (Museum am Friedrichsplatz), Erbprinzenstr. 13, Postfach 4045, D-7500 Karlsruhe 1, Federal Republic of Germany.

Burkhard Stephan

Remarks on Reconstruction of *Archaeopteryx* Wing

Abstract

Reconstruction of the wing is of significance for assessment of the evolutionary position of *Archaeopteryx* and may serve as a model case. The feathers are of particular importance. Attention is payed to number and attachment of primaries and to the function of the remicle. Reconstruction of the wing goes hand in hand with assessment of its functions. Functions commonly attributed to the claws are questioned.

Zusammenfassung

Die Rekonstruktion des Flügels hat Einfluß auf die Bewertung der Stellung der Urvögel im System. Sie hat Modellcharakter. Besondere Bedeutung hat dabei die Befiederung. Einige Details werden diskutiert, insbesondere Zahl und Ansatz der Handschwingen sowie die funktionelle Bedeutung des Remicle. Die Rekonstruktion des Flügels läßt Schlüsse auf die Funktion der Vorderextremität zu. Die den Krallen zugeschriebenen Funktionen werden in Frage gestellt.

Introduction

Whether *Archaeopteryx* is in or off the line of descent of modern birds, resemblances in wing architecture are such as to allow reconstruction of *Archaeopteryx* partly by consulting the modern version. Such reconstructions, in turn, serve as models encompassing our present knowledge and are of significance in phylogenetic research and teaching. Reconstruction in the case of *Archaeopteryx* has been dealt with elsewhere (see STEPHAN 1974, 1979). Nonetheless, more recent attempts continue to repeat a number of mistaken points: The fingers are rendered as protruding beyond the wing. Creatures of such design would hardly have been viable and definitely incapable of crucial functional activities. There is no point in taking them as models.

Feathering of *Archaeopteryx* forelimb

As evidenced by the slabs (see Berlin specimen) and in accordance with modern birds the remiges are attached to hand and forearm. They are primaries and secondaries and not, as presented by reconstructions which render the hand free of feathers, secondaries and "tertiaries". The remiges of *Archaeopteryx* were rather lengthy and thus in need of some support, which for the primaries could be provided only by the hand, including the second finger. This however implies that digit II and III did not protrude beyond the feathers. The Berlin, "Maxberg", and Eichstätt specimens present the evidence that the second, and thus the third finger had feathers. "Tertiaries" presuppose rather long humerus – the extreme case being the large Procellariiformes. "Tertiaries" however are not remiges for they do not provide for upward propulsion. Their function is one of coverage when the wing is closed and one of bridging the gap between body and secondaries when in flight, thereby providing for aerodynamic perfection. Short-winged birds need no "tertiaries". With all birds forward propulsion is brought about by the primaries. In the case of flapping flight, the "tertiaries" would become a handicap. Wing shape in *Archaeopteryx* indicates flapping flight. There is no evidence of "tertiaries" on the slabs. Approximately two thirds of the upper arm (see crista lateralis) was covered by muscles. The row of remiges stretches from the next to last phalanx of the second digit to shortly behind the elbow.

The remicle is attached (distally beyond the outermost primary) to phalanx 2 of digit II and stabilizes the feathers of this part of the wing. The distal portion of the remicle is covered by the outer vane of the outermost primary. Whenever the remicle is missing, its function is taken over by the outer vane of this primary. Overlapping of the wing feathers is not only of advantage when spreading and folding the wings (gliding function of the feathers), but is also in support of each other during flight, therefore the minor size of the outermost feather of the row of primaries in the ancestral engineered wing. The remicle has been maintained in various groups of modern birds. For the same reason the primaries increase in length one by one from the remicle inwards, so that maximal length is attained from the fifth primary onwards. This condition is present in the Berlin specimen and similarly in modern birds with rounded wings. The remicle however is always short, and therefore surely must also have been the case in *Archaeopteryx*, and not protruding beyond the coverts of this part of the wing allows the conclusion that the outermost primary (of the Berlin specimen) is not a remicle (see HELMS 1982).

The number of primaries differs among modern birds, whether or not a remicle is present (STRESEMANN & STEPHAN 1968, STEPHAN 1970, 1974). Variations occur between 9 and 11 and are all due to the carpometacarpus. There is no deviation from the number of 4 digital primaries: *Podiceps, Ciconia, Phoenicopterus* 7 + 4 = 11, Indicatoridae 5 + 4 = 9, all other flying birds 6+ 4 = 10. The remicle, when present, is given extra credit, e.g. 11 + 1. Three of the four digital primaries are attached to the basal phalanx and one to the last phalanx (or to the one before last, when a rudimentary claw phalanx is present as in *Phoenicopterus*) of digit II. This was taken into account in my reconstruction of *Archaeopteryx'* wing (STEPHAN 1974, fig. 45; 1979, fig. 54). A remicle was added to show its stabilizing function which, however, doesn't reach beyond the coverts and claw, and thus actually isn't to be seen. The remicle is attached to the phalanx 2 of the second finger. According to this, among modern flying birds primaries are attached to the carpometacarpus, phalanx 1 and phalanx 2 of digit II in the order 5 − 7, 3, 1 + 0 or 1 R. There is no primary and no remicle attached to the claw phalanx. According to HELMS (1982) *Archaeopteryx* possessed 12 primaries. He suggests that they are attached to the bone elements in this order 6, 3, 2, 1; that is, 2 to phalanx 2 and one to phalanx 3 (the claw phalanx). This however, is improbable, even if the claws were not used for clinging. The rather lengthy second phalanx would provide room for 3 primaries, if the angle of attachment invoked by HELMS (sketch) were correct. HELMS takes the outermost of the visible primaries to be a remicle. He denotes it a functional remicle because it shares in bringing about forward propulsion. The stabilizing function of the remicle which induces its smallness (see above) remains thus out of consideration.

Among modern birds the innermost primary is attached rather close to the wrist. If the shafts of primaries outlined in HELMS' picture (1982 p. 188) are placed a little more proximally (because in this picture the gap between primaries and secondaries is too large, even when forearm and hand are approximated) the place of attachment of the primaries also will alter. Room would become available for the attachment of 7 primaries to the metacarpus as well as 3 to the basal phalanx of the second finger and 2 to its second phalanx. This would be in better accordance with the curvature of the shafts of all primaries and in particular to that of the outer one which is attached to the bone at a closer angle. Primary no. 12 cannot be taken as a remicle.

As mentioned before, reduction of the remicle takes place independently of the number of primaries because the latter alters only on the carpometacarpus. However, a reduction of primary no. 12 would have been necessary because of the shortening of phalanx 2 which didn't effect the claw phalanx. Primary no. 12 has the shape of a primary and not of a remicle. Stabilization of this wing area here was necessary too. This allows the assumption, that *Archaeopteryx* also was in possession of a remicle: 12 H + R.

HELMS, taking into account the degree of fossilization, arrived at 12 primaries. The bird (Berlin specimen) sank upside down touching the bed of the lagoon with its back. The underlying ooze was covered by a clayey coat which enabled present feather disconnection. The edges of the feathers were bent upwards slightly as a result of the bird sinking into the lime and thus the overlapping of the feathers hasn't been completely preserved. While the skeleton was preserved by its enclosure in lime, the feathers desintegrated thereby leaving cavities in the principal slab which finally left their imprints on the underlying slab. The underlying slab, however, already contained the imprints of the dorsal side of the

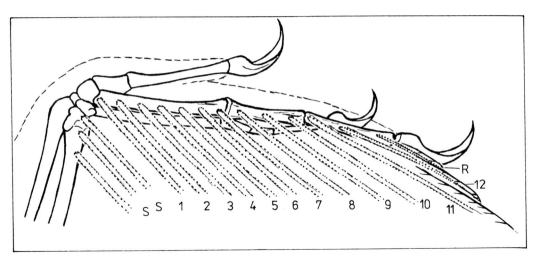

Fig. 1
Structure of the right hand.
Digit I with two, digit II with three and digit III with four phalanges. Only the first finger could be moved freely. It is shown at the front margin of the wing and the attachment of primaries (1–12), remicle (R) and outer secondaries (S).

wings so that a double imprint arises. This is why the imprints of the shafts differ, which hitherto has been interpreted in various ways. The slab with the fossil split mainly along the clayey coat so that most of the bones remained in the principal slab, the imprints of the feathers on the principal and underlying slab correspond, and there is a negative of the shaft on the principal slab, and a positive on the underlying slab.

My considerations (STEPHAN 1976, 1979) therefore need correcting and clearer interpretation. The bird was on its back and the splitting occurred above the animal, e.g. also above the bones of the wings. However, as a result of the embedding (its position in the coat of clay), only the imprints of a few shafts have been preserved as far as the dorsal side of the wings is concerned and it is the ventral side which is more clearly preserved in the principal slab. Normally, when spreading the wings each of the primaries glides over the next distal primary in such a manner that in the spread state only the ends of the primaries become free. From the dorsal side of the wing only the outer vanes are to be seen and from the ventral side only the inner vanes of the remiges. From above, the shafts are free and from below they are covered by the overlapping inner vanes. This is because the inner vanes are much broader in width than the outer vanes. In the case of the Berlin specimen the shafts became uncovered because of the slight upward curvature of the vanes imposed by the clay. It was HELMS (1982) who gave the plausible story to the somewhat muddled details of the imprints. However, he also maintained that the main slab is the hanging slab.

The functions of the wings

Opinions on the functions of the front extremities in *Archaeopteryx* differ (see STEPHAN 1974, 1979, OSTROM 1979, FEDUCCIA & TORDOFF 1979). The rather long primaries and secondaries presuppose an efficient manner of closing the wings and of holding them tightly to the body which suggests that terrestrial or arboreal locomotion may have been impeded. Therefore, the gliding function of the feathers mentioned above has to be presupposed apart from elasticity and strength for flight. These rounded wings were suitable for flapping and powered flight (Kraftflug). The architecture of the humerus and remiges support this view. However, this view is weakened by the absent sternum.

The shoulder articulation has shifted close to the vertebral column. This crucial change as compared with the quadrupeds clearly effects the locomotion of the front extremities. The proximal part of the humerus was covered in *Archaeopteryx* by a strong muscular system, as is indicated by the crista lateralis. The wings were capable of circular movement. The forearm remained at a slight angle even when the wing was spread, as is to a lesser extent also the case in modern birds. The hand points more or less to the side when the wing is spread and when closed it points to the back. A hand pointing to the side when the wing is spread remains in this position when the forearm (including the attached feathers) is directed foreward by bending the wrist accordingly. However, there is no possibility for a hand arranged in this manner to point foreward as can be done in tetrapods. This kind of arrangement of the wing bones, as a condition for the performances of the feathers (in flying, terrestrial and arboreal locomotion) excluded any kind of employment of the wings other than those known in modern birds.

The architecture of the skeleton and the feathering of the wings no longer allowed employment of the claws for clinging or preying. Employment of the wings was possible in active flight, parachute jumps, and during display. Similar to the contour feathers of the body, the coverts had an aerodynamic function; furthermore, they served in thermoregulation and as camouflage.

The claws in the Berlin specimen which are also rather long (FISCHER & KRUEGER 1979), may have been without specific function. Functional change of a system of organs, as witnessed in the front extremities, need not effect all elements at once. The elimination of the claws leads to reconstruction of the finger tips, which surely was a complex evolutionary step.

Acknowledgements

I am grateful for the discussions with Dr. J. HELMS, director of the Paläontologisches Museums of the Museum für Naturkunde der Humboldt-Universität zu Berlin, and with Dr. K. FISCHER, head of the vertebrate department of this Paläontologisches Museum, which contributed to this paper, and Dr. P. BEURTON, Berlin, for the translation.

References Cited

BAKKER, R. T. (1975): Dinosaur Renaissance. – Sci. Amer. **232**, 4: 58–78.
DABER, R. & HELMS, J. (1983): Fossil Treasures. – 231 pp. Leipzig (Edition).
FEDUCCIA, A. & TORDOFF, H. B. (1979): Feathers of *Archaeopteryx*: Asymmetric vanes indicate aerodynamic function. – Science, **203**: 1021–1022.
FISCHER, K. & KRUEGER, H.-H. (1979): Neue Präparation am Berliner Exemplar des Urvogels *Archaeopteryx lithographica* H. v. Meyer, 1861. – Z. geol. Wiss. **7**, 4: 575–579; Berlin.
HELMS, J. (1982): Zur Fossilisation der Federn des Urvogels (Berliner Exemplar) – Wiss. Z. Humboldt-Univ. Berlin, Math.-Nat. R. **31**, 3: 185–198; Berlin.
OLSON, S. L. & FEDUCCIA, A. (1979): Flight capability and the pectoral girdle of *Archaeopteryx*. – Nature **278**, No. 5701: 247–248; London.
OSTROM, J. H. (1974): *Archaeopteryx* and the origin of flight. – Quart. Rev. Biol. **49**: 27–47.
OSTROM, J. H. (1979): Bird flight: how did it begin? – American Scientist **67**: 46–56.
STEPHAN, B. (1970): Über Vorkommen und Funktion des Remicle. – Beitr. z. Vogelk. **16**, 1/6: 372–385; Leipzig.
STEPHAN, B. (1974): Urvögel Archaeopterygiformes. 167 pp. Wittenberg Lutherstadt (A. Ziemsen Verlag).
STEPHAN, B. (1976): Zeigt das Berliner Exemplar des Urvogels die Ventral- oder die Dorsalseite von Schwanz und Flügeln? – Orn. Jber. Mus., Hein. **1**: 71–77; Halberstadt.
STEPHAN, B. (1979): Urvögel Archaeopterygiformes. 2. Ed., 182 pp. Wittenberg Lutherstadt (A. Ziemsen Verlag).
STRESEMANN, E. & STEPHAN, B. (1968): Zahl und Zählung der Handschwingen bei den Honiganzeigern (Indicatoidae). – J. Orn. **109**, 2: 221–222; Berlin.

Addendum

It is very significant, that Dr. RIETSCHEL arrived at similar or equal conclusions as I did. For comparison I cite the concerned pages of 1st and 2nd edition of my monograph "Urvögel" (Wittenberg Lutherstadt 1974 and 1979) and refer to my contribution in the "Proceedings".

1. Primaries. In comparison with recent birds I mentioned 10 primaries and 1 remicle, now, after HELMS 1982, I think, *Archaeopteryx* has had 12 p. and 1 remicle. The remicle can not be seen, it is under the coverts and has a function of stabilization in this wing part (1 ed. p. 55 ff., 2 ed. p. 63 ff., Proc.).
2. Secondaries. 12 and more (1 ed. p. 55 ff., 2 ed. p. 63 ff.).
3. Finger 2 and 3. Not free (1 ed. p. 122, 128 ff., 2 ed. p. 133 ff., 140 ff.).
4. Claws. No use for locomotion (climbing) and grasping (1 ed. p. 122 ff., 128 ff., 2 ed. p. 133 ff., 140 ff.). But *Archaeopteryx* could not use the claws also for grooming, because the flexibility of the wing was reduced (see construction of the wing and the long feathers).
5. Flight. Active flight (1 ed. p. 117 ff., 2 ed. p. 128 ff.).
6. Carina. I have compared *Archaeopteryx* with bats (1 ed. p. 119, 2 ed. p. 130).
7. Tail. Rudder, but also as bearing plane and for display (1 ed. p. 149–150, 2 ed. p. 162).
8. Proavis. Bipedal, small animal with endothermy (1 ed. p. 139 ff., p. 98, 2 ed. p. 151 ff., p. 110–111).
9. Origin of flight. Flight as a prolonged jump (1 ed. p. 120, 126–127, 2 ed. p. 132, 137–139). Hopping is different from jumping with both feet. Only a little part of birds can hop (see Fringillidae, Ploceidae), most birds can walk, run, and jump.
10. Fossilisation. Now (Proc.) after HELMS 1982. In ed 1 (p. 152–153) after SCHÄFER 1955, in ed. 2 (p. 164–166) I have considered SCHÄFER 1976 and RIETSCHEL 1976.

Author's address: Dr. Burkhard Stephan, Museum für Naturkunde der Humboldt-Universität in Berlin, Invalidenstr. 43, 1040 Berlin, German Democratic Republic.
Translation: Dr. P. Beurton, Berlin.

Russell P. Balda, Gerald Caple, William R. Willis

Comparison of the Gliding to Flapping Sequence with the Flapping to Gliding Sequence

Abstract

The arboreal parachuting, gliding pathway to powered flight has been accepted by many as the most logical route for the evolution of flight in *Archaeopteryx*. We have investigated the physics and aerodynamics of this pathway. A model, with only flat plate drag, shows that an animal the size of *Archaeopteryx* would not enter the aerial environment as a parachuter. Using coefficient of lift divided by wing loading, lift to drag ratio, and vertical sink speed, we developed a three dimensional graph from which it was possible to closely estimate the gliding capabilities of virtually any animal. An animal the size of *Archaeopteryx* would need considerable morphological change (flattening) to enter the gliding niche at sub-lethal speeds. Although this is a smaller morphological change than would be required for parachuting, it is large enough that more than a small fringe of feathers would be needed to be useful. From this we conclude there is no compelling reason for a gliding precursor to develop a modern bird-like wing. A high aspect ratio wing was assumed and then we investigated the requirements for adding powered, flapping, flight capabilities. First a flapping motion was added without a twisting motion added to the flap. This allowed us to calculate a vertical flap speed at which wing tip stall occurs. Now, by using dihedral losses and an estimation of loss caused by a recovery stroke, we calculated the relationship between the maximum dihedral angle and the frequency of the wing beat to maintain the original glide path. Again, there is a very specific set of conditions that must be met before a wing beat would be useful to improve the original glide path. We conclude that the parachuting, gliding, to powered flight scenario is a highly improbable evolutionary route for *Archaeopteryx*.

Zusammenfassung

Der Weg zum aktiven Flug über ein Fallschirm-Gleitflugstadium ist von vielen als der logischste Verlauf der Evolution des Fluges von *Archaeopteryx* angenommen worden. Wir haben die physikalischen und aerodynamischen Verhältnisse dieses Weges untersucht. Ein Modell (mit nur flachem Plattenwiderstand) zeigt, daß ein Tier von der Größe von *Archaeopteryx* in das Medium Luft nicht als Fallschirmtier eintreten würde. Unter Verwendung des Koeffizienten Auftrieb durch Flächenbelastung, des Verhältnisses von Auftrieb zu Strömungswiderstand und der vertikalen Sinkgeschwindigkeit, entwickelten wir eine dreidimensionale Grafik, von der es möglich war, die Gleitfähigkeiten von faktisch jedem Tier genau zu schätzen.

Ein Tier wie *Archaeopteryx* benötigte eine beträchtliche morphologische Änderung (Abflachung), um die Gleitnische bei sublethalen Geschwindigkeiten besetzen zu können. Obwohl dies eine kleinere morphologische Veränderung ist als für das Fallschirmspringen nötig wäre, ist sie doch groß genug, daß mehr als ein schmaler Federsaum nötig wäre, um von Vorteil zu sein. Daraus schließen wir, daß es keinen zwingenden Grund für einen gleitenden Vorläufer gab, einen modernen vogelartigen Flügel zu entwickeln.

Es wurde ein Flügel mit hohem Streckungsverhältnis angenommen. Dann untersuchten wir die Bedingungen für zusätzliche aktive Schlagflugfähigkeiten. Zunächst wurde eine Schlagbewegung ohne Drehbewegung zum Flügelschlag hinzugefügt. Dies erlaubte nun, eine vertikale Flügelschlaggeschwindigkeit zu berechnen, bei der eine überzogene Fluglage an den Flügelspitzen besteht. Jetzt berechneten wir die Beziehung zwischen dem maximalen Flächenwinkel und der Flügelschlagfrequenz, um den ursprünglichen Gleitweg einzuhalten, indem wir die Neigungswinkelverluste und eine Schätzung des Verlustes, der durch den Aufschlag verursacht wird, benutzten. Wiederum gibt es eine Reihe spezifischer Bedingungen, die erfüllt sein müssen, bevor ein Flügelschlag den ursprünglichen Gleitweg verlängern könnte.

Wir kamen zu dem Schluß, daß ein Scenario von einem Fallschirm- und Gleitstadium zum aktiven Flug ein höchst unwahrscheinlicher evolutionärer Weg für *Archaeopteryx* war.

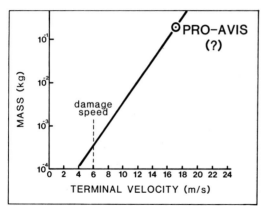
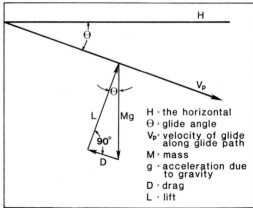

Fig. 1
Terminal velocities for organisms flattened so their thickness equals 0.1 of their diameter. Organisms were assigned a maximum coefficient of drag.

Fig. 2
A stable glide where lift is at right angles to glide path (P) when drag (D) is parallel to the glide path and weight (mg) is at right angles to the horizontal. The triangle describes the conditions for a glide down the path Vp.

Introduction

Ancestors of *Archaeopteryx*, according to arborealists, passed through two critical transitional phases on the route to becoming airborne. A parachuting organism developed gliding capabilities and this glider became the intermediate form for the development of powered, flapping flight. The physics and aerodynamics of these transitions have never been critically evaluated. These two transitions are only necessary in a scenario in which the ancestors of *Archaeopteryx* are tree dwellers. The purpose of this paper is to critically evaluate the physical parameters involved in making these transitions.

The three key factors we will address are: (1) The physical requirements for a parachuting animal; (2) The physical and biological constraints on the gliding mode of locomotion; (3) The physical and biological requirements necessary to add a useful, powered flap to an airfoil with gliding capabilities.

Methods

Our procedure is to develop a model the size and shape of *Archaeopteryx* and then apply aerodynamic laws to it. These laws delineate limits within which the model must operate. Thus, Newton's Laws of motion specify the relations between forces and between various velocities in gliding and flapping. It is characteristic of this approach that it is broadly applicable (although we have specified a model like *Archaeopteryx*) and that it cannot always be used to gain definitive answers to questions about the past. This approach, however, can be used to evaluate the likelihood of an event occurring in the past.

Results and Discussion

Parachuting:
In the arboreal theory, tree dwellers supposedly could leap between trees. This led to a parachuting stage (MARSH 1880, BOCK 1965). Here Pro-avis dropped through the air using air resistance to slow its descent. This animal purportedly had no capacity to generate lift. Parachuting has been arbitrarily defined as falling in a descent path of greater than 45° from the horizontal (OLIVER 1951, RAYNER 1981), however, by this definition the animal has the capability to generate lift. Since true parachuting employs

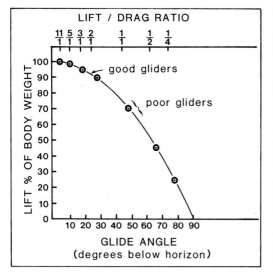

Fig. 3
The percent lift needed to travel down a given glide path. Also shown is the necessary lift to drag ratio for a given glide angle. Entry into the gliding realm begins where the L/D ratio is 0.70, and percent lift about 60%.

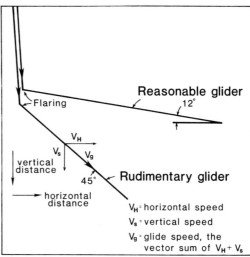

Fig. 4
Entry into the gliding realm occurs when an organism drops to gain speed and then flares into the glide path. Rudimentary gliders achieve a glide angle of about 45° which reduces V_s by 30%.

only forces of drag, we define any motion utilizing lift as a gliding motion, albeit often poor gliding. By our definition proponents of the arboreal theory espoused a poor gliding stage rather than a strict parachuting stage.

A parachuter is a very unlikely precursor to *Archaeopteryx*. In our model of an efficient parachuter, we assumed a flattened animal with a thickness equal to 1/10 its diameter and a density of approximately one. We gave this animal a maximum coefficient of drag (equal to that of a hemispherical parachute) and demanded a terminal velocity of no more than 6 m/sec, above which we assume injury will occur (TEMPLIN 1979). No animal with a mass greater than 8.0 gm can meet these criteria (Figure 1). If the falling distance is short enough Pro-avis will not obtain terminal velocity. Any animal with the mass postulated for *Archaeopteryx*, in the implausible shape we define, can parachute only about 2.0 m without sustaining injury. No small fringe of feathers (*contra* SAVILE 1962) will significantly improve parachuting performance because this fringe will have only a minuscule effect on the coefficient of drag. Thus, the concept of parachuting as a functional stage in the evolution of flight should be dropped, as parachuters are highly unlikely precursors to gliders.

Gliding:
By definition, gliding requires an airfoil that generates lift. If a glider, including any potential ancestor to *Archaeopteryx*, could not enter the gliding mode via parachuting another route must be found. Clearly, gliding is a successful means of transport as evidenced by the numerous, diverse kinds of organisms that have become gliders (RAYNER 1981).
In order to discuss the aerodynamics of gliding we must define the physical relationships between weight, lift, drag, and glide angle. These relationships are as follows:

$$\% \text{ Lift} = 100 \, L/W = 100 \cos \Theta = (100 \, L/D)/[1+(L/D)^2]^{1/2} \quad (1)$$

where L is lift, D is drag, W is weight, and Θ is the glide angle.

We define a stable glide as one with a constant speed and glide angle (Figures 2 and 3). The implications of Figure 3 are threefold. First, and most surprising, is that an animal gliding at 45° generates over 70%

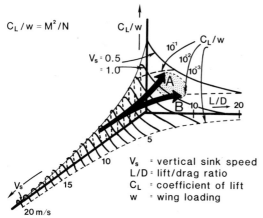
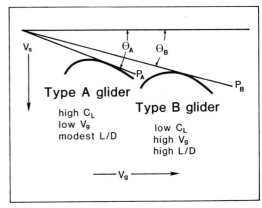

Fig. 5
Vertical sink speed (V_s), lift to drag ratio (L/D) and the ratio of coefficient of lift to wing loading are combined to show how these factors interact on an airborne organism. A glider the size of *Archaeopteryx* could operate in the stippled area. *Archaeopteryx* could have entered the gliding realm as shown in A with a high C_L/w, low sink speed and a modest L/D ratio. Evolution of a glider from a parachuter would have started along the V_s line, and involved a "jump" over the ridge.

Fig. 6
Conventional glide polar diagram for Type A and B gliders. Type A glider has a slow glide speed (V_g) a low sink speed, and a steep glide angle (θ_A) along path P_A. Most modern arboreal gliders are at or close to Type B gliders. *Archaeopteryx*, if it were arboreal would have been a Type A glider.

of its body weight in lift. RAYNER (1981) reports that flying geckos (*Ptychozoon* spp.) rather poor gliders, have this capability. A flying snake (*Chysopelia* sp.) reported to be gliding at an angle of 30° will generate a lift force greater than 85 % of its body weight. Second, at very small increments of lift there is little change in glide path or vertical speed. We have previously shown (CAPLE et al. 1983) that small increments of lift do not appreciably lengthen a jump path. Third, for a modest glider, with a glide angle of 10° to 20°, the lift to drag ratio changes dramatically for small changes in glide angle. The lift to drag ratio can only be improved by substantial morphological changes. We conclude, based on the three interpretations given above, that modern arboreal gliders did not enter the gliding realm via small increments of lift. Thus, we believe that the first gliders were already equipped (preadapted) with a substantial amount of lift. We propose that gliding evolved in the following sequence. First animals dropped to gain speed. Once this speed was high enough to gain substantial lift, control was used to assume a stable glide angle (Figure 4). The animal must be of a shape that is conducive to generating lift such as having a flattened body or a patagium.

If an ancestral glider performs this drop – first flare, later behavior – as depicted in Fig. 4, it would reduce its vertical impact speed by 30 % if the flare angle were 45°. The ability to flare and become a glider is a behavioral change as previously suggested by OLIVER (1951), and SAVILE (1962), and requires the animal to have the ability to flatten its body during the drop. This flattening need not be as extensive as in a true parachuter. The red squirrel *(Tamiasciurus hudsonicus)*, capable of flattening its body, can achieve a glide angle of 30° by this mechanism (SAVILE 1962). OLIVER (1951) showed that the Carolina Anole *(Anolis carolinensis)* when dropped from a height of 10 m, flared when about 5 m from the ground and landed safely. We therefore conclude, that if Pro-avis entered the air through a gliding stage it would have started with a highly flattened body, capable of generating substantial amounts of lift.

This drop-flare mechanism requires the animal be at a certain elevation above the ground. The maximum height of the flora found in the Solnhofen Limestone that preserved the fossils of *Archaeopteryx*, precludes the arboreal evolution of a glider in this environment (VIOHL, this volume).

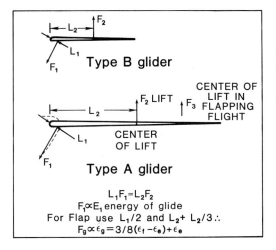
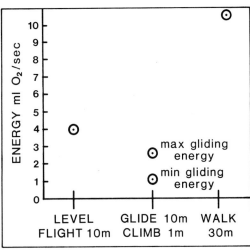

Fig. 7
Muscle requirements for holding a wing level in Type A and B gliders. If L_1 and F_2 are equal in both gliders the force necessary to hold the wing horizontal is larger in a Type A glider than in a Type B one. In flapping flight the center of lift is distal (F_3) and L_2 is longer.

Fig. 8
Energy consumption for a 200 gm *Archaeopteryx* flying, gliding, and walking (climbing included) to a new perch 10 m from the starting perch but at the same elevation. Gliding is the most energy efficient mode of transport.

If Pro-avis entered the air as suggested above, then additional increments of lift would allow it to extend its glide path. For Pro-avis to have been a precursor to *Archaeopteryx* these incremental changes must have occurred in the forelimbs and tail, thus changing the aspect ratio (relationship of airfoil length to its width). Changing from a low aspect ratio wing (less than one) to a high aspect ratio one of 7.2 as in *Archaeopteryx* (HEPTONSTALL 1970) results in large, multiple, aerodynamic changes in the animal. We will attempt to show the consequence of these changes in a gliding Pro-avis.

In order to study the changes in aerodynamic parameters that would occur on the evolutionary route to powered flight, if in fact, a gliding stage was intermediate, we have developed a 3-dimensional graph. Figure 5 shows how vertical sink speed, lift to drag ratio, and the coefficient of lift (C_L) divided by wing-loading (w) are integrated into a single curved surface. This graph depicts the characteristics of any gliding object or animal that is performing a stable glide.

The left axis (V_s) where lift to drag is zero, i.e. no lift is generated, represents true parachuting. There is a very pronounced ridge on the surface of the graph at a constant lift to drag ratio of 0.7. Small changes in C_L (refinements of the airfoil) at this low lift to drag ratio would not result in appreciable lengthening of the glide path. This strengthens our earlier conclusion that the evolution of flight did not begin with a parachuting stage, as a parachuter would have to "climb" this steep ridge – an unlikely biological event. If *Archaeopteryx* did not enter the gliding realm as a parachuter, then (to follow the arboreal scenario) it must have been a primitive glider.

A glider the size of *Archaeopteryx* would have occurred to the right of the ridge at a C_L/w of approximately .005. If *Archaeopteryx* was a gliding only animal it would occur somewhere within the shaded area shown on the surface in Figure 5. We estimated the maximum coefficient of lift to be less than 1.6 but greater than 0.4. Our upper value is about the maximum achieved by modern birds (PENNYCUICK 1978) and our lower value is a very modest one. Arrows A and B trace the potential evolutionary pathways to different types of gliders. Gliders on Path A would increase their coefficient of lift, or decrease their wing-loading, or both. Both of the above result in increased drag, thus the animal will have a relatively low glide speed and a higher relative sinking velocity (low L/D). Path B gliders maintain a nearly unchanged C_L/w and refine their body shape, which results in a lower drag and thus a larger lift to drag ratio. A glide performed by such an animal would have a fast glide speed and could have a relatively lower sinking velocity (high L/D). Animals on Path A must make substantial morphological changes to achieve these slow glide speeds. This is so, because for Path A animals the wing area must be substantially enlarged and would probably be accompanied by an increased aspect ratio. Path B animals need make smaller morphological changes which lead to a high L/D ratio. These changes mainly involve the streamlining of the airfoil and body. Both Paths A and B can be taken as extremes for entering the gliding realm. A and B paths cannot be taken as mutually exclusive, as an animal with the capabilities of A can, with behavioral modifications, also perform as in B. In Figure 5, it should be noted that the larger the lift to drag ratio, the longer the glide path. This is also shown in a more conventional glide polar diagram (Figure 6). HUMMEL (1980) described how gliding birds can change their wing area, and C_L, thus making the analogous change from a Type A to a Type B glider which would result in lengthening the glide path. The B type glider with its faster glide would probably land at a faster speed than an A type glider, however, this landing problem was resolved upon entry into the gliding realm. For an animal that becomes an arboreal glider there are no compelling aerodynamic reasons why it should follow Path A. Among modern arboreal gliders there are no examples of Type A, high aspect ratio gliders.

RAYNER (1985) comments that the absence of the larger aspect ratio wings in contemporary gliders presents a possible objection to these animals being precursors of powered flight. The glider most often cited (RAYNER 1985) as an intermediate is the Colugo because the flight membrane is incorporated into the hand. The Colugo has the lowest aspect ratio of all modern membrane type gliders and therefore is the glider furthest from the necessary precursor suggested by RAYNER (1985). JEPSEN (1970) suggested that the Colugo is the end result of a bat that adopted the glide, climb-up route. The lower aspect ratio has a minor effect on the glide performance, but greatly enhances surface mobility.

There is a potentially large mechanical difference between Type A and B gliders in the structure of the wing and shoulder. Type B gliders have a much shorter distance from the shoulder joint to the center of lift on the wing than do Type A gliders. Type A gliders would be required to develop either a strong mechanical stop or large pectoral muscles (Figure 7) to keep the wings in a horizontal position. Both of these solutions involve large morphological changes.

Transition from Gliding to Flapping

If a gliding Pro-avis were to initiate a flap as stated in the arboreal scenario then major energetic, morphologic, and aerodynamic consequences would result. We will discuss the physical limitations of these three constraints in this transition.

Any glider that increases its aspect ratio does so at increased energy consumption (even if no mechanical work is done). The increased aspect ratio will also decrease surface mobility. While the increased aspect ratio can improve a glide path, it is an adaptation for a slow speed glide.

We calculated the energy consumption of a 200 gm animal in the shape of *Archaeopteryx* moving from a position 10 m up in one tree to the same height in a second tree 10 m from the first. We assumed a glide and flight speed of 9 m/sec and a walking speed of 0.84 m/sec (TAYLOR 1977). The most primitive and energetically costly way to do this is to climb down, walk to the adjacent tree and then climb to the original takeoff elevation. A second method is to glide to a position in the second tree and climb up to recover the lost elevation (we chose a one meter climb). A third method is direct, powered flight between the two trees. Experimental data are available for energy consumption during running, climbing, and flapping flight (TAYLOR 1979, NACHTIGALL 1980). Gliding energy requirements have not been reported. Here, we estimate a minimum and maximum value.

Using the energy value for flapping flight (NACHTIGALL 1980) we calculated the energy requirements for gliding flight based on the following two facts: 1) as a flapping wing moves from the horizontal the pectoral muscle must exert a larger force to attain the same torque; 2) the energetic requirement to exert an impulse (force × time) is proportional to the impulse. From the above we argue that the ratio of the energies involved in flapping and gliding is the same as that of the force involved to invoke the same torque. Since the force in flapping flight includes that force necessary to produce thrust to overcome drag the estimate of the gliding energy obtained above will be too large. Two factors are involved in estimating torque in flapping flight; the lever arm associated with the muscle used in the flap stroke decreases as the wing moves from the horizontal, and torque required for a given lift increases as the center of lift moves outward. We find that the energy above existence energy required for gliding is ⅜ that required for flapping. We calculated the maximum energy of a glide by using the equation given in Figure 7. The minimum estimate was assumed to be existence energy (TAYLOR 1977).

Our energy calculations for the three methods of transport (surface travel, gliding, and powered flight) show that a glide combined with a climb is the most economical means of traversing the 10 m between trees without losing altitude. Any reasonable glider with an L/D ratio greater than one is operating with a lower energy consumption than the climber and flapper (Figure 8). Energy, thus cannot be used as an inducement for the transition from gliding to flapping flight.

Critical to the arboreal theory is the addition of a flap to a gliding airfoil. We hasten to point out that flapping any airfoil can be detrimental as well as advantageous to the flapper, since changes in the dihedral angle and unloading on the upstroke cause loss of lift. Only if the gain in lift on the downstroke is greater than these combined losses will a flap be advantageous.

As a wing flaps on its downstroke, it is moving fastest at the tip. This additional velocity is directed downward and is directly proportional to the distance from the shoulder joint. Since the angle of attack of the wing is determined by the direction of the relative wind, it will also increase toward the wing tip. The additional angle of attack due to the flap can be found as,

$$\tan \sigma = 3lf\,\Phi/V \qquad (2)$$

where σ is the increase in attack angle, V is the speed along the path, Φ is the maximum angle of the wing from the horizontal (maximum dihedral angle), l is the distance along the span of the wing, and f is the frequency of the beat.

For *Archaeopteryx* gliding at its best L/D (small C_L), a beat frequency of about 3/sec would lead to stalling of the wing tips. Any animal gliding near its maximum C_L would stall at very low beat frequencies unless the wings were twisted to keep the angle of attack below stalling. Thus, a transitional Pro-avis attempting to beat a gliding airfoil, would be required to incorporate the complex twisting of the wing (seen in modern birds) in the very first stages of flapping. In our analysis of a flap we ignore these changes in angle of attack and simply assume the animal is maintaining the optimal value of C_L along the wing by proper configurational changes.

The loss of lift caused by dihedral angle changes during a beat cycle easily can be found by calculating the specific dihedral angle at which lift is the same as the average lift over the downstroke. Assuming that the angular speed of the wing during the beat is constant, the average dihedral is α where:

$$\alpha = \Phi/1.75 \qquad (3)$$

We assume the wing is unloaded during the upstroke which requires ⅓ of the total flap cycle time. Thus, the total loss of lift due to flapping can be found as the following fractions of gliding lift:

$$\text{Unloading Loss} = 1/6 \text{ (half the wing is unloaded)} \qquad (4)$$
$$\text{Dihedral Loss} = 1 - \cos\alpha \qquad (5)$$

When a flap is added to a glider with a stable equilibrium glide, the equilibrium is disturbed and a new path will be followed. In this case we will start with a stable glide at a speed of 10 m/sec and a sink speed of 2 m/sec. This glide is perturbed by a complete wing beat cycle. Downstroke time (T_1, Fig. 9) will be twice the time of the upstroke (PENNYCUICK 1978, RICHNER 1980). The amplitude of the wing beat will be based upon the dihedral change and length of the wing, the air velocity at the wing tip will be taken as twice that of the average air velocity over the wing.

The condition for returning the now flapping glider to its original path with a forward speed of 10 m/sec and a sink speed of 2 m/sec depends on correcting for two losses. The first of these is dihedral loss

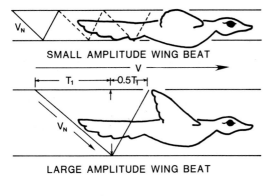
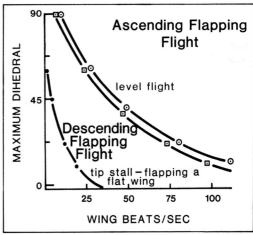

Fig. 9
Geometry necessary to calculate wing beat frequencies needed to maintain the original flight path when flapping is initiated. Two amplitudes are illustrated.

Fig. 10
The relationship between wing beat frequency and amplitude to maintain the original flight path after flapping is initiated.

(Eq. 5), the second is the upstroke or recovery loss, when part of the wing will be unloaded and generating no lift. Any reciprocating motion involving a downstroke must be followed by an unloading upstroke if there is to be forward motion.

The losses during a flap are recovered by increasing the air velocity (V_n) over the wing. This new air velocity (V_n) was calculated and used to generate lift during the downstroke of the wing beat (Figure 9). The amplitude of this wing beat is determined by the total angular deflection above and below the horizontal. Simple geometry allows a calculation of the relationship between wing beat frequency and angular displacement necessary to restore the original path. Although there are some assumptions used in this analysis, the calculated beat frequency at an angular deflection of 90° is close to those calculated by PENNYCUICK (1983) for a 200 gm bird with the aspect ratio of *Archaeopteryx*, and observed for birds by GREENEWALT (1961).

The calculated value for the beat frequency of about 9 hz at a speed of 10 m/sec agrees with that calculated by RAYNER (this volume) and with his observation that the energy expenditures for flapping flight via the vortex ring gait or constant circulation gait are comparable for level flight.

We also calculated conditions necessary for level flight (Figure 10). We conclude that a nearly complete wing beat cycle comparable to that used by modern birds is necessary before an extension of the original glide path is obtained. For low amplitude wing beats high frequency is required to generate comparable amounts of lift, but this results in greater energy consumption than occurs at the original beat frequency. In its initial stages a glider learning to flap does not have a low amplitude, low energy wing beat. This amplitude-frequency effect has been directly observed in modern birds (RUPPELL, 1978).

By modifying an equation developed by PENNYCUICK (1978) to allow for the effects of gravity, we can write:
$$f = [kmg \cos \theta / Ar]^{1/2} \qquad (6)$$
where f is the beat frequency, θ is the descent angle, and Ar is the aspect ratio. Equation (6) predicts a sizeable wing beat frequency to flap down a reasonable path. It also predicts a higher wing beat frequency for the low aspect ratio wing, and since increasing frequency means a higher energy, it would be more difficult to develop a flap in a shorter wing. As we have shown earlier, the shorter wing however, may give improved gliding performance as a Type B glider.

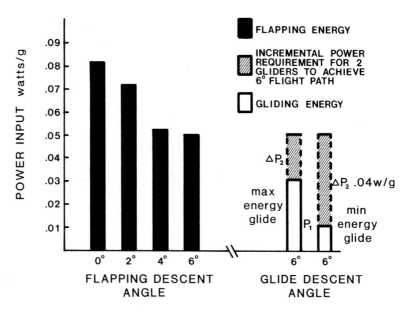

Fig. 11
The power relationship between level and descending flapping flight in the fish crow. Open bars are our estimates to glide at a 6° descent angle. Hatched bars show the additional energy needed to flap down the same flight path.

A wing beat requires an energy input (PENNYCUICK 1969, RAYNER 1977). Power requirements for level flight have been measured for many birds and NACHTIGALL (1980) has determined power requirements for the fish crow in flapping, descending, flight. As the fish crow is a bird in the size range of *Archaeopteryx*, we will use these data, along with an estimate of gliding energy to determine the extra power needed to add a useful flap to a glider. As no gliding energies are available, we will estimate the maximum gliding energy as a fraction of the energy required for level flapping flight. This is a lever arm calculation from Fig. 7 and gives an overestimation of the gliding power, as the power requirements for increased drag and inertial changes during the flap have not been subtracted.

It is possible that the fish crow could attain a six degree glide path, using a reduced wing span, a low C_L, but a high L/D ratio (HUMMEL, 1980). Thus the energy requirements are estimated for gliding and known for flapping down this descending path. We argue that any incremental addition of power to gliding power P_1 that is less than P_2 will only cause a descent path steeper than six degrees (Figure 11). In fact, an additional power requirement of P_2 is required for the gliding fish crow to flap down its previous glide path. The generalization we make here is that the more efficient a glider becomes in terms of its glide path and energy consumption, the larger the increment P_2 becomes. The gliding to flapping transition is more difficult and, therefore, less likely.

Our calculations strongly suggest that in any transition from a glider to a powered flier there is a non-adaptive transition zone. With primitive flaps of a gliding airfoil the glide path is not extended, in contrast, it is degraded. The question then becomes, why would a glider flap its' airfoil to increase its' angle of descent? There are far simpler means available to the glider to increase its' angle of descent. RAYNER (1985, this volume) has postulated that there does exist a specific set of frequency-amplitude parameters describing how a gliding airfoil, when flapped, can increase the glide length. It must be noted, however, that these parameters apply only to a sophisticated, high speed gait postulated to be used by only a few modern day birds. Such a gait, if it exists, is in all probability too complicated to be ascribed to a primitive flapping animal.

A powered flyer can easily become a glider, but a glider would face a most difficult transition to a powered flyer. In fact, this latter transition demands the loss of the original glide path. The postulates of the arboreal theory indicate this would be non-advantageous. The power requirement arguments made above should be verifiable by experimental work.

For a specific muscle type there is a direct relationship between muscle mass and power generated. For this reason arguments presented above using small additional increments of power are directly applicable to small additions of mass. The additions of mass emphasize an underlying assumption in the previous arguments, i.e. that power could be added at constant mass. It is more probable that the addition of power was concommitant with an increase in mass. Such an increase in mass will lead to a degradation in glide path or an increase in speed. In the first of these the length of the glide is decreased and in the second the problems of landing have been exacerbated.

So far we have said nothing about the role of the tail of *Archaeopteryx*. It would appear to have a large drag, as it has a large area. We speculate that it has low lift because of the associated problems with balance. Its action is that of a small force through a large distance. For the glider, the large surface, low lift combination would serve to slow the glide and decrease the glide distance. For the running *Archaeopteryx*, drag would still be there, but small movements of the tail can be used for control during a run-jump sequence. For the glider these same movements only serve to degrade the glide path.

Summary

We have investigated the physical requirements for the arboreal evolution of flight in *Archaeopteryx*. We predict that entry into the gliding niche would have been by a highly flattened precursor, with the morphology and behavior capable of generating over 50 % of the body weight in lift. There also was no parachuting stage in the evolution of any arboreal gliders. The gliding capabilities of *Archaeopteryx* have been assessed. It was difficult to find any compelling reason why a glider would have evolved with the morphology of *Archaeopteryx*. In fact, it appears *Archaeopteryx* would have been a "better" glider, that is, one with an extended glide, if it had a smaller surface area, a lower coefficient of lift, but retaining a high lift to drag ratio and a high speed (Type B glider).

Assuming an *Achraeopteryx*-like glider, we investigated the requirements for adding powered flight to this glider. Here only a complete wing beat cycle was considered, as this is a requirement for powered flight. In the transition from gliding to flapping flight there are no evolutionary advantages to be gained by performing a partial wing beat, and the wing must move with sophisticated twisting motions. Power requirements for adding a flapping motion to the wing indicate that small increments of power (and therefore small increments of muscle mass) would only lead to a more sharply descending flight path. We conclude that a glider is not a likely precursor to *Archaeopteryx* and that *Archaeopteryx* and its wing is the end result of flapping precursors, with any gliding capabilities being a useful, but secondary adaptation of the flapping wing. A powered flier will always have some gliding capabilities but an animal limited to gliding is unlikely to flap.

References Cited

Bock, W. J. (1969): The origin and radiation of birds. – Ann. N.Y. Acad. Sci. **167**: 147–155.
Caple, G., Balda, R. P. and Willis, W. R. (1983): The physics of leaping animals and the evolution of preflight. – Am. Nat. **121**: 455–476.
Greenewalt, C. H. (1962): Dimensional relationships for flying animals. – Smithson Misc. Collect. **144** (2): 1–46.
Heptonstall, W. B. (1970): Quantitative assessment of the flight of *Archaeopteryx*. – Nature **228**: 185–186.
Hummel, D. (1980): The aerodynamic characteristics of slotted wing-tips in soaring birds. – Acta XVII Congressus Internationalis Ornithologicus 391–396.
Jepsen, G. L. (1970): Bat origins and evolution. – In: Biology of bats (W. A. Wimsatt, ed.): 1–64; New York (Academic Press).
Marsh, O. C. (1880): Odontornithes: A monograph on the extinct toothed birds of North America. – Rep. U.S. Geological Exploration of the 40th Parallel, No 7. Washington, D.C.
Oliver, J. A. (1951): "Gliding" in amphibians and reptiles, with a remark on an arboreal adaptation in the lizard, *Anolis carolinensis carolinensis* Voight. – Amer. Nat. **85**: 171–176.
Pennycuick, C. J. (1968): Power requirements for horizontal flight in the pigeon. – *Columba livia*. J. Exp. Biol. **49**: 527–555.

Pennycuick, C. J. (1978): Fifteen testable predictions about bird flight. – Oikos **30**: 165–176.

Rayner, J. (1977): The intermittent flight of birds. – In: Scale effects in animal locomotion (T. J. Pedley, ed.): 437–443; New York (Academic Press).

Rayner, J. (1981): Flight adaptations in vertebrates. – Symp. Zool. Soc. Lond. **48**: 137–172.

Rayner, J. (1985): Vertebrate flapping flight mechanics and aerodynamics and the evolution of flight in bats. – Biona Report **5**: in: Fledermausflug, ed. W. Nachtigall. Heidelberg (Gustav Fischer).

Rayner, J. (1985): Mechanical and ecological constraints on flight evolution. – This volume.

Richner, H. (1981): Funktionsmorphologische Untersuchungen am Flugapparat verschiedener Gänse- und Entenvögel [Anatidae]. – Dissertation, Universität Zürich.

Rothe, H. J. and Nachtigall, W. (1980): Physiological and energetic adaptation of flying birds, measured by the wind tunnel technique. A Survey. – Acta XVII Congressus Internationalis Ornithologicus: 400–405.

Ruppell, G. (1977): Bird Flight. – 191 pp., Van Nostrand Reinhold Co. N. Y. 191 pp.

Savile, D. B. O. (1962): Gliding and flight in the vertebrates. – Am. Zool. **2**: 161–166.

Templin, R. J. (1979): Size limits in flying animals. – In: Major patterns in vertebrate evolution (Hecht, M., K., Goody, P. C., Hecht, B. M. ed.): 411–421; New York (Plenum Press).

Taylor, C. R. (1977): The energetics of terrestial locomotion and body size in vertebrates. In: Scale Effects in Animal Locomotion. (T. J. Pedley ed.): 127–141; New York (Academic Press).

Viohl, G. (1985). Geology of the Solnhofen Limestones and the Environment of *Archaeopteryx*. – This volume.

Yalden, D. W. The flying ability of *Archaeopteryx*. Ibis; **113**: 349–356.

Authors' address: Prof. Dr. Russell P. Balda, Prof. Dr. Gerald Caple and Prof. Dr. William R. Willis, Northern Arizona University, Flagstaff, Arizona 86011, USA.

Jeremy M. V. Rayner

Mechanical and Ecological Constraints on Flight Evolution

Abstract

Consideration of mechanical and ecological constraints on the evolution of flapping flight in birds and bats shows that objections to the gliding model are largely unfounded, and calculation for a gliding proto-microbat reveals significant ecological advantages. Owing to the mechanical requirements of flapping flight gaits, gliding is pre-adapted for the evolution of flapping; pectoral girdle structure in *Archaeopteryx* is consistent with this hypothesis. In a discussion contribution I demonstrate that flapping flight is far more likely to have evolved from an arboreal glider than from a gliding or running cursor.

Zusammenfassung

Die Betrachtung der mechanischen und ökologischen Zwänge bezüglich der Entwicklung des Schlagfluges bei Vögeln und Fledermäusen zeigt, daß Einsprüche gegen die Gleitflughypothese grundlos sind, und die Modellrechnung für eine gleitende Ur-Fledermaus offenbart wichtige ökologische Vorteile. Infolge der mechanischen Erfordernisse von Schlagflugstilen ist der Gleitflug an den Schlagflug vorangepaßt; die Brustgürtelmorphologie in *Archaeopteryx* ist mit dieser Hypothese vereinbar. In einem Diskussionsbeitrag zeige ich, daß der Schlagflug sich wahrscheinlicher bei einem baumlebenden Gleitflieger als bei einem gleitenden oder rennenden Läufer entwickelt hat.

Introduction

Flapping flight in vertebrates is a complex interaction of many factors reflecting the relationship of flight performance to many different facets of the animal's biology. A discussion of the evolution of flapping flight must consider not only the mechanics of flight, but also the correlates of flight with morphology, physiology, behaviour, and ecology. In this paper I describe a number of important constraints on flight performance in birds and bats, and demonstrate why gliding is attractive as an intermediate stage in the evolution of flapping flight. The evidence I present does not on its own prove that flight evolved through gliding (although – on mechanical and ecological grounds – it certainly c o u l d have), any more than the efficacy of cursorial activity (e. g. CAPLE et al. 1983) can i n i s o l a t i o n prove that the proto-bird was a cursor: independent corroboration must be adduced before drawing any such conclusion. In a discussion contribution to this symposium I present a tentative mechanical model consistent with a cursorial habit in pro-avis, but an arboreal (or at least a gliding) origin of avian flight is at least mechanically more attractive. There is however overwhelming evidence that flying bats originated from arboreal gliders (RAYNER 1985, SCHOLEY 1985b): the problem is simpler for bats than for birds, since there is little doubt about what constitutes a bat, no material doubt about bat ancestry, and very little fossil evidence to invite misinterpretation. Bats evolved from quadrupedal mammals which could thermoregulate and could sustain the aerobic metabolism needed for active flight; the flight organs involve all four limbs and are directly homologous to quadrupedal structure, and the behaviour and ecology of the proto-bat can be deduced from comparisons with contemporary gliding mammals. For birds some of these points are – at the least – arguable. I feel that it is preferable not to risk misinterpretation as a result of the diversity of opinion as to what *Archaeopteryx* was and how it was able to fly, and so have presented this discussion in general terms, with calculations for a proto-microbat. There is nonetheless no difficulty in applying these general mechanical and ecological principles equally well to a proto-bird as to a proto-bat.

Models of flight evolution

Three physical models have been advanced for the development of flapping flight in birds and bats. First, a cursorial (terrestrial) animal makes small jumps from the ground – to catch prey or to escape predators – and these gradually lengthen as the forelimbs develop into wings, either to catch more prey or to enhance running stability. Second, an animal (cursorial or arboreal) suddenly flutters its wings and flies to catch insects. Thirdly, an arboreal animal first jumps and parachutes, then glides between trees, and then extends its glides by flapping.

An adaptive model for the evolution of flight must be consistent with the (scanty) fossil evidence, and must predict a sequence of stages in which structural and behavioural locomotor adaptations allow the animal to be ecologically successful. Moreover, these adaptations must be bounded by physical constraints imposed by the animals' structure and phylogeny (BOCK 1965; RAYNER 1981; PADIAN 1982, 1983). For our purposes it is valuable to define these constraints more specifically in the ways in which they act on the propulsion system of a flying animal. At every phase in the evolution of flight the forces and moments produced by the wing muscles, the geometry (kinematics) of the wingbeat, and the overall mechanical and metabolic power requirements must remain consistent with the structure of the flight organs. The aerodynamic performance of the forelimb must necessarily improve as flight develops, and as far as possible the changes must be consistent with the structure of the flight organs, must be gradual, and must result in improved ecological – or ultimately reproductive – performance so that retention and further enhancement of the proto-flight adaptations may be favoured. Previous scenarios and models have failed to satisfy completely these requirements, and in particular have not explained satisfactorily the final phases in the development of flapping from gliding or running. CAPLE et al. (1983) have rightly drawn attention to the importance of stability in running animals and have shown how some form of gliding aerofoil may have evolved in a cursor, but they fail to solve adequately the most significant problem, namely the development of a thrusting wingstroke. In this paper I explore the relation of these mechanical and ecological constraints to the origin of avian flight: they are most closely consistent with the evolution of flapping flight from gliding, and with an arboreal habit for pro-avis, and appear to rule out alternative hypotheses.

The cursorial model has been invoked by many authors (NOPCSA 1907, 1923; OSTROM 1974, 1979; PADIAN 1982; CAPLE et al. 1983). There are various reasons to doubt its application to birds (RAYNER 1981, U. M. NORBERG 1985): the main source of thrust in running – the legs – is removed as soon as the animal becomes airborne, there is no explanation for the appearance of a wingstroke capable of producing thrust, and running speeds may be well below those at which flapping wings (particularly the primitive wings of a proto-flier) are efficient. The cursorial model also ignores reverse selection due to the loss of other forelimb functions and to the impairment of terrestrial locomotion by wing drag, and the fact that a cursor evolving flight must possess well-developed pelvic and pectoral muscles. Undoubtedly improved foraging performance and stability contribute to selection in a running animal for the development of wings, but precisely the same may be said for a gliding proto-flapper.

The proto-bat could not have been cursorial since the hind limb, integral with the flight membrane, must have inhibited terrestrial progression. The related fluttering hypothesis (JEPSEN 1970, PIRLOT 1977, CAPLE et al. 1983) has arisen in response to unfounded, and often illogical, objections to the gliding model, but it is untenable since it ignores the animal's intermediate appearance and behaviour, which must have fitted it neither for running, climbing nor fluttering. As seems inevitable also with the cursorial model, the first powered flights – with poorly developed wings – would have been in hovering or at low speeds, requiring exceptional metabolic and sensory performance and a highly complex wingbeat; since slow flight is so taxing it is unreasonable to suppose that evolution was able to solve these problems before sustained fast flight became possible (CLARK 1977; U. M. NORBERG 1985).

The gliding model, first proposed by DARWIN (1859) for bats and by MARSH (1880) for birds, has been widely accepted (BOCK 1965; YALDEN 1971; SMITH 1977; RAYNER 1981, 1985; U. M. NORBERG 1985). It is intuitively attractive as it allows gravity to help rather than hinder the development of flight, it assumes that flapping first developed at high flight speeds, and it assigns more importance to the relatively simple requirements of forward flight than to control, manoeuvrability or prey capture.

Flapping flight mechanics

To understand the evolution of flapping flight it is essential to define what a flying animal achieves by beating its wings. Space does not permit detailed consideration here: I have given fuller discussions elsewhere (RAYNER 1979, 1980, 1985).

The wings of flying animals generate lift by aerofoil action. Lift is associated with drag due to friction on the wing surface and to the trailing vortices behind the wing which are responsible for the lift; in addition there is drag owing to friction on the body. In a gliding animal lift balances the weight but the drag is uncompensated, and the animal must descend. To fly level the wings must be flapped so that they generate a horizontal thrust component of lift to balance drag. At first sight it may appear paradoxical that the purpose of wing flapping is not to support the weight, but the shallow glide angles of most birds confirm that wings readily support the weight, but energy is always lost to drag, and the absence of a horizontal thrust when the wings are not flapped is the cause of a descending glide.

The need for thrust constrains the configuration of the wingbeat: simply flapping the wings up and down with constant lift gives no thrust since any horizontal force from the downstroke is cancelled by the effect of the upstroke. Some form of asymmetry – geometric or aerodynamic – between the two stroke phases is essential. The flapping gait (frequency, amplitude, geometry) is selected according to wing design, speed and to other criteria such as stability and control. The possible gaits are distinguished largely by the aerodynamic function of the upstroke and by the structure of the vortex wake; preferred gait also correlates with pectoral girdle anatomy owing to the varying pattern of aerodynamic force generation (RAYNER 1985).

Two particularly important gaits which we have demonstrated experimentally are useful to clarify wing function under different conditions (SPEDDING et al. 1984, RAYNER 1985). In slow flight, and in birds with low aspect ratio wings, upstroke weight support is mechanically inefficient, and the wing is flexed and moved close to the body to minimize drag. All weight support and thrust are generated by the downstrokes, and the wake is a chain of ring vortices. In fast flight and in birds and bats with larger aspect ratio the upstroke is used for lift. In the downstroke the wing moves perpendicular to the flight path, while in the upstroke it is swept or flexed, and the wingtips move back and slightly inwards in order to provide geometric asymmetry: thrust from the downstroke more than balances retardation from the upstroke. This is a much simpler gait than flight with unloaded upstroke: the wake is a continuous undulating vortex pair, and weight is supported throughout the wingbeat. It is typical of many birds and bats, while the vortex ring gait appears to be confined to slow flight and to birds with sufficiently large wings to support the weight on the downstroke alone.

These gaits can be modelled theoretically by considering momentum and energy transport in the vortex wake (RAYNER 1979, 1985). The energy costs of the two gaits are comparable, but for mechanical and anatomical reasons the continuous lift gait is more attractive for fast flight. A typical mechanical power curve (for the noctule bat *Nyctalus noctula*, which is known to fly with continuous lift) is shown in figure 1. The U-shaped curve shows the advantages of flight at speeds which minimize power or maximize range (see also figure 4). Although the structure of the gait used is based on analysis of high speed film, an important feature of this model is that it estimates frequency and amplitude by an equilibrium condition, and in the noctule, as in many other animals, estimated kinematics agree well with our results from film analysis, so that other predictions derived with it may be accepted with some confidence. Most important is the pattern of time variation of the aerodynamic moments at the wing root, since these indicate the action of the thoracic muscles (figure 2). The most striking features are that wing inertia dominates the

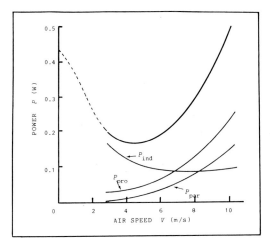

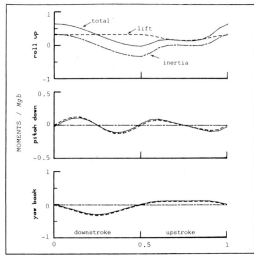

Fig. 1

Estimated mechanical power consumption for a noctule bat (*mass* = 0.0275 kg) flying at different speeds, with continuous lift and with optimum wingbeat kinematics. For full details of the model used and for experimental justification see RAYNER (1985). Noctules are observed to fly at 6–8 m/s, close to the maximum range speed.

Fig. 2

Moments of aerodynamic forces (mainly lift and wing inertia) at the root of one wing for a noctule flying at 6 m/s, normalized by semispan times weight. The wings are flexed in the upstroke and remain perpendicular to the flight path so there are no pitch or yaw components of inertia. Roll corresponds to dorso-ventral action of the pectoralis. The yaw and roll components from the two wings cancel and have no net effect on the body, but pitch demands either compensatory tail movements or some horizontal action of the pectoralis to ensure stability. Similar moment patterns are calculated in other flying animals.

vertical roll moment (inertia has generally been ignored hitherto), that pectoralis action begins and peaks in the final phases of the upstroke, and that (at relatively high speeds at least) during the upstroke aerodynamic lift is sufficient to raise the wing so that there is therefore no need for an elevator muscle. The rather smaller pitch and yaw moments are balanced by variations in wing incidence, by tail movements, and by other wingroot muscles (coracobrachialis, deltoid group). There is limited experimental confirmation that muscle action corresponds to this pattern (HERMANSON & ALTENBACH 1983, RAYNER 1985).

Since it is mechanically much simpler, fast flight is far easier to achieve than slow flight or a ground take-off: moreover the continuous lift gait is aerodynamically very similar to gliding, with only the instantaneous geometrical configuration of the wing being varied. The forces in gliding and flapping are also comparable: a gliding animal must apply a roll moment of magnitude 0.23 Mgb (RAYNER 1985; M = mass, g = gravity, b = wing semispan) at the wingroot to balance lift on the forelimb. This is only slightly smaller than the average moment in flapping in fast flight, and is comparable to the moments in climbing: if the animal hangs on a vertical surface with its forelimbs at an angle of 45° it can support its weight with a moment 0.36 Mgb', where b' (which may be smaller than b) is the semispan of the hands; much larger moments are needed in active climbing. In fast flight the predominantly dorso-ventral (roll) moment is best served by a pectoral muscle lying beneath the wing. In birds the supracoracoideus muscle and the posterior attachment of the pectoralis and supracoracoideus to the sternum are associated with slow flight or take-off, where the wing must also be moved horizontally and where the upstroke is unloaded and must be raised actively. Birds habitually flying fast with continuous lift have comparatively small supracoracoideus muscles since they rarely need to elevate the wing. The musculature for slow flight, for fluttering or for take off from the ground is unlike that for any other mode of locomotion, and appears to be more derived than that needed for simple flapping. Gliders – without elevator muscles – already possess the musculature for flapping at normal flight speeds (although the muscles themselves may not be sufficiently large), and therefore are pre-adapted for the development of true flapping flight. Once level fast flight is achieved, pressures for greater control, for slower flight or hovering, for more advanced manoeuvres, and for improved landing would favour development of elevator muscles and of greater upstroke control.

The proto-flappers and the gliding niche

The likely sequence of behavioural and morphological events as gliding develops into flapping has been described in birds by BOCK (1965) and in bats by SMITH (1977). I follow these models to explore the mechanical developments in the later stages between gliding and flapping, but shall first demonstrate how gliding itself is successful, and can therefore form a viable platform for the appearance of flapping. It is reasonable to model the proto-bat ecologically as a quadrupedal arboreal glider, which had gradually evolved a flight membrane spread by the four limbs as jumps around a forest canopy lengthened into glides. It was undoubtedly related to the Insectivora, and recent evidence suggests that Chiroptera and Dermoptera share many similarities in membrane anatomy and are sister groups (NOVACEK 1982). The Dermoptera are the only gliding mammals to have a membrane spanning the fingers as well as the limbs; the order today comprises a single genus, the colugo (*Cynocephalus*, mass = 1.5 kg), but it is an ancient group which included many smaller species. The relationship of bats to Dermoptera suggests at the least a predisposition to gliding. The proto-bat is thought to have been insectivorous, although if the order Chiroptera is diphyletic (SMITH 1977) a proto-megabat evolving flight independently may have been frugivorous (SCHOLEY 1985b). The diet, however, is not important: all animals benefit from the ability to move quickly and safely away from the forest floor and from a locomotor strategy which increases the distance traveled and enlarges the area searched for food.

I cannot be as specific about the status of the proto-bird or its relationship to *Archaeopteryx*. These mechanical arguments remain valid for a small climbing biped with proto-wings formed from the forelimbs: the details of its running and climbing gaits will differ – but not materially – from those of a mammalian quadruped, but once in the air, aerodynamic – as opposed to morphological – differences between birds and bats, and equally between proto-birds and proto-bats, are slight (RAYNER 1985). For this reason I prefer not to discuss the possible arboreality of pro-avis or of *Archaeopteryx* in detail.

Gliding vertebrates occupy a specialized ecological niche whereby – as long as they can climb – they exploit their environment efficiently without the need for powered flight. All small animals in trees or in the canopy of forest, benefit by being able to move away from the obstacles and risk of predation on the forest floor. Trees are often too widely spaced or irregular for running or leaping through the canopy to be reliable, and this may have formed the initial pressure for leaping and jumping between trees. The true gliding niche is more sophisticated, and conveys distinct ecological advantages for both insect and fruit (or foliage) eaters; that it is particularly successful is indicated by its frequency in terrestrial vertebrates: gliding has evolved at least twice in amphibians, five times in reptiles and eight times in mammals, quite apart from in birds or pterosaurs (RAYNER 1981). Most gliders are arboreal, and while it is natural to assume the same to be true of a gliding proto-bird or proto-bat this is not absolutely essential: the woolly flying squirrel *Eupetaurus* of the Himalayas lives above the tree line and glides among rocks.

A gliding frugivore in a rain forest may need to cover large distances to locate dispersed food sources (SCHOLEY 1985a, b). By successively climbing trees to gain height and then gliding the animal remains away from the forest floor, but most importantly makes significant savings in the energy cost of locomotion; *Petaurista petaurista* (mass 1.3 kg) reduces its cost of transport (energy to transport unit weight of the animal through unit distance) by 25 % compared with level unimpeded running (SCHOLEY

1985a), and hence gains a considerable increase in foraging radius. For smaller gliders the savings are much greater, and moreover the maximum size at which this strategy is viable (around 1.5 kg) coincides with the largest contemporary gliding mammals and with the largest megabats (RAYNER 1981, 1985). The strategy is economical because the energy cost of the gliding phase is extremely low and so the energy for horizontal transport is little more than that for gaining height; the savings improve with shallower glide angle, and the main limits on performance and size relate to tree-climbing. The strategy is directly comparable to the undulating flight of many birds (including coucals and magpies, both often compared to *Archaeopteryx*) in which glides are interspersed with wing flaps in order to save energy (RAYNER 1985; see also discussion).

The distribution of fruit supplies is probably the main stimulus for climbing and gliding in frugivores, but this is not the sole reason for the strategy. Any regular movement or migration through a forest favours gliding in order to save energy, while gliding has obvious advantages in more open or diverse habitats. Arboreal insectivores generally have small ranges, but some also glide: they then descend to the base of the same, or another tree, to enlarge the area they can search for prey (RAYNER 1981). This strategy is adopted by the flying lizard *Draco* and some smaller gliding mammals *(Glaucomys, Acrobates)*, which do not catch flying insects, but find them on the surfaces of trees and foliage. R. Å. NORBERG (1981, 1983) has shown how climbing and gliding is energetically advantageous for insectivorous birds (nuthatch, treecreeper, woodpeckers). Climbing and gliding may have been adopted by proto-bats or proto-birds; not only is the strategy itself an efficient way of exploiting a diverse habitat, but also there could then be strong selection favouring enhancement of foraging performance as glide angle is reduced by flapping.

The evolution of flapping flight in bats

Even if early bats were poor fliers, able only to glide and to flap their wings to generate weak thrust, they would still enjoy efficient locomotion provided that they could climb, as of course can contemporary bats. Landings, on a vertical surface, would have been stable and similar to those of their gliding ancestors, with a progressive rise in incidence as speed drops until the wing stalls as the animal lands (SCHOLEY 1985a). As nocturnal aerial competitors were absent the first bats would not need exceptional performance: they did not need great agility or manoeuvrability, and were certainly unlikely to have caught flying insects, which demands all of the adaptive resources of contemporary microbats.

Small wingbeats with continuous lift and swept or flexed upstroke superposed on gliding are ideal for the production of thrust, provided the animal is travelling at normal glide speeds. As I have shown there is no need for an upstroke elevator muscle or for complex control of wing movements. Shortening of the forearm by flexing the elbow and wrist joints is a natural movement for any quadruped (OSTROM 1976), and is precisely the movement used by gliding mammals to control membrane tension. This gait performs best when the handwing is short – as at an early stage of wing growth – since then upstroke flexure can be greatest; this would necessarily have been the case in the first flappers, and is exemplified by *Rhinopoma*, thought to be the most primitive extant microbat, which has a strikingly short handwing, and is one of the few microbats to glide (RAYNER 1985). Gliding represents strong aerodynamic and morphological pre-adaptation for flapping flight.

To quantify the ecological advantages of flapping to a glider I have modelled the gliding to flapping sequence in a relatively large microbat (the noctule, mass = 0.027 kg), by assuming that it evolved from a dermopteran glider (aspect ratio around 1.5) by progressive growth of wingspan and wing area, without change of mass. This is a purely hypothetical example, and implies neither that the first bat was this size, nor that the noctule is primitive, nor that wing growth was regular or gradual. I choose the noctule because its flight mechanics are well understood, and it is probably around the smallest size at which flapping could feasibly develop from gliding. I have given full details of these calculations elsewhere (RAYNER 1985). As wing area increases, air speed reduces, so that the lift coefficient is constant, and the animal always flies at the shallowest possible descent angle. The wings are flapped to produce increasing thrust so that the flight angle gradually reduces below the glide angle until the animal can fly

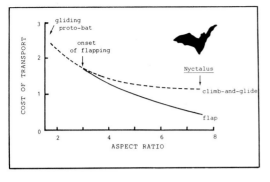

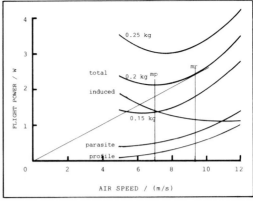

Fig. 3
The fall in cost of transport as aspect ratio increases as a gliding proto-microbat evolves into a typical bat *(Nyctalus noctula)*. The glider has aspect ratio 1.5, is generally similar to contemporary gliding mammals, and uses the climbing and gliding strategy. Even without flapping (dashed line) the reduction in cost is considerable, but flapping improves performance still further, and is probably essential to compensate for loss of agility as the forelimb lengthens. See RAYNER (1985) for full details.

Fig. 4
Estimated flight performance for *Archaeopteryx* and components of flight power, with dimensions estimated by YALDEN (1971; $M = 0.20$ kg), and also higher (0.25 kg) and lower (0.15 kg) body mass with unchanged wing size. The model assumes that the animal was capable of flying with the continuous lift gait with optimum kinematics at all speeds. Maximum range speed is 9.3 m/s; the shallowest glide angle is 6.4° at 8.8 m/s.

horizontally, and I have made the conservative assumption that the animal can only fly level when its aspect ratio reaches that of the noctule. The calculated reduction in cost of transport for climbing and gliding or flapping as aspect ratio rises is shown in figure 3. When aspect ratio is small – below about 3 or 4 – induced drag is large and the wings cannot produce sufficient thrust for level flight. But, at this stage only small amounts of thrust are needed (of the order of 1 % of weight), and thereafter with small, modest amplitude wingbeats, thrust rises rapidly as larger aspect ratio is evolved. Modest increases in power output (and muscle size) bring large reductions in flight angle and dramatic increases in foraging range: increased foraging input produces strong positive feedback favouring further increases in aspect ratio and flapping performance. In addition, the much extended flight path and ultimately the possibility of level flapping allow the animal to exploit more open habitats. The estimated cost for level running (at 3 m/s) for an animal of this size is about 4, and the cost of transport for a flying noctule may be as little as one-tenth that for a comparable non-flying quadruped. The savings in cost from flapping compared to running remain significant for heavier animals up to about 1.5 kg (SCHOLEY 1985b).

I have already shown that wingroot moments in flapping are comparable to those in gliding or climbing. Gross energy consumption is also comparable: the power for a 27 g mammal to climb a vertical tree at 0.7 m/s (SCHOLEY 1985a) is about 0.18 W, virtually equal to the flapping power at minimum power speed for a noctule (Figure 1). Moreover, the same group of thoracic muscles are employed in both climbing and in flapping. Even at an early stage in wing development when aspect ratio is low, physiological and mechanical constraints permit small amplitude flapping provided that there are modest increases in pectoralis size and the pectoralis moment arm remains proportional to wing length.

The evolution of flapping flight in birds

Discussion of bird flight evolution naturally revolves around *Archaeopteryx*, although it is unclear whether it was a glider or flapper, a cursor or a tree-dweller. Nonetheless, the parallel with bat evolution, where gliding permits ready exploitation of food sources and is a pre-adaptation for flapping at high speeds, is strong, and the only major obstacles to its acceptance in birds are the apparent cursorial habit and the bipedality of *Archaeopteryx*. The gliding model offers a possible explanation for the development of the surpracoracoideus in birds: *Archaeopteryx* appears not to possess an elevating supracoracoideus, and to have a pectoralis without a posterior extension along the sternum as in modern birds, but rather lying beneath the humeral joint with origin on the furcula and coracoid (OLSON & FEDUCCIA 1979); that this structure is precisely appropriate for gliding or for simple flapping in fast flight, strongly suggests that *Archaeopteryx* was capable of flapping, which had been acquired through a gliding phase, but could not control slow flight and was restricted to landings on vertical surfaces, probably using its claws to climb. Subsequent development of upstroke control and of wing pitch adjustment could encourage changes in supracoracoideus configuration similar to those suggested by OSTROM (1976) as a horizontal component to the line of action of the pectoralis develops.

I have calculated the aerodynamic performance of *Archaeopteryx* by using the model described above, assuming the flight dimensions to be: mass $M = 0.200$ kg, wingspan $2b = 0.58$ m, wing area 0.0479 m^2 (YALDEN 1971); slightly smaller than estimated by YALDEN (1984). The results are shown in figure 4. This estimate is likely to be more accurate than previous calculations since the wake structure is correctly accounted for, and there is no reason to believe that the aerodynamic performance of *Archaeopteryx* differed markedly from comparable contemporary birds. The aspect ratio of 7.0 is relatively large for modern arboreal birds (it is around 5.6 for a magpie), and from both contemporary analogues and the known pectoral girdle structure it is realistic to assume that at high speeds (greater than 5 m/s) the animal flew with continuous lift. The estimated wingbeat frequency at the maximum range speed (9.3 m/s) is 9.3 Hz. If the flight muscles comprised 15 % of the total weight (a modest figure), and could produce 200 W/kg (see RAYNER 1979), the maximum power output is 6 W, and so is comfortably above the required mechanical power of 2–3 W, even allowing for the fact that *Archaeopteryx* may not have achieved the ideal areodynamic performance of modern birds which I have assumed.

The problem remains that it may be hard to accept whether a bipedal proto-bird could have been arboreal. There is no doubt that such an animal with clawed wings could climb, but its bipedality may have constrained other movements in the trees. Arboreality is perhaps not essential for the gliding to flapping model: bipeds are not confined to level ground, and any cursor has opportunities to jump which it might extend to glides (see discussion). Provided that glides are fast and that there is pressure for the glides to lengthen, selection favours the evolution of flapping. There are no extant bipedal gliders (but this does not indicate that they could not have existed in the past), but such an animal would have had the advantage that its forelimbs may have developed a higher aspect ratio than a quadrupedal glider since the wings have less use in terrestrial movement. I venture the hope that future morphological and palaeontological evidence will help to clarify the behaviour of *Archaeopteryx* and thereby illuminate the ecology of the proto-bird.

Discussion

The gliding model offers strong selective pressures for the evolution of powered flight: gliding conveys clear ecological benefits from a behaviour pattern which has a clear parallel in contemporary vertebrates; moreover the mechanical and anatomical developments envisaged are fully consistent with both the aerodynamics of flapping flight and with archosaur morphology. This analysis has shown how a limited form of flapping flight could have evolved from gliding: beyond this stage birds or bats are no longer dependent on gravity for forward momentum, they could exploit a wide range of habitats and could radiate to a range of ecological strategies as performance improvements are coupled with developing control of the flight surfaces, manoeuverability, the ability to make metastable landings, and the sophisticated morphology to control the upstroke in slow flight or hovering; only then do they adopt the vortex ring gait with unloaded upstroke. It is certainly false to assume that primitive fliers necessarily had the same remarkable agility and control ability as contemporary animals; equally, gliding ancestry of flapping flight does not mean that contemporary gliding and soaring birds are in any way primitive.

Opponents of the gliding model have argued that gliders show no adaptations for flapping and that gliding is an adaptive plateau (JEPSEN 1970, PADIAN 1982, CAPLE et al. 1983); this is of course true, although as I have shown, gliders are p r e - a d a p t e d for flapping, and it is precisely because gliding is so successful that there were animals extant in which flapping could evolve. There is no reason to argue that the success of gliding p r e v e n t s progression to flapping, while to argue that gliders should have disappeared through competition with flappers is as illogical as to argue that dinosaurs became extinct because one group of them evolved into birds. Perhaps a more serious objection to the model is the paucity of intermediate gliding forms with larger aspect ratio (say between 1.5 and 4). Nevertheless it must be remembered that conditions have altered greatly since birds and bats evolved, and an adaptation which may once have been a successful plateau may have later become non-viable through competition with true flappers. In any case similar objections apply to the cursorial hypothesis: there are no extant equivalents to a cursorial bird ancestor, and there are very few non-flying vertebrates sufficiently able to catch flying insects for this to form a major factor in their ecology or selection.

The relevance of the gliding model to *Archaeopteryx* must remain unproven. Even as a running biped it may have been secondarily arboreal, and might have climbed well with its forelimb claws, although its agility in the trees could hardly have approached that of small mammals. The apparent absence of an elevator muscle and the ventral action of the pectoralis suggest that it was certainly capable of gliding, and there is no reason to think that the pectoralis was not sufficiently large for it also to have allowed simple flapping; the muscle architecture suggests though that it was not capable of the more horizontal wingbeat essential for slow flight and for take-off from the ground. These factors and the ecological arguments applied above to proto-bats combine to make gliding an attractive explanation for the evolution of avian flight and to argue against the cursorial model. However, evolutionary hypotheses of this kind are untestable: it can never be demonstrated that any model can be even approximately correct, and the evidence presented in this paper is largely circumstantial. Nevertheless, although details of this model or of the assumptions adopted are questionable, it is hard to see a more successful explanation of the mechanics and evolution of vertebrate flight evolution from any alternative to the gliding hypothesis.

Acknowledgements

I wish to acknowledge the financial support of the Royal Society of London, and to thank many friends and colleagues for discussions and criticism during this work, and in particular COLIN PENNYCUICK, ULLA and AKE NORBERG, KEVIN PADIAN and KEITH SCHOLEY.

References Cited

BOCK, W. J. (1965): The role of adaptive mechanisms in the origin of higher levels of organization. – Syst. Zool. **14**: 272–287; Washington.
CAPLE, G., BALDA, R. P. & WILLIS, W. R. (1983): The physics of leaping animals and the evolution of preflight. – Am. Nat. **121**: 455–467; Chicago.
CLARK, B. D. (1977): Energetics of hovering flight and the origin of bats. – In: Major patterns in vertebrate evolution, ed. M. K. HECHT, P. C. GOODY & B. M. HECHT, pp. 423–425. New York (Plenum Press).
DARWIN, C. (1859): The origin of species. – 502 pp. London (John Murray).
HERMANSON, J. W. & ALTENBACH, J. S. (1983): The functional anatomy of the shoulder of the pallid bat, *Antrozous pallidus*. – J. Mammal. **64**: 62–75, Lawrence, Ks.
JEPSEN, G. L. (1970): Bat origins and evolution. – In: Biology of bats, ed. W. A. WIMSATT, vol. 1, pp. 1–64. New York (Academic Press).
MARSH, O. C. (1880): Odontornithes. – 201 pp. Washington (Government Printing Office).
NOPCSA, F. (1907): Ideas on the origin of flight. – Proc. zool. Soc. pp. 222–236; London.
NOPCSA, F. (1923): On the origin of flight in birds. – Proc. zool. Soc. pp. 463–477; London.
NORBERG, R. Å. (1981): Why foraging birds should climb and hop upwards rather than downwards. – Ibis **123**: 281–288; London.
NORBERG, R. Å. (1983): Optimum locomotion modes for birds foraging in trees. – Ibis **125**: 172–180; London.
NORBERG, U. M. (1985): Evolution of vertebrate flight: an aerodynamic model for the transition from gliding to active flight. – Am. Nat. (in press); Chicago.

Novacek, M. J. (1982): Information for molecular studies from anatomical and fossil evidence on higher eutherian phylogeny. – In: Macromolecular sequences in systematic and evolutionary biology, ed. M. Goodman, pp. 3–41. New York (Plenum Press).

Olson, S. L. & Feduccia, A. (1979): Flight capacity and the pectoral girdle of *Archaeopteryx*. – Nature **278**: 247–248; London.

Ostrom, J. H. (1974): *Archaeopteryx* and the origin of flight. – Q. Rev. Biol. **49**: 27–47; New York.

Ostrom, J. H. (1976): Some hypothetical anatomical stages in the evolution of avian flight. – Smithson. Contr. Paleobiol. **27**: 1–21.

Ostrom, J. H. (1979): Bird flight: how did it begin? – Am. Scient. **67**: 46–56; New Haven.

Padian, K. (1982): Macroevolution and the origin of major adaptations: vertebrate flight as a paradigm for the analysis of patterns. – Proc. 3rd North American Paleontological Convention, vol. 2, pp. 381–392.

Padian, K. (1983): A functional analysis of flying and walking in pterosaurs. – Paleobiology **9**: 218–239.

Pirlot, P. (1977): Wing design and the origin of bats. – In: Major patterns in vertebrate evolution, ed. M. K. Hecht, P. C. Goody & B. M. Hecht, pp. 375–410. New York (Plenum Press).

Rayner, J. M. V. (1979): A new approach to animal flight mechanics. – J. exp. Biol. **80**: 17–54; Cambridge.

Rayner, J. M. V. (1980): Vorticity and animal flight. – In: Aspects of animal movement, ed. H. Y. Elder & E. R. Trueman, pp. 177–199. Semin. Ser. Soc. exp. Biol. **5**. London (Cambridge University Press).

Rayner, J. M. V. (1981): Flight adaptations in vertebrates. – Symp. Zool. Soc. Lond. **48**: 137–172; London.

Rayner, J. M. V. (1985): Vertebrate flapping flight mechanics and aerodynamics, and the evolution of flight in bats. – In: Biona Report **4**, Fledermausflug, ed. W. Nachtigall. Stuttgart (Gustav Fischer).

Scholey, K. D. (1985a): The climbing and gliding locomotion of the giant red flying squirrel *Petaurista petaurista*. – In: Biona Report **4**, Fledermausflug, ed. W. Nachtigall. Stuttgart (Gustav Fischer).

Scholey, K. D. (1985b): The evolution of flight in bats. – In: Biona Report **4**, Fledermausflug, ed. W. Nachtigall. Stuttgart (Gustav Fischer).

Smith, J. D. (1977): Comments on flight and the evolution of bats. – In: Major problems in vertebrate evolution, ed. M. K. Hecht, P. C. Goody & B. M. Hecht, pp. 427–437. New York (Plenum Press).

Spedding, G. R., Rayner, J. M. V. & Pennycuick, C. J. (1984): Momentum and energy in the wake of a pigeon *(Columba livia)* in slow flight. – J. exp. Biol. **111**: 81–102; Cambrdige.

Yalden, D. W. (1971): The flying ability of *Archaeopteryx*. – Ibis **113**: 349–356; London.

Yalden, D. W. (1984): What size was *Archaeopteryx*? – Zool. J. Linn. Soc. **82**: 177–188; London.

Author's address: Dr. Jeremy M. V. Rayner, Department of Zoology, University of Bristol, Woodland Road, Bristol BS8 1UG, United Kingdom.

Jeremy M. V. Rayner

Cursorial Gliding in Proto-Birds
An Expanded Version of a Discussion Contribution*

This conference has provided a unique opportunity to discuss the mechanisms by which flapping flight has evolved in vertebrates. This is of fundamental importance for the interpretation of the ecological performance – and to some extent also the phylogenetic status – of *Archaeopteryx* and the early birds.

Palaeontological opinion and evidence appears roughly equally divided between a cursorial and an arboreal origin of flight, although arguments against an arboreal habit in proto-birds have been weakened by YALDEN'S demonstration that the forelimb claws of *Archaeopteryx* are adapted for climbing. It is valuable therefore to explore whether a biomechanical argument can help to resolve the problem. I find it helpful to rephrase the distinction between the cursorial and arboreal hypotheses in terms of "upwards" and "downwards" pathways to flight, distinguished by the need to overcome gravity or the opportunity to take advantage of it. In separate papers in this conference ULLA NORBERG and I have demonstrated that a downwards origin of flight, in which the wings of a gliding proto-flier gradually enlarge and the animal begins to flap, is fully self-consistent on mechanical, morphological, and ecological grounds, and that the intuitive attraction of the gliding model can be justified by relatively straight forward physical arguments. There remain non-mechanical limitations with the "downwards" model: the behavioural repertoire of a bipedal, arboreal proto-bird is unclear (what posture did it adopt on the trunk or among the branches, particularly preceding the jumps and glides?), and *Archaeopteryx* itself shows few definite arboreal adaptations (but in view of the relatively small size and the uncertainty of its relation to the proto-bird this may be irrelevant). It is vital therefore to compare the two hypotheses with a similar mechanical analysis to that with which the arboreal model is justified. This I attempt to do here.

It is far from easy to suggest the likely sequence of events in the development of a flapping wing from an arm with little or no locomotory function in a running biped. CAPLE et al. (1983) have given a convincing explanation for the development of a high-aspect ratio gliding or stabilizing wing in a running biped, but their argument for the s i m u l t a n e o u s appearance of flapping of this wing is unconvincing, and no realistic explanation for subsequent development of flapping has been suggested. A possible solution is that the proto-bird at this stage entered the trees (or perhaps jumped among rocks or cliffs) and then used gravity to glide before starting flapping. Mechanically this is a fully reasonable explanation; it is arguably unreasonable to demand such a significant change of behaviour in the bird's ancestor, although the access to new niches it would permit might be sufficient encouragement for the adaptation to survive. However, the absence of a mechanically-consistent evolutionary scenario does not mean that flight did not evolve cursorially, although it remains intuitively hard to see how this could have occurred. To model the upwards evolution of flapping I consider a parsimonious situation in which a gliding wing has evolved for the purposes of stability at high running speeds; if flapping can further improve running performance usefully then the cursorial model may be viable; moreover the intermediate stage of gliding w h i l e s t i l l running will be more advantageous than simple running. In this way a direct comparison between cursorial and arboreal evolution is possible since the same mechanical and physiological assumptions can be applied to each. If running and gliding is feasible then flapping may have evolved in a cursor without the need for some form of gliding. It is necessary therefore to show whether a glide is possible within the context of level running.

I have made various assumptions about the energy cost of running and other mechanical limitations on performance; as far as possible these are identical to the assumptions I made in considering arboreal gliding. The power P_r in watts for mammals and birds (bipeds and quadrupeds) running at speed V_r (m/s) has been measured as
$$P_r(V_r) = 6.0\, M^{0.7} + 10.7\, M^{0.68}\, V_r,$$
where M is the mass in kilogrammes (TAYLOR & HEGLUND 1982). I have chosen c o s t o f t r a n s p o r t as a yardstick to compare performance: I do not favour an evolutionary response to predation (either by or on the animal) as a source

---

* Editors' Note: This supplementary paper was developed by Dr. RAYNER at the request of the Editors as a consequence of the lengthy discussions that followed his original paper.

of the evolution of major locomotor adaptations (RAYNER 1981), but cost of transport (energy to transport the whole animal through unit distance) is often important to selection, and cost of transport is a useful currency of selection of locomotor adaptations; in certain conditions it is inversely proportional to foraging range, and it measures indirectly the efficiency of predation. To minimize cost ($C = P/MgV$) a running animal must run at the highest available a e r o b i c power output P_m; anaerobic metabolism is rarely relevant to normal locomotor activity since it can be sustained only for short durations.

Suppose that the proto-bird has already developed a feathered wing of moderate aspect ratio (perhaps as high as 7, by comparison with *Archaeopteryx*). If it runs with wings outstretched, and raises its centre of gravity by jumping or leaping or by using aerodynamic lift, can it reduce energy consumption for given distance? The outstretched wings increase profile drag, but based on values for flying birds I estimate the effect of this to be small, resulting in a reduction of at most 5 % in running speed, and so I have neglected it. If any useful aerodynamic lift is to be generated the running speed must be high since otherwise the wing will stall or the amount of lift will be limited by induced drag. By neglecting profile and induced drag the adverse effects of the outstretched wings are neglected, and the comparison will favour the cursor. The energy used to raise the centre of gravity is more substantial, and detracts significantly from running speed. If the animal gains height at angle Φ during the running phase (vertical speed = $V_r \tan \Phi$), the power may be written

$$P_{rc}(V_{rc}) = P_r(V_{rc}) + 4MgV_{rc} \tan \Phi,$$

where the factor 4 in the gravity term represents the muscular efficiency. Note that the pelvic musculature would be responsible both for work done while running and for any height gain. At maximum (aerobic or anaerobic) power P_m the running speed must be reduced; this is one of the reasons why predation – demanding the ability to jump or leap at insects – is a poor explanation for cursorial wing and flight evolution. For a 0.2 kg animal, at the small angle Φ of 10° the running speed falls by 28 %, while at 20° it falls by 43 %. Even if level running speed V_r is sufficient (and I demonstrate below that it is not), this brings the running and jumping speed V_{rc} well below the range at which flapping flight is efficient and must make the animal both more vulnerable to predation and a less successful predator. Nonetheless, the animal may use potential energy gained at the expense of speed against drag in a glide, and since the animal only does work for part of the time the cost of transport may reduce.

If the wing loading of proto-bird equals that of *Archaeopteryx*, and if no energy is used in gliding, then the ratio of the costs for level running and for a strategy of running and jumping and then gliding at angle γ may be compared from the above formulae as

$$C_{rg}/C_r = V_r / [(1 + \tan \gamma/\tan \varphi) V_{rg}];$$

therefore running with intermittent gliding is energetically attractive – and will be favoured by selection – if $\tan \gamma$ is less than $10.7M^{0.68} / (4gM)$, that is if $\tan \gamma < 0.273M^{-0.32}$, which is independent of maximum power and of running speed. For an animal of mass 0.2 kg the glide angle need only be shallower than about 24°, which is well within the performance of even a poor glider, and the saving becomes more attractive for smaller animals. Interspersing of glides into running is e n e r g e t i c a l l y attractive, although the savings obtained are considerably smaller than those for the comparable climbing and gliding strategy of an arboreal proto-flier. However, this simple energetic theory does not consider whether the animal is mechanically capable of attaining such glides. In particular, can it run fast enough, and can it jump sufficiently high to be able to flap its wings?

The flight speeds of birds are far higher than the speeds attained by most running animals (RAYNER 1985). These high flight speeds are imposed primarily by the need to support the weight with wings of limited size, but speed can be advantageous since it permits a dramatic reduction in cost of transport compared to running. (In fact, the slowest flying birds are aerial insectivores such as swifts and hirundines – fast flight is a hindrance to them since their insect prey fly much slower.) The typical flight speed for modern birds of mass 0.2 kg is at least 7 m/s (in flapping or gliding), and it is probable that a poorly-adapted proto-flier would if anything fly faster. As I have discussed elsewhere in this symposium, flapping flight is most unlikely to have evolved from gliding at a speed lower than this since in slow flight the energy demands are large and the wingbeat gait is functionally complex and is inconsistent with the anatomy of *Archaeopteryx*. Thus, a running animal would have needed to attain a speed of around 6 m/s before it could glide in a way which could evolve into flapping. At lower speeds the glide angles are high and the above energetic argument rules out intermittent gliding.

The closest contemporary parallels to a running bipedal proto-bird are small lizards such as *Basiliscus* and *Callisaurus*. The normal locomotion speeds of these species are not recorded, but anaerobic burst speeds reach 2.6 m/s in *Basiliscus* (HIRTH 1963) and 8.1 m/s in *Callisaurus* (BELKIN 1961); it seems therefore that a small biped might attain sufficiently high speeds. However, burst speeds are only attained for very short times (1–2 s): sustained speeds in lizards are very much slower, often less than 0.5 m/s (BENNETT 1982). Since gliding r e d u c e s speed it is hardly likely to be of use in escape, although it may permit enhanced escape distance. Sustained speeds in homeotherms are larger than those of lizards, but still do not reach the values required: maximum aerobic metabolic power has been estimated as $18.1M^{0.68}$ W in small mammals (LECHNER 1978), and this corresponds to a scale-independent maximum running speed of 1.1 m/s. ALEXANDER (1976) has estimated the speeds of running bipedal dinosaurs to be between 1.0 and 3.6 m/s (and hence perhaps anaerobic), and the likely maximum running speed of *Archaeopteryx* is around 2.5 m/s (RICHARD THULBORN, discussion this symposium). The fastest running speed of a flying bird is in the roadrunner (*Geococcyx californianus*; mass = 0.285 kg); although there are records as high as 6.7 m/s when chased (COTTAM et al. 1942), FEDAK & SEEHERMANN (1979) could not obtain sustained speeds above about 1 m/s. The derivation of a definite range of running speeds for proto-bird is clearly not easy, but all this evidence indicates that the sustained speed is unlikely to have been more than about 2 m/s unless the animal was blessed with exceptional performance. The proto-bird would need to run three times faster than this in order to fly.

Not only must this hybrid running and gliding gait be fast and energetically attractive, it must also be geometrically consistent. In particular, is the time taken to gain height comparable with the stride rate? Suppose that the animal runs and glides at 7 m/s and has a shallow glide angle of 6° (compare *Archaeopteryx*), and that the glides are sufficiently long for three wingbeats to be superposed (three or four beats are typical in undulating flapping birds) at the rate of 9 Hz. The glide must therefore be at least 2.3 m in length, and the centre of gravity must be raised through 0.25 m. For an animal at most 0.50 m in height this is substantial, and likely to be achieved by jumping or hopping rather than aerodynamically. Comparable experimental data are unavailable, but I believe it is somewhat larger than might be expected of any small biped. The pelvis height of *Archaeopteryx* is about 0.12 m (Berlin specimen) and the maximum stride length will be of the order of 0.50 m (ALEXANDER 1976): the height gain must therefore be at an angle of around 45°, and the angle can be usefully reduced only at exceptional running speed: energetic limitations therefore rule out the possibility of jumping sufficiently high. Jumping performance is weakly scale-dependent, with smaller animals being more highly favoured, but for size to alter the main conclusion the proto-bird must have been as small as 30 g; this appears inconsistent with the fossil record.

I conclude that – while remaining adaptively a cursor – a running biped can in principle benefit from short glides, which could usefully be lengthened by wing flaps. The prerequisites for this to be satisfactory argue against this adaptation in an animal the size of *Archaeopteryx*: extremely high running speed, high metabolic power and the ability to gain height in a single stride must all be beyond the levels reached by contemporary animals, so the parsimonious running and gliding scenario becomes exactly the opposite. These demands are far greater than the climbing ability required in the downwards scenario, where potential energy assists the generation of sufficient glide speed for flapping to appear. The probability is therefore strong that avian flapping flight evolved through an arboreal gliding phase.

The most conspicuous omission of this model is some consideration of whether flapping could have accompanied the evolution of the wing – as in the arboreal model – rather than following it as I have proposed here. A mechanism related to stability as loosely proposed by CAPLE et al. (1983) is feasible, and must be explored further. Provided that the e c o l o g i c a l advantage gained increases progressively I see no objection to flapping appearing more rapidly than I have assumed, but the running speeds are unlikely to have attained the necessary values at which flapping – or gliding – are advantageous unless the flight is gravity assisted, while flapping to take off at slow running speeds is a highly complex manoeuvre and appears inconsistent with the simple flight musculature of *Archaeopteryx*. A major limitation of the cursorial model is that no clear ecological advantage has been assigned to the running-leaping-gliding strategy: it is unlikely to be encouraged by escape from predators since direct running is faster, while no contemporary cursor travels at such high speeds while commuting to foraging sites. There remains finally the problem which has beset all cursorial models of flight evolution. For a running animal, selection is on the hind-limb, and favours large pelvic musculature; for a flapping flier the pectoral muscles are enlarged, but the hybrid flapping runner would have required both sets of muscles, with the weight penalty and all of the physiological problems this would entail. The arguments given here confirm that selection would act a g a i n s t such excess design. There are no mechanical or morphological arguments against the arboreal or gliding model for the origin of flapping flight: by comparison the cursorial model remains inadequate.

References Cited

ALEXANDER, R. McN. (1976): Estimates of speeds of dinosaurs. – Nature **261**: 129–130: London.

BELKIN, D. A. (1961: The running speeds of the lizards *Dipsosaurus dorsalis* and *Callisaurus draconoides*. – Copeia pp. 223–224; Gainsville, Fla.

BENNETT, A. F. (1982): The energetics of reptilian activity. – In: Biology of the Reptilia, volume 13, Physiology D, Physiological Ecology, ed. C. GANS & H. POUGH, pp. 155–199. New York (Academic Press).

CAPLE, G. BALDA, R. P. & WILLIS, W. R. (1983): The physics of leaping animals and the evolution of preflight. – Am. Nat. **121**: 455–467; Chicago.

Cottam, C. Williams, C. S. & Sooter, C. A. (1942): Flight and running speeds of birds. – Wilson Bull, **54**: 121–131.

Fedak, M. A. (1979): Reappraisal of energetics of locomotion shows identical cost in bipeds and quadrupeds including ostrich and horse. – Nature **282**: 713–716; London.

Hirth, H. F. (1963): The ecology of two lizards on a tropical beach. – Ecol. Monogr. **33**: 82–111; Ithaca, N.Y.

Lechner, A. J. (1979): The scaling of maximal oxygen consumption and pulmonary dimensions in small mammals. – Resp. Physiol. **34**: 29–44; Amsterdam.

Rayner, J. M. V. (1981): Flight adaptations in vertebrates. – Symp. zool. Soc. Lond. **48**: 137–172; London.

Rayner, J. M. V. (1985): Speeds of flight. – In: A new dictionary of birds, ed. B. Campbell & E. Lack, pp. 224–226. Calton. T. & A. D. Poyser).

Taylor, C. R. & Heglund, N. C. (1982): Energetics and mechanics of terrestrial locomotion. – A. Rev. Physiol. **44**: 97–107; Palo Alto.

Author's address: Dr. Jeremy M. V. Rayner, Department of Zoology, University of Bristol, Woodland Road, Bristol BS8 1UG, United Kingdom.

Ulla M. Norberg

Evolution of Flight in Birds: Aerodynamic, Mechanical, and Ecological Aspects

Abstract

A model, based on quasi-stationary areodynamics, shows that even at very low flapping speeds a net horizontal thrust force will be produced, which does not exist in equilibrium gliding, while the vertical lift force remains unchanged, balancing the weight. A constraint is that the flapping amplitude remains within certain limits, dictated (from the model) by the particular flapping speed chosen. A corollary is that the flapping frequency must exceed some threshold value (which may be rather low, however). The net thrust force can be used to flatten out the flight path and hence to lengthen the glide. This strongly supports the arboreal theory of the evolution of flight in birds, bats, and pterosaurs, which involves a gliding stage before powered flight originated. The early proto-birds might have foraged in trees and used gliding as a cheap and rapid mode of locomotion between trees during foraging. Maximization of net energy gain during foraging might have been a reason for strong selection for increased gliding performance. Increased wing area reduced wing loading and hence gliding speed, allowing safer landings. Evolution of longer wings (higher aspect ratio), more powerful flapping with coordinated movements, and asymmetric wingstrokes eventually led to horizontal flight. Evolution of more sophisticated wing characters improved the aerodynamic performance.

Zusammenfassung

Ein auf quasistationärer Aerodynamik basierendes Modell zeigt, daß sogar bei sehr niedrigen Schlaggeschwindigkeiten eine horizontale Netto-Schubkraft erzeugt wird, die im ausgeglichenen Gleiten nicht existiert. Dagegen bleibt die vertikale Auftriebskraft, die dem Körpergewicht entspricht, unverändert. Ein (vom Modell) durch die gewählte besondere Schlaggeschwindigkeit diktierter Zwang ist es, daß die Schlagamplitude innerhalb gewisser Grenzwerte bleibt. Eine Folge ist, daß die Schlagfrequenz einen gewissen Grenzwert überschreiten muß, der allerdings ziemlich niedrig sein kann. Die Netto-Schubkraft kann dazu benutzt werden, den Flugweg zu verflachen und somit das Gleiten zu verlängern.

Dies spricht stark für die "Arboreal"-Theorie der Flugentstehung bei Vögeln, Fledermäusen und Flugsauriern, die vor der Entstehung des aktiven Antriebsfluges ein Gleitstadium annimmt. Die frühen Vorvögel dürften ihre Nahrung in Bäumen gesucht haben. Sie benutzten dabei das Gleiten als eine "billige" und schnelle Fortbewegungsweise zwischen den Bäumen. Die Maximierung des Netto-Energiegewinnes während der Nahrungssuche könnte ein Grund für eine starke Selektion verbesserter Gleitleistungen gewesen sein. Eine vergrößerte Flügelfläche reduzierte die Flächenbelastung und somit die Gleitgeschwindigkeit, was ein sichereres Landen erlaubte. Die Entwicklung längerer Flügel (höheres Streckungsverhältnis), kraftvoller Flügelschläge mit koordinierten Bewegungen und asymmetrischer Flügelschläge führten schließlich zum horizontalen Flug. Die Evolution von verfeinerten Flügelkriterien verbesserten die aerodynamischen Fähigkeiten.

Introduction

There are two major, contradicting, hypotheses on the origin of flapping flight in birds. The arboreal theory suggests that powered flight evolved via gliding in tree-living proto-birds (e.g. DARWIN 1859; MARSH 1880; HEILMANN 1926; BOCK 1965, 1969, 1983; PARKES 1966; PENNYCUICK 1972; FEDUCCIA 1980). The cursorial theory, on the other hand, states that birds evolved from ground-running proto-birds which ended up as active fliers without a gliding intermediate (e.g. NOPSCA 1907, 1923, OSTROM 1974, PADIAN 1982, CAPLE et al. 1983). In recent years the arboreal theory has been criticized on some fundamental points.

CAPLE et al. (1983) claimed that powered flight cannot develop from gliding, since flapping would reduce lift in a gliding animal. They further stated that animals with no control of their bodies could not successfully evolve true flight, a fact that should impede gliding animals to evolve flapping flight, but support the cursorial theory of flight. They suggested that the forelimbs in ground-running proto-birds were extended and moved for stability and control of the body. Better stability and lift increment increased the ability of insect capture. Wing-stroke-like limb motions for control and stability became more and more rapid and eventually produced enough lift and thrust to permit powered flight.

These objections and other arguments against the arboreal theory are treated by U. M. NORBERG (1985a, b), who also presented an aerodynamic model showing that a transition from gliding to active flight is mechanically and aerodynamically quite feasible. I will here discuss the major objections against the arboreal theory and will then consider possible pathways of the evolution of flight in birds, from tree-living proto-birds to birds with powered flight. I will also give a brief summary of my aerodynamic model (U. M. NORBERG 1985a).

Tree-climbing and gliding

Proto-birds might have been tree-living insectivores or omnivores. BOCK (1983) suggested that climbing and clinging among branches and leaves would require better control of movements with concomitant improvements in sense organs, neuro-muscular control and external morphology. He also suggested that feathers developed as insulating devices, that development of elongated feathers would have reduced the rate of fall and promoted safe landings, and that elongated feathers on the forelimbs later could have acted as gliding surfaces and eventually been developed to flapping wings (BOCK 1965). Another theory is that feathers evolved directly for flight, as first proposed by HEILMANN (1926) and further discussed and promoted by FEDUCCIA (1980).

R. Å. NORBERG (1981, 1983) showed that gliding from one tree to another and climbing upwards during foraging is a way of minimizing energy costs of locomotion among modern birds. SCHOLEY (1983, 1985) showed that the cost of transport for climbing a tree and then gliding in a gliding mammal *(Petaurista)* is less than that for unimpeded running. Furthermore, gliding between trees takes less time than climbing down and then running to the next tree. So, maximization of net energy gain during foraging in trees might have been a reason for strong selection for increased gliding performance in the early birds.

Once a glide surface had evolved, the proto-bird increased its foraging efficiency and reduced time and energy required for locomotion during foraging. The first glides were certainly steep, since the aspect ratio and wing area were small and hence lift to drag ratio low in the initial stages. But even a steep parachuting jump would be important for energy and time reduction, and might also have facilitated escape from predators. Elongation of the forelimb and tail feathers decreased surface loading and hence gliding speed, with safer landings as a result.

The selection pressure for good control of movements must have been high already in the tree-climbing proto-bird, and higher still for the early glider. Better control would have allowed the glider to change course to reach a particular destination. Good stability and control of movements were probably achieved before powered flight evolved, and may have evolved progressively along with the ability to glide. There might have been no difficulties in a gliding proto-bird to achieve stability and control of movements by means of simple wing movement coordination (twisting, retraction, control of dihedral angle etc.). For instance, a long tail (as in *Archaeopteryx*) could be used for pitch and yaw stability and control, and dihedral of the wings and wing twisting for roll and yaw stability and control, as in modern birds. Stable landing on tree-branches probably did not occur until powered flight was established. Landing on a tree-trunk after a glide in the early birds was certainly preceded by a flattening out of the glide followed by a terminal climb, which highly reduce speed, as in modern gliding animals. The statement by CAPLE et al. (1983) that the demands for stability and control of the body for stable landings would argue against the theory of gliding as a pre-stage to flight therefore is dubious.

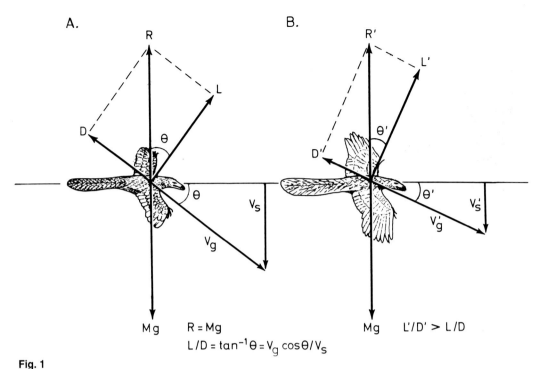

Fig. 1
Aerodynamics of gliding.
A. the glide angle θ is determined by the lift to drag ratio, L/D. R is the resultant of L and D and must balance the body weight Mg. V_g and V_s are the gliding and sinking speeds.
B. An increase of aspect ratio (longer wings) results in a larger L/D ratio and hence in a smaller gliding angle and lower sinking speed. Starting from a given height the animal will cover more ground in a glide the better its L/D ratio (glide ratio) is. A decrease of wing loading, as a result of increased wing span and area, reduces the minimum gliding speed.

Gliding and powered flight

I have shown (U. M. NORBERG 1985a, b) in a model, based on quasi-stationary aerodynamics, that thrust is produced even during very slight flapping in a gliding animal while lift remains unaffected, provided that the flapping frequency exceeds some threshold value. The transition between gliding and flapping does not involve any problems with transition of the shed vortex pattern (U. M. NORBERG 1985a, RAYNER 1985). This highly supports the arboreal theory on the evolution of flight in vertebrates, which involves a gliding stage before powered flight originated.

From the beginning slight flapping was probably used as movements for stability and for manoeuvres in turning and landing. When flapping became more powerful it could also be used to increase the glide path length. This would make locomotion in forest more efficient both in terms of speed and energy cost. SAVILE (1962) and BOCK (1969) stated that small increments of lift will increase the length of the glide path, but CAPLE et al. (1983) claimed that it adds only minute distance to it. The glide path length does not increase as long as the lift to drag ratio (L/D) remains unchanged. But if the L/D ratio increases, so does the length of the glide. In stable gliding (Fig. 1) the ratio between the gliding speed V_g and the sinking speed V_s is

$$V_g/V_s = L/(D \cos \theta), \qquad (1)$$

where θ is the glide angle. The glide path length increases only if θ is made smaller, i.e. only if the L/D ratio increases. This can be obtained by an increase of the wing length (increase of aspect ratio), which results in larger increment of lift than of drag. This is discussed further in U. M. NORBERG (1985a). A

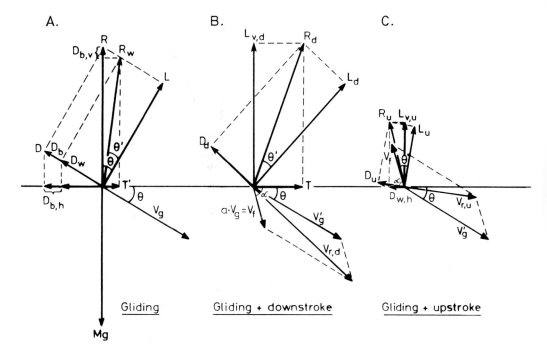

Fig. 2
Force diagrams for equilibrium gliding and partially powered gliding flight.
A. Forces acting on a hypothetical proto-bird during gliding at angle θ to the horizontal. L and D are the lift and drag forces and R their resultant. R equals body weight Mg. The total drag, D, is the sum of wing drag, D_w, and body drag, D_b. R_w is the resultant of the forces acting on the wings only. $D_{b,h}$ and $D_{b,v}$ are the horizontal and vertical components, respectively, of body drag. T' is the horizontal component of R_w and equals the horizontal component of body drag, T' = $D_{b,h}$. V_g is gliding speed.
B. Forces acting on the wings during the downstroke in a proto-bird flapping at angle α to the horizontal in partially powered gliding flight. $L_{v,d}$ is the lift component, perpendicular to the long wing axis, of the instantaneous resultant force, R_d, and T the horizontal thrust component of R_d. V'_g is the gliding speed vector, V_f the instantaneous flapping speed, and $V_{r,d}$ their resultant.
C. Forces acting on the wings during the upstroke in a proto-bird flapping at angle α to the horizontal in partially powered gliding flight. $L_{v,u}$ is the lift component, perpendicular to the long wing axis, of the instantaneous resultant force R_u, and $D_{w,h}$ the horizontal drag component of R_u. $V_{r,u}$ is the resultant of V'_g and V_f.
(From U. M. NORBERG 1985 b).

successive increase of the L/D ratio would be accompanied by a proportional increase of the glide path length, and this increase should not be regarded as unimportant.

CAPLE et al. (1983) stated that powered flight cannot be developed from gliding, since flapping would dramatically reduce lift. But my model shows that thrust can be produced without any associated loss of the vertical lift needed to balance the weight. When flapping amplitudes and velocities are large, as they are in slow forward flight and hovering, unsteady effects become important. But at higher forward speeds, such as in gliding, the wing stroke speed is low relative to the forward speed, and the air flow more stationary. The quasi-stationary aerodynamic theory used here is then appropriate. A brief summary of the model is given below.

The model is so constructed that, regardless of the amount of flapping, the various flight variables always combine in such a way that the vertical lift force produced during one complete wing stroke in partially powered flight by a hypothetical animal does always equal the weight of the animal (as in gliding with no flapping). With this constraint the model then explores whether there is any net horizontal thrust force, over and above that needed to balance body drag. Any such thrust can be used to flatten out the glide. When $L_{v,d}$ is the vertical lift force produced during the downstroke and $L_{v,u}$ that for the upstroke, T the horizontal thrust produced during the downstroke, D_h the horizontal drag produced during the upstroke, and **t** the time for an entire wing stroke, then

$$\int_{t=0}^{t=t/2} L_{v,d}\, dt + \int_{t=t/2}^{t=t} L_{v,u}\, dt = \int_{t=0}^{t=t} Mg\, dt \qquad (2)$$

and

$$\int_{t=0}^{t=t/2} T\, dt > \int_{t=t/2}^{t=t} D_h\, dt. \qquad (3)$$

The initial stage is a hypothetical proto-bird gliding at an angle θ to the horizontal (Fig. 2a). The resultant force R of the lift and drag forces is vertical and must balance the weight Mg in stable gliding, R = Mg. Drag D (= Mg sinθ) is the sum of wing drag D_w and body drag D_b. R is thus the sum of the vertical component of the wing's resultant force $L_{v,w}$, and of the vertical component of the body drag $D_{b,v}$. Similarly the horizontal drag D_h is the sum of the horizontal components of wing drag $D_{w,h}$ and body drag $D_{b,h}$. Equations (2) and (3) can then be written as

$$\int_{t=0}^{t=t/2} L_{v,d}\, dt + \int_{t=t/2}^{t=t} L_{v,u}\, dt + \int_{t=0}^{t=t} D_{b,v}\, dt = \int_{t=0}^{t=t} R\, dt \qquad (4)$$

and

$$\int_{t=0}^{t=t/2} T\, dt > \int_{t=t/2}^{t=t} D_{w,h}\, dt + \int_{t=0}^{t=t} D_{b,h}\, dt. \qquad (5)$$

(Fig. 2b, c).

The wings' lift to drag ratio, L/D, is assumed to equal that in gliding and to remain constant throughout the wingbeat cycle. The downstroke and upstroke are further assumed to be of equal duration, and the wings to supinate during the upstroke so that the air meets the ventral sides of the wings during the entire wing stroke, making the wing-tip trailing vortex always rotate in the same direction.

The lift and drag force components of the wing as well as the local gliding and flapping speed vectors change with the instantaneous wingbeat angle (due to dihedral and anhedral of the wings). This must be accounted for (see U. M. NORBERG 1985a).

The vertical and horizontal forces obtained during flapping vary with time and the wing's positional angle during the wingstroke. The lift force index, $L_v/(1/2)\varrho C_L S V_g^2 t$, and net horizontal force index, $(T-D_{w,h}-D_{b,h})/(1/2)\varrho C_L S V_g^2 t$, where ϱ is air density, C_L lift coefficient, and S wing area, were calculated for different positions of the wings in the stroke plane when glide angle was taken to be 30°, wing-stroke-plane angle 90° to the horizontal, the root-mean-square flapping speed V_f half the gliding speed, $V_f = 0.5V_g$, and the ratio (wing drag)/(body drag) 5 (Fig. 3). The vertical lift produced during flapping must equal that produced during gliding – shaded area above the force index line for gliding must equal that below it in Fig. 3. Furthermore, the net horizontal thrust produced during the downstroke must exceed the horizontal drag produced during the upstroke – hatched area above the zero line must exceed that below it in Fig. 3 – for the animal to be able to flatten out the glide. The outcome of the model is that a horizontal net thrust force will be generated during partially powered flight, given certain flight variables.

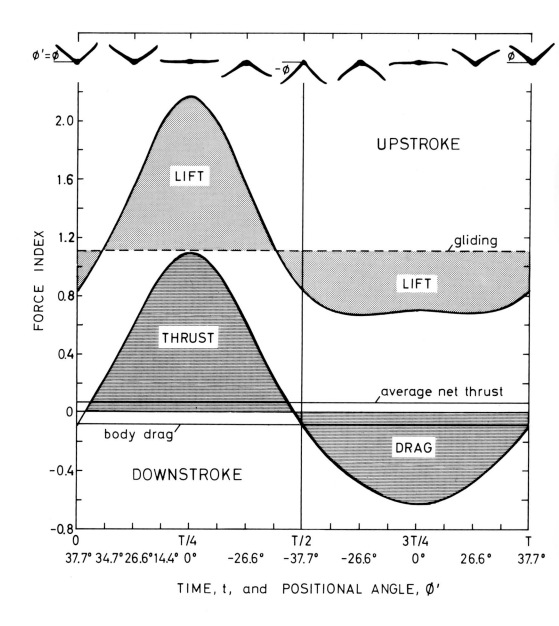

Fig. 3
Diagram showing how the lift force index $L_v/(1/2)\varrho C_L SV_g^2 t$ and net horizontal force index $(T-D_{w,h}-D_{b,h})/(1/2)\varrho C_L SV_g^2 t$ vary with time and position angle during one complete wingstroke in a gliding-flapping hypothetical proto-bird, when the glide angle θ is 30°, wingstroke plane angle is 90°, the mean flapping speed V_f is 0.5 times the gliding speed V_g, and the ratio (wing drag)/(body drag) is 5. For these values the vertical lift produced during flapping (solid curves) equals that produced during gliding when the wingstroke amplitude $\emptyset$ is 37.7° ($\emptyset'_{max} = \emptyset$; shaded areas above and below the broken line for gliding being equal). The thrust produced during the downstroke is larger than the drag produced during the upstroke (hatched area above zero line larger than that below). This net thrust can be used to flatten out the glide. (From U. M. NORBERG 1985b).

Fig. 4 shows how the net thrust index varies with the flapping speed. Higher lift to drag ratios than the 1.73 ratio used in Fig. 3 (corresponding to a glide angle of 30°), i.e. smaller θ, would give lower net thrust. But smaller stroke plane angles (more forwardly tilted stroke planes) would increase the thrust because the resultant airspeed becomes larger. When the stroke plane angle becomes larger than 90° (about 100° in the example illustrated in Fig. 3) then the wings' net thrust ($T-D_{w,h}$) would decrease and approach body drag (D_b). When flight becomes more or less horizontal, the resultant force during the upstroke becomes tilted strongly backwards relative to the vertical in an extended wing, and the wing drag then becomes larger than the thrust produced during the downstroke. The animal then has to flex its wings and/or rotate the wings as to make the angle of attack very small during the upstroke to reduce wing drag.

Incipient wing flapping in a gliding proto-bird would have carried larger benefits (in terms of flattened flight paths) the better the L/D ratio of the wings. No matter how small the wing flapping speed then is, it will provide a net thrust force as well as the lift necessary to balance the body weight, provided that the stroke amplitude is small enough and the flapping frequency reaches a threshold value: For every value of the flapping speed the model identifies a particular value of the stroke amplitude that must be used by the animal to provide exactly the necessary lift (Fig. 4). This amplitude increases almost linearly with the flapping speed, and is small at low speeds. This means that the flapping frequency is rather constant irrespective of the flapping speed. A particular flapping frequency thus has to be attained for the proto-bird to obtain thrust and the lift necessary to balance body weight, but even very small flapping speeds and amplitudes are sufficient. For a proto-bird the size of *Archaeopteryx* with a wing span of 60 cm and a glide speed of about 7 ms^{-1}, the wingbeat frequency must be about 6.5 with vertical wing strokes and about 6 when the stroke plane is tilted forwards relative to the vertical as in modern birds, if the duration of the downstroke is taken to equal the duration of the upstroke. This is rather a high flapping frequency. But if the proto-birds beat the wings faster during the downstroke than during the upstroke the stroke frequency could be reduced to 2 strokes per second (U. M. NORBERG 1985a). So, it would be quite possible for the proto-bird to produce a net thrust force as well as the lift necessary to balance body weight even with very slight flapping.

Increased aspect ratio of the wings (longer span) and more efficient wing profiles, giving larger L/D ratio and hence thrust, and increased abilities of wing flexion and movement coordination eventually led to horizontal flight. More sophisticated wing characters evolved, such as wing slots and a complex structure and asymmetry of the feathers. These features occurred already in *Archaeopteryx*, which indicates that the feathers of this early bird had an aerodynamic function (FEDUCCIA and TORDOFF 1979, FEDUCCIA 1980, R. Å. NORBERG this volume).

Increased manoeuvrability permitted the birds to forage in the air. Radiation to different flight habits led to different wing forms, specialized for different habitats and flight modes.

Running, jumping, and powered flight

CAPLE et al. (1983) proposed that birds evolved from cursorial proto-birds that jumped for flying insects. The forelimbs were extended and moved for stability and control of the body. Their study rightly indicates that small increments of lift will increase body orientation control. Better stability and lift increment would increase the animal's ability of insect-catching. The authors suggested that the forelimb motions were similar to wingstrokes in modern birds, and that rapid movements and twisting of the forelimbs for control and stability gave more lift and thrust and eventually led to powered horizontal flight.

This latter step from a ground-running jumping mode of life to powered flight seems very difficult to make, since the animal had to work a g a i n s t gravity. In a running hypothetical proto-bird the drag on body and wings would increase with speed squared, as in flight. The drag would retard the animal, and it would have to do still more work to reach the speed needed to generate the lift and thrust required for take-off. Longer forelimbs and feathers would produce still larger drag and increase the need of more work for the same speed. Such proto-birds would thus have to produce not only the power necessary for

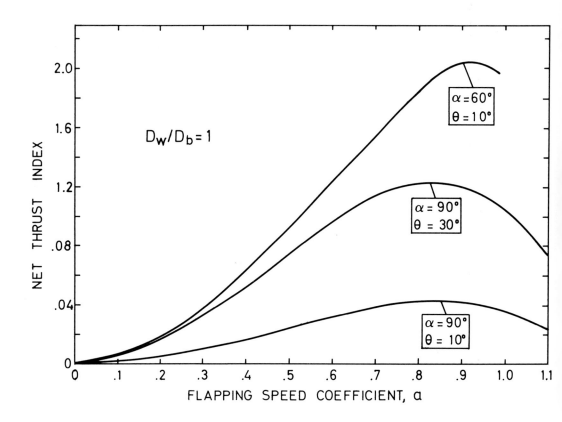

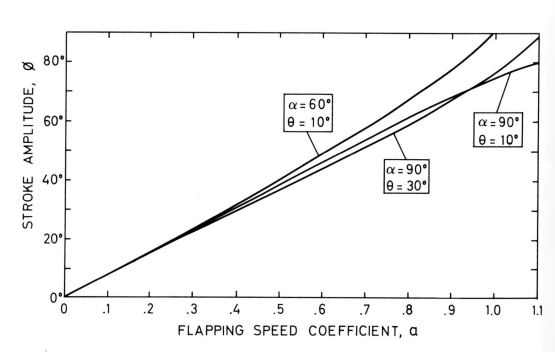

Fig. 4
Net horizontal thrust index $(T-D_{w,h}-D_{b,h})/(1/2)\varrho C_L S V_g^2 \mathbf{t}$, for various values of the flapping speed coefficient **a** ($\mathbf{a} = V_f/V_g$ and named a d v a n c e r a t i o in aerodynamic theory) at glide angles of 10° and 30°, stroke plane angles of 60° and 90°, and (wing drag)/(body drag) ratios of 1 and 10 (top figure). For every value of **a** the model indentifies a particular value of the stroke amplitude that must be used by the proto-bird to provide the lift necessary to balance body weight (bottom figure).
(Modified from U. M. NORBERG 1985a).

flight but also the power needed to run near take-off speed during the transient evolutionary stages. Since active flight is very power-consuming, it would then be very difficult for the animal to provide the power required. This power argument identifies a formidable obstacle to the evolution of flight from a cursorial mode of life as sketched by CAPLE et al. (1983) (U. M. NORBERG 1985a). A gliding animal does not suffer from this negative feed-back system, since it works w i t h gravity, and hence reaches a high forward speed almost free of cost before it begins to flap.

Still, it could be possible that a running-jumping-gliding behaviour was used by proto-birds as an energy-saving mode of locomotion, and that powered flight evolved from these short glides. Fast running with r e t r a c t e d wings intermitted by regular leaps, followed by shallow glides on extended wings, is a commuting mode that is energetically cheaper than continuous running with the same average speed (RAYNER, this volume). This is thus a possible pathway for the origin of flight in birds, although I think it is unlikely that the proto-birds had reason to move so fast except for escape from predators (modern cursorial animals seldom travel so fast in commuting runs). But jump-and-glide would involve some deceleration and would then be slower than continuous running. The high speeds needed for flight are reached in gliding from some elevation without the high power output demanded from a cursorial proto-flier. In light of this, the arboreal theory seems a much more attractive hypothesis than the running-leaping-jumping one, both in terms of energy and time saving arguments, as well as for aerodynamic reasons (U. M. NORBERG 1985a).

Summarizing, it is quite clear that gliding must have preceded flapping, either from some elevation, such as trees, or followed upon leaps from the ground. My aerodynamic model applies to the transition from gliding to powered flight regardless of whether gliding occurred from some elevation or followed upon leaps in fast runs.

Acknowledgements

I am grateful to ÅKE NORBERG and JEREMY RAYNER for valuable discussions, and to ÅKE NORBERG for critically commenting on the manuscript. This work was supported by a research grant from the Swedish Natural Science Research Council (B-BU 4455-106).

References Cited

BOCK, W. J. (1965): The role of adaptive mechanisms in the origin of higher levels of organization. – Syst. Zool., **14**: 272–287.
BOCK, W. J. (1969): The origin and radiation of birds. – Ann. N. Y. Acad. Sci., **167**: 147–155.
BOCK, W. J. (1983): On extended wings. Another view of flight. – The Sciences **23**: 16–20.
CAPLE, G., BALDA, R. P. & WILLIS, W. R. (1983): The physics of leaping animals and the evolution of preflight. – Am. Nat. **121**: 455–467.
DARWIN, C. (1859): The origin of species. – 386 pp. London (JOHN MURREY).
FEDUCCIA, A. (1980): The age of birds. – 196 pp. Cambridge (Mass.) and London (Harvard University Press).
FEDUCCIA, A. & TORDOFF, H. B. (1979): Feathers of *Archaeopteryx*: asymmetric vanes indicate aerodynamic function. – Science, **203**: 1021–1022.
HEILMANN, G. (1926): The origin of birds. – London (Witherby).

MARSH, O. C. (1880): Odontornithes: a monograph on the extinct toothed birds of North America. – Rep. U. S. Geological Exploration of the 40th Parallel, no. 7. Washington, D. C.
NOPSCA, F. VON (1907): Ideas on the origin of flight. – Proc. Zool. Soc. Lond., **1907**: 223–236.
NOPSCA, F. VON (1923): On the origin of flight in birds. – Proc. Zool. Soc. Lond., **1923**: 463–477.
NORBERG, R. Å. (1981): Why foraging birds in trees should climb and hop upwards rather than downwards. – Ibis, **123**: 281–288.
NORBERG, R. Å. (1983): Optimal locomotion modes of foraging birds in trees. – Ibis, **125**: 172–180.
NORBERG, R. Å. Function of vane asymmetry and shaft curvature in bird flight feathers; inference on flight ability of *Archaeopteryx*. – This volume.
NORBERG, U. M. (1985a): Evolution of vertebrate flight: an aerodynamic model for the transition from gliding to active flight. – Am. Nat. **126**: 303–327.
NORBERG, U. M. (1985b): On the evolution of flight and wing forms in bats. – Biona Report, **4**, Fledermausflug, (ed. W. NACHTIGALL). Stuttgart (Gustav Fischer). (In press).
OSTROM, J. H. (1979): Bird flight: How did it begin? – Am. Sci., **67**: 46–56.
PADIAN, K. (1982): Running, leaping, lifting off. – The Sciences, May/June: 10–15.
PARKES, K. C. (1966): Speculations on the origin of feathers. – Living Birds, **5**: 77–86.
PENNYCUICK, C. J. (1972): Animal flight. – 68 pp. London (EDWARD ARNOLD).
RAYNER, J. M. V. (1985): Vertebrate flapping flight mechanics and aerodynamics, and the evolution of flight in bats. – Biona Report **4**, Fledermausflug, (ed. W. NACHTIGALL). Stuttgart (Gustav Fischer). (In press).
RAYNER, J. M. V. (1985): Cursorial gliding in proto-birds. – This volume.
SAVILE, D. B. O. (1962): Gliding and flight in the vertebrates. – Am. Zool., **2**: 161–166.
SCHOLEY, K. D. (1983): Developments in vertebrate flight: climbing and gliding of mammals and reptiles, and the flapping flight of birds. – PhD. Thesis, University of Bristol.
SCHOLEY, K. D. (1985): An energetic explanation for the evolution of flight in bats. – Biona Report **4**, Fledermausflug, (ed. W. NACHTIGALL). Stuttgart (Gustav Fischer). (In press).

Author's address: Dr. Ulla M. Norberg, Department of Zoology, University of Göteborg, Box 25059, S-40031 Göteborg, Sweden.

R. Åke Norberg

Function of Vane Asymmetry and Shaft Curvature in Bird Flight Feathers; Inferences on Flight Ability of *Archaeopteryx*

Abstract

The structural asymmetry of bird flight feathers is essential for their bending and twisting behaviour throughout the wing-beat cycle. The interaction between feather structure and dynamics is much more complicated than suggested in previous investigations. Based on high-speed photographs of flying birds, wind-tunnel tests on wings and feathers, and data on feather flexural stiffness, a new mechanical and aerodynamical theory is developed on the function of vane asymmetry and feather curvature. It explains how feather angles-of-attack are adjusted automatically throughout the wing-beat cycle, despite continuously varying directions and velocities of the relative wind. All structural characteristics necessary for this passive maintenance of proper angles of attack were present in *Archaeopteryx*. This suggests that *Archaeopteryx* flew with true powered flight.

Zusammenfassung

Die strukturelle Asymmetrie der Flugfedern von Vögeln ist wesentlich für Biege- und Drehverhalten während des Flügelschlag-Zyklusses. Die Wechselwirkung zwischen Federstruktur und Kräftespiel ist komplizierter als früher angenommen. Gegründet auf extrem kurz belichtete Fotografien fliegender Vögel, Windkanaltests an Schwingen und Federn sowie Daten über die Biegefestigkeit von Federn wird eine neue mechanische und aerodynamische Theorie über die Funktion der Asymmetrie der Federfahnen und die Krümmung der Federn entwickelt. Sie erklärt, wie die Anstellwinkel der Federn automatisch während des Flügelschlag-Zyklus angepaßt werden trotz ständig variierender Richtungen und Geschwindigkeiten des Flugwindes. Alle strukturellen Merkmale, die notwendig für das passive Beibehalten geeigneter Anstellwinkel sind, waren bei *Archaeopteryx* vorhanden. Dies läßt darauf schließen, daß *Archaeopteryx* ein echter aktiver Flieger war.

1. Introduction

There are various opinions on whether *Archaeopteryx* was a cursorial animal, which used its wings essentially for non-flight purposes, or whether it could just glide, using gliding descents for horizontal progression among trees (which is cheap in energy; R. Å. NORBERG 1983), or, finally, whether it was even capable of true, powered flight (review in FEDUCCIA 1980). I address this problem here by developing a new mechanical and aerodynamical theory on the function of flight feathers in a modern bird, and then comparing flight feather structure in modern birds with the excellent fossil record of *Archaeopteryx*.

Based on high-speed photographs of modern birds in flight [Pied Flycatcher, *Ficedula hypoleuca* (Pallas)] and wind-tunnel tests on wings and feathers, I here describe the behaviour of flight feathers when subjected to the continuously varying relative wind throughout the wing-beat cycle. Using information also on feather flexural stiffness, I will develop a theory on the functional interaction between vane asymmetry (narrower vane in front of than behind the shaft) and feather shaft curvature. These are actually two structural means of achieving the same goal – the automatic adjustment of feather pitch. Comparisons between feather structures in *Archaeopteryx* and in modern birds reveal advanced feather adaptations in *Archaeopteryx*, suggestive of flapping, powered, flight.

The overall aerodynamic function of separated primary feathers in bird wings is to increase lift by the leading-edge slat effect, and, possibly, to reduce induced drag (review in WITHERS 1981). These and other reasons why the primary feathers separate in the first place are not pursued here. Emphasis is instead placed on the mechanics of the various feathers.

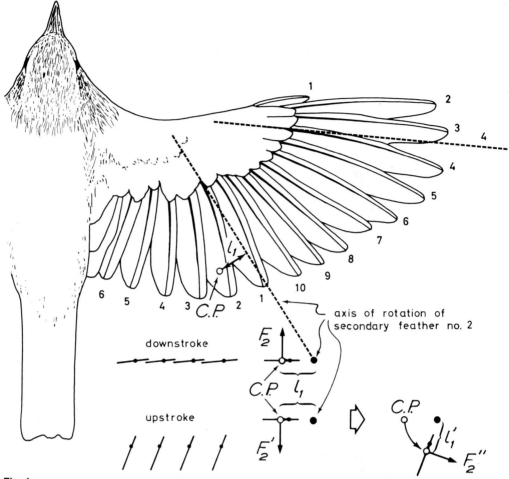

Fig. 1
Dorsal view of Pied Flycatcher with right wing extended laterally as in the middle of the downstroke. The longitudinal axis through the feather base (calamus), at its attachment to the wing, is shown in broken line for primary feather no. 4 and secondary feather no. 2. Planform figure drawn from a photograph of a first-year bird held by hand. The wing feathers are not subjected to any aerodynamic load, but exhibit the curvature typical of the resting state.
The lower diagrams show end-on views of secondary feathers as seen from behind, in the plane of the upper, planform figure.

Readers are referred to R. Å. NORBERG (1972, 1973) for information on movements of the aerodynamic centre of pressure along the wing chord, following upon changes in angles of attack, and for explanations of mechanisms for automatic control of wing pitch angles.

2. Rotational characteristics of the feather base

The anatomy of feather attachment in the wing is such that both primary and secondary flight feathers are rather free to rotate in the nose-up sense. This "in socket" rotation is possible through about 90°, from the normal orientation with the feather vane in the plane of the wing (Figure 2). During the upstroke the entire feather may thus rotate as a rigid unit about its axis of rotation through its base (calamus) with little or no deformation (Figures 1 and 2).

Fig. 2
Pied Flycatcher in hovering flight. The wings are half-folded and near the middle of the upstroke. The primary as well as secondary feathers are rotated in the nose-up sense through about 90° relative to their normal orientation, letting air through the wing. Photo: R. Å. NORBERG.

But from the feather's normal orientation, with its vane in the plane of the wing, there is no freedom of rotation in the nose-down sense at the feather's attachment to the wing. Any nose-down rotation of the vane must therefore be by torsional, elastic, deformation of the feather shaft (Figures 3 and 4).

3. Upstroke

3.1. Relative wind in upstroke

The resultant relative wind comes down on the dorsal side of the wing, meeting it from obliquely in front at some negative angle of incidence. The direction of the relative wind varies along the wing and changes throughout the upstroke. But the details are not important for feather mechanics in the upstroke.

3.2. Mechanics of primary and secondary feathers in upstroke

The following account of feather mechanics in the upstroke refers to secondary feather no. 2 in Figure 1. The aerodynamic force F'_2, generated by the distal part of secondary feather no. 2, acts downwards

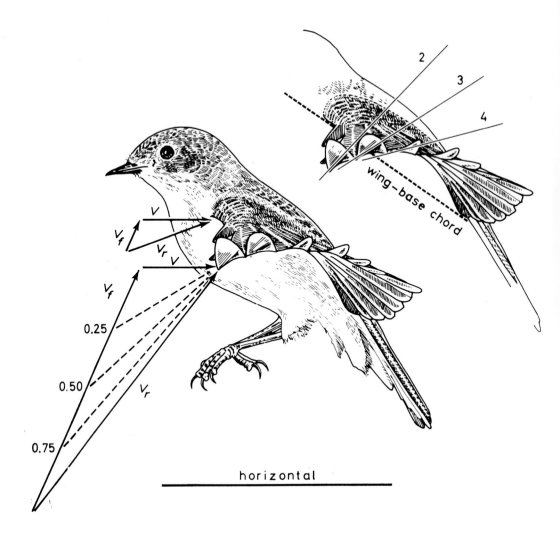

Fig. 3
Pied Flycatcher in slow horizontal, near-hovering, flight. Diagrams of relative air speeds are shown for a spanwise position near the wing base and for the outer ends of the separated primary feathers. The local resultant velocity V_r of the relative wind is the vector sum of the horizontal flight speed V and the local, tangential flapping speed V_f. Both diagrams refer to mid-downstroke when the flapping speed V_f, and hence also V_r, are largest. The broken lines show directions and magnitudes of the resultant speed V_r when the flapping speed takes the intermediate values 0.25, 0.50, and 0.75 of its maximum.
Inset figure at upper right shows chord orientation of the wing-base and of individual primary feathers number 2, 3, and 4 in their separated region near the tip.
Bird drawn strictly from a high-speed photograph of a Pied Flycatcher in near-hovering flight. Inclination of the wing-stroke plane to the horizontal is taken to be 65°, horizontal flight speed 2.4 ms^{-1}, and the maximum flapping speed at the wing tip 8 ms^{-1} (based on Table 1 in U. M. NORBERG 1975).

through the chord-wise centre of pressure C.P., which lies near the feather's mid-chord point. Whenever the relative wind is more or less perpendicular to the wing plane, the section C.P. is located at or near the mid-chord point. It is only when the relative wind meets the leading edge at some angle of incidence smaller than 90° (relative to the airfoil chord) that the C.P. moves ahead of the airfoil's mid-chord point.

Force F'_2 acts with moment arm l_1 about the feather's axis of rotation. During the upstroke an initial moment $F'_2 l_1$ therefore tends to rotate the feather in the nose-up sense. The section C.P. then moves along a circular arc about the axis of rotation through the feather base, as shown at bottom right in Figure 1, and the entire feather rotates in the nose-up sense as a single unit. In the process, the feather's angle of attack becomes smaller than 90°, resulting in a lift force component so that the resultant aerodynamic force F''_2 rotates laterally (counter-clockwise in Figure 1, lower right). The section C.P. moves forwards into a position between the mid-chord point and the quarter-chord point, and the moment arm changes from l_1 to l'_1. The more the feather chord becomes aligned along the relative wind, i.e. the smaller the feather's angle of attack, the smaller the aerodynamic force F''_2 becomes. So, as the feather chord approaches a "feathered" position along the direction of the relative wind, the resultant force F''_2 diminishes, and the torsion moment vanishes. There will of course always remain some friction drag on the feather (along the direction of the relative wind), but when the feather has rotated into a downwind position relative to the feather axis of rotation, there is no longer any torsion arm l_1 for the aerodynamic drag to act on.

The above description for secondary feather number 2 applies in principle also to the other secondary feathers and to all primary feathers as well. During the upstroke they all rotate in a similar manner through about 90°, each feather rotating as a rigid unit with little or no twisting (Figure 2). With this nose-up rotation of the flight-feathers, air spills through the wing and little or no useful aerodynamic forces will be elicited. The upstroke therefore is largely a recovery stroke in slow flight and hovering. It is the downstroke that is the power stroke which supplies most of the useful aerodynamic forces in slow flight and hovering. It will be considered next.

4. Downstroke

4.1. Relative wind in downstroke

The resultant relative wind comes up against the ventral side of the wing, meeting it from obliquely in front, at some positive angle of incidence. It is composed of two main speed vectors.

In horizontal flight there will always be a horizontal flight speed vector (V in Figure 3), which remains constant along the entire wing span and throughout the wing-beat cycle. In additon, there is in powered flight a flapping speed vector (V_f in Figure 3), which increases linearly from the wing base outwards and whose magnitude also varies continuously throughout the wing-beat cycle. Therefore, both the direction and magnitude of the resultant relative wind (V_r in Figure 3) vary throughout the wing-stroke cycle and do so in different but characteristic ways for every position along the wing span.

The most vertical direction of the relative wind is at the wing tip in mid-downstroke, as indicated in Figure 3. Since a wing operates with the most efficiency at a certain angle of attack to the relative wind, there is a need of continuous adjustments of wing and/or feather twisting throughout the wing-beat cycle. It is apparent from Figures 3 and 4 that the strong nose-down orientation of the tips of the anteriormost primary feathers in relation to the wing-base chord is due partly to their forward location in the antero-posteriorly cambered wing, partly to twisting of the entire wing, and partly to twisting of the separated feathers themselves.

The mechanics of feather bending and twisting in the downstroke will be considered next. Because the dynamics is much simpler for secondary feathers than for primaries, let us begin with the former. The dynamics of the outermost primary feathers will then be discussed in some detail.

Fig. 4
Pied Flycatcher in near-hovering flight. The wings are fully extended and near the middle of the downstroke. The anteriormost primary feathers are bent forwards-upwards and twisted in the nose-down sense in response to the aereodynamic load. Photo: R. Å. Norberg.

4.2. Mechanics of secondary feathers in downstroke

Also this account refers to secondary feather no. 2 in Figure 1. The aerodynamic force F_2, generated during the downstroke, acts through the same chord-wise centre of pressure as does force F'_2 during the upstroke. The moment arm l_1 about the feather's axis of rotation is also the same. During the downstroke there will thus be a moment $F_2 l_1$, which tends to rotate the feather in the nose-down sense.

Because of the curvature of the feather shaft and the overlap between adjacent feathers, the aerodynamic pressure on the vane of one feather will be taken up largely by the nearest inner neighbour feather. In their zone of overlap, adjacent feathers will therefore be pressed together much more than if each feather were straight and carried its own aerodynamic load rather than much of that of its neighbour. This pressing together of adjacent feathers, together with the action of a specialized friction surface of the vane in the zone of overlap, tend to prevent the flight feathers from slipping apart, and make them form a closed wing surface. The friction zone owes its character to feather microstructures (Graham 1930; Sick 1937; Oehme 1963; Lucas and Stettenheim 1972, pp. 260). The modified barbules were termed "friction barbules" by Lucas and Stettenheim (1972).

Fig. 5
Living Pied Flycatcher whose left wing is manipulated by its wrist to be in the same position as in Figure 4. The backward curvature of the unloaded primary feathers contrasts sharply with their anterodorsal bending when they are loaded aerodynamically in the downstroke (Figure 4). Photo: U. M. NORBERG.

4.3. Mechanics of primary feathers in downstroke

The mechanics of the anteriormost primary feathers differs strongly from that of the secondary feathers in the downstroke. From the inner secondaries outwards, there is a gradual change in the orientation of the feather shafts in relation to the resultant relative wind. Whereas the relative wind comes more or less in the lengthwise direction of the secondaries, it comes nearly in the direction of the chords of the anteriormost primaries, i.e. at about 90° to the feathers' leading edge. Furthermore, the anteriormost primary feathers are separated near their tips, each feather therefore acting as an airfoil of its own. The separation is due partly to the spreading apart of the feathers at the wing tip, partly to some feathers being emarginated both in their anterior and posterior vane.

The following mechanical treatment is divided into three sections (numbered 4.3.1., 4.3.2., and 4.3.3.), each treating a specific combination of feather curvature and flexural stiffness.

4.3.1. Curved feather; equal flexural stiffness dorso-ventrally and antero-posteriorly

Figure 6 shows the response to aerodynamic force F exhibited by a primary feather which is rather curved and which is equally resistant to bending in the dorso-ventral and antero-posterior directions. When the relative wind meets the feather at some near-optimal angle of attack, such as shown in Figure 3, the section C.P. is located near the one-quarter or one-third chord points (as measured from the feather's leading edge). This is behind the feather shaft which acts as the local axis of rotation of the feather section (Figures 6–8).

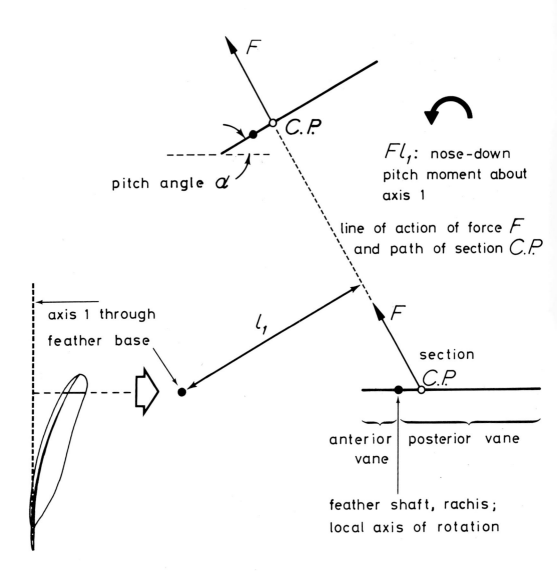

Fig. 6
Bending and twisting behaviour of a primary feather subjected to aerodynamic forces in the downstroke. At lower left is a planform view of a primary feather in its resting state. The longitudinal, torsion, axis, aligned along the feather base, is indicated in broken line. The resultant relative wind is taken to come from the left and to be perpendicular to the longitudinal axis through the feather base. The distal chordwise section, marked out along the direction of the relative wind, is shown enlarged in the right diagram as it appears when the feather is seen in end-on view along the line of longitudinal torsion axis 1. The angle of view of the right diagram thus is as if the feather were tipped 90° out of the plane of the figure. The relative wind is taken to meet the feather profile at lower right at some positive angle of incidence, from below and in front (from the left) as in Figure 3. The resultant aerodynamic force is therefore tipped forwards, providing a thrust component.
The feather is rather curved in its resting state and is assumed to possess equal flexural stiffness dorso-ventrally and antero-posteriorly. At lower right is a cross-sectional view of the feather in its resting position. Under the influence of force F the feather bends and the feather section moves to the upper forward position and rotates in the nose-down sense due to the pitching moment Fl_1.

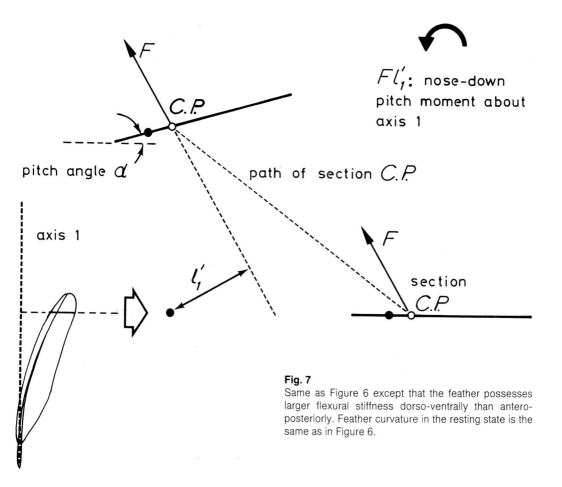

Fig. 7
Same as Figure 6 except that the feather possesses larger flexural stiffness dorso-ventrally than antero-posteriorly. Feather curvature in the resting state is the same as in Figure 6.

The resultant aerodynamic force F bends the feather. Because the feather shaft exhibits equal flexural stiffness in all directions (under the present assumption), the section C.P. will move along the line of action of force F. Bending increases until sufficient resistance to further bending has built up as a result of elastic deformation of the feather.

Force F also sets up a torsion moment Fl_1 about the axis through the feather base (Figure 6). The distal part of the feather therefore rotates in the nose-down sense until sufficient resistance to further rotation has built up as a result of elastic twisting of the feather shaft. There is also a nose-down pitching moment about the local axis of rotation (the shaft) within the profile itself, owing to the asymmetry of the vane.

When F increases the feather bending also increases, which is unavoidable. The load-extension curve of the shaft is nearly linear for Pigeon primary feathers (*Columba livia* GMELIN; PURSLOW and VINCENT 1978, p. 258). Since the feather bends in the direction dictated by the aerodynamic force F, the moment arm l_1 (about the torsion axis through the feather base) remains the same regardless of the degree of bending of the feather (so long as the force direction remains constant). The nose-down pitching moment Fl_1 therefore also increases as F increases.

This constancy of the moment arm l_1, due to equal flexural stiffness of the shaft in all directions, is probably detrimental. To see this, assume that the velocity of the relative wind increases strongly, while its direction remains the same. When force F is then strongly increased as, for instance, during flight acceleration or manoeuvres, the nose-down pitching moment Fl_1 may become unduly large so that the

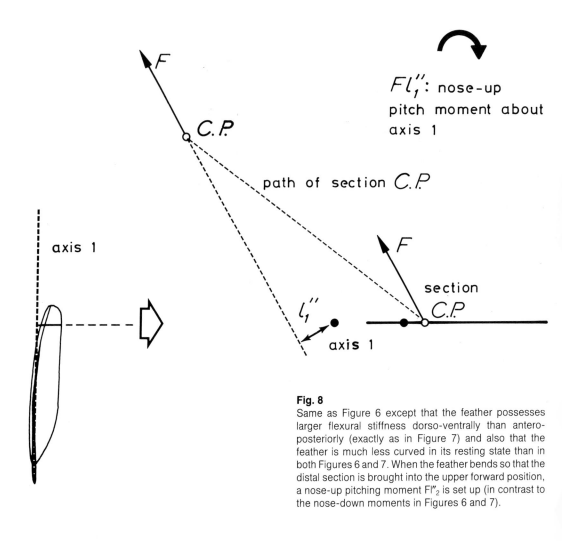

Fig. 8
Same as Figure 6 except that the feather possesses larger flexural stiffness dorso-ventrally than antero-posteriorly (exactly as in Figure 7) and also that the feather is much less curved in its resting state than in both Figures 6 and 7. When the feather bends so that the distal section is brought into the upper forward position, a nose-up pitching moment Fl''_2 is set up (in contrast to the nose-down moments in Figures 6 and 7).

section angle of attack approaches zero. Then the aerodynamic force will become much less than if the nose-down pitching moment had remained smaller. But this linear increase of moment Fl_1 with increasing F is avoidable with a different stiffness distribution of the feather shaft, as will be explained in the next section.

4.3.2. Curved feather; larger flexural stiffness dorso-ventrally then antero-posteriorly

The situation is the same as in the previous section and in Figure 6 except that the feather is taken to be more resistant to bending dorso-ventrally than antero-posteriorly (Figure 7). This is actually the case in real feathers. Thus PURSLOW and VINCENT (1978, p. 257) showed that the feather stiffness is much larger dorso-ventrally than laterally for flight feathers of the Pigeon (*Columba livia*; except for the anteriormost primary, which is equally stiff laterally as dorso-ventrally). The generally larger dorso-ventral stiffness can easily be demonstrated by simple manipulation of flight feathers. The ventral groove along part of the feather shaft is likely to add to the feather's dorso-ventral stiffness (HERTEL 1966, pp. 51–54).

Because the flexural stiffness is lower laterally than dorso-ventrally, the aerodynamic force on a distal feather section results in the feather bending in such a way that the section C.P. moves along a path inclined ahead of the line of action of the force (Figure 7). Therefore the moment arm l_1 about the torsion axis through the feather base becomes progressively shorter the more the feather bends. As force F increases, the moment arm l_1 therefore decreases, which tends to prevent the nose-down pitching moment Fl_1 from becoming unduly large (Figure 7). Instead the angle of attack may be maintained at near-optimal values despite large changes of force F following upon large variations of the relative air speed. This mechanism is statically stable; it tends towards self-adjustment of the angle of incidence at the outer, free parts of the anteriormost primary feathers.

4.3.3. Straight feather; larger flexural stiffness dorso-ventrally than antero-posteriorly

A function of primary feather curvature was described above. The obvious alternative to a curved feather is a straight, or near-straight one. The function (or, rather, malfunction) of such a feather is considered below as a contrast to the static stability of angles of attack that applies to a curved feather. The following is a hypothetical example that is presented to further illustrate the function of primary feather curvature.

The situation is taken to be the same as in the previous section and as shown in Figure 7 except that the feather is rather straight already in its resting state (Figure 8). It is assumed to be stiffer dorso-ventrally than antero-posteriorly, just as in the previous case (4.3.2.).

If the feather were thus too straight in its resting state, before being subjected to the aerodynamic force F, then bending in response to force F would tend to move the line of action of force F ahead of the torsion axis through the feather base (Figure 8). Then a moment Fl''_1 would give a nose-up pitching moment, tending to flip the feather over in the nose-up sense, bringing the vane into a vertical attitude with the posterior vane lowest. (This nose-up pitching moment obtains for the entire feather despite the nose-down pitching moment present with respect to the local axis of rotation within any individual chordwise section.) This flip-over would occur easily because there is no resistance at the feather base, or from adjacent, overlapping feathers, against rotation in the nose-up sense, which occurs during every upstroke in slow flight or hovering as described above. But its occurrence in the downstroke would be aerodynamically disastrous.

So there is an urgent need for the anterior primary feathers to be curved if they are stiffer dorso-ventrally than laterally. And this they should be for the automatic maintenance of near-optimal angles of incidence, despite continuous variation of the direction of the relative wind and the magnitude of the aerodynamic forces. (The same malfunction described above would occur also if the feather were normally curved but had too little flexural stiffness antero-posteriorly as compared with its dorso-ventral stiffness.)

5. Summary on feather planform, structure, and function in modern birds

It will have been apparent from the foregoing that the feather vane asymmetry and feather curvature are two structural means of achieving the same goal, namely the creation of appropriate pitching moments along the feather. One function of the vane asymmetry is to make sure that the local pitching moment at every chordwise feather section does always act in the same direction as does the overall pitching moment set up about the longitudinal axis through the feather base. This applies to the valve-like feather function in the upstroke as well as to the elastic twisting in the downstroke.

Throughout this paper emphasis has been laid on moments about the longitudinal torsion axis through the feather base. Nose-up pitching rotation about this axis is also possible for each individual flight feather. But it should be observed that nose-down rotation of feathers is hampered by the nearest inner neighbour feather in the zone of overlap (Figures 1, 3–5), as well as by the structure of the feather's attachment to the wing. This does not disturb the previous arguments. It only shows that nose-down twisting is not only counteracted by the torsional stiffness of the feather itself, but also by the presence of a more or less overlapping inner feather.

The vane asymmetry and feather curvature jointly tend towards self-adjustment of the angles of incidence of the relative wind at the outer free parts of the primary feathers. This mechanism is statically

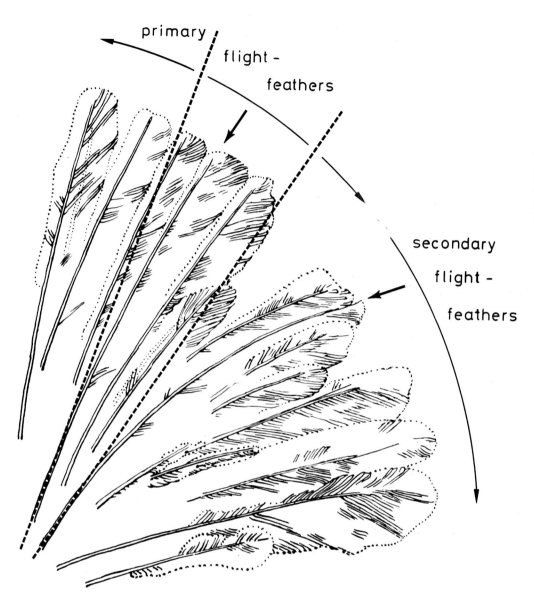

Fig. 9
Drawing of feather impressions of the right wing of the British Museum (London) specimen of *Archaeopteryx lithographica*, main slab. The wing's leading edge is on the left. The orientation of the longitudinal axis through the proximal part of the feather shaft (calamus) is shown in broken line for one primary and one secondary feather, the tips of which are indicated by arrows. The pronounced shaft curvature is apparent with reference to these axes.
Photographic copy of the drawing serving as key to plate 11 in DE BEER (1954) but with all the original labels and label lines removed. The two broken lines, the arrows, and the labels "primary flight-feathers" and "secondary flight-feathers" are my additions. The separation between primaries and secondaries is according to DE BEER (1954).

and dynamically stable. The inherent aeroelastic stability of the feather is achieved passively by an appropriate combination of three main feather characteristics: (1) the vane asymmetry, (2) the curved shaft, and (3) the greater flexural stiffness dorso-ventrally than antero-posteriorly. The latter is achieved structurally, at least in part, by the ventral furrow of the shaft, which is likely to add strongly to dorso-ventral stiffness.

6. Feather structure in *Archaeopteryx*

The fossil record of *Archaeopteryx* exhibits remarkably fine details of feather structure (DeBeer 1954; Feduccia 1980; Helms 1982; Rietschel 1985). It is therefore eminently possible to compare *Archaeopteryx* feathers with those of modern birds. Focus is here placed on those particular characters in a modern bird that have been explored in this paper from mechanical and aerodynamical points of view.

6.1. Feather vane asymmetry

The vane asymmetry is caused by the anterior (or outer) vane being narrower than the posterior (or inner) one. Feduccia and Tordoff (1979) were the first to draw attention to this feather asymmetry in *Archaeopteryx*. Its presence in *Archaeopteryx* is abundantly evidenced by the fossil record; the first-discovered *Archaeopteryx* feather was thus asymmetrical and the flight feathers of the London as well as Berlin specimen are about as asymmetrical as those in modern birds (Figures 9 and 10).

6.2. Feather curvature

Backward curvature of flight feathers, essentially like that in modern birds, occurred already in *Archaeopteryx*. The first-discovered fossil feather impression of *Archaeopteryx* is strongly curved backwards (illustrations in for instance DeBeer 1954, plate 15; Feduccia and Tordoff 1979; Feduccia 1980). It is not obvious from where on the bird the feather comes. But in modern birds the secondaries are usually more curved than the primaries. Judging from its strong curvature, the first-discovered *Archaeopteryx* feather impression may therefore well be from a secondary wing feather.

The primary and secondary wing feathers of the London *Archaeopteryx* are all curved very much like those in modern birds (Figure 9). In the Berlin specimen the first four primary feathers are also curved backwards, whereas the eight remaining (inner) primaries are bent the wrong way in their outer parts, i.e. forwards (Figure 10; Helms 1982, p. 192). The secondary feathers seem all to have the usual backward curvature. The very prominent ridges at the base of the hand-wing of the Berlin specimen may mark the orientation of the basal part (calamus) of the primary feathers, covered by wing coverts (Figure 10; and much more obvious in Figure 1 in Helms 1982). If so, the eight primary feathers with a distal, forward bend do still have their outer part well behind the longitudinal axis through their respective base. And this is what matters functionally, as explained in this paper. Moreover, the distal, forward bend in the eight inner primaries of the Berlin specimen may be the result of flattening of the wing during the early sedimentation process (Helms 1982, p. 191; Rietschel, pers. comm.).

Taken together, the fossil evidence thus shows that *Archaeopteryx* had backwardly curving flight feathers with respect to longitudinal axes through the feather bases, very much as in modern birds.

6.3. Ventral furrow of the feather shaft

According to S. Rietschel (1985; and pers. comm.) there is a longitudinal furrow along the ventral side of all flight feathers in *Archaeopteryx*, similar to that in modern birds.

7. Flight Ability in *Archaeopteryx*

The aerodynamic function of the various feather features in *Archaeopteryx* will now be evaluated for gliding flight and for active, flapping flight.

(1) The vane asymmetry in *Archaeopteryx* indicates an aerodynamic function (Feduccia and Tordoff 1979; Feduccia 1980). The pitching moments, created because of the vane asymmetry, are necessary

Fig. 10
Left wing of the Berlin specimen of *Archaeopteryx lithographica*, main slab. The wing's leading edge is on the top of the figure. The upper broken lines and arrows show approximate orientation of the longitudinal axis through the base of an outer primary wing feather (the third one from the leading edge, according to HELMS 1982). The orientation of the feather base (calamus) is according to my best estimate from the photograph. My judgement is shared by SIEGFRIED RIETSCHEL (pers. comm.). The shaft curvature is obvious with reference to the broken line which is interrupted along the feather, not to obscure its structure.
The lower broken line shows approximate course of the shaft of a secondary feather according to SIEGFRIED RIETSCHEL'S interpretation (pers. comm.). The bases of adjacent secondary feathers are similarly oriented, suggesting they have similarly curved shafts. Photograph supplied by JOCHEN HELMS. Photo: BARTHEL and HELMS.

for gliding flight as well as for flapping flight (for pressing the feathers together in their zones of overlap to keep the wing surface closed, and for automatic pitch angle control.).

(2) The feather curvature likewise indicates an aerodynamic function. The pitching moments, created because of the feather curvature, strongly enhance the above effects of the vane asymmetry – important for gliding as well as for flapping flight. The curvature of flight feathers is important for their valve function in the upstroke and for the automatic adjustment of angles of attack of separated primaries in the downstroke. Feather curvature is thus essential for flapping flight, but obviously has beneficial effects also for gliding flight.

(3) The furrow along the ventral side of the feather shaft adds to structural strength in the dorso-ventral direction. This is an obvious adaptation to make the flight feather resistant to aerodynamic loads, which act predominantly in the dorso-ventral direction in gliding as well as in flapping flight. A larger flexural stiffness in the dorso-ventral direction than in the antero-posterior direction is also a necessary precondition for the automatic pitch angle control of primary feathers in the downstroke.

All these three features are essential for flapping flight. But since they are beneficial also for gliding flight, they cannot be used to decide for certain if *Archaeopteryx* was only a gliding animal or whether it was also capable of active, flapping, flight. It seems unlikely, however, that the flight feathers would have been so strongly curved in a purely gliding animal. The curvature of the flight feathers strongly facilitates their valve function during the upstroke when this is an unloaded, recovery stroke. The curvature is also a prerequisite of the automatic adjustment of angles of incidence at the primaries throughout the downstroke, provided that they do separate. And this they seem to have done, as judged by the following arrangement at the leading edge.

The leading edge of the hand-wing of *Archaeopteryx* is made up of four primary feathers of increasing length. The four feathers do each make up a remarkably similar portion of the length of the leading edge (Figure 10). This is an ideal arrangement for their separation towards the tip to function as leading-edge slats, which are high-lift devices. They are particularly important at, and shortly after, take-off, when the forward speed is still low so that most of the relative air speed at the wings has to be achieved by flapping. So, even though the evidence from the feather structure in *Archaeopteryx* is not entirely conclusive, it does strongly indicate that *Archaeopteryx* was capable of active flight. The striking similarity in feather structure between *Archaeopteryx* and modern birds even makes it likely that *Archaeopteryx* flight feathers had exactly the same function as described in this paper for a modern flying bird. Or, to talk in falsification terms, the structures of wings and flight feathers give no grounds whatsoever for rejecting the hypothesis that *Archaeopteryx* flew with true powered flight.

It should also be evident from this account, and from U. M. NORBERG (1985 a and b), that flight feathers, the wing, and flight in birds must have had a long and advanced evolutionary history before *Archaeopteryx*.

8. Acknowledgments

I thank Professor SIEGFRIED RIETSCHEL for his expert judgements on feather curvature in *Archaeopteryx* (Berlin specimen; Figure 10) and for his information that the feather shaft of *Archaeopteryx* wing feathers has a ventral furrow as in modern birds. I also thank Professor JOCHEN HELMS for providing me with excellent photographs taken by him and BARTHEL of *Archaeopteryx* (Berlin specimen) and for allowing me to publish one in this paper (Figure 10). My thanks also to Dr. GÜNTER VIOHL and his co-editors for their forbearance with me for the delay in submitting the manuscript. Financial support from the Swedish Natural Science Research Council (grant No. B 4450) is gratefully acknowledged.

9. References Cited

DEBEER, G. (1954): *Archaeopteryx lithographica*. A study based upon the British Museum specimen. – 68 pp., 15 pl. London (British Museum).
FEDUCCIA, A. (1980): The age of birds. – 196 pp. Cambridge (Mass.) and London (Harvard University Press).
FEDUCCIA, A. & TORDOFF, H. B. 1979. Feathers of *Archaeopteryx*: Asymmetric vanes indicate aerodynamic function. – Science, **203**: 1021–1022.

Graham, R. R. (1930): Safety devices in wings of birds. – Brit. Birds, **24**: 2–21, 34–47, 58–65.
Helms, J. (1982): Zur Fossilisation der Federn des Urvogels (Berliner Exemplar). – Wiss. Z. Humboldt-Univ., math.-naturwiss. R., **31**: 185–199; Berlin.
Hertel, H. (1966): Structure, form, movement. – New York (Reinhold Publ. Corp.).
Lucas, A. M. & Stettenheim, P. R. (1972): Avian anatomy: Integument, Part 1. – Washington, D. C. (U. S. Govn. Printing Office).
Norberg, R. Å. (1972): The pterostigma of insect wings an inertial regulator of wing pitch. – J. Comp. Physiol., **81**: 9–22.
Norberg, R. Å. (1973): Autorotation, self-stability, and structure of single-winged fruits and seeds (samaras) with comparative remarks on animal flight. – Biol. Rev., **48**: 561–596.
Norberg, R. Å. (1983): Optimal locomotion modes of foraging birds in trees. – Ibis, **125**: 172–180.
Norberg, U. M. (1975): Hovering flight of the Pied Flycatcher *(Ficedula hypoleuca)*. – In: Wu, T. Y.-T., Brokaw, C. J. & Brennen, C. (Eds.): Swimming and flying in nature. Vol. **2**: 869–881. New York (Plenum Press).
Norberg, U. M. (1985a): Evolution of vertebrate flight; An aerodynamic model for the transition from gliding to active flight. – Amer. Nat. **126:** 303–327.
Norberg, U. M. (1985b): Evolution of flight in birds: Aerodynamic, mechanical and ecological aspects. – This volume.
Oehme, H. (1963): Flug und Flügel von Star und Amsel. – Biol. Zentbl., **82**: 413–454, 569–587.
Purslow, P. P. & Vincent, J. F. V. (1978): Mechanical properties of primary feathers from the Pigeon. – J. exp. Biol., **72**: 251–260.
Rietschel, S. (1985): Feathers and wings of *Archaeopteryx*, and the question of her flight ability. – This volume.
Sick, H. (1937): Morphologisch-funktionelle Untersuchungen über die Feinstruktur der Vogelfeder. – J. f. Ornithol., **85**: 206–372.
Withers, P. C. (1981): The aerodynamic performance of the wing in Red-shouldered Hawk *Buteo linearis* and a possible aeroelastic role of wing-tip slots. – Ibis, **123**: 239–247.

Author's address: Dr. R. Åke Norberg, Department of Zoology, University of Göteborg, Box 25059, S-40031 Göteborg, Sweden.

Samuel F. Tarsitano

The Morphological and Aerodynamic Constraints on the Origin of Avian Flight

Abstract

The problem of the origin of the birds and the development of flight has two aspects. The first depends upon the morphological starting point and the second pertains to the constraints aerodynamic considerations have on how flight may evolve. The morphology of *Archaeopteryx* indicates that at this level of avian evolution there was no development of a wing elevator pulley system. According to the works of Brown (1963) and Sy (1936) a prerequisite for the terrestrial origin of flight is the supracoracoideus pulley system. An arboreal proavis is not subject to this constraint since creating an air flow for takeoff is accomplished merely by falling or jumping from an elevated position. Selection for feather evolution can be traced to at least two sources: 1) thermoregulation by virtue of increased surface area and 2) the penetration of the boundary layer by scales.

Zusammenfassung

Das Problem des Ursprungs der Vögel und der Entstehung des Fluges hat zwei Aspekte. Der erste hängt vom morphologischen Ausgangspunkt ab, der zweite betrifft die Zwänge, welche mit aerodynamischen Überlegungen zur möglichen Entstehung des Fluges verbunden sind.

Die Morphologie von *Archaeopteryx* zeigt an, daß es in diesem Stadium der Vogelevolution keine Entwicklung eines Zugsystems für den Flügelaufschlag gab. Nach den Arbeiten von Brown (1963) und Sy (1936) ist das Supracoracoideus-Zugsystem eine Voraussetzung für den terrestrischen Ursprung des Fluges. Ein arboricoler Proavis ist diesem Zwang nicht unterworfen, da der für den Start erforderliche Luftstrom hauptsächlich durch den Fall oder Sprung von einer erhöhten Position erzeugt wird. Eine Selektion für die Federentwicklung kann auf wenigstens zwei Faktoren zurückgeführt werden: 1) Thermoregulation dank vergrößerter Oberfläche und 2) die Durchdringung der Grenzschicht durch Schuppen.

Introduction

There are two aspects to the theory of the origin of the birds. They are: 1) the morphological considerations and 2) the development of flight. The two are inseparable insofar as the morphology provides the primary constraint as to how flight may evolve. Aerodynamic considerations through selection could refine and mold the morphology through time to produce a flying organism. *Archaeopteryx* represents a stage in the evolution of birds and the development of flight, one of the few we have. This Urvogel need not be considered the ancestor of all later birds to be of use in evaluating the development of avian flight. It is possible that *Archaeopteryx* is a holdover from an earlier level of organization in avian history. For many of us *Archeaopteryx* is a stage of avian morphological development and flying capability which may help us determine whether birds evolved from an arboreal habitat or required Jurassic Startbahn.

Aerodynamic Considerations

While all is not known concerning the aerodynamic phenomenon of avian flight there are basic constraints operating on organisms moving through a fluid medium such as air. As a body moves through air its movement relative to the medium creates a resistance the value of which is dependent on its form, size, velocity, the nature of the fluid through which it travels, and the profile the body presents to the on-flowing stream of air (NACHTIGALL, 1977). These parameters are reflected in the Reynolds Number at which the organism operates. The resistances an organism encounters are usually divided up into: 1) parasite drag – which is related to the cross-sectional area of the body and its forward velocity (PENNYCUICK, 1972; NACHTIGALL, 1977; LIGHTHILL, 1974); 2) friction drag – due to the surface architecture (NACHTIGALL, 1977); and 3) profile drag – a resistance produced by the movement of the wings through the air (PENNYCUICK, 1972).

According to NACHTIGALL (1977) an airflow about the body will depart from that body at the place where the cross-sectional area is the greatest. From this observation it is important in the evolution of the flying organism to streamline the body and to place as far caudally as possible the widest and highest portion of the body. To do otherwise invites the early departure of the laminar flow from the body surface. The result of a poorly streamlined body form would be large scale turbulence over the body surfaces thereby increasing the parasite drag (W_{pa}). This phenomenon is indicative of the behavior of a sphere in an airflow (NACHTIGALL, 1977) and has a strong bearing on how an unmodified proavian limb would react to an airflow, whether this flow was generated by running or jumping or falling from an elevated area. An aerodynamically unmodified limb would have a turbulent boundary layer, inducing a departure of laminar flow at an earlier point from the limb's surface than would an aerodynamically shaped limb. Thus the cross-sectional area of a non-feathered and unstreamlined proavian limb would create large value profile drag (W_{pr}) in the opposite direction of the limbs angle of attack (Fig. 1).

Another consideration is the friction drag and its relation to the flow of air in the boundary layer. It is clear that a proavis must have had a scale covering since feathers are derived from a portion of the reptilian scale (HEILMANN, 1926; REGAL, 1965). These scales would automatically penetrate the boundary layer flow and create friction which could lead to widescale turbulence and early departure of the laminar flow from the limb's surface. What this means in terms of the early stages of flight is twofold. First, a non-aerodynamically shaped limb with protruding scales should cause widescale turbulence and therefore drag on the limb. The same can be said for the body shape and body scale covering. The result of this is a dominating resistance (W_{pr}) + (W_{pa}) which would act to slow the animal down. One may then consider how this would affect a proavis starting from a terrestrial or an arboreal setting. The resistance has a strong retarding influence on movement of either a terrestrial or an arboreal proavian archosaur. The resistance tends to keep a terrestrial proavis on the ground whereas it would act to slow the descent of an arboreal proavis. It is clear that in the early stages of proavian evolution this resistance benefits an arboreal proavis that has fallen or lept from a tree but resists the forward movement of a terrestrial proavis and its ability to become airborne. The drag created by an unmodified body and limbs would slow an animal's descent particularly if the animal had a large surface to volume ratio which is indeed true of very small tetrapods (JAGER, pers. comm.). This resistance production is automatically produced merely by having the animal fall from an elevated position and allowing the airflow created essentially by gravity to act on the body surfaces. This coincides with the very first stage of flight, namely parachuting or a steep angled glide, where the animal must create as much drag as possible. The terrestrial animal by running to create an aircurrent must actually expend energy to create the resistance it doesn't want. Thus there is little doubt that the early body and limb form and its scale covering would create large scale turbulence about the body surfaces. This works for an arboreal proavis but works against a terrestrial one. The amount of drag on a terrestrial proavis would perhaps select for the reduction of scales protruding from the body surface (and therefore against the evolution of feathers) whereas selection in the early stages of arboreal flight (parachuting) would be for an ever increasing scale length in order to affect the boundary layer flow, culminating in the evolution of feathers.

The evolutionary aspects of feather formation have been discussed by many (PARKES, 1966; REGAL, 1975; GOULD and VRBA, 1983; etc.). It is their opinion that feathers were formed as thermoregulatiory or insulating devices that were only later modified for flight. This has led to the interpretation of feathers as exaptations or secondary adaptations to flight. None of these authors has considered the effect a protruding protofeather has on the boundary layer flow. NACHTIGALL (1977) has described the effects of unmodified small projections in the boundary layer and the turbulence they produce. Upon entering the boundary layer (which adheres to the body surface), the scales of a proavis would affect the air flow

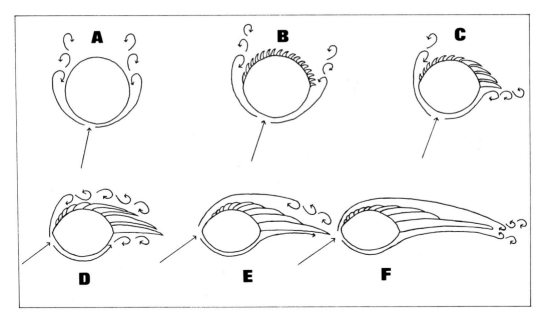

Fig. 1
The hypothetical stages in the evolution of the avian wing in cross section. A. = the pre-proavian limb; B. = the beginning of scale elongation in the proavis during the parachuting stage. C. = further elongation of the scales to increase the component of drag. D. = proto-feather elongation sufficient to allow the limb to act as an airfoil – beginning of gliding phase. E. = further elongation and development of the feathers, beginning of weakly powered flight phase. F. = formation of a well developed feathered wing. Arrows indicate direction of air flow.

about the limbs and body and therefore must be considered to contribute to the aerodynamic picture. However, scales or protofeathers also create more surface area which could act to increase the heat reception from the sun (REGAL, 1975). Thus it is clear that feathers had at least a dual positive selection: for flight and for thermoregulation. The evolution of contour feathers can also be explained initially by boundary layer disturbances (friction drag) and secondarily (in the phylogenetic sense) by a change over to turbulence dampening devices during flight. We can conclude that the concept of exaptation as it relates to the evolution of feathers is incorrect.

Limb girdle morphology and parachuting

Parachuting requires that the proavis be small, lightly built, and able to extend its limbs to present as much surface area as possible to the airflow. The shoulder joint morphology presents no difficulties whether one begins from a crocodylomorph, theropod, or avimorph thecodontian proavis. The pectoral girdle in these forms allows the limbs to be held in a horizontal (or nearly so) position. In contrast, the pelvic girdle has a limited mobility. Theropods have an offset femoral head associated with a hip roller joint which would make it impossible for these dinosaurs to extend the thigh laterally. If there were arboreal theropods (as yet unknown) then they would be unable to perform the required lateral position for the hindlimb. During a fall their hindlimbs would be more vetically placed and thus offer less resistance to the airflow. Crocodylomorphs do not have this problem. According to the data of TEMPLIN (1977) they are, however, much too large and heavily constructed to survive an arboreal fall. In addition, they are not known to have clavicles. While this is negative evidence, it suggests that if crocodylomorphs are closely related to birds then this relationship must be at a stage earlier then the loss of clavicles. Therefore, the supposed branching point must be earlier than the known Sphenosuchia. Furthermore,

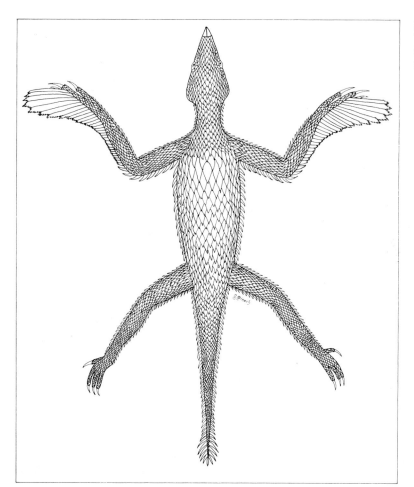

Fig. 2
The arboreal proavis during the parachuting stage.

there are no known arboreal types within the crocodylomorph lineage. Thecodonts have a primitive pelvic girdle with no offset femoral head. Therefore lateral movement of the femur is not only possible but close to the usual position of the femur. A more lateral femoral position also provides more surface area for climbing and holding on to tree trunks and branches. What is clearly required by the arboreal theory is a very small proavis on the order of size of a flying gecko with a high surface to volume ratio. The parachuting stage therefore requires a new lineage of small thecodonts, whose adaptations are for an arboreal life style (Fig. 2).

The change to gliding

The increase in the size of the scale-proto-feather and the flattening of the limb to present as much resistance surface to the oncoming air flow would lead the proavis to the next level of flight, gliding. The transition from parachuting to gliding is easily understood in terms of scale patterning in the necessary aerodynamic positions. In parachuting the air flow will mainly affect the edges of the limb more than the limb's upper surface. Considering the distribution of scales on the limbs of reptiles the elongation where present is directed postero-laterally. Therefore even in nonflying amniotes the scale patterns are preadapted to the morphological requirements of forming a wing. Selection could favor the elongation of the scale along the proto-wing's edges (Fig. 1). When the scale-proto-feathers had elongated sufficiently so they could act collectively as airfoils, the rotation of the limb at the shoulder would present an entirely new morphology to the air flow (Fig. 1). The change-over from parachuting to gliding could be accomplished by manipulating the angle of attack from the resistance-dominating realm to the inertial-dominating realm. This brings the air flow from an above critical angle of attack on the leading edge of the proto-wing to an under-critical air flow due to a lower angle of attack. The proto-wing now using the inertial qualities of the air may now be defined as a wing or airfoil (Fig. 1). It is likely that such a limb morphology would present to the on-flowing air a rounded and slightly bumpy surface (composed of surface scales on the limb's upper surface) with a tapering trailing edge made up of proto-feathers. Selection could then act to refine the upper surface of the wing to lower the REYNOLDS Number at which the limb must operate and therefore postpone the point of departure of the airflow from the wing's surface (GREENWALT, 1962; NACHTIGALL, 1977). Selection would also favor the reduction of sharp leading edges on the wing to reduce turbulence generated by changing the angle of attack above or below an angle which is parallel to the direction of the air flow. However, small disturbances on the leading edge of a wing may actually lead to a beneficial small-scale turbulence within the boundary layer flow (PENNYCUICK, 1972). The small-scale turbulence introduced into the boundary layer flow may actually keep the boundary layer on the wing's surface for a longer period of time. This would act to postpone, in the direction of the trailing edge of the wing, the transition from a mainly laminar flow to large-scale turbulence and flow departure from the wing's surface.

PENNYCUICK (1972) has also provided an aerodynamically sound scenario for the evolution of the avian feather from an archosaurian scale. He points out that the lifting force acting on the proto-feathers must be transmitted to the limb and body to act as a buoyant force. According to PENNYCUICK (1972), it was inevitable that a rachis (or functional strut) formed which could also reduce the weight of the proto-feather. The hollowing out of the rachis would still provide the necessary strength to transmit the forces acting on the wing and reduce the weight of the airfoil as a whole.

The functional reason behind the evolution of the feather vane, ie., the evolution of a system of barbs and hooklets versus a solid feather, is perhaps still open to debate. According to PENNYCUICK (1972), forces generated about the edges of a scale during a glide would tend to bend the scale upward. Notches developing along the scale's edges could reduce such turbulence inducing bends and energy wasting conditions along the wing's edge (vortex formation and an increase in induced drag). These corregations could lead inward toward the rachis to form a modern feather. From what we know of the avian feather a complex network of barbules and barbicels was necessitated as this transition from scale to feather took place (PENNYCUICK, 1972). Another viewpoint put forth by NACHTIGALL (1977) is that feather formation may be related to turbulence dampening along the proto-wing's surface. Small-scale turbulences could be kept from inducing a thickening of the turbulent boundary layer and eventual stalling by making it difficult to suck off the boundary layer from the proto-wing's surface through a slightly porous feather vane. Thus it is possible that a combination of these two evolutionary pathways (PENNYCUICK, 1972; NACHTIGALL, 1977) in addition to the hypothesis of REGAL (1975), provided the basis for the selection of the avian feather.

Gliding and the refinement of the flight apparatus

Gliding can actually lead to the further refinement of the flight morphology of the proavis. The lifting forces on the proavian wings would have to be countered by the downward pull of the pectoralis and stabilized by the supracoracoideus, deltoids etc. Therefore, it is possible that there was a positive selection for increased size of the pectoralis in the pre-supracoracoideus pulley system stage. The ventral elongation of the coracoids would also tend to elongate the pectoralis and supracoracoideus muscle which would increase the excursion of their insertion points (GANS and BOCK, 1965) and increase the amount of available contractile force. Another aspect of coracoid elongation is the lowering of the center of gravity of the thorax making it more difficult for wind forces to overturn the proavis while gliding. The need for holding the limbs outward and upward during gliding would also select for the upward migration of the glenoid fossa.

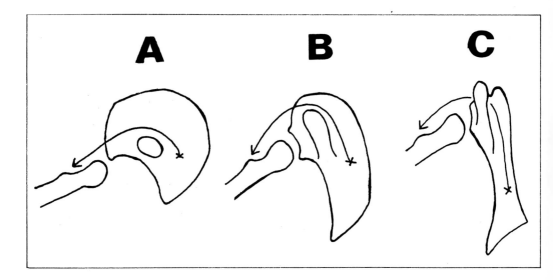

Fig. 3
The evolution of the supracoracoideus pulley system. A. = the coracoid of *Archaeopteryx*; B. = the ventral elongation of the coracoid forming the supracoracoideus pulley system using the biceps tubercle; C. = an example of the modern avian wing elevator pulley system.

A proto-wing which has a greater lift than drag component can be elevated merely by using the force of the air passing over it which is generated by the forward motion of the proavis. If the glenoid-humeral complex has been so modified as to allow vertical movement of the limb then the passive upstroke could set the stage for a greater power downstroke. Thus the evolution of a supracoracoideus pulley system need not have arisen during the early gliding phase, but could have developed gradually as OSTROM (1976a) indicated. The elongation of the coracoid would gradually lower the origin of the M. supracoracoideus, changing its direction of pull from the anterio-medial direction to the dorso-ventral direction. The "biceps tubercle" (WALKER, 1972) could then automatically function as the incipient pulley surface (Fig. 3). This could have been brought about in the following way: 1) primitively the biceps tubercle lies fortuitously below the tendon of the M. supracoracoideus; 2) as the coracoid elongated ventrally, the tendon of the supracoracoideus would be forced to loop over the biceps tubercle and form a pulley system (OSTROM, 1976a); 3) further coracoid elongation and backward migration of the supracoracoideus onto the sternum would further enhance the applied force of this muscle to the humerus. Thus, the primitive archosaurian morphology was preadapted for the formation of the wing elevator system.

Weight considerations

According to PIRLOT (1977) and TEMPLIN (1977) the aspect ratio of the wing versus the body mass can be of use in evaluating the flying capabilities of animals whether they are active flyers or gliders. This is true since an animal's mass must be supported by the lift on the wings. PIRLOT (1977) has shown that large body sizes require alterations in the aspect ratio to compensate for the different airflows that will be encountered or created by the gliding angle or wing beat. The early proavis whether it was terrestrial or arboreal must have had a narrow wing due to the lack of feather development. Although the aspect ratio is not an absolute determining factor, it is clear from figure 5 of PIRLOT (1977) that the proavis must have been very small and lightly built. TEMPLIN (1977) feels that an organism must be able to come back to earth at a speed of less than 6 meters per second to avoid injury. While this speed would certainly vary depending upon the organism, it is an important factor to take into account. The bone width and size of even the smallest coelurosaurs such as *Microvenator* and *Compsognathus* are incongruent with the above stated requirements. The small size required by the arboreal hypothesis necessitates a new group of archosaurs. The discovery of representatives of this group, which I refer to as the Avimorpha (TARSITANO, in press), may have been accomplished by CALZAVARA, MUSCIO and WILD (1981). The upper Triassic *Megalancosaurus* was apparently arboreal with a large manus and claws similar in morphology to those of pterosaurs. WILD (pers. comm.) recognized the importance of this form as a possible link to the group of thecodonts which may have given rise to birds. *Megalancosaurus* though only partially preserved was much smaller and more lightly built than any coelurosaur. It is interesting that CAPLE et al. (1983), who support a terrestrial origin of flight, also require a small archosaur. Interestingly, even the smallest coelurosaur is much too large to fit their model. The heavy bone construction and size of the hindlimbs and pelvis of even the smallest theropods would preclude them from successfully completing an arboreally launched flight. The bone width of *Archaeopteryx* is clearly different from that of the slightly larger Munich specimen of *Compsognathus*. It is evident that *Compsognathus* has a heavily built pelvic and hindlimb morphology which is adapted to a terrestrial environment. In comparable morphological areas it appears that *Compsognathus* is differently proportioned than *Archaeopteryx* and was therefore evolving in quite a different direction. From the preserved aspect ratio of the wing of the Berlin example of *Archaeopteryx* it is clear that its increase in body size and mass in comparison to an early arboreal proavis was tied to the development of the wing. For this reason *Archaeopteryx* is phylogenetically a substantial distance from the early stages of the avian lineage. If one is forced to work with a theropod ancestor it becomes clear from PIRLOT's assessment of the aspect ratio and body mass that coelurosaurs are much too heavy and their forelimbs too short relative to the body size to even have the possibility of lifting off. The arguments used by some that ornithomimids have relatively long forelimbs is of little value since these theropods are much too large to attempt flight and their pectoral girdle (the all important coracoid) is reduced.

Archaeopteryx and the origin of avian flight

While *Archaeopteryx* may not be on the main line of avian evolution it nevertheless could indicate a similar level of organization of a comparable stage leading to the evolution of modern birds. The feather structure and distribution of the primaries and other retrices in *Archaeopteryx* are similar to modern birds (STEPHAN 1979). This has led many (HEILMANN, 1926; GEORGE and BERGER, 1966; YALDEN, 1970; OLSON and FEDUCCIA, 1979; FEDUCCIA and TORDOFF, 1979) to assign aerodynamic function to the feathers of *Archaeopteryx*. The asymmetrical morphology of the feathers of *Archaeopteryx* is consistant with reducing turbulence on the upper surface of the wing (NACHTIGALL, 1977) and cambering the wing profile to create an airfoil. What is interesting is that *Archaeopteryx* does not have a vertically elongated coracoid and therefore no supracoracoideus pulley system. SY (1936) has shown that when the supracoracoideus tendon of pigeons is surgically cut, these birds are unable to take off from level ground as they are deprived of their clap and fling mechanism. They could, however, actively fly if they were thrown from an elevated position (the wing elevation mainly provided by the lift generated by the wing and the deltoids). In a similar manner *Archaeopteryx* could have flown from trees or elevated points. The coracoid of *Archaeopteryx* was not ventrally elongated and therefore no supracoracoideus pulley system was present. Thus the upward movement of its wings was limited by the ventro-laterally directed glenoid. However, *Archaeopteryx* has a seemingly modern feather structure and well developed primaries. According to MAREY (1890), the wing's of ducks are accelerated upward during the upstroke in the course of takeoff and slow flight. The same observations were made by BROWN (1948, 1963) in his studies of pigeon flight. BROWN concluded that "one must accept that the upstroke is an important power phase in slow flight". Since takeoff is an acceleration process it is clear that a terrestrial takeoff also requires an upstroke acceleration – something which *Archaeopteryx* was not biomechanically able to perform. The use of a passive upstroke will create a good deal of profile drag and hamper the efforts of such an aerodynamic system to get into the air from the ground. This is one of the major arguments against the evolution of a terrestrial origin for avian flight. The pectoral girdle morphology of

Archaeopteryx demonstrates that arboreal gliding must have come first and that only after there was a functional supracoracoideus pulley system correlated with a ventrally expanded thorax for greater stability about the body axes could birds begin flight from the ground up. Thus it is likely that *Archaeopteryx* was mainly a glider which could utilize weakly powered flight once airborne gliding was underway. A highly vertical wing beat as would be necessary for a terrestrial takeoff is not required for an arboreal proavis since gliding produces the airflow over the wing, thus reducing the value of the induced velocity necessary for flight (BROWN, 1948, 1963). In a terrestrial takeoff as proposed by CAPLE et al. (1983) the reduction in the required induced velocity is supplied by the legs. As stated above, an arboreal animal need only fall.

According to OLSON and FEDUCCIA (1979), *Archaeopteryx* had the possibility of powered flight since the M. pectoralis has a partial origin on the furcula. It has been pointed out by PENNYCUICK (1972) that in bats the M. pectoralis muscle originates in part along a midventral raphe extending dorsally to the sternum. Likewise *Archaeopteryx* could have had a larger M. pectoralis than has previously been assumed by using a midventral raphe connected to a cartilaginous sternum. The stout furcula could have functioned to resist the compressive forces of the wing's downstroke and in the early stages of flight, to resist the lift forces generated by the wings during gliding. The thoracic vertebrae of *Archaeopteryx* are unfused and there is little or no development of a synsacrum. These are characteristics of gliding animals (CAPLE et al., 1983). *Archaeopteryx* has a ventro-laterally directed glenoid (TARSITANO and HECHT, 1980) which should limit the upward mobility of the wing. The head of the humerus is well rounded and large which may have enabled *Archaeopteryx* to raise its wing above the level of the glenoid only by some small angle.

Digital homologies

The manus of *Archaeopteryx* still retains three digits which are similar in morphology to the first three topographic digits of some coelurosaurs and pterosaurs. There has been a growing debate as to the correct numbering of the digits of *Archaeopteryx* and birds. Paleontologists generally consider the digits to be I-II-III for theropods and *Archaeopteryx* (HEILMANN, 1926; ROMER 1966; OSTROM, 1976b) whereas modern embryologists (HINCHLIFFE, 1977) believe the digits of birds and thus *Archaeopteryx* are represented by digits II-III-IV (HINCHLIFFE, this volume; HINCHLIFFE and HECHT 1984). The similarity of the carpus and digital proportions in coelurosaurs and *Archaeopteryx* (HEILMANN, 1926; OSTROM, 1976a, b), argues strongly for the same numbering system in both groups. This view is supported by the apparent reduction pattern in theropods and prosauropods (HEILMANN, 1926; ROMER, 1966; OSTROM, 1976b; TAQUET, 1977). Furthermore, in crocodilians the size of the phalanges and metacarpals in digits 4 and 5 are reduced relative to the first three digits. The reduction pattern is a stepwise reduction from the postaxial side. If the embryological interpretation is correct then the similarity between the theropod and avian manus is the result of homoplasy. If the paleontologists are correct then this similarity indicates that theropods and *Archaeopteryx* (and perhaps birds) shared a distant common ancestry.

At the present time both hypotheses are plausible but the morphological resemblance between the theropod and *Archaeopteryx* manus indicates that their digits represent the first three topographic digits of the pentadactyl plan. Of interest is the fact that the first metacarpal in *Megalancosaurus* (WILD, pers. comm.) is shorter than its proximal phalanx. This morphology had been used (PADIAN, 1982) as a key character between *Archaeopteryx* and theropods. Such a synapomorphy is now completely invalid because it is primitive at the thecodont level of the Avimorpha. Furthermore, in many pterosaurs the first three topographic digits bear a resemblance to the theropodan plan (TARSITANO and HECHT, 1980). This condition can be interpreted as homoplasy but it reflects the archosaurian manner of forming grasping digits. The similar formation of the first three digits in these groups may represent an underlying genetic program for a similar grasping manus. Considering the morphological differences between theropods and birds (WHETSTONE and MARTIN, 1979; TARSITANO and HECHT, 1980; MARTIN et al., 1979), the similarity of the manus need not indicate an ancestor-descendant relationship as proposed by OSTROM (1976a, b). What the manus of *Archaeopteryx* demonstrates is that a bipedal arboreal landing was not yet necessary

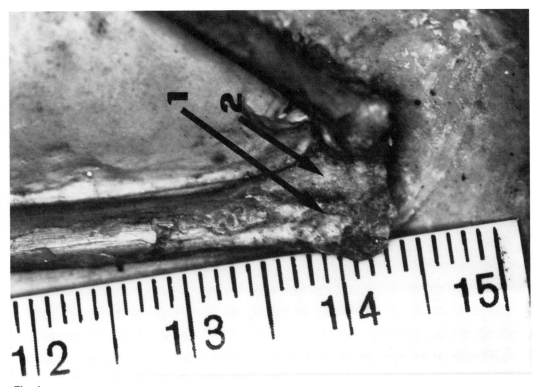

Fig. 4
The left tibiotarsus of the London example of *Archaeopteryx* demonstrating that the appearences of an ascending process of the astragalus is due to the breakage of the distal end of this bone. Arrow one indicates the area of breakage and arrow two indicates the supposed ascending process.

for *Archaeopteryx*. The manus of *Archaeopteryx* is sufficiently developed for use in a four point landing or for climbing.

The tibiotarsus

The tarsus of *Archaeopteryx* consists of fused proximal tarsals to form the distal tibial condyles as seen in the right tibiotarsus of the London specimen. There have been reports of separate proximal elements (OSTROM, 1976b; WELLNHOFER, 1974), but none as such exist (TARSITANO and HECHT, 1980) in the material under consideration. The Berlin specimen purportedly shows a right and left astragalus which have been interpreted as having ascending processes. These ascending processes are not mirror images of each other and are best interpreted as breakage of the tibial walls due to calcite precipitation within the bones. This is true for the left tibiotarsus in the London specimen which reportedly has a well developed astragalar process (Fig. 4). McGOWEN (1984) has studied the development of the tarsus in modern birds. He found that there are three tarsal conditions. According to McGOWEN, Tinamous have both an astragalar process and a pretibial bone, a process of the calcaneum. Other ratites have only an astragalar process whereas carinates have only a pretibial bone. *Archaeopteryx* as presently known either lacks any tarsal processes or has a fused condition whereby the processes would be completely masked. If *Archaeopteryx* lacks tarsal processes then the astragalar and pretibial processes in living birds could be interpreted as neomorphs. The Tinamous are thus interesting because they have both processes present in their development. There are perhaps then two questions to be answered. They are: 1) do Tinamous represent the primitive condition for birds? 2) have other ratites merely lost the pretibial bone or have they modified the pretibial bone to connect to the astragalus? The ascending process of the astragalus found in theropods must yet be found in *Archaeopteryx*. There is still the possibility of functional homoplasy in the development of various locking devices between the proximal tarsals and the tibia in order to resist the torques generated by a bipedal gait.

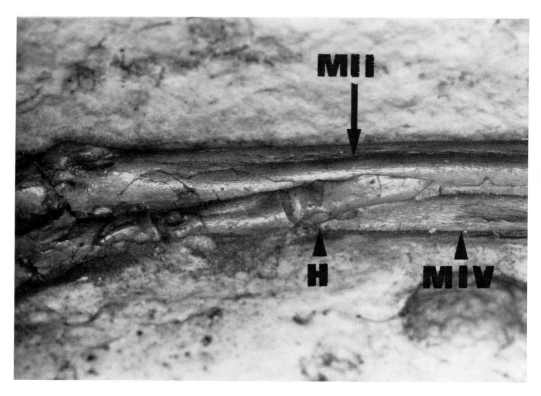

Fig. 5
The preserved position of the hallux in the Munich specimen of *Compsognathus*. Note the reduced morphology of the hallux, its high placement on metatarsal II and its unnatural association with metatarsal IV. H = hallux; M II = metatarsal II; M IV = metatarsal IV.

The pes

Many authors such as HEILMANN (1926) and WELLNHOFER (1974) have noted that *Archaeopteryx* has a reflexed hallux and have associated this character with an arboreal habitat. The morphology of its hallux is consistent with this interpretation. The hallux ungual of *Archaeopteryx* is nearly as large as that of the middle digit. In addition, metatarsal I is placed near the end of metatarsal II, allowing opposability with the other digits. When one compares the hallux of *Archaeopteryx* with those of theropods it becomes apparent that the theropod condition is distinctly a reduction character (TARSITANO, 1983) . This is evident for two reasons. First, the ungual of the theropod hallux is reduced in comparison with those of the other pes digits. Second, metatarsal I is placed nearly at the midpoint of metatarsal II precluding its opposability to the other digits (Fig. 5). If one views the morphology of the theropod hallux in terms of reducing the inertia of the limb it is clear that this reduction is an adaptation for cursorial locomotion. The theropod pes that were found in articulation supports this interpretation. The pes of *Ceolophysis*, *Saurornithoides* and *Velociraptor* were found in articulation with a hallux parallel to digit II. WHYBROW (pers. comm.) also came to the same conclusion for the hallux of *Deinonychus* during its preparation. *Compsognathus* also has a well preserved pes with a parallel hallux. There has, however, been some dispute as to the natural position of the hallux in this form. Metatarsal I does not line up with its phalanges, which are offset approximately ninety degrees from metatarsal I. In the Munich specimen metatarsal I abutts against metatarsal IV – a most unnatural condition (Fig. 5). The displacement of metatarsal IV has flattened and distorted metatarsal I. Therefore it is likely that the position of metatarsal I has been altered by metatarsal IV. In contrast, the phalanges of metatarsal I do not contact metatarsal IV and thus have been unaffected by the displacement of metatarsal IV against metatarsal I. In conclusion the position of the hallux phalanges are more reliable than that of metatarsal I in this specimen of *Compsognathus*. In all aspects of pes morphology theropods are adapted to a terrestrial environment and do not have a reflexed hallux.

Conclusion

Long ago HEILMANN (1926) hypothesized the steps taken by a small archosaur toward the evolution of flight. He recognized that the archosaur must have been small and although beginning as a terrestrial form soon adopted an arboreal habitat. Leaping from branch to branch and then between trees the proavis would develop over the course of its evolution the morphology and neural development necessary for flight. The one flaw in HEILMANN's scenario was that the terrestrial ancestor of the proavis was already a biped. Based on the studies of BRINKMAN (1981) and TARSITANO (1983) pseudosuchian thecodonts were not true bipeds as HEILMANN had supposed. From the data presented in this paper it is clear that even pseudosuchians (sensu TARSITANO, 1983) are too large and specialized to be a direct ancestor of birds. The same is true of theropods, crocodilians and paracrocodylians.

If the first stage of flight was indeed parachuting or steep-angled gliding then the hindlimbs by the necessity to create more drag, must be held close to a horizontal attitude. Thus, the early members of the avian lineage must have been associated with quadrupedal forms early in the arboreal stage. From studies of boundary layer phenomenon a terrestrial proavis would reduce the scalation (to reduce friction drag) and emphasize the development of the hindlimb as opposed to the forelimb. This should be true since it is the hindlimbs of a terrestrial proavis which must produce the induced velocity by running and leaping. It is easily seen that the development of the hindlimbs would dominate over the forelimbs which at best would be used for grasping in a terrestrial proavis. This may be the reason why theropods have reduced coracoids. Therefore, in a terrestrial proavis the positive selection would be on the wrong limb and girdle for the evolution of flight. In conclusion, the terrestrial scenario suffers on both anatomical and aerodynamic grounds.

Selection for an arboreal proavis would first favor the development of proto-feathers to create drag. When their elongation was sufficient, they could then act as airfoils by rotating (changing the angle of attack) the limb. If an arboreal proavis merely falls, thus creating an air flow, then selection would favor the elongation of scales along both its limbs, its tail and along the contours of its body. This then was the first stage in the development of feathers and of flight. If an animal leaps, however, from a branch perhaps to avoid becoming the main course of another's dinner then it is the forelimbs which first encounter the airflow (aside from the head) and thus scale elongation should proceed to a greater degree on the forelimb than on the hindlimb. BROWN (1963) has described the various means of correcting pitch, roll and yaw using the forelimbs as the main source of lift. Thus, hindlimb airfoils would perhaps cause problems of rotation about the body axes and it seems clear why hindlimb wings were not evolved. Another aspect germane to the problem of which limb would be developed for flight is the mobility of the manus versus the pes. A proavian archosaur could spread its metacarpals and digits much more than is anatomically possible for the pes. Therefore it was inevitable that wing formation proceeded with the forelimb rather than with the hindlimb.

As HEILMANN (1926) and BOCK (1965) hypothesized the stages of flight would proceed gradually from parachuting to gliding as the proto-feathers elongated (Fig. 6). The act of gliding would be sufficient to ensure the enhancement of the neural capabilities of muscular coordination and eyesight. Gliding will positively select for: 1) feather development; 2) enlarged pectoralis muscles; 3) deeper thorax forming a supracoracoideus pulley system and 4) fusion of the clavicles and interclavicle to form a furcula (this would act to stabilize the scapula and coracoid against the lifting forces generated on the limbs and consequently at the shoulder joint). The development of an ossified sternum and its coracoid connections in actively flying birds has probably usurped the above stated primal function of the furcula. The furcula may now function to reduce the compressive force exerted by the downstroke and as OLSON and FEDUCCIA (1979) and GEORGE and BERGER (1966) point out, as an attachment site for the pectoralis muscle.

The aerodynamic analysis presented above weighs heavily in favor of an arboreal origin of flight. In each stage: parachuting, gliding, weakly powered flight, and powered flight the morphology of the scales, girdles and limbs are congruent and adaptive. There is no need for macroevolutionary events. In contrast, the terrestrial theory of flight (OSTROM, 1976a, b; CAPLE et al., 1983) requires a jump between

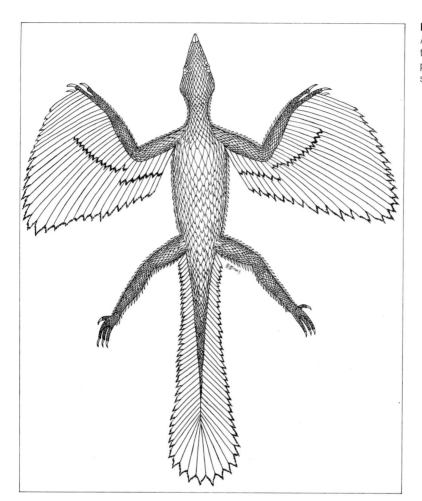

Fig. 6
A diagrammatic representation of the advanced proavis during the gliding stage.

leaping into the air and active flight. Such a step requires a major overhaul of the integument, pulley systems, girdle morphology, muscle bulk and physiology (GEORGE and BERGER, 1966), neural control and sensing systems, pneumatic system, and the balance system of bipedality. For lifting off the ground all of the above systems must function harmoniously and thus must be evolved simultaneously. Thus it is very unlikely that tetrapods could originate flight from the ground up (TUCKER 1938). The running, jumping, and lifting off advocated by CAPLE et al. (1983), PADIAN (1982) and LEWIN (1983) would most certainly be followed by crashing.

The morphological considerations point to a phylogenetic position earlier than the common ancestor of the Crocodylomorpha, Theropoda, and Pseudosuchia. An examination of the cranium of crocodilians and birds demonstrates that there are: 1) an intertympanic sinus, the floor of which is partly made up by an epiotic bone, housing much of the semicircular canals; 2) a basicranial sinus within the basisphenoid and basioccipital which communicates with the tympanic cavity and the pharynx (MULLER, 1963) and often misnamed the "median eustachian tube" (TARSITANO, in press); 3) a lateral head vein-stapedial artery anastomosis above the tympanic cavity; 4) fenestra rotundum lying below and at nearly right angles to the fenestra ovalis; and 5) definitive eustachian tubes.

Character 1 is at least incipiently present in the Crocodylomorpha. It is not yet known in theropods or thecodonts. In this regard the crania of *Megalancosaurus* and *Longisquama* are not preserved in such a way as to permit the identification of this character. A basicranial sinus system (character 2), extending from the pituitary fossa to the pharynx is known in theropods (OSBORN, 1906; TARSITANO, in press) and crocodylomorphs, but has not been studied in thecodonts. Character 3, the lateral head vein – stapedial artery anastomosis, is known in birds (SAIFF, 1979) and in eusuchian crocodilians (TARSITANO, in press). Its presence in fossils would be extremely difficult to determine. Character 4, the presence of a fenestra rotundum, is interesting but has been incompletely studied. According to MARTIN (1983), theropod dinosaurs do not have a fenestra rotundum despite claims to the contrary. Much of what has been called the fenestra rotundum in theropods and thecodonts is probably not this structure. In theropods the structure called the fenestrae rotundum has not been demonstrated to have the possibility of communicating with the fenestra ovalis. The thecodontian otic capsule is not well known since much of the otic capsule walls were apparently not ossified. Character 5, distinct tubular eustachian tubes, is found in birds and the Crocodilia. The morphology of the true lateral eustachian tubes in these two taxa are not similar (TARSITANO, in press) and probably are not homologues. Eustachian tubes have not as yet been identified in theropods, thecodonts, or paracrocodilians. There is need of further investigation of these characters in all archosaurian taxa. It still remains that there is no intertympanic sinus or foramen aereum in the articular bone in theropods. If one combines this fact with the lack of clavicles and small size in paracrocodylians it becomes clear that the ancestry of birds must lie at a level earlier than the Paracrocodilia and therefore within the Thecodontia. I have therefore come to a similar conclusion as that of HEILMANN (1926) who illustrated the type of proavis which probably existed in the Triassic. The groups which have a small mosaic of seemingly similar characters, such as the theropods, Crocodilia and Paracrocodylia are too specialized and large to be ancestors of birds. This conclusion is consistent with the known archosaurian morphologies and aerodynamic principles.

Acknowledgements

I wish to thank the National Science Foundation for Grant No. BSR 8307345 for my support as a Research Associate. I also wish to thank the following people for friendly discussions and for making available material in their care for study: A. CHARIG, M. CLUVER, A. CROMPTON, A. ELZANOWSKI, E. FREY, E. GAFNEY, C. GOW, M. HECHT, K. JAGER, J. KITCHING, W. NACHTIGALL, J. OSTROM, M. RAATH, W. REIF, J. RIESS, B. RUBIGE, B. STEPHAN, G. VIOHL, C. WALKER, P. WELLNHOFER, F. WESTPHAL and R. WILD. I thank J. WINSCH for his skillful preparation of the figures and J. BENSKO for valuable aid in editing this manuscript.

References Cited

BOCK, W. J. (1965): The role of adaptive mechanisms in the origin of higher levels of organization. – Syst. Zool. **14**: 272–287.
BRINKMAN, D. (1981): The hindlimb step cycle of *Caiman sclerops* and the mechanism of the crocodilian tarsus and metatarsus. – Can. J. Zool., **58**: 2187–2200.
BROWN, R. H. J. (1948): The flight of birds. I. The flapping cycle of the pigeon. – J. Exp. Biol., **25**: 322–333.
BROWN, R. H. J. (1963): The flight of birds. II. Wing function in relation to flight speed. – J. Exp. Biol., **30**: 90–103.
CALZAVARA, M., MUSCIO, G. & WILD, R. (1981): *Megalancosaurus preonensis*, n. g., n. sp., a new reptile from the Norian of Friuli, Italy. – Atti Mus. Friul. Storia. Nat., **2**: 49–64.
CAPLE, G. R., BALDA, R. P. and WILLIS, W. R. (1983): The physics of leaping reptiles and the evolution of pre-flight. – Amer. Nat. **121** (4): 455–476.
FEDUCCIA, A. & TORDOFF, H. B. (1979): Feathers of *Archaeopteryx*: asymmetric vanes indicate aerodynamic function. – Science, **203**: 1021.
GANS, C. & BOCK, W. (1965): The functional significance of muscle architecture: A theoretical analysis. Ergeb. Anat. Entwicklungsges., **38**: 115–142.
GEORGE, J. C. & BERGER, A. J. (1966): Avian Myology. – 500 pp., New York & London: (Academic Press).
GOULD, S. J. & VRBA, E. S. (1982): Exaptation – a missing term in the science of form. – Paleobiology, **8**: 4–15.
GREENWALT, C. H. (1962): Dimensionless relationships for flying animals. – Smithsonian. Misc. Coll., **144** (2): 1–46.
HEILMANN, G. (1926): Origin of the Birds. – 210 pp. London (Wittherby).
HINCHLIFFE, J. R. (1977): The chondrogenic patterns in chick limb morphogenesis. In: D. A. EDE, J. R. HINCHLIFFE & M. BALLS (Eds.), Vetebrate Limb and Somite Morphogenesis: 293–308. – Cambridge (University Press).

HINCHLIFFE, J. F. & HECHT, M. K. (1984): Homology of the Bird Wing Skeleton: Embryological versus Paleontological Evidence. – Evol. Biol., **18**: 21–39.
LEWIN, R. (1983): How vertebrates take to the air? – Science, **221**: 38–39.
LIGHTHILL, M. J. (1974): Aerodynamic aspects of animal flight. – Bull. Inst. Maths. Appls., **10**: 369–393.
MAREY, E. J. (1890): Vol des Oiseaux. – 249 pp. Paris.
MARTIN, L. D., STEWARD, J. D. & WHETSONTE, K. N. (1980): The origin of birds: structure of the tarsus and teeth. – The Auk, **97**: 86–93.
MARTIN, L. D. (1983): The origin of birds and avian flight. – Curr. Ornithol., **1**: 105–129.
McGOWAN, C. (1984): Evolutionary relationships of ratites and carinates: evidence from the ontogeny of the tarsus. – Nature (**307**) **5953**: 733–735.
MULLER, H. J. (1963): Die Morphologie und Entwicklung des Craniums von *Rhea americana* Linné, Part II. Visceralskelett, Mittelohr und Osteocranium. – Zeit. Wiss. Zool., **168**: 40–118.
NACHTIGALL, W. (1977): Zur Bedeutung der Reynoldszahl in der Schwimmphysiologie und Flugbiophysik. – Physiology of movement-biophysics. – Fortschr. Zool., **24** (2/3): 13–56.
OLSON, S. L. & FEDDUCIA, A. (1979): Flight capability and the pectoral girdle of *Archaeopteryx*. – Nature (London) **278**: 247–248.
OSBORN, H. F. (1906): *Tyrannosaurus*, Upper Cretaceous carnivorous dinosaur (Second communication). – Bull. Amer. Mus. of nat. Hist., **22**: 281–296.
OSTROM, J. H. (1976a): Some hypothetical anatomical stages in the evolution of avian flight. – Smiths. Contr. Paleobiol., **27**: 1–21.
OSTROM, J. H. (1976b): *Archaeopteryx* and the origin of birds. – Biol. J. Linn. Soc., **8**: 91–182.
PADIAN, K. (1982): Macroevolution and the origin of major adaptions: vertebrate flight as a paradigm for the analysis of patterns. – Proc. Third North Amer. Paleo Conv., **2**: 387–392.
PARKES, K. C. (1966): Speculations of the origin of feathers. – The Living Bird, **V**: 77–86.
PENNYCUICK, C. J. (1972): Animal Flight. – Studies in Biology no. **33**. Southampton, The Camelot Press Ltd., 68 pp.
PIRLOT, P. (1977): Wing design and the origin of bats. In: M. K. HECHT, P. C. GOODY & B. M. HECHT (Eds). Major patterns in vertebrate evolution, NATO A.S.I. **14**: 375–426. – New York and London (Plenum Press).
REGAL, P. J. (1975): The evolutionary origin of feathers. – Quart. Rev. Biol., **50**: 35–65.
ROMER, A. S. (1966): Osteology of the Reptiles. – 987 pp. Chicago (Universtiy of Chicago Press).
SAIFF, E. I. (1978): The middle ear of the skull of birds: the Pelecaniformes and Ciconiformes. – Zool. J. of Linn. Soc., **63**: 315–370.
SAVILE, D. B. O. (1957): Adaptive evolution in the avian wing. – Evol., **11**: 212–224.
STEPHAN, B. (1979): Urvogel. – 196 pp. Wittenburg (A. Ziemsen Verlag).
SY, M. (1936): Funktionell-anatomische Untersuchungen am Vogelflügel. – Jour. Ornith., **84**: 199–296.
TAQUET, P. (1977): Variation ou rudimentation du membre anterieur chez les Theropodes (Dinosauria)? In: Mécanismes de la rudimentation des organes chez les embryons de vertébrés, pp. 333–339, Colloques Internationaux CNRS, No. **266**, Paris.
TARSITANO, S. & HECHT, M. K. (1980): A reconsideration of the reptilian relationships of *Archaeopteryx*. – Zool. J. Linn. Soc. Lond., **69**: 149–182.
TARSITANO, S. F. (1983): Stance and gait in theropod dinosaurs. – Acta Palaeont. Polonica., **28**: 251–264.
TARSITANO, S. F. (1985): Cranial metamorphosis and the origin of the Eusuchia. – N. Jb. Geol. Palaeont. Abh. **170**, 1: 27–44.
TEMPLIN, R. J. (1977): Size limits in flying animals. In: M. K. HECHT, P. GOODY & B. HECHT (Eds) Major patterns in vertebrate evolution, NATO A.S.I. **14**: 411–423.
TUCKER, B. W. (1938): Functional evolutionary morphology: The origin of birds. In: Sir G. DE BEER (Ed) Evolution: 321–336. Oxford (Clarendon Press).
WALKER, A. D. (1972): New light on the origin of birds and crocodiles. – Nature, London, **237**: 257–263.
WALKER, A. D. (1982): The pelvis of *Archaeopteryx*. – Geol. Mag., **117**: 595–600.
WELLNHOFER, P. (1974): Das fünfte Skelettexemplar von *Archaeopteryx*. – Palaeontogr. Abt. A. **147**: 169–216.
WHETSTONE, K. N. & MARTIN, L. D. (1979): New look at the the origin of birds and crocodiles. (Nature, London, **279**: 234–236.
YALDEN, D. W. (1970): The flying ability of *Archaeopteryx*. – Ibis **113**: 349–356.

Author's address: Dr. SAMUEL F. TARSITANO, Queens College of the City University of New York, Dept. of Biology, 65–30 Kissena Boulevard, Flushing, New York 11367–0904, USA.

Dietrich Schaller

Wing Evolution

Abstract

The different known wing types are compared with each other and classified with respect to their ecological background that offers the selective stimulus towards evolution of flight.

All extant forms can thus easily be classified. It thereby becomes obvious, that there is no need to look for new parameters in order to understand the extinct forms.

Zusammenfassung

Die unterschiedlichen bekannten Flügeltypen werden miteinander verglichen und vor dem ökologischen Hintergrund beleuchtet, der den selektiven Anreiz zur Flügelwerdung gibt.

Alle rezenten Formen können dadurch sehr einfach klassifiziert werden. Es zeigt sich deutlich, daß keine neuen Parameter gesucht werden müssen, um auch die fossilen Formen zu verstehen.

In order to comprehend an animal wing in full, all known extant and extinct volant groups are investigated and grouped according to topology, function, and evolutionary patterns. The interactions between these forms and their determining surroundings are discussed with respect to flight and are then categorized. Some aspects can be demonstrated:

– Gliding-wing evolution occurs quite frequently and is accomplished by relatively simple means.
– Flapping-wing evolution in contrast occurs quite rarely. With but one exception all flapping-wings had a gliding-wing as precursor.

In systematic comparisons some principal bauplans can be derived enabling a better understanding of the multitude of wings. It can be shown that some bauplans occur analogously while others represent isolated phenomena. In the overall view, the single lines can very well be judged against their physical background conditions. It becomes obvious that only two realms allow the evolution of wings

– the obstruction-free water surface
– the association of trees – the forest

Three principle requirements have to be met in combination

– the environment must supply food
– the specific mode of locomotion must always be possible
– the flight must always end in the home biotope

The obstruction-free water surface is the starting point of wing evolution for water-breathers. They can only dive into the airspace, comparable to air-breathers diving into water. Water-breathing volants dive into the airspace in order to flee, not to feed. The energy for the gliding-flight is obtained by swimming. A gliding-flight fish, for instance, lives close beneath the water surface and finds its food there. It can jump out of the water in any direction and flee gliding, but as a water-breather it must always land in its home biotope – water. A repeated start is only possible from the water surface.

The association of trees – the forest – is the starting point of wing evolution for air-breathers, which therefore are permanent residents in the medium air. Air-breathing gliders travel through the obstruction-free airspace either to get to food or to flee. The energy for the gliding-flight is obtained by climbing. A gliding squirrel for instance lives in the association of trees and finds its food there. It can climb to an elevated point from where it can launch a glide, but it must reach its home biotope, tree, again for security, food, and because a repeated start is only possible from a raised point.

The path from tree to tree within the home biotope leads over the contact of leaves, and within the alien biotope over the diverse, obstructed ground. A third path leads over the uniform obstruction-free airway. It can be bridged

- over short distances via jumping
- over greater distances via gliding
- over great distances via flying

Quite a few animals of the arboreal realm are capable of jumping, far less are able to glide, and only pterosaurs, bats, birds and insects can bridge airspace via flight. The airpath is only possible when body-structures have evolved especially for this mode of locomotion. Principally they represent aerofoils which in the first phase have the task of converting "external" falling-energy into horizontal progression and in the second phase the task of transforming "internal" muscular energy by power-generators into motion.

It is apparent that gliding is a comparatively simple mode of locomotion and is very efficient in its arboreal realm, but is also restricted to it due to the necessity of launching the glide. Flying on the other hand liberates from such constraints, but necessitates evolutionary complicated design alterations.

Terrestrial and aerial locomotion is accomplished by very different locomotive systems.

- The locomotion of an animal across a 2-dimensional surface is by a walking-device, the limbs or the walking-apparatus.
- The locomotion of an animal through the 3-dimensional airspace is by a flying-device, the wings or the flying-apparatus.
- The locomotion of some animals in water is by a sculling-caudal – a thrust-generator of specific importance for water-breathing gliders.

The principal problems at the beginning of flight evolution are the generation of lift and the control of space-orientation. Therefore any set of wings must be able to produce lift and to control roll, pitch and yaw. Devices counteracting gravity are lift-generators and those counteracting drag are thrust-generators. A gliding-wing is a lift-generator and a flapping-wing is a lift- and a thrust-generator combined.

The two basic modes of locomotion through airspace therefore can be categorized according to their means of accomplishment.

Gliding-wings
- lift-generator independent of thrust-generator
  lift-generator:
  water- and air-breathers employ gliding-wings to transform "external" energy into lift (flying-apparatus of gliders)
  thrust-generator:
  water-breathers obtain their "external" energy by swimming (sculling-caudal of thrust-gliders)
  air-breathers obtain their "external" energy by climbing (walking-apparatus of gravity-gliders)

Flapping-wings
- lift-generator combined with thrust-generator

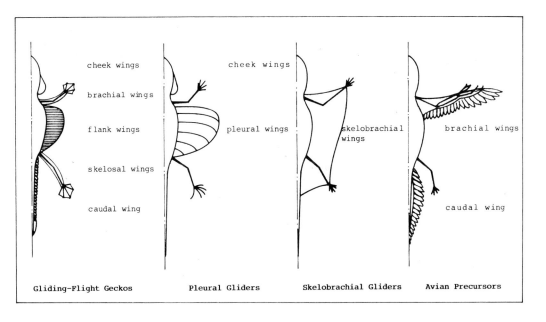

Fig. 1
Evolution as conversion utilizing preadaptive structures of the tetrapod bauplan for the wing evolution.

power-generator:
water and air-breathers employ flapping-wings to transform "internal" energy into lift and thrust (flying-apparatus of flappers)

Wing types

The following wing types are distinguished (see Fig. 1):

Jointless wings
chitin-membrane wings
(self-supported aerofoils; unstowable when not in use)
– chitin gliding-wings   (Gliding-Flight Insects)

muscle-membrane wings
(muscle-supported aerofoils; flapped away when not in use)
– cheek-wings   (Ptychozoon, Draco)
– flank-wings   (Ptychozoon)
– ear-wings   (Gliding Mammals, Bats)

rib-membrane wings
(brace-supported aerofoils; swung away when not in use)
– pleural wings   (Draco, Icarosaurus)

Jointed wings
single-jointed gliding-wings
(self-supported aerofoils; swung away when not in use)
– pectoral-pelvic gliding-wings   (Exocoetidae; Thoracopteridae)

single-jointed flapping-wings
(self-supported aerofoils; swung away when not in use)
- pectoral flapping-wings (Gasteropelecidae)
- chitin flapping-wings (Flapping-Flight Insects)

Limb wings
skelobrachial gliding-wings
double-mast, 2-membered armleg-wings
(brace-supported aerofoils; foldable when not in use)
flying-apparatus combined with walking-apparatus
(Pterosaur Precursors,
Chiropteran Precursors,
Podopteryx, Gliding Mammals)

skelobrachial flapping-wings
double-mast, 3-membered armleg-wings
(brace-supported aerofoils; foldable when not in use)
flying-apparatus combined with walking-apparatus
(Archaetype Pterosaurs, Bats)

brachial gliding-wings
single-mast, 3-membered arm-wings
(muscle-, brace-, self-supported aerofoils; foldable when not in use)
flying-apparatus either used as, or independent of walking-apparatus
(*Rhacophorus, Ptychozoon*,
Avian Precursors)

brachial flapping-wings
single-mast, 3-membered arm-wings
(self-supported aerofoils; foldable when not in use)
flying-apparatus independent of walking-apparatus
(Derived Pterosaurs, Birds)

skelosal wings
single-mast, 3-membered leg-wings
(brace-supported aerofoils; foldable when not in use)
flying-apparatus used as walking-apparatus
(*Rhacophorus, Ptychozoon*,
Primary Derived Pterosaurs)

Unpaired wings
lift- and steering-wings
(self-supported aerofoils; unstowable when not in use)
- caudal-wings (*Ptychozoon, Hylopetes*, Birds)

The control of space-orientation is facilitated by:

Momentum-rudders
Highly agile bipedal animals require a sophisticated terrestrial space-controlling device. Bipedal dinosaurs and kangaroos developed an unpaired momentum-rudder – the tail; man utilizes a paired momentum-rudder – the arms.

Drag-rudders
Some gliders utilize a drag-rudder to orient their body-axis towards the aerial pathway-axis. *Anomalurus* and *Petauroides* have an unpaired drag-rudder – the cylindrical tail.

Flow-rudders

Other gliders employ a flow-rudder as an aerial steering-device and unpaired lift-wing combined. *Ptychozoon, Hylopetes* and birds have an unpaired flow-rudder – the flat caudal-wing.

Wing evolution reflects basic design principles which either offer possibilities or impose limitations. On the one hand the evolutionary course usually proceeds along the most simple routes by utilizing preadapted structures. Bats for example use their arms and legs as bracers and their skin as an aerofoil, following the most frequently repeated bauplan. Insects in contrast represent an isolated phenomenon. Their wing is a new development altogether. On the other hand the basic ecological demand has to be met that the "apparatus not in use" should be packed away as well as possible: The wing should not impede the foot while walking and the foot not impede the wing while flying. A wide range of solutions can be observed, some very promising and others quite restricting. At one end of the scale the skelobrachial gliders and skelobrachial flappers have their walking-apparatus fully combined with their flying-apparatus. Intermediates like gliding-flight frogs and gliding-flight geckos use their walking-apparatus as a flying-apparatus as well. At the other end of the scale the – "Pegasus-like" – insects, pleural gliders, brachial gliders and brachial flappers have their walking-apparatus totally independent of their flying-apparatus.

A checklist of all known extant and extinct volant forms is compiled:

Water-breathers

Gliding-Flight Fishes
- Thoracopteridae (Triassic)
- Exocoetidae (Cretaceous-Extant)

Flapping-Flight Fishes
- Gasteropelecidae (Extant)

Air-breathers

Gliding-Flight Frogs
- *Rhacophorus* (Extant)

Gliding-Flight Geckos
- *Ptychozoon* (Extant)

Pleural Gliders
- *Draco* (Extant)
- *Kuehneosaurus* (Triassic)
- *Icarosaurus* (Triassic)
- *Daedalosaurus* (Permian)
- *Weigeltisaurus* (Permian)

Skelobrachial Gliders
Skelobrachial gliding lizards
- *Podopteryx* (Triassic)

Skelobrachial gliding mammals (Extant)
- *Acrobates* : feather-tail gliding possums
- *Petaurus* : squirrel gliding possums
- *Petauroides* : greater gliding possums

- Anomalurinae   : scaly-tail gliders
- Petauristinae   : gliding squirrels
- Dermoptera   : gliding lemurs

Postulated skelobrachial gliders
- Pterosaur Precursors   (Triassic)
- Chiropteran Precursors   (Tertiary)

Skelobrachial Flappers
- Archaetype Pterosaurs   (Triassic-Jurassic)
- Chiroptera   (Tertiary-Extant)

Brachial Gliders
Postulated brachial gliders
- Avian Precursors   (Jurassic)

Brachial Flappers
- Derived Pterosaurs   (Triassic-Cretaceous)
- Aves   (Jurassic-Extant)

Insects
- Gliding-Flight Insects   (Devonian-Carboniferous)
- Flapping-Flight Insects   (Carboniferous-Extant)

Only volant forms with aerofoils are listed; gliding-flight snakes, ballistic-flight squids and gossamer-parachute spiders are not counted.

A brief description of the different volant groups is presented.

Gliding-Flight Fishes
In gliding-flight fishes lift-generators and thrust-generators are morphologically and functionally independent of each other, a situation prevailing in the dense medium – water, but not observable in the attenuate medium – air. Lift is induced by the pectoral gliding-wings and in some species by the pelvic wings as well, and thrust is induced by the hypobatic lobe of the caudal sculling in the water, whilst the rest of the body is airborne. Thrust-gliders can bridge enormous distances, because the sculling propulsion accelerates the fish with ease. At the end of a glide a repeated start can be accomplished simply by dipping the hypobatic lobe back into the water again to scull. Gliding-flight fishes are very good and stable gliders. However they can control space-orientation and pathway-parameters only within certain limits. Gliding-flight in fishes has arisen a few times independently.

Flapping-Flight Fishes
In flapping-flight fishes lift and thrust are induced by the same apparatus, the pectoral flapping-wings. This is the only case in which flapping-flight evolved without an intermediate gliding stage, because the same device operating in the same manner preexisted. Fishes have evolved two different modes of fin motion. Paddling-fins are arranged perpendicular to the motion axis and paddle parallel to it. Flapping-fins are arranged parallel to the motion axis and flap perpendicular to it. For transgression from the dense medium – water – into the 770 times attenuate medium – air – the flapping-fins just had to be enlarged and their flapping frequency increased in order to obtain flapping-wings.

Flapping-flight fishes are very unstable flyers, their flapping-wings barely being able to control space-orientation and pathway-parameters. The fish "hangs" between the whirring pectoral flapping-wings until it is exhausted and drops back into the water. Flapping-flight in fishes seems to have arisen only once.

In contrast to water-breathing gliders which are thrust-gliders all air-breathing gliders are gravity-gliders.

Gliding-Flight Frogs

Rhacophorus has two pairs of limb-wings, brachial and skelosal gliding-wings. The brachial wing is 3-membered. The proximal wing section consists only of the propatagium and the distal wing section of the large webbed hand-wing. To the rear of the ulna and continued along the outer finger is a muscular flap. The skelosal wing is 4-membered. The two inner members, femur and shin, are not embedded in a membrane. The two outer members, the elongated tarsal section and the large webbed foot-wing support the aerofoil. There is a muscular flap to the rear of the tarsal section and continued along the outer toe. The existing cephalocaudal shortness necessitates a very sensitive pitch control which can be enhanced by shifting the foot-wing with the help of the two proximal spanmembers. The wing areas of the different *Rhacophorus* species are very diverse and thus their gliding ability must differ accordingly. Gliding-flight in frogs may have arisen more than once independently.

Gliding-Flight Geckos

Of all known volant forms *Ptychozoon* possesses the highest number of morphologicaly and functionally independent wings, thereby exhibiting a mosaic of patterns which reoccur in different combinations within other volant groups (see Fig. 1). They have four paired wings and one unpaired wing, altogether nine. Counted from head to tail these are the paired cheek-wings, the brachial wings, the flank-wings, the skelosal wings, and the unpaired caudal-wing. The two pairs of 3-membered limb-wings have muscular membranes extending to the front and the rear of the two inner spanmembers. These membranes are flapped away when not in use. The hand and the foot are fully webbed. The cheek-wings and flank-wings are self-cambering muscle-membrane aerofoils and are only extended when in use. The cheek-wings are highly cambered and channel the air around the neck and onto the elbow-wings and the flank-wings. They serve as a lift-wing and a steering-device combined. The flank-wings spread wide and constitute the largest aerofoils. The system cheek-wing/flank-wing of *Ptychozoon* very much resembles the system cheek-wing/rib-wing of *Draco* in topology and function. The unpaired caudal-wing is assembled of small metameric self-supported areas, that expand the tail laterally forming a lift-wing and a flow-rudder combined. Gliding-flight geckos are poor but stable gliders. Gliding-flight in lizards has arisen a few times independently.

Repeated Wing-Bauplans

Two different bauplans of gliders have repeatedly evolved independently. One bauplan, the pleural glider, is realized by which greatly elongated ribs brace and camber the rib-membrane – the pleuropatagium – thus forming the rib-wing – the pleuropteron (see Fig. 1). The flying-apparatus is therefore morphologically and functionally independent of the walking-apparatus. Pleural gliders have evolved at least five times independently. The other bauplan, the skelobrachial glider, is realized in such a manner that the walking-apparatus, the front and hind limbs, brace and camber the tensile flight-membrane – the tenopatagium –, thus forming the armleg-gliding-wing – the 2-membered skelobrachiopteron (see Fig. 1). The flying-apparatus is therefore morphologically and functionally combined with the walking-apparatus. Skelobrachial gliders have evolved at least nine times independently.

Pleural Gliders

The extant pleural glider *Draco* is an extremely good and dextrous glider with large aerofoils. Its pleuropteron is manipulated with great adroitness and when not in use can be swept back alongside. It also has highly cambered cheek-wings which serve as a lift-wing and steering-device combined. Whether the extinct forms had cheek-wings or not is not indicated by their remains, but the very obvious analogy of the cheek-wing/flank-wing system of *Ptychozoon* and the cheek-wing/rib-wing system of *Draco* make it probable. They did however have comparatively larger pleuroptera. By comparison with the extant form one might speculate about their volant abilities. Evolutionary conversions of ribs to wing-bracers are characteristic of pleural gliders. Being hard parts they are predestined to good fossilization and when unearthed are as such indicative. More forms of this bauplan probably await discovery because conversions of arboreal lizards to pleural gliders seem evolutionarily simple for the lizard bauplan.

Skelobrachial Gliders

Developing a skelobrachial gliding-wing seems to be the most simple and most frequently repeated bauplan of converting an arboreal animal into a volant. All prerequisites are preadaptively present. The limbs are stretched out laterally forming the bracers of the armleg-gilding-wing – the 2-membered skelobrachiopteron. The skelobrachial gliding-wing thus folds twice: in the shoulder joint and the elbow joint. The evolutionary spreading outwards of the body-skin converts it into a frame-supported formlabile tensile flight-membrane – the tenopatagium. The aerofoil of the double mast elbow-wing – the angkalepteron – is generally subdivided into the tenopatagial propatagium, -plagiopatagium and -uropatagium, and is cambered by rotating the elbow and the knee out of the patagial tension-plane. Some gliding mammals and bats also use their auricles – the ear-wings – as a lift-wing and a steering-device combined.

Different evolutionary stages and designs of the skelobrachial gliding-wing can be observed in extant and extinct gliders.

1. Early Forms
– *Acrobates*
In this form the patagia have not as yet reached the wrist and ankle, demonstrating an early stage in the evolution of the skelobrachial gliding-wing. All evolutive options are still open.

2. Late Forms
– Dermoptera
"A little more of the same", overextended proximal spanmembers limit arboreal adroitness, and "no innovations" lead to stagnation.

3. Closed Forms
– Anomalurinae
– *Petauroides*
– *Podopteryx*
In these forms the further general development of the skelobrachial gliding-wing seems inhibited, but its specialization is enhanced. In Anomalurinae the outward pointing elbow-strut which subdivides the diapatagium from the plagiopatagium seems to prevent further evolutionary lateral expansion of the wing, but it offers a very effective additional cambering-device by rotating span member-II around the axis span member-I/elbow-strut. In contrast both *Petauroides* and *Podopteryx* exhibit a skelobrachial gliding-wing with a 1-membered front spanmast and a dominant 2-membered rear spanmast. The possum utilizes the inward flexed forearm and hand combined as a slotted fore-wing forming a very effective lift- and cambering-device. The lizard seems to camber its wing additionally by some self-cambering structure within its greatly enlarged uropatagium. A 1-membered front spanmast is evolutionarily not very promising and a dominant rear spanmast incorporated in a compound-wing makes life on foot quite problematic.

4. Developed Forms
– *Petaurus*
In this form the patagia – propatagium, plagiopatagium, and uropatagium – have reached wrist and ankle forming the skelobrachial gliding-wing proper. A slight lateral wing expansion beyond the wrist is already noticeable indicating that only digit V will become the single bracer of the newly developing distal wingpart and the third spanmember to be of the front spanmast, but a third joint – the wing-folding joint – has not started to develop as yet.

5. Open Forms
– Petauristinae
In this form the evolutionary expansion of the proximal wingpart, the angkalepteron, beyond the wrist can be observed in an advanced stage. The 2-membered front spanmast is being extended beyond the wrist by a finger-like third span member – the d a c t y l o i d – and made foldable backwards around a wing-folding joint thus forming the single bracer for the newly developing distal wingpart. The 3-membered skelobrachiopteron of gliding squirrels therefore folds zig-zag three times, but a cambering-device for the distal aerofoil is not obvious as yet.

6. Throughway Forms
– Pterosaur Precursors
– Chiropteran Precursors
In these forms the 2-membered front spanmast of the skelobrachial gliding-wing was extended beyond the wrist by a third span member and made foldable backwards around a wing-folding joint, thus folding the 3-membered skelobrachiopteron zig-zag three times. Along with this a cambering-device for the distal aerofoil was developed and power-generators applied, thereby creating the skelobrachial flapping-wing. These forms are not yet documented by fossils but their descendents are well known.

Skelobrachial gliders are extremely good and dextrous gliders controlling space-orientation and pathway-parameters with great adroitness. At the same time they are very agile arboreal animals. This bauplan requires no principal alterations and has repeatedly evolved independently during geological times. Therefore more forms of this bauplan probably await discovery. Because the modifications pertain mostly to soft parts, their evidence can only be recognized on extremely good fossil records showing the gliding membranes as well. It therefore is not astonishing, that only one fossil piece of proof has been found as yet, the skelobrachial gliding-lizard *Podopteryx mirabilis*. Extant however a whole palette of skelobrachial gliding mammals are known. Skelobrachial gliders were also the base from which two groups of skelobrachial flappers evolved.

Skelobrachial Flappers

The conversion of a skelobrachial gliding-wing to a skelobrachial flapping-wing is achieved by expanding the preadapted proximal wingpart, the angkalepteron, outwards beyond the wrist, thereby developing a new distal wingpart composed of a bracer made foldable backwards around a wing-folding joint and of an aerofoil with its cambering-device. The 3-membered skelobrachiopteron therefore folds zig-zag three times: in the shoulder joint, the elbow joint and the wing-folding joint. The application of power-generators thus creates the skelobrachial flapping-wing.

The most simple conversion of this kind was that of the chiropteran precursors to flapping bats. All prerequisites were preadaptively present. The webbed hand – as demonstrated in the angkalepteron of Dermoptera – only needed to be extended beyond the wrist as the third span member of the front spanmast and made foldable backwards around a wing-folding joint, thus forming the bracer and the cambering-device of the newly developing distal wingpart, the hand-wing of Chiroptera – the c h i r o p t e r o n. The 3-membered skelobrachiopteron of bats therefore folds zig-zag three times. The employment of power-generators thus created the skelobrachial flapping-wing of bats. It consists of only one type of flight-membrane, the formlabile tensile tenopatagium. Their angkalepteron assembles the propatagium, plagiopatagium, and uropatagium and their chiropteron composes the dactylopatagium brevis, -minor, -medius and -major.

The conversion of gliding pterosaur precursors to flapping archaetype pterosaurs was achieved by expanding the angkalepteron beyond the wrist by elongating their (at that time) outermost digit IV, and making it foldable backwards around a wing-folding joint to become the single bracer of the newly developing distal wingpart, the fingerwing of pterosauria – the d a c t y l o p t e r t o n. Because the single bracer can neither brace nor camber a formlabile tensile flight membrane a new kind of f o r m s t a b i l e s e l f - s u p p o r t i n g s e l f - c a m b e r i n g r a y - s t r u c t u r e d f l i g h t - m e m b r a n e – the a k t i n o - p a t a g i u m – had to be invented. This pterosaur-specific aktinopatagium supplied the single-braced

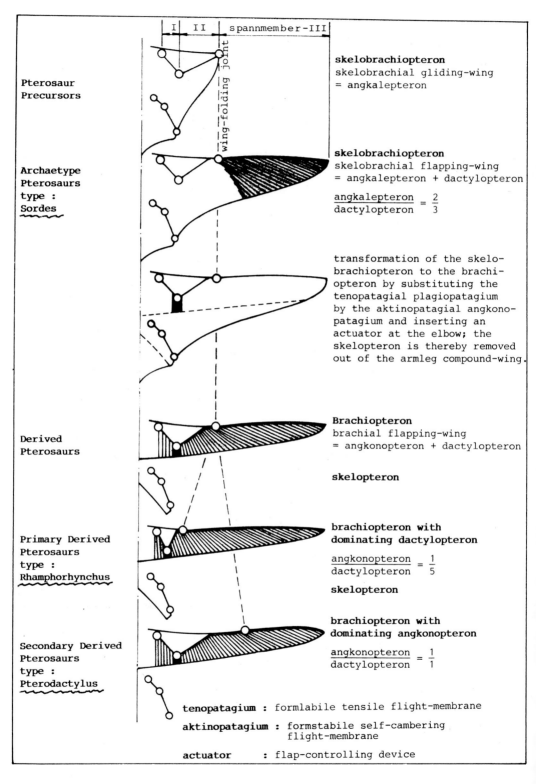

dactylopteron with a highly sophisticated camber-controlled flight membrane – the d a c t y l o p a t a -
g i u m. The 3-membered skelobrachiopteron of pterosaurs therefore folded zigzag three times. The employment of power-generators thus created the skelobrachial flapping-wing of pterosaurs. It consisted of two structurally and functionally very different types of flight-membranes. Their angkalepteron assembled the formlabile tenopatagial propatagium, -plagiopatagium, and -uropatagium, whereas their dactylopteron consisted of the formstabile aktinopatagial dactylopatagium. It is obvious that a double-braced angkalepteron can only function within a formlabile tensil compound because of the metachronous beat of the two limb pairs, whereas a single braced dactylopteron can only function when the dactylopatagium is of a self-supporting self-cambering structure. It was the property of this pterosaur-specific aktinopatagium that enabled the conversion of the 2-membered skelobrachial gliding-wing of pterosaur precursors to the 3-membered skelobrachial flapping-wing of archaetype pterosaurs. Subsequently the same aktinopatagial structure also allowed the transformation of the double-mast skelobrachial flapping-wing of archaetype pterosaurs to the single-mast brachial flapping-wing of derived pterosaurs (see Fig. 2).

Extant skelobrachial flappers, the bats, are very good and dextrous flyers and it can be assumed that the extinct skelobrachial flappers, the archaetype pterosaurs, had similar abilities. Pterosaurs originally bipedal and equipped with a highly sophisticated momentum-rudder, the tail, for controlling the space-orientation of their terrestrial bipedal locomotion remained biped throughout their entire phylogenetic history. Bats originally quadrupedal continue being so. They never used their tails as rudders.

Brachial Gliders

Extant brachial gliders, the gliding-flight frogs and the gliding-flight geckos, share with the gliding avian precursors the single-mast arm-wing – the b r a c h i o p t e r o n, but in contrast to emanate birds their flying-apparatus is used as a walking-apparatus as well. They also exhibit such diverse bauplans, that they are treated independently.

The course of bird evolution is quite complex. The avian precursors originate as small bipedal dinosaurs that venture into the arboreal realm, then transform to brachial gliders and subsequently to brachial flappers. Preadaptively they were equipped with two independently operating circuits: one controlling the forelimbs for grasping, the other controlling the hindlimbs for walking. Their refined unpaired momentum-rudder, the tail, was devised for the control of space-orientation of their terrestrial bipedal locomotion. It progressively transformed to a flow-rudder and a lift-wing combined, the unpaired caudal-wing.

The leg was ready-made and remained independent of the flying-apparatus throughout their phylogenetic history. The arms with their very long fingers needed modifications while transforming to the paired 3-membered brachial gliding-wings. Their at that time 3-fingered hand became functionally 2-fingered by dermally fusing the second and third digits up to their claws and transforming them to become the third span member of the front spanmast and making them foldable backwards around a wing-folding joint. The 3-membered brachiopteron therefore folded zig-zag three times.

The hand had a slight supinated orientation with respect to the flexing plane of the wing-folding joint and the claws were directed accordingly in a forward downward orientation. This arrangement made a very dextrous climbing-hand without interfering with the rearward oriented feathers. The mobile first digit could hook up independently of the fused second and third. The paired armwings and the unpaired caudal wing form the 3-winged brachial glider.

Birds and their emanates evolved two structurally and functionally quite different aerofoils – dermal sublayers and feathered coverlayers. The evolution to the primary 3-winged gliding avian precursor was achieved by creating membraneous aerofoils, the tenopatagial metapatagium for the posteriorly gaping shoulder joint and the tenopatagial propatagium for the anteriorly gaping elbow joint. Besides that

Fig. 2
Wing evolution of Pterosaurs. Transformation of the skelobrachiopteron to the brachiopteron.

muscle membrane flaps – the ala membrana – developed to the rear of the second and third spanmembers. Muscle membrane flaps also formed laterally alongside the caudal. The transition to the secondary 3-winged gliding avian precursor was achieved by transforming cover-scales to feathers, thus forming feathered aerofoils. The remiges were created to greatly expand the dermal aerofoils beyond their trailing edges, whereby the tertiaries expanded the metapatagium, the secondaries expanded the proximal part of the ala membrana, and the primaries expanded the distal part, thus creating the feathered brachial gliding-wings. This arrangement avoided problems of tensions in the dermal plane to the rear of the elbow joint and of structural interferences between the primaries- and secondaries-plane at the wing-folding joint. The feathered brachiopteron therefore folded zig-zag three times. In addition the rectrices were devised to greatly expand the caudal-flaps laterally, thus creating the feathered caudal-wing. Altogether they formed a very dextrous and stable 3-winged glider with especially effective devices for controlling roll, pitch, and yaw. The gliding avian precursors

- walked with the hind limbs
- climbed with four limbs
- glided with the brachial and caudal wings
- landed on tree trunks head up and hands first

This was exactly what the flapping archaetype bird still did. But with strengthening and refinement of the paired brachial flapping-wings the climbing-hands successively lost their function. Continued sophistication of the pitch-control and increased effectiveness of the airbrakes progressively transferred the initial landing-gear of the gliding avian precursors – the hands – to the final landing-gear of the derived flapping birds – the feet.

Brachial Flappers
The brachial flapping stage, which is characterized – "Pegasus-like" – as having the flying-apparatus totally independent of the walking-apparatus, was reached twice via very different evolutionary pathways; birds developed a f e a t h e r e d b r a c h i o p t e r o n whereas pterosaurs evolved a m e m - b r a n o u s b r a c h i o p t e r o n.

Birds on the one hand converted to a brachial flapping stage by expanding the three feathered wings of their gliding precursors and applying power-generators to the brachial gliding-wings, thereby transforming them to brachial flapping-wings. The continued evolution from a dynamically stable flyer to a dynamically labile flyer, a general tendency observable in all volant groups, is essentially a cephalocaudal compression correlated with a marked lateral wing expansion. Most of the functions of the caudal-wing, lift and steering, are thereby progressively transferred to the structurally and functionally refined brachial-wings.

Pterosaurs on the other hand reached the brachial flapping stage via transformation of their double-mast skelobrachial flapping-wing to the single-mast brachial flapping-wing. This conversion was achieved by evolutionarily spreading the distal single-braced aktinopatagial dactylopatagium bodywards, thereby progressively substituting the proximal double-braced tenopatagial plagiopatagium and transforming it to the single-braced aktinopatagial a n g k o n o p t a g i u m . Because the inward progression of the ray-structured actinopatagium interfered with the metachronous wingbeat of the tensile tenopatagium, the leg had to be removed out of the double-braced elbow wing, the angkalepteron, thus creating the single-braced elbow wing – the a n g k o n o p t e r o n . This transformation was only possible because of the self-supporting and self-cambering properties of the aktinopatagial angkonopatagium.

Spreading inwards the aktinopatagium had to extend around the elbow joint, the most critical point in a membranous brachiopteron. Because of extreme tensions at this place a tongue-like flap-controlling device – the a c t u a t o r – had to be inserted into the wing at the rear of the elbow joint pointing backward in order to facilitate the camber-control of the patagial angkonopteron. The newly devised angkonopatagium together with the preadapted dactylopatagium compose the complete aktinopatagial arm-wing membrane – the b r a c h i o p a t a g i u m – which runs along the entire length of the 3-membered front spanmast. The tenopatagial propatagium retains its original position, structure, and function in front of the

Fig. 3
Flightsilhouettes.

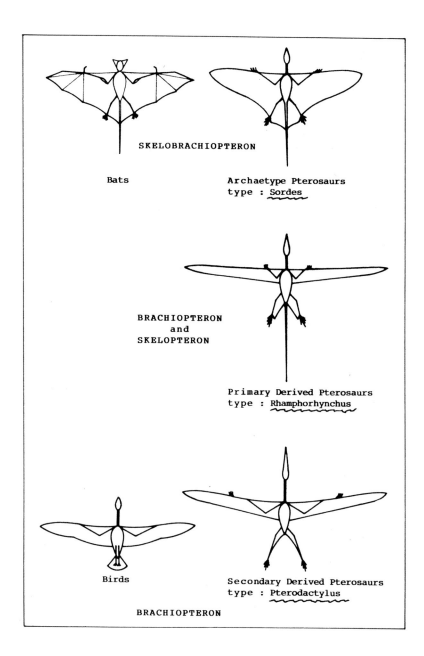

spanmast. The membraneous brachiopteron therefore folded zig-zag three times. The leg which was removed out of the armleg compound-wing becomes the leg-wing – the s k e l o p t e r o n – of derived pterosaurs. Its tenopatagial flight-membrane – the s k e l o p a t a g i u m – is a derivate of the tenopatagial uropatagium of archaetype pterosaurs. It is obvious that a double-braced skelobrachiopteron can only function within a formlabile tensile compound because of the metachronous beat of the two limb pairs, whereas a single-braced brachiopteron can only function when the brachiopatagium is of a self-supporting self-cambering structure (see Fig. 2). It becomes evident that during their very long phylogenetic history pterosaurs evolved two quite different types of wings – one a derivate of the other – which present themselves in three characteristic flightsilhouettes (see Fig. 3).

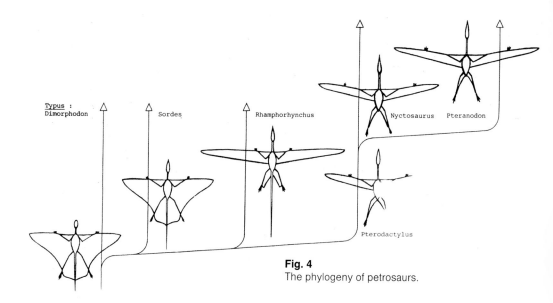

Fig. 4
The phylogeny of petrosaurs.

The continued evolution of pterosaurs brought forth a side line of successful, but highly specialized flyers – the primary derived pterosaurs. They transformed the brachial wing to become the specialized high-frequency flapping brachiopteron with dominating dactylopteron thereby utilizing their skelosal wings very much for the same purpose that birds use their caudal-wing. They retained the highly sophisticated momentum-rudder, the tail, for controlling the space-orientation of their terrestrial bipedal locomotion. In contrast the main line led to very generalized and extremely successful adaptively radiating flyers – the secondary derived pterosaurs. They transformed the brachial wing to become the very generalized brachiopteron with dominating angkonopteron thereby reducing the skelosal wings and transfering all their functions to the brachial wings. Linked with this the head-axis was cranked away from their neck-axis thus forming an effective new momentum-equalizer, the shifting head/body-system. It compensated for the reduction of the momentum-rudder, tail, and took over all its controlling functions of space-orientation for their terrestrial bipedal locomotion.

To summarize it becomes obvious that the wing evolution of pterosaurs passed through three stages (see Fig. 2):

Stage I Development of the double-mast 2-membered skelobrachial gliding-wing of pterosaur precursors. Walking- and flying-apparatus are morphologically and functionally coupled. This wing topologically resembles the wing of gliding squirrels.

Stage II Expansion of it to the double-mast 3-membered skelobrachial flapping-wing of archaetype pterosaurs. Walking- and flying-apparatus remain morphologically and functionally coupled. This wing topologically resembles the wing of bats.

Stage III Transformation of it to the 3-membered single-mast brachial flapping-wing of derived pterosaurs, thereby separating the walking-apparatus morphologically and functionally from the flying-apparatus. This wing topologically resembles the wing of birds.

The phylogeny of pterosaurs can be briefly described as follows (see Fig. 4):
The skelobrachial gliding-wing of the pterosaur precursors transformed to the skelobrachial flapping-wing of the archaetype pterosaurs. The basic group with a lizard-like skull form the primary archaetype pterosaurs (type *Dimorphodon*).

Close to their base a group branched off which modified its skull but retained the skelobrachial flapping-wing. They form the secondary archaetype pterosaurs (type *Sordes*).

Shortly thereafter a new offshoot transformed the double-mast skelobrachial wing of the archaetype pterosaurs to the single-mast brachial wing of the derived pterosaurs. The leg was thereby removed out of the skelobrachial compound-wing and became the skelosal wing. This offshoot again split into two different evolutionary pathways:

One group conserved its skeletal features and retained the skelopteron. They developed the brachiopteron with the dominating dactylopteron, the wing of the primary derived pterosaurs (type *Rhamphorhynchus*).

The other group altered its skeletal features and reduced the skelopteron. They developed the brachiopteron with the dominating angkonopteron, the wing of the secondary derived pterosaurs (type *Pterodactylus*).

Pterosaurs having an entirely closed scapulocoracoid-ring form a natural entity. This unique structure serves as a continuous longeron in the wings of advanced pterosaurs (type *Pteranodon*).

The remaining progressive pterosaurs without a closed scapulocoracoid-ring have to be carefully studied for their systematic arrangement, also using criteria other than flight parameters. The creation of the brachial wing by removing the leg out of the skelobrachial compound-wing is a step of great evolutionary significance, comparable to the conversion of the skelobrachial gliding-wing to the skelobrachial flapping-wing. The evidence shows that the principal phylogenetic bifurcation does not lie between the tailed and tailless pterosaurs but between the skelobrachiopteral and brachiopteral pterosaurs. This implies that the tailed brachiopteral *Rhamphorhynchus* is more closely related to the tailless brachiopteral *Pterodactylous* than to the likewise tailed, but skelobrachiopteral *Sordes* (see Fig. 4).

Insects

The most ancient and most successful volant animals are the insects. With respect to their wings they are also exceptionally polymorphous. From the outset their flying-apparatus was – "Pegasus-like" – strictly independent of their walking-apparatus. Original nonvolant insects had a locomotion-tagma consisting of the three metamerous segments each of which had one pair of legs but no wings.

Initially three pairs of chitin membrane duplications evolved creating three pairs of gliding-wings, one on each segment of the tagma. Gliding-flight insects therefore had three pairs of gliding-wings.

In due course only wing-pair-II and wing-pair-III continued to evolve to become flapping-wings, whereby wing-pair-I was successively reduced. Archaetype flapping-flight insects therefore had two pairs of flapping-wings. The further evolution brought forth some truly 4-winged insects which remain rare by insect standards (Odonata). Many groups show a tendency towards a functional 2-wingedness by coupling the smaller wing-pair-III to the dominating wing-pair-II (Hymenoptera). Few groups have become truly 2-winged either by reduction (wing-pair-III, Diptera) or transformation (wing-pair-II, Cetoniinae). And other groups have lost their wings altogether and become derived wingless (Siphonoptera).

Fossil Documents
Skelobrachyopteron
(archaetype pterosaurs)
Sordes pilosus Sharov, 1971.
Upper Jurassic, Karatau, Kasachstan, USSR
Palaeontological Institute, Moscow, USSR
PIN No 2585/3

Brachyopteron with Dominating Dactylopteron
(primary derived pterosaurs)
Rhamphorhynchus muensteri (GOLDFUSS, 1831)

Upper Jurassic, Eichstätt, FRG
Bayer. Staatssamml. Paläont. Geol., München, FRG
No 1880 II 8

Brachyopteron with Dominating Angkonopteron
(secondary derived pterosaurs)
Pterodactylus kochi (WAGNER, 1837)
Upper Jurassic, Altmühlalb, FRG
Nat. Hist. Mus. Wien, Austria
No 1975/1756

References Cited

BAKKER, R. T. (1975): Dinosaur renaissance. – Sci. Amer. **232** (4): 48–78; San Francisco, USA.
BENES & BURIAN (1980): Tiere der Urzeit. – (Artia) Prha, Hanau, (Werner Dausien), FRG.
BRAMWELL, C. D. & WHITFIELD, G. R. (1974): Biomechanics of *Pteranodon*. – Phil. Trans. Roy. Soc. London (B) **267**: 503–581; London, UK.
BROWER, J. C. (1980): Pterosaurs: How they flew. – Episodes **1980** (4): 21–24; USA.
CAPLE, G. et al. (1983): The physics of leaping animals and the evolution of preflight. – Am. Nat. **121**: 455–467; Chicago, USA.
FREY, E. & RIES, J. (1981): A new reconstruction of the pterosaur wing. – N. Jb. Paläont. Abh., **161** (1): 1–27; Stuttgart, FRG.
HAUBOLD, N. & KUHN, O. (1977): Lebensbilder und Evolution fossiler Saurier. – Neue Brehm Bücherei, **509**; Wittenberg Lutherstadt (A. Ziemsen), GDR.
HERZOG, K. (1969): Anatomie und Flugbiologie der Vögel. – Stuttgart (Gustav Fischer), FRG.
HOLST, E. V. (1975): Der Saurierflug. – Paläont. Z., **31** (1/2): 15–22; Stuttgart, FRG.
KINGDON, J. (1974): East African Mammals. – Academic Press, Vol. II A+B; London, UK.
LEKAGUL, B. & MCNEELY, J. (1977): Mammals of Thailand. – Kurusapha Press; Bangkok, Thailand.
LORENZ, K. (1963): Die "Erfindung" von Flugmaschinen in der Evolution der Wirbeltiere. – Therapeutischer Monat, **13**: 137–195.
MOHR, E. (1954): Fliegende Fische. – Neue Brehm Bücherei, **133**; Wittenberg Lutherstadt (A. Ziemsen), GDR.
PADIAN, K. (1983): A functional analysis of flying and walking in pterosaurs. – Paleobiology, **9** (3): 218–239; Berkeley, Cal. USA.
RÜPPELL, G. (1980): Vogelflug. – Hamburg (Rowolt Taschenbuch), FRG.
SCHALLER, D. (1983): Neubeschreibung des Pterosaurierflügels. – Zoologisches Institut der Universität München; München, FRG.
SCHMIDT, H. (1960): Der Flug der Tiere. – Senckenberg-Buch **39**; Frankfurt (Senckenberg, Nat. forsch. Ges.), FRG.
SCHMIDT-NIELSON, K. (1971): How Birds Breathe. – Scient. Am. **225** (6): 72–79; San Francisco; USA.
SHAROV, A. G. (1971): New Flying Reptiles from the Mesozoic Deposits of Kazakhstan and Kirgizin. – Akad. Nauk. SSSR Trudy Paläont. Inst. **130**: 104–113; Moscow, USSR.
SMITH, J. D. (1977): Comments on flight and the evolution of bats. – NATO Adv. Stud. Inst. Ser. A: Life Sci.: 427–437.
STEFAN, W. (1974): Urvögel. – Neue Brehm Bücherei, **465**; Wittenberg Lutherstadt (A. Ziemsen), GDR.
WALKER, E. P. (1975): Mammals of the world. – Baltimore (John Hopkins University Press), USA.
WELLNHOFER, P. (1975b): Die Rhamphorhynchoidea (Pterosauria) der Oberjura-Plattenkalke Süddeutschlands. I: Allgemeine Skelettmorphologie. – Palaeontographica, A. **148**: 1–33; Stuttgart, FRG.
WELLNHOFER, P. (1975c): Die Rhamphorhynchoidea der Oberjura-Plattenkalke Süddeutschlands. II: Systematische Beschreibung. – Palaeontographica, A. **148**: 132–186; Stuttgart, FRG.
WELLNHOFER, P. (1978): Pterosauria. – In: Handbuch der Paläoherpetologie (**19**): 82 pp.; Stuttgart (Gustav Fischer), FRG.
WELLNHOFER, P. (1980): Flugsaurier. – Neue Brehm Bücherei, **534**; Wittenberg Lutherstadt (A. Ziemsen), GDR.
WILD, R. (1978): Die Flugsaurier (Reptilia, Pterosauria) aus der oberen Trias von Cene bei Bergamo, Italien. – Boll. Soc. Pal. Ital. **17** (2): 176–256; Modena, Italia.
WILD, R. (1984): Flugsaurier aus der Obertrias von Italien. – Nat. Wiss. **71**: 1–11; Stuttgart, FRG.
YALDEN, D. W. & MORRIS, P. A. (1975): The Lives of Bats. – London (David & Carles), UK.

Author's address: DIETRICH SCHALLER, Landshuter Allee 156, D-8000 München 19, Federal Republic of Germany.

Günter Viohl

CARL F. and ERNST O. HÄBERLEIN, the Sellers of the London and Berlin Specimens of *Archaeopteryx*

Fig. 1
CARL F. HÄBERLEIN (1787–1871), the seller of the London specimen of *Archaeopteryx*.

Abstract

Photos of Dr. CARL F. HÄBERLEIN and ERNST O. HÄBERLEIN, the sellers of the London and the Berlin *Archaeopteryx* and some new details about them are presented here. Both gained fame when they instantly recognized the scientific importance of these specimens and sold them to public institutions together with their rich collections of Solnhofen fossils.

Zusammenfassung

Fotos von Dr. CARL F. HÄBERLEIN und ERNST O. HÄBERLEIN, den Verkäufern des Londoner und des Berliner *Archaeopteryx*, werden zusammen mit einigen neuen Details über sie veröffentlicht. Beide haben sich dadurch verdient gemacht, daß sie die wissenschaftliche Bedeutung dieser Stücke sogleich erkannten und sie an öffentliche Institutionen verkauften, zusammen mit ihren reichen Sammlungen Solnhofener Fossilien.

During the International *Archaeopteryx* Conference Mr. FRIEDRICH HINGKELDEY from Erfurt (German Democratic Republic), a great-grandson of CARL F. HÄBERLEIN, presented photos of his great-

Fig. 2
The house of CARL F. HÄBERLEIN at Pappenheim.

grandfather, the latter's house, and son ERNST O. HÄBERLEIN, a family tree as well as a copy of a newspaper article about ERNST HÄBERLEIN, to the Jura-Museum. He had heard of the meeting from a press report and despite his great age of 91 years he dared the long journey to personally hand the aforesaid documents over to the author, complaining that in no museum visited by him were the real discoverers of the first two *Archaeopteryx* specimens, given credit or mentioned.

Dr. CARL F. HÄBERLEIN and ERNST O. HÄBERLEIN played a decisive role in the history of the *Archaeopteryx* findings, but hardly a photo of them can be found in the literature. Only a photo of ERNST O. HÄBERLEIN in the book of ROECK (1973) is known to the author. Therefore the photos presented by Mr. HINGKELDEY are published in this volume.

In the years 1862/63 Dr. CARL HÄBERLEIN sold the first find of *Archaeopteryx* along with his complete collection of Solnhofen fossils for £ 700 to the British Museum as is generally known. The details of this purchase and the story of the early discussions about that specimen are reported by DE BEER (1954), HELLER (1959), and WELLNHOFER (1983, and this volume), and need not be repeated. These authors, however, quote the first name of Dr. HÄBERLEIN as "Karl". That is not quite correct. His full name was CARL FRIEDRICH HÄBERLEIN and obviously he was called "CARL". Details about his biography are not known. Only very little additional information is given by the family tree. According to it he was born at Solnhofen on October 28, 1787 and died at Pappenheim on February 20, 1871. On March 29, 1813, he married in a second marriage, (about his first wife nothing is known), KATHARINA FRIEDERIKA WEIGEL, who died on February 26, 1848. She gave birth to eleven children, two sons and nine daughters, three of whom had already died when their father came into possession of the *Archaeopteryx*. Dr. HÄBERLEIN was

Fig. 3
ERNST O. HÄBERLEIN (1819–1896), the seller of the Berlin specimen of *Archaeopteryx*.

Royal Bavarian District Medical Officer and according to the communication of his great-granddaughter, Mrs. GERTRUD KUTSCHER (Augsburg), he too was an obstetrician. This was no easy profession at that time. By day and night and in all kinds of weather Dr. HÄBERLEIN had to ride by coach or sleigh to his patients. As Mrs. KUTSCHER wrote, he also suffered from gout so that he always had to be heaved in and out by two strong men.

The sale of *Archaeopteryx* brought Dr. CARL F. HÄBERLEIN into ill repute as a clever businessman greedy for money. This judgment is, however, not right. It must be considered, that many of his patients were poor quarrymen from whom he could not take money. They payed with fossils. In this way Dr. HÄBERLEIN acquired a unique collection of Solnhofen fossils, comprising, after DE BEER (1954), 1703 specimens. One cannot blame him for utilizing this dead capital, since he had a large familiy to care for. The proceeds of the sale provided one of his daughters with a dowry (DE BEER 1954). Concerning the price, it must be admitted that it was not excessive even at that time, to say nothing of the present-day prices of fossils. Thus in 1864 the market value, among private buyers, of C. F. HÄBERLEIN'S collection including *Archaeopteryx* was estimated by JOHN RUSKIN to have been "some thousand or twelve hundred pounds" (DE BEER 1954). It was strange that CARL F. HÄBERLEIN allowed neither a description nor an illustration of *Archaeopteryx* before its sale. Probably he feared that thereby the specimen would have lost its interest for scientists and through that its value, too.

The son of Dr. CARL F. HÄBERLEIN, ERNST OTTO HÄBERLEIN, acquired the second *Archaeopteryx*. He was born at Pappenheim on July 3, 1819, and died likewise at Pappenheim on February 18, 1896. In between he lived at Weidenbach near Ansbach (Mittelfranken), as can be drawn from his address in the first communication of the new *Archaeopteryx* (HÄBERLEIN 1877). The family tree indicates as his profession "Aufschlagnehmer" which means tax consultant (kind communication of Mr. HINGKELDEY). In this function he probably had good contacts with quarry owners and in consideration of his counsel he might have obtained many a fossil.

For the *Archaeopteryx*, however, ERNST HÄBERLEIN had payed 2000 Marks according to his granddaughter ELMIRA KINDLER (Grönenbach). This is reported in the newspaper article left to us by Mr. HINGKELDEY (G-D 1973). Interesting details can also be borrowed from a contemporary newspaper article (Sch. 1877). According to it the discovery of the second *Archaeopteryx* actually occurred in the autumn of 1876, and not in 1877, as is usually presented. The fossil still covered by a limestone layer, stood out but indistinctly against the slab. The quarry owner regarded it as a *Pterodactylus*. HÄBERLEIN uncovered a part of the tail with feathers. This revealed to him the great significance of the find and he offered it (for the price of 15000 fl = 25710 Marks), first to the Bavarian State Collection in München, which, however, could not buy it because of the high sum. Thereupon HÄBERLEIN continued his preparation work which brought to light the complete animal. He now offered the specimen along with his rich collection of Solnhofen fossils for 36000 Marks (compare OSTROM: "The Yale *Archaeopteryx*...", this volume). For six months he left the specimen to "Freies Deutsches Hochstift" for the arrangement of a purchase by a German public institution, but this could not be realized. After a reduction of the price the Prussian State acquired the *Archaeopteryx* for 20000 Marks, Dr. WERNER VON SIEMENS advancing the sum. The proceedings, until the incorporation of the specimen into the collections of the Humboldt University in Berlin, have repeatedly been reported (W. DAMES 1884, R. DAMES 1927, HELLER 1959, WELLNHOFER 1983). Later on the Prussian State also bought the rest of E. O. HÄBERLEIN's collection.

The ill repute of CARL F. and ERNST O. HÄBERLEIN in some publications does not seem justified after a detached examination of all facts. Sure, both were good businessmen, but both deserve credit for collecting an extensive amount of fossils of the Solnhofen Lithographic Limestones and selling them to scientific institutions. Their names will be forever linked with the first two specimens of *Archaeopteryx*. They were the first to acquire them and to make them accessible to science. The price they asked was high, but appropriate, in view of the importance of the fossils as representatives of a "connecting link" between reptiles and birds. Indeed even then the museums had financial difficulties, and as nowadays, the public authorities were not willing to spend too much money for documents of the history of life.

Acknowledgements

I am especially indebted to Mr. FRIEDRICH HINGKELDEY for leaving the above mentioned material to the Jura-Museum. Many thanks also to Dr. WINFRIED WERNER (Bayerische Staatssammlung für Paläontologie und historische Geologie München) for sending me the newspaper article of 1877 and Mr. DUANE HARRIS for correcting the English text.

References Cited

BEER, SIR G. DE (1954): *Archaeopteryx lithographica*. – Brit. Mus. (Nat. Hist.), 68 pp., 16 pls., 9 figs.; London.
DAMES, R. (1927): WERNER VON SIEMENS und der *Archaeopteryx*. – Nachr. Ver. Siemens-Beamt. e. V. Sportver. Siemens Berlin e. V., **19**: 233–234; Berlin.
DAMES, W. (1884): Über *Archaeopteryx*. – Palaeont. Abh., **2**, 3: 199–198, 1 pl., 5 figs.; Berlin.
G – D (1973): Ein Pappenheimer fand den Urvogel. – Memminger Zeitung, **24** (5); Memmingen.
HÄBERLEIN, E. (1877): *Archaeopteryx lithographica* v. MEYER. – Leopoldina, **13**, 9–10: 80; Dresden.
HELLER, F. (1959): Ein dritter *Archaeopteryx*-Fund aus den Solnhofener Plattenkalken von Langenaltheim/Mfr. – Erlanger geol. Abh., **31**, 25 pp., 15 pls., 2 figs.; Erlangen.
ROECK, B. (1973): Spuren im Stein. – 64 pp.; Augsburg.
SCH. ,TH. (1877): Die Räthseleidechse oder der Urvogel. – Neue Frankfurter Presse, **262** (8. 11. 1877); Frankfurt.
WELLNHOFER, P. (1983): Solnhofer Plattenkalk: Urvögel und Flugsaurier. – 59 pp., 58 figs.; Maxberg (Freunde des Museums beim Solnhofer Aktienverein).

Author's address: Dr. GÜNTER VIOHL, Jura-Museum, Willibaldsburg, 8078 Eichstätt, Federal Republic of Germany.

Peter Wellnhofer

The Story of Albert Oppel's *Archaeopteryx* Drawing

Abstract

In the Bavarian State Collection of Palaeontology and Historical Geology in Munich an old drawing of the London *Archaeopteryx* specimen is preserved. It is supposed to have been drawn by ALBERT OPPEL in 1861, the year in which OPPEL travelled to Dr. HÄBERLEIN in Pappenheim to see the famous specimen. It is said that he made the drawing after his return to Munich from memory. It is supposed that the drawing was indeed made by OPPEL, but probably using a sketch which he managed to make in Pappenheim.

Zusammenfassung

In der Bayerischen Staatssammlung für Paläontologie und historische Geologie in München wird eine alte Zeichnung des Londoner *Archaeopteryx*-Exemplares aufbewahrt. Sie soll von ALBERT OPPEL stammen und 1861 nach dem Studium des Fossils bei Dr. HÄBERLEIN in Pappenheim nach seiner Rückkehr in München aus dem Gedächtnis angefertigt worden sein. Es wird vermutet, daß es sich tatsächlich um die Zeichnung OPPEL'S handelt, der aber wahrscheinlich eine in Pappenheim gefertigte Skizze zugrunde lag.

As is generally known the first description of the first *Archaeopteryx* skeleton, the London specimen, was published by JOHANN ANDREAS WAGNER in 1861. This was a printed version of a paper read before the session of the mathematical-physics class of the Royal Bavarian Academy of Sciences in Munich on November 9th, 1861, only five weeks before Professor WAGNER passed away.

At that time, WAGNER was conservator, what would today be director of the Palaeontological Collections of the Bavarian State in Munich, and professor of palaeontology at the university. During the summer of the same year, 1861, an amateur palaeontologist, O. J. WITTE, Law-Councillor in Hannover, had visited WAGNER in Munich and reported on the feathered fossil from the Lithographic Stone he had seen in the possession of the District Medical Officer, Dr. HÄBERLEIN, in Pappenheim (WITTE 1863).

WITTE himself mentioned later that he had urged WAGNER to purchase the specimen for the Munich collection. But WAGNER regarded the fossil as being a feathered reptile rather than a bird because he considered geology and palaeontology to be in close agreement with the biblical interpretation of divine creation, and there was no place for a Jurassic bird.

Only when WITTE told him that HERMANN VON MEYER, the famous Frankfurt palaeontologist, would be preparing a publication on a fossil bird feather from Solnhofen, did WAGNER finally decide to send his assistant, ALBERT OPPEL, to Pappenheim in order to investigate the mysterious fossil.

WAGNER's description of 1861 is solely based on the report brought back by OPPEL which follows in his own words (p. 148):

"Ich habe nämlich von einem meiner Freunde, der vollständiger Sachkenner ist... einen Bericht über dieselbe Platte, welche Herr O. J. WITTE einzusehen Gelegenheit hatte, mitgetheilt bekommen. Wenn derselbe auch keine Zeit hatte, eine umständliche Vergleichung der Platte vorzunehmen, so reichte dieselbe ihm doch aus, um wenigstens über die Hauptstücke derselben eine sichere Auffassung zu gewinnen."

WAGNER avoided citing his informant by name, and also he left open to question what OPPEL'S opinion of the fossil was, although giving the impression of agreement with his own suggestion of a feathered reptile which he proceeded to name *Griphosaurus*.

The story is reported in detail by FERDINAND VON HOCHSTETTER (1866), then professor of geology and mineralogy in Vienna and a close friend of OPPEL'S. He wrote in his obituary on OPPEL, who had died after only four years in office at the age of 34:

"Im Winter 1860 (sic!) war OPPEL nach Pappenheim gegangen, um das wunderbare fossile Federthier von Solnhofen zu sehen, das damals noch im Besitze Dr. HÄBERLEIN'S war. Der Eigenthümer erlaubte nicht einmal eine Zeichnung davon zu machen. OPPEL entwarf aber die Zeichnung bei seiner Rückkehr nach München aus dem Gedächtnis. WAGNER, der sie mit Erstaunen betrachtete, hielt das Fossil für ein befiedertes Reptil gegen OPPEL'S Meinung, der dasselbe sogleich richtig als Vogel mit reptilienartigem Schwanze erkannte. WAGNER aber wollte den Ruhm für sich haben, das merkwürdige Thier zuerst beschrieben zu haben, und OPPEL überließ ihm dazu auf's Bereitwilligste seine Zeichnung. So ist WAGNER'S *Griphosaurus* (Räthselechse), der jetzige *Archäopteryx*, entstanden, und OPPEL'S Zeichnung, die heute noch aufbewahrt ist, stimmt wunderbar mit den seither nach dem Original publicirten Abbildungen."

Obviously based on this obituary WILHELM VON GÜMBEL, the Munich geologist, wrote about ALBERT OPPEL in 1887:

"Schon im Winter 1860 (sic!) hatte OPPEL die berühmte Versteinerung eines Vogels im lithographischen Schiefer von Solnhofen richtig erkannt, aber in seiner liebenswürdigen Bescheidenheit die Beschreibung dieses wichtigen Überrestes WAGNER überlassen, der ihn aber grundsätzlich zu einer Eidechse stempelte."

It is surprising that both HOCHSTETTER and GÜMBEL gave 1860 as being the year of OPPEL'S visit to Pappenheim. This, however, disagrees with the paper of WAGNER (l.c.) read before the Academy on November 9th, 1861 where he stated:

"Im Laufe dieses Sommers hatte ich das Vergnügen von Herrn Oberjustizrath WITTE in Hannover... einen Besuch zu erhalten, bei welcher Gelegenheit er mir alsogleich mittheilte, daß er, im Besitze des Herrn Landarztes HÄBERLEIN zu Pappenheim, eine aus dem lithographischen Schiefer von Solnhofen stammende Platte gesehen habe, auf welcher ein Skelet mit einer Combination von Merkmalen, wie man sich dieselben nicht befremdlicher und abenteuerlicher denken könne."

Only thereafter, could WAGNER have sent his assistant, OPPEL, to Pappenheim, that is in summer 1861, at its earliest.

Should the travel already have been undertaken in the winter of 1860, as mentioned by HOCHSTETTER and GÜMBEL, the year of discovery of the London specimen would have to be dated back to 1860 as well. This, however, would be contradictory to the letters written by HERMANN VON MEYER to the Neues Jahrbuch für Mineralogie etc. dated August 15th and September 30th, 1861 (H. V. MEYER 1861a; 1861b) in which he reported on the discovery of the feather a n d of the skeleton respectively.

Although an exact date for the discovery of the London specimen was never published, the year 1860 was probably suggested by HOCHSTETTER and GÜMBEL erroneously.

In the Bavarian State Collection of Palaeontology and Historical Geology in Munich an old drawing is kept which is supposed to be OPPEL's original drawing of 1861. No other records exist in the archives of the State Collections about it. According to personal communications of Professor DEHM, director emeritus of the State Collection, the drawing was exhibited in the geology department of the Deutsches Museum in Munich, until 1950. It was then taken over by the State Collection, together with palaeontological and geological objects. It is unknown how and when the drawing was obtained by the Deutsches Museum, which opened in 1925. Nevertheless, the question we are faced with is, whether it really is the famous drawing of OPPEL.

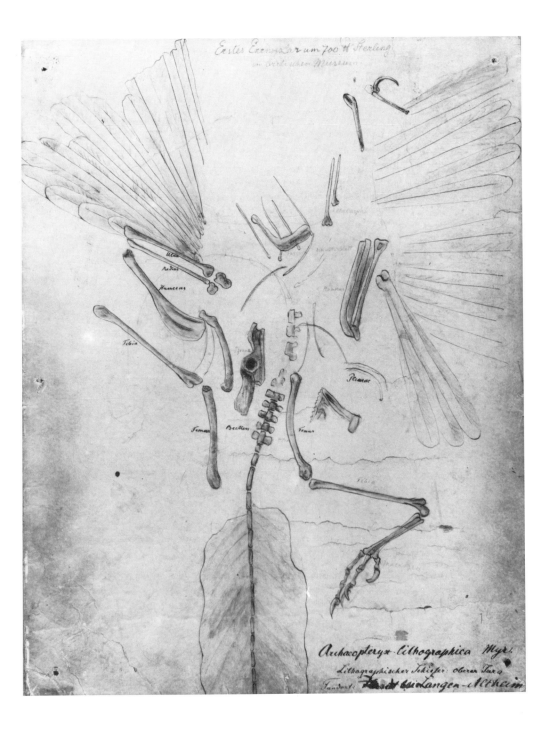

Fig. 1
Drawing of the London *Archaeopteryx* specimen supposed to have been made by ALBERT OPPEL in 1861. The drawing shows the situation of the skeleton on the main slab but in reverse. Bayerische Staatssammlung für Paläontologie und historische Geologie, München. Size 32,5 × 41 cm. Photo: F. HÖCK.

The size of the rather thin and yellowed paper is 32,5 × 41 cm. It is mounted on cardbord which apparantly was once framed. The line drawing is done in brown sepia ink, superimposed on pencil lines, and finished with pencil in order to give the bones and the feathers a three dimensional effect. On the sheet, there are several inscriptions, carried out first in pencil and then followed partly in sepia ink. They seem to have been written all by the same writer. The inscriptions read as follows. Middle top margin: "Erstes Exemplar um 700 £ Sterling im britischen Museum"
Bottom right:
"*Archaeopteryx lithographica* MYR., Lithographischer Schiefer: oberer Jura. Fundort: Haardt bei Langenaltheim."
In small lettering several bones are named: "Ulna, Radius, Humerus, Becken, Pfanne, Femur, Tibia, Pleurae, Schulterblatt, Furcula, Metacarpus, Phalangen."
Certainly, the handwriting is not WAGNER'S, for the specimen went to London in the autumn of 1862, after the death of WAGNER. It can not be the handwriting of OPPEL either, because only after the discovery of the second specimen, the Berlin specimen, in 1877, one could distinguish the London specimen as the "first" specimen. OPPEL who succeeded WAGNER in 1862 died already in 1865, only 34 years old. In 1866 OPPEL was replaced by KARL ALFRED VON ZITTEL who was in office until his death in 1904. So, the writing could have been added by ZITTEL between 1877 and 1904.

The drawing itself shows the specimen in natural size, but surprisingly, in reverse! One could argue that it showed the counterpart slab, but there are elements figured not present as bones but just as impressions on the counterpart slab, such as the furcula and the right ilium. The skull fragment with the natural mould of the brain, located at the left margin of the main slab is not figured and was obviously not taken for a part of the skeleton, whereas the fragment of the snout which is preserved mainly on the counterpart slab (DE BEER 1954: 10, pl. III), is drawn in detail but not signed nor mentioned by WAGNER.

Apart from being reversed, the drawing appears to be surprisingly close to the original. The positions, the angles and the lengths of the skeletal elements and the feather imprints differ only minimally from the real fossil. So it is almost incredible that OPPEL, according to HOCHSTETTER (p. 354) should have made the drawing from memory after his return to Munich. This is even more so if one takes into account that in 1861 a journey from Pappenheim to Munich by coach and train took almost a whole day, after which OPPEL is supposed to have made a correct but reverse drawing of the specimen!

In his description of 1861, WAGNER did not mention the drawing although he drew upon OPPEL's personal report. Are there any hints to the drawing?

WAGNER (1861: 149) gave some comments on the positions of bones, for example: "Of the pelvis, only the r i g h t half is preserved"... or: "On the l e f t side the complete hind leg, on the r i g h t side only the thigh and the shank are present" This corresponds to the situation met with on the m a i n slab. If WAGNER had the drawing at his disposal, he knew that it was a reverse.

After all, there is, in my opinion, little doubt about the authenticity of the drawing and the authorship of ALBERT OPPEL. However, no reason can be given why he should have made the drawing in reverse, so far from Pappenheim without the original fossil.

One explanation could be that OPPEL somehow managed to make a sketch of the specimen, maybe using the counterpart slab with additions from the main slab. This would explain the reverse image. This pencil drawing was completed with sepia ink after his return to Munich. Otherwise he must have used a mirror for the drawing in Pappenheim which seems to be quite unlikely because according to HOCHSTETTER, the owner of the fossil would not give permission for preparations for drawings. OPPEL would have had to make the drawing quickly and secretly, thus WAGNER could not mention the drawing in his publication.

No personal comments or letters of OPPEL on this story are known. He does not refer to it in his paper on *Ichnites lithographicus* from Solnhofen which he regarded as a bird's trackway (OPPEL 1862).

So, the real circumstances of the origin of this drawing are obscure and will probably never be known. But we are glad to have preserved it as documentation of the history of our science.

References Cited

DE BEER, G. (1954): *Archaeopteryx lithographica*. – Brit. Mus. (Nat. Hist.) London: 1–68, 9 figs., 16 pls.; London.
EVANS, J. (1865): On Portions of a Cranium and Jaw, in the Slab containing the Fossil Reamains of the *Archaeopteryx*. – Nat. Hist. Rev., **5**: 415–421; London.
GÜMBEL, C. W. v. (1887): Über ALBERT OPPEL. – Allg. Deutsche Biographie, **24**: 388–390; Leipzig.
HOCHSTETTER, F. v. (1866): Zur Erinnerung an ALBERT OPPEL. – Jb. Geol. Reichsanst., **16**: 59–67; Wien.
MEYER, H. v. (1861a): Vogel-Federn und *Palpipes priscus* von Solenhofen. – N. Jb. Miner etc., **1861**: 561; Stuttgart.
MEYER, H. v. (1861b): *Archaeopteryx lithographica* (Vogel-Feder) und *Pterodactylus* von Solenhofen. – N. Jb. Miner. etc., **1861**: 678–679; Stuttgart.
OPPEL, A. (1862): Ueber Fährten im lithographischen Schiefer *(Ichnites lithographicus)*. – Pal. Mitt. Mus. Kgl. Bayer. Staat., **2**: 121–125, Taf. 39; Stuttgart (Ebern & Seubert).
WAGNER, J. A. (1861): Ein neues, angeblich mit Vogelfedern versehenes Reptil aus dem lithographischen Schiefer. – Sitz.-Ber. Bayer. Akad. Wiss., math.-phys. Cl. vom 9. Nov. 1861: 146–154; München.
WITTE, O. J. (1863): Briefl. Mitt. betreffend *Archaeopteryx lithographica*. – N. Jb. Min. etc., **1863**: 567–568; Stuttgart.

Author's address: Dr. PETER WELLNHOFER, Bayerische Staatssammlung für Paläontologie und historische Geologie, Richard-Wagner-Str. 10, 8000 München 2, Federal Republic of Germany.

John H. Ostrom

The Yale *Archaeopteryx*: The One that Flew the Coop

Introduction

There is a legend at my university that Yale might have become the prestigeous caretaker of one of the prized specimens of *Archaeopteryx*, but until recently there was no documentary evidence to support that rumor. Some readers will understand why that legend had special interest for me because it is well known that I have long been intrigued with the handful of specimens of *Archaeopteryx*. That intrigue, together with a surprising discovery in the Yale archives, has resulted in this contribution. What follows was presented to the International *Archaeopteryx* Conference in Eichstätt solely for purposes of historical entertainment.

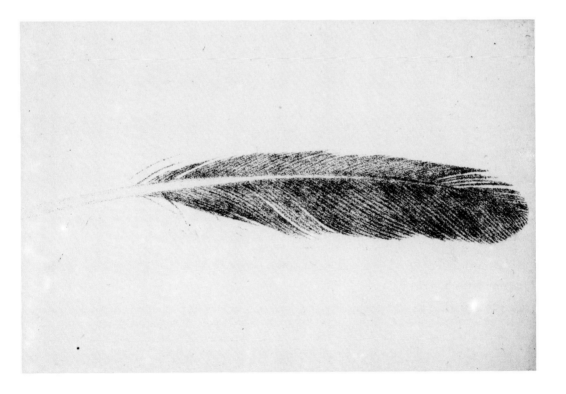

Fig. 1
Lithograph print of the feather impression preserved in a slab of Solnhofen Lithographic Limestone of Bavaria and reported by HERMANN VON MEYER (1861a). This illustration appeared in VON MEYER's article of 1862.

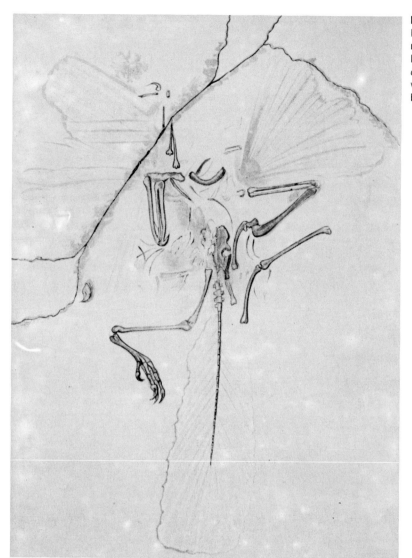

Fig. 2
HENRY WOODWARD's colored woodcut print of the London specimen of *Archaeopteryx*. This print was published in December, 1862.

Archaeopteryx and Art

We are so accustomed to modern illustrative techniques – high quality black and white or color photographs, artistic line or shaded drawings, micro and stereo photographs, X-ray and scanning electron microscopic images – that we often fail to appreciate the workmanship and quality of illustrations published during the 19th century and before. The early illustrations published of the first specimens of *Archaeopteryx* provide an interesting reminder and also served as an appropriate diversion for the conferees, as well as an introduction to Yale's *Archaeopteryx*, which you will see in due time.

As most readers will know, this (Fig. 1) is the first specimen to be recognized and ultimately attributed to *Archaeopteryx*. This fine lithograph of the isolated feather imprint was published by HERMANN VON MEYER in 1862. It has often been regarded as t h e first published illustration of a specimen of *Archaeopteryx*,

Fig. 3
Sir RICHARD OWEN's lithograph of the London specimen which was included in his 1863 description of that specimen.

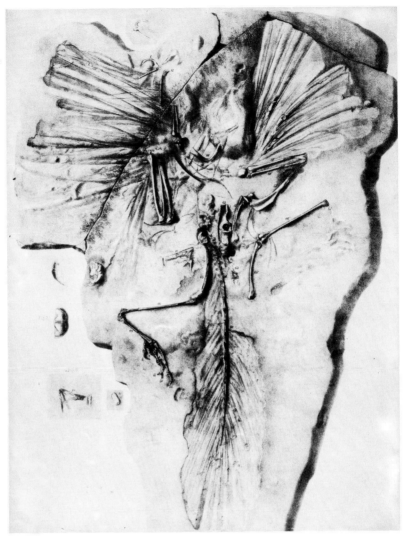

but that is not correct. To demonstrate how remarkably accurate this artistic rendition of the actual specimen really is, compare this Figure with Figure 1 of the "Introduction to *Archaeopteryx*" in this volume.

Only two months after VON MEYER (1861a) reported the discovery of this single feather imprint illustrated in Fig. 1, VON MEYER (1861b) announced the discovery of the first skeletal remains of *Archaeopteryx* associated with impressions of feathers. To this specimen VON MEYER applied the now famous name *Archaeopteryx lithographica*. Fig. 2 is a reproduction of the f i r s t published illustration of that specimen, which is now widely known as the London specimen. It was published in December, 1862, slightly more than a year after the specimen was found, in an article by HENRY WOODWARD of the British Museum of Natural History that was entitled "On a feathered Fossil from the Lithographic Limestone of Solnhofen".

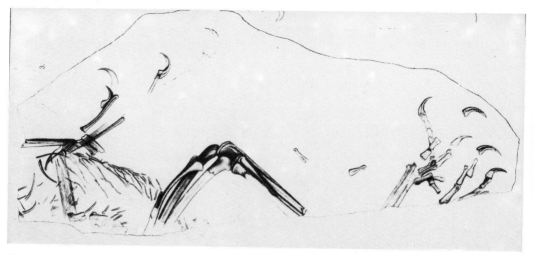

Fig. 4
HERMANN VON MEYER's 1859 lithograph of the Teyler (or Dutch) specimen of *Archaeopteryx*. It was published two years before the discoveries of the better known feather and London specimens.

Fig. 5
The author's photograph of the Teyler specimen for comparison with VON MEYER's lithograph (Fig. 4).

It is a woodcut, rather than a lithograph, in which each published print was handcolored. The crudeness of this woodcut print is evident, especially when it is compared with the lithograph reproduced in Fig. 3.

In 1863, Sir RICHARD OWEN described this first skeletal specimen and included the remarkably detailed lithograph presented here in Fig. 3. It is noteworthy that this Figure is of the Yale copy of OWEN'S

Fig. 6
CARL VOGT's photograph of the Berlin specimen, the first published photograph of any of the specimens of *Archaeopteryx*.

lithograph, which probably is the only unfolded print of that lithograph still in existence because all other copies known-to-date were folded to fit the size format of the Philosophical Transactions of the Royal Society of London in which OWEN'S 1863 paper "On the *Archaeopteryx* of VON MEYER" was published. This lithograph is Yale's *Archaeopteryx* archival treasure # 1.

These three illustrations would appear to be the first three published illustrations of the original specimens of *Archaeopteryx*, but that turns out not to be true. The very first published illustration of a specimen of *Archaeopteryx* turns out to be that shown in Fig. 4, which was published in 1859 by HERMANN VON MEYER. At that time, this specimen (discovered in 1855 – six years before the feather) was thought to represent a pterosaur, one of the flying reptiles, and VON MEYER (1857) established a new species, *Pterodactylus crassipes*, on the basis of this specimen. Figure 5 is a photograph by the author of that specimen portrayed in VON MEYER'S 1859 lithograph. Compare them. In 1970, 115 years after its original discovery, this specimen was finally recognized as a fragmentary specimen of *Archaeopteryx*. Comparison of Figs. 4 and 5 is clear testimony of the skill and accuracy of at least one 19th century lithographer.

The most famed of all the specimens of *Archaeopteryx* is the so called Berlin specimen that was found in 1877 (HÄBERLEIN, 1877). Figure 6 is the first published illustration of that specimen. Moreover, it is the

Fig. 7
The pencil tracing of the Berlin specimen of *Archaeopteryx* in the Yale Archival Collection. Notice that it is labeled at upper right as Urvogel/ "Rätseleidechse" – which means original bird/ "riddle lizard".

first published p h o t o g r a p h of any of the specimens of *Archaeopteryx*. Figure 6 is a reproduction of Yale's print of the original photograph that was published by CARL VOGT in 1879. This is the s e c o n d archival treasure on *Archaeopteryx* at Yale University.

While VOGT'S photograph of the Berlin specimen is the oldest previously known illustration of that specimen, it may surprise readers to learn that it was not the first image made of that specimen, nor was it the last – by many thousands.

The Yale Legend Confirmed

In 1940 a biography was published by CHARLES SCHUCHERT and CLARA LEVENE of the 19th century Yale paleontologist O. C. MARSH. In that work they recorded that MARSH first learned of the discovery of the Berlin specimen in June 1877. (It had been publicly announced by HÄBERLEIN in May of that year.) That prompted MARSH to write to a friend, a Professor GEINITZ of Dresden, requesting his assistance in purchasing it for Yale College. His friend made the effort on behalf of Yale, but the owner of the specimen, ERNST HÄBERLEIN, refused GEINITZ' offer of 1000 Deutsche Marks. Except for a few letters between MARSH and GEINITZ preserved in Yale Archives, no other tangible evidence was known at Yale to document MARSH'S legendary efforts to acquire this most famous of all fossil specimens. That is – not until the summer of 1983.

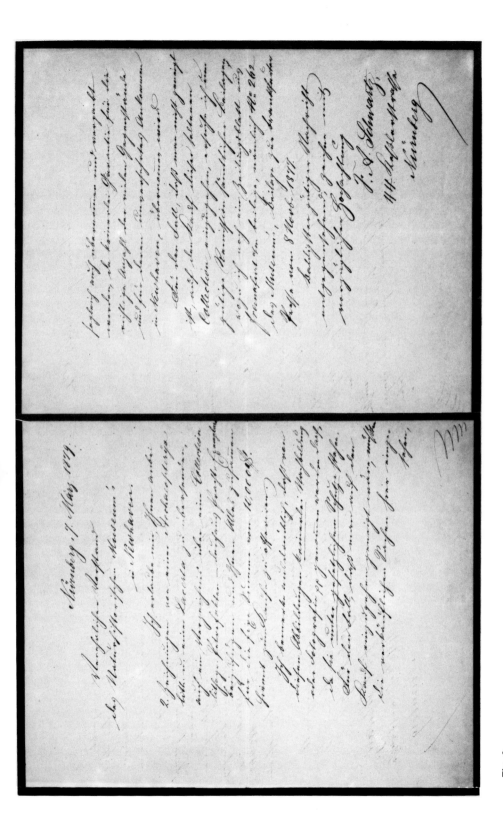

Fig. 8
Reproduction of F. A. SCHWARTZ's 1879 letter to Yale College offering the Berlin specimen of *Archaeopteryx* for sale at $ 10,000.

In August, 1983, MIRIAM SCHWARTZ, my secretary and right hand assistant for more than twenty years, came across a large and obviously old manila envelope while reorganizing Yale's reprint library on vertebrate fossils. That envelope contained two letters, a newspaper article from a Nürnberg newspaper dated November 8, 1877 and two tissue paper tracings of fossil specimens from the Solnhofen Limestone. One of the tracings was that of a lizard-like animal, probably the rhynchocephalian *Homoeosaurus*. The other tracing was that of what we now know as the Berlin specimen of *Archaeopteryx*. That tracing is reproduced in Fig. 7. This is Yale's *Archaeopteryx* archival treasure number three, and certainly the most valuable of all.

This tracing was done by an unknown "artist", perhaps the owner ERNST HÄBERLEIN, sometime before March 7, 1879 because the most important letter accompanying it is clearly dated March 7, 1879. From this I conclude that this tracing was on MARSH'S desk at Yale before he ever saw the photograph (Fig. 6) published by CARL VOGT in September, 1879. So we may conclude that this tracing was the first image ever made of the famous Berlin specimen of *Archaeopteryx*.

But that is not all of the story on the "Yale *Archaeopteryx*". The letter that accompanied the tracing was written by an F. A. SCHWARTZ of Nürnberg. His letter is reproduced here in Fig. 8 and translates as follows:

Nürnberg, 7 March, 1879

To the Managers of the Natural History Museum in Newhaven I take the liberty of sending you the two drawings enclosed of an *Archaeopteryx lithographica* and of a Lacerta, and add also a list of a collection of lithographic stones which are first class examples, and offer all of this to you herewith for the sum of $ 10,000 for purchase. I remark expressly you may not reproduce or photograph these drawings in any way as they are under the protection of the law.
In the case that you would be inclined to buy it, the material for sale would have to be looked over here and packed, as no guarantee will be given for the correct number of the many objects and for their safe arrival in Newhaven.
In case you are not inclined to purchase this rare collection, I ask for your kind return of all enclosures to which I further add a newspaper page from Frankfurt a/m no. 262 of the museum extra sheet to the Frankfurt Press of November 8, 1877.
Looking forward to a prompt answer from you I sign with sincere greetings.

F. A. Schwartz
44 Kessler Street
Nürnberg

The astonishing fact is that this appears to have been a *bona fide* offer to sell the HÄBERLEIN (or Berlin) specimen to Yale for a mere $ 10,000 well before it was offered elsewhere. It was ultimately purchased by WERNER SIEMENS, who donated it to Humboldt University in Berlin, for the much lower price of 20,000 Deutsche Marks.

So far, we have not found any documentary evidence to explain why MARSH failed to follow through on this opportunity. It may have been because some of his friends in Europe, particularly Professors GEINITZ of Dresden and ZITTEL of Munich, had advised him that his original offer in 1877 ($ 1,000 via Professor GEINITZ) was outrageous, or it may have stemmed from genuine doubts about the authenticity of this and the two earlier finds. The Nürnberg newspaper article noted that some experts considered this latest find to be a fraud and cited a Professor GIEBEL in Halle. Perhaps a more likely explanation as to why the Berlin specimen is not at Yale is the well-documented fact that O. C. MARSH was not a spendthrift. He may simply have been trying to bargain the price downward, and he lost out to a higher bidder.

It is easy for us to say that the $ 10,000 price quoted to Yale in 1879 was a cheap price. At least it seems cheap considering how important and famous the Berlin specimen has become. Nevertheless, that was a considerable sum then. Today, $ 10,000 1879 dollars are worth somewhere between 3 and 5 million dollars of 1984 buying power.
This is the curious story of what m i g h t have been the YALE specimen of *Archaeopteryx* – the one that flew the coop.

References Cited

HÄBERLEIN, E. (1877): Neue Funde von *Archaeopteryx*. – Leopoldina, **13**: 80; Halle.

MEYER, H. VON (1857): Beiträge zur näheren Kenntniss fossiler Reptilien. – N. Jahrb. Min. Geol. Paläont. **437**: 532–543; Stuttgart.

MEYER, H. VON (1859): Zur Fauna der Vorwelt. Reptilien aus dem lithographischen Schiefer des Jura in Deutschland und Frankreich. – pp. 64–66; Frankfurt.

MEYER, H. VON (1861a): Vogel-Federn und *Palpipes priscus* von Solnhofen. – N. Jahrb. Min. Geol. Paläont. **1861**: 561; Stuttgart.

MEYER, H. VON (1861b): *Archaeopteryx lithographica* (Vogel-Feder) und *Pterodactylus* von Solnhofen. – N. Jahrb. Min. Geol. Paläont. **1861**: 678/679; Stuttgart.

MEYER, H. VON (1862): *Archaeopteryx lithographica* aus dem lithographischen Schiefer von Solnhofen. – Palaeontographica **10**: 53–56; Stuttgart.

OWEN, R. (1863): On the *Archaeopteryx* of VON MEYER, with a description of the fossil remains of a long-tailed species from the lithographic stone of Solnhofen. – Philos. Trans. Roy. Soc. London **153**: 33–47; London.

SCHUCHERT, C. & LEVENE, C. M. (1940): O. C. MARSH, Pioneer in Paleontology. – Yale University Press, New Haven. 541 pp.

VOGT, C. (1879): *Archaeopteryx*, ein Zwischenglied zwischen den Vögeln und Reptilien. – Naturforscher **42**: 401–404; Berlin.

Author's address: Prof. Dr. JOHN H. OSTROM, Peabody Museum of Natural History, Yale University, 170 Whitney Avenue, P. O. Box 6666, New Haven, Conn. 06511, USA.

Eric Buffetaut

The Strangest Interpretation of *Archaeopteryx*

Abstract

The strangest interpretation of *Archaeopteryx* was given in 1897/1898 by the French writer ALFRED JARRY (1873–1907). In his satirical play *Ubu cocu, ou l'Archéoptéryx*, he compared the earliest known bird to various birds and mammals and used it as a symbol of the unusual. This shows how quickly *Archaeopteryx* had become one of the most famous fossil animals.

Zusammenfassung

Die merkwürdigste Deutung von *Archaeopteryx* wurde 1897/1898 vom französischen Schriftsteller ALFRED JARRY (1873–1907) gegeben. In seinem satirischen Stück *Ubu cocu, ou l'Archéoptéryx*, verglich er den Urvogel mit verschiedenen Vögeln und Säugetieren und benutzte ihn als Symbol des Ungewöhnlichen. Das zeigt, wie schnell *Archaeopteryx* eines der berühmtesten fossilen Tiere geworden war.

"Il nous paraît préhistorique, croisé vampire-archéoptéryx, ichthyornis, avec de nombreuses qualités des chéiroptères, léporides, rapaces, palmipèdes, pachydermes et porcins" (it seems prehistoric to us, a cross between a vampire and an archaeopteryx, ichthyornis, with many qualities of the chiropters, the leporids, the birds of prey, the palmipedes, the pachyderms and the pigs). This unusual, to say the least, interpretation of the affinities of *Archaeopteryx*, which is somehow reminiscent of recent speculations on the mammalian affinities of ancestral birds, appears in a play written in 1897 or 1898 by the French writer ALFRED JARRY (1873–1907). JARRY, an eccentric playwright and novelist, is mainly remembered for the famous character he created, "Père Ubu", a personification of "everything grotesque in the world". Ubu is a stupid, greedy and bloodthirsty coward, whose absurd adventures as usurper of the throne of Poland, and then as "maître des phynances" (master of finance) in France, are the subject of several satirical plays *(Ubu roi, Ubu cocu, Ubu enchaîné)* written by JARRY in the 1890s in a rather peculiar, comical and somewhat dirty, language. The description of *Archaeopteryx* quoted above is taken from *Ubu cocu, ou l'Archéoptéryx* (Ubu cuckolded, or, the *Archaeopteryx*), a play written in 1897 or 1898, parts of which were published only a long time after JARRY'S death. In this play, the *Archaeopteryx* is the son of Ubu's wife (a character as unpleasant and ridiculous as Ubu himself) and of her lover, a certain Barbapoux ("Lousybeard"), and the description is given by Ubu himself when he discovers his misfortune. *Archaeopteryx* appears several times in the play (measurements of it are even given by Ubu's wife, who calls it a "beautiful bird"), but it never talks. Incidentally, both Ubu and his wife state that *Archaeopteryx* can fly through the air (although this can hardly be used as evidence in the debate about the palaeobiological interpretation of *Archaeopteryx* and the origin of avian flight!).

The use of *Archaeopteryx* as a character in a play, less than 40 years after the discovery of the London specimen and 20 years after that of the Berlin specimen, is interesting as an example of the ease with which some spectacular fossil vertebrates can become famous outside scientific circles. JARRY was apparently interested in palaeontology and evolution. In his *Almanach illustré du Père Ubu* ("Father Ubu's illustrated almanac") for 1901, a work in the same comically nonsensical vein as his plays, he lists "Gaudry d'Asson, professeur au Muséum" as a commander in the imaginary Order of the Gidouille, of which, of course, Ubu was supposed to be Grand Master. ALBERT GAUDRY (1827–1908), who had become famous for his work on the Miocene mammals from Pikermi, was then professor of palaeontology at the Muséum national d'Histoire naturelle. He was one of the main representatives of evolutionary palaeontology in France at the time, and JARRY may have read some of his works (for instance, the *Enchaînements du monde animal*, published in three volumes between 1878 and 1890).

That JARRY was seriously interested in palaeontological problems is also shown by a review he published in 1900 of a French translation of a memoir by ERNST HAECKEL on "the present state of our knowledge on the origin of man". This memoir was HAECKEL'S contribution to the 4th International Zoological Congress, held in Cambridge in 1898. In his brief review, published in the literary magazine *La revue blanche*, JARRY also mentions the famous reconstruction of *Pithecanthropus erectus* by EUGÈNE DUBOIS, which was on display in the pavilion of the Dutch East Indies at the Paris World Exhibition of 1900; his conclusion is that "the descent of man from a series of extinct Tertiary primates is no longer a vague hypothesis, but a real historical fact".

JARRY was not serious, of course, about *Archaeopteryx*. He seems to have used the earliest known bird, with its intermediate characters, as a kind of nonsensical symbol of the bizarre and the hybrid (it is, after all, the result of the illicit love affair between Ubu's wife and Barbapoux). This is not the only instance in French literature of prehistoric animals being used as symbols of the abnormal or strange. In 1909, for example, two years after JARRY'S early death, GUILLAUME APOLLINAIRE introduced a talking ichthyosaur in his prose poem *L'enchanteur pourrissant* ("The rotting enchanter"), a variation on the Merlin theme. There, the ichthyosaur appears among a crowd of other animals, some real, some mythical (dragons, werewolves, sphynxes, and the Biblical Leviathan and Behemoth), and some invented by the author, which are all doomed to extinction. In APOLLINAIRE'S work, Mesozoic reptiles are thus assimilated to the fabulous creatures of myths and legends – a role they probably still play today, to some extent, in the popular imagination, as exemplified by their frequent use in science-fiction books and films.

To return to *Archaeopteryx*: by the time ALFRED JARRY raised it to literary fame by using it as an improbable character in one of his bizarre plays, its scientific fame was already well established, and it must have become relatively familiar to the public (at least to the intellectuals who were JARRY'S public). There can be no doubt that *Archaeopteryx* had rapidly become one of the most famous fossils ever found, as well as the archetype of the "no longer missing" link between groups of apparently unrelated or widely separated animals. JARRY'S fanciful comparison with various birds (including *Ichthyornis*) and mammals, besides its comical effect, also expresses the intermediate and problematical nature of *Archaeopteryx*. More than a hundred years after the discovery of the first specimen of the earliest known bird, the phylogenetic significance of this animal is still the subject of scientific debates and controversies. Seen in this light, ALFRED JARRY'S contribution to the problem of the affinities of *Archaeopteryx* is decidedly a minor one, but there is little doubt that no stranger interpretation has ever been given!

Acknowledgements

I thank MICHEL MARTIN for his help in locating some of the "palaeontological" references in JARRY'S works.

References Cited

APOLLINAIRE,, G. (1909): L'enchanteur pourrissant. – Paris (Kahnweiler). (A recent edition has been published by Gallimard, Collection Poésie, Paris, 1972).

GAUDRY, A. (1878–1890): Les enchaînements du mode animal dans les temps géologiques. – 3 vol., 293 pp., 317 pp., 322 pp. Paris (Hachette and Savy).

JARRY, A. (1968): Tout Ubu. – 470 pp. Paris (Fasquelle). (This comprises all the plays of the Ubu cycle, including *Ubu cocu, ou l'Archéoptéryx*, as well as JARRY'S other writings on the Ubu theme).

JARRY, A. (1969): La chandelle verte. – 696 pp. Paris (Le livre de poche). (A collection of articles by JARRY, including his review of HAECKEL'S work on the origin of man).

Author's address: Dr. ERIC BUFFETAUT, U. A. 720 du C.N.R.S., Laboratoire de Paléontologie des Vertébrés, Université Paris VI, 4 place Jussieu, 75230 Paris Cedex 05, France.

Siegfried Rietschel

False Forgery*

> "The whole idea is quite ridiculous" he said.
> Marlowe shook his head.
> "This comes of reading science fiction".
> FRED HOYLE 1958

A few months after the Eichstätt-Symposium *Archaeopteryx*, and especially it's plumage, became the centre of extraordinary public interest. This was not a result of the symposium itself, but of a well orchestrated campaign to challenge the credibility of *Archaeopteryx* as a proof of Darwin's theory of evolution. A group of physicists proposed, in a series of papers, that the fossil's feathers had been faked. Their information was based on the result of new photographs of the London specimen which they regard as exceptional. However their studies exhibited a distinct lack of paleontological knowledge and experience; but, nevertheless, the features and arguments on which the forgery-hypothesis has been erected must be critically examined, a task I will attempt in the following points.

1. In March 1985 the British Journal of Photography reported (BJPa: 265**) that the most well know specimens (London and Berlin) were "unique" because of their "unmistakable feather imprints". The BJP (d: 694) also pretended that only the finds of the Häberlein-era, have feathers. This is not true. Descriptions of the Teyler- (OSTROM 1972), Maxberg- (HELLER 1959), and Eichstätt-specimens (WELLNHOFER 1974) report and figure "impressions" of feathers. We have to admit, that these are of a different quality of preservation, but they are present at topographic positions comparable to the Berlin- and London-specimens' feathers. It is likely that different stages of decay and slight differences in the sediment are responsible for the poorer preservation. But in each specimen there are at least the outlines of feathers and the Maxberg-specimen has clear feather structures with rachis and barbs visible. To illustrate this I took a series of new photographs from a plaster cast of the Maxberg-specimen, as the original is not available for research at the moment. These casts, which are authentic (made by myself and two preparators in 1973), show clear feather structures in areas where the covering thin sediment film has broken away (Fig. 1). HELLER (1959: 14–15) gives a good description of the structure of these feathers and the covering sediment, which had been removed by preparation in only a few places. Because his photographs (pl. 3, taken by W. STÜRMER from the original) show only the major structures but no details, new photographs are presented here.

So it can be definitely stated that feather structures are present in all know specimens, independent of their date of discovery, collector, preparator or scientific author.

---

*Editors' Note: This supplementary paper was developed by Professor RIETSCHEL at the request of the Editors. Our request to Prof. Rietschel was prompted by the "*Archaeopteryx*-fraudulent-feather-hypothesis" advanced by FRED HOYLE et al., which provoked considerable public interest just a few months after this Conference.

**The four 1985 papers of HOYLE, RABILIZIROV, SPETNER, R. S. & J. WATKINS, and WICKRAMASINGHE, published in Brit. J. Photogr. are henceforth referred to as "BJPa-d". References to the authors' paper on "Feathers and wings of *Archaeopteryx*" in this volume are shortened to "RL". – The author thanks MIKE HOWGATE, London, for brushing up the English of the original manuscript.

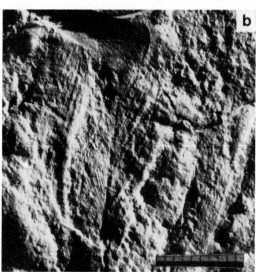

Fig. 1.
Archaeopteryx lithographica, Maxberg-specimen, main-slab (plaster cast of 1973); scale 10 mm. – a: Large feathers, presumably primaries of the left wing, near the claw of leftside 2nd digit; they are only visible, where a thin sheet of covering (originally underlying) sediment has broken off (locally by preparation). – b: little feathers of about 2 cm length between left tibia and imprint of left scapula (upper edge), which HELLER (1959: 14, pl 6–7) supposed to belong to the "feather-trousers".

2. The feather structures of the *Archaeopteryx* specimens are generally regarded by the BJP-authors as "impressions" (e. g. BJPa: 265); but elsewhere the authors speak of "featherlike material" (e. g. BJPd: 693). While residues of the organic material of feathers are undoubtedly present in the first find of an isolated feather, the five skeletal *Archaeopteryx*-specimens exhibit only casts of feathers.

The forging of the feather structures on these specimens could have been managed by several different techniques, such as by mounting artificial casts (or impressions) or original feathers upon an original (or faked) fossil. However the BJP-authors seem to believe, that the feather structures were faked directly upon the originals. They have thus to answer the question, how can all the feathers of all the *Archaeopteryx*-specimens be faked? (In fact, one, two or more "artificial" feather structures on the *Archaeopteryx*-specimens would not harm the "feathered reptile". The history of geosciences would be enriched by a new anecdote, and nothing more).

3. A forgery of all the feathers in the five specimens – each represented by main-slab and counterslab – consequently suggests that all plumage structures were faked

a) on the main-slabs
b) as 3-dimensional exactly matching negatives on the counterslabs
c) using the same catalogue of details for very different specimens and
d) adapting this catalogue to the particular specimen.

In that case, we would not only be dealing with a forger of feathers, but with a forger of technically and systematically identitical wings, tails and body plumage from 1857 (Teyler specimen) to the present day.

a) The main-slab of the London specimen not only shows feathers, but an intricate arrangement of them, which forms wings and a tail. The tail feathers are arranged in a way, which is typical of *Archaeopteryx*, but has never been found in any fossil or recent bird. The wing feathers are in their turn arranged the same way as in modern birds; the slight dissarrangement can be interpreted as due to decay of the carcass before being covered by sediment. This dissarrangement is not present in the Berlin specimen, where we can see the complete undersides of the wings with their different kinds of feathers, looking very similar to a modern birds' wing, but with an expanded area of coverts (RL: pl. 1–2 and fig. 4).

A faker would not only have had to produce whole wings, similar to those of modern birds, as well as body feathers (single and connected ones) for such a forgery, but he would had to invent a unique type of a tail (initially mistaken in BJPa: 265, as a "large and conspicuous tail feather"). Even a perfect genius would have needed many years to do that – and we may ask, how he could have managed it without a prototype and the knowledge of scientists of some generations later.

b) The counter-slab of the Berlin specimen is a 95 % true negative copy of the main-slab, and the structures of the plumage are even better preserved than those of the main-slab, which has suffered by research work. Differences between main-slab and counterslab result mainly from the original splitting of the limestone-slabs and subsequent preparation. Those differences which, occuring on the counter-slab of the London specimen, furnished, according to BJP (a: 266, b: 358 f.) proof of the forgery, are just those features which helped in the interpretation of the mode of preservation of feather structures (RL: fig. 1). Very thin sediment layers may be lost in fragments, when a block of Solnhofen limestone is split. With these layers superficial structures are lost too (some still remain present adhering to the counterpart slab) and other structures of a deeper layer may be visible. Here we have a reason for misinterpreted differences between the two slabs in BJP (e. g. b: 359, fig. 2–3).

Generally speaking it would be possible to cast a feather impression onto one slab especially using modern artificial resin techniques. But it seems just impossible, to produce an identical negative copy of such a cast exactly at the right place on the counter-slab. In the *Archaeopteryx*-specimens the total BJP-postulated forgery of the plumage must necessarily have been copied in 3 dimensions of the counter-slabs (including the mentioned "faults") if the forgery-hypothesis is correct.

c) There is a very complicated arrangement of structures on both main- and counter-slab (RL). What we need is a theory which explains step by step the fossilisation process and its results concerning the preserved feather structures always in keeping with actualpaleontological observations. For a long time there has been no convincing theory which explains all the details of the feather structures, as shown by the different explanations for the so-called „double struck feather impressions". A detailed analysis of the Berlin specimen's plumage (RL) provides a rich catalogue of complex structures, such as shaft shadows, split vanes, zigzag-structures, preservation of only feather undersides and of several feather layers etc. It has been shown, that all these structures may result from a natural succession of postmortal and diagenetic processes.

Therefore it is not surprising, that we also find most of the detail structures in the plumage of the London and Maxberg specimens. The lack of detail in the other specimens is unquestionably depedant on the quality of preservations (see d.).

It is impossible that a faker could have utilized such a variety of detailed structures in order to imitate the result of a natural process without making a mistake, especially as the known faked fossils from Solnhofen and other sites are all done in a more or less primitive way. How could a faker, for instance, get the idea, to produce the feather impressions as casts of the undersides on both slabs when even well experienced paleontologists would have expected the upper surfaces of the wings to be displayed on the counter-slab (see 5.)?

d) The catalogue of feather structures is not restricted to one specimen, nor is it complete in all specimens, and the quality of the detail differs distinctly from specimen to specimen. There is an evident relationship between the preservation of the feather structures and disarrangement of bones by decay. We know, that preservation of feathers depends on several factors such as: how rotten the carcass was when it reached its final resting place, the chemical and physical conditions of the surrounding water and of the substrate and the mode of sedimentation. From a study of the above we can today elucidate general conditions from the complex of factors involved and it is once again pertinent to ask, how the faker could have faked a consistent and complex taphonomy.

4. The last point leads to the Eichstätt specimen, wich shows – seemingly in contradiction to it's marvellously preserved skeleton – very poorly preserved plumage. Only faint outlines of primaries and the tail feathers are visible. But these shadowlike features clearly demonstrate, that the plumage was present, when the bird started to become a fossil! The only question is, was it during early sedimentation or later diagenesis, that the feathers disappeared? It has been proposed that algal mats played an important role in the preservation of delicate feather structures (RL). The *Archaeopteryx*-specimens come from different places and different layers within the Solnhofen limestone (see WELLNHOFER 1974: fig. 1, not BJPa: 264, as criticized by HOWGATE 1985), so we might reasonably expect differences in environment and sedimentation locally and it is probable that algal mats were not involved where or when the Eichstätt specimen was fossilized.

Algae, bacteria and fungi very often cover recent subaquatic animal carcasses, coating feathers or fur. The sediment they build and/or agglutinate, is often much finer in texture than the surrounding sediment. This may serve as an explanation for the BJP-statements, that feather structures of the London specimen are characterized by finer sediment (compared by BJPa: 367 with "chewing gum"), finer than the surrounding limestone, which generally belongs to another bedding plane.

So the difference of fine and coarser – but still fine – matrix around the bones and underneath the feathers of the fossil may have two reasons, a natural and an artificial: apparently there is very fine sediment upon the feathers ("upon" concerning orientation during sedimentation, what is preserved as a cast of the underside of the feathers!). But also we have to distinguish from the smooth surfaces formed by preparation, which can easily be mistaken by a layman for finer grained material.

These structures are hard to elucidate in *Archaeopteryx* originals, and only a series of thin sections can resolve all the questions. Quite rightly most people would hesitate to destroy precious fossils, such as *Archaeopteryx* by sectioning and other damaging analytic methods, especially as differences in the sediment are no proof of a fakers activity.

5. As another point it should be mentioned, that the London and the Berlin specimens have been in the possession of scientific institutions for over 100 years. During this time several casts were taken of the specimens. That meant, that the original specimens had to be treated with chemicals, and especially their surface was effected. FISCHER & KRUEGER (1979) describe how often preparations and casts of the Berlin specimen were done, the surface of which has been described as a "chemical laboratory". In any case, casting and preparation should have brought to light, if there had been any, artificially added feathers or feather structures!

6. Last but not least (BJPd), the authors have found a motive for the postulated fakers. Concerning this hypothesis, we must again remember that the plumage structures of the fossilized specimens are very complicated. The Häberleins knew how to make money out of fossils, and they did it by the best, which was also the easiest, way. They were not such outstanding intellectuals as to have discovered and faked the processes of sedimentation and fossilization – processes which were are still trying to elucidate now, more than 100 years later. And OWEN as a fellow conspirator?, as attributed to him by BJPd. As a competent morphologist he gave a very accurate description of the new fossil, especially regarding the skeletal elements. But it has to be mentioned, that he misunderstoodt the special preservation of the feathers as casts; this is shown by the fact, that in his description the left side of the body is mistaken for the right one and vice-versa. This only could happen, because we see on the main slab the bones as solid elements from the back, but the wings as casts from the underside! Even HEINROTH (1923) wondered about this little riddle, which results from the fact, that there is no impression of the upper side of the wing, but only cast of the underside – such as we would see, if the bones of a hand were lying upon a cast of the palm when the flesh disappeared. A palaeontologist of the 19th century, would never have used the underside of the feathers for a forgery on both main- and counter-slab!

7. It makes no sense to argue about the quality and usefulness of the photographs, which the BJP-authors used in their interpretation of the *Archaeopteryx* feathers. Photographs are a very important aid in elucidating detail which might otherwise be difficult to locate and identify. But serious reasons are rare, for example in the case of lost originals, to base a research work only on photographs. Compare the plates of RL with the BJP-illustrations and see to what extent paleontological research was "limited by the techniques available in the past" (BJPa: 265).

There are still many more reasons why we cannot believe in a forgery, but:
The striking argument f o r the forgery-hypothesis would be, if somebody successfully f a k e s a bird's plumage with all the characters of the *Archaeopteryx* originals, using Solnhofen limestone material. During unpublished experiments (1976–78) on the mode of preservation of *Archaeopteryx* feathers, I failed, when I tried to get casts of bird wings identical to those of *Archaeopteryx*. I may predict, that it will not be possible, to produce feather casts or impressions which could be confused with the *Archaeopteryx* originals. For only the first step of embedding the feathers in the sediment is reproducible, and not the subsequent steps.

What can be concluded on the whole?
It undoubtedly helps in finding a new hypothesis and unusual arguments, if one is not loaded down with the knowledge from too much literature and material. But once having formulated a new hypothesis, only critical reading of the literature and research on the fundamental material can establish it as a serious theory. The BJP-authors apparently didn't study much more than the photographs before bursting into print. There is no statement by the BJP-authors, which cannot be explained better by classical and modern paleontological knowledge, and we don't need to invoke the activity of generations of fakers to make *Archaeopteryx* a feathered reptile, respectively an archaic bird. However in a popular-science dispute it is always much harder to counter unscientific arguments with scientific ones, especially when cosmology and/or religious belief is introduced and common sense and scientific experience are neglected. In this discussion we want to be free from dogmatic arguments about *Archaeopteryx* and evolution. Our scientific work depends on experience, knowledge, facts, and ideas, only story-tellers can replace these with pure phantasy.

So we can only ask, is the BJP-hypothesis a joke, is it a science fiction story, is it, contrary to this paper, a serious scientific attempt – or is it just a new version of a popular fairy tale, telling us that, once upon a time, Mother Carey was plucking her chickens for *Archaeopteryx* from space?

References Cited

FISCHER, K. & KRUEGER, H.-H. (1979): Neue Präparationen am Berliner Exemplar des Urvogels *Archaeopteryx lithographica* H. v. Meyer, 1861. – Z. geol. Wiss., **7**(4): 575–579; Berlin.

HELLER, F. (1959): Ein dritter *Archaeopteryx*-Fund aus den Solnhofener Plattenkalken von Langenaltheim/Mfr. – Erlanger geol. Abh., **31**: 25pp., 15 pl, 2 fig.; Erlangen.

HOWGATE, M. E. (1985): *Archaeopteryx* counterview. – Brit. J. Photogr., **132** (29. 3. 85): 348; London.

HOYLE, F. (1958): The Black Cloud. – after: SCHWARZBACH, M. (1983), Eiszeit-Probleme. Diesmal gelöst von Fred Hoyle. – Naturwiss. Rdsch., **36** (5): 219–222; Stuttgart.

Hoyle, F., Wickramasinghe, N. C. & Watkins, R. S. (1985): *Archaeopteryx.* – Brit. J. Photogr., **132** (21.6.85): 693–695, 703, 3 fig.; London. (BJPd)

Ostrom, J. H. (1972): Description of the *Archaeopteryx* specimen in the Teyler Museum, Haarlem. – Proc. k. nederl. Akad. Wet. (B), **75** (4): 289–305; Amsterdam.

Rietschel, S. (1985): Feathers and wings of *Archaeopteryx,* and the question of her flight ability. – This volume.

Watkins, R. S., Hoyle, R., Wickramasinghe, N. C., Watkins, J., Rabilizirov, R. & Spetner, L. M. (1985a): *Archaeopteryx* – a photographic study. – Brit. J. Photogr., **132** (8.3.85): 264–266, 5 fig.; London. (BJPa)

Watkins, R. S., Hoyle, F., Wickramasinghe, N. C., Watkins, J., Rabilizirov, R. & Spetner, L. M. (1985b): *Archaeopteryx* – a further comment. – Brit. J. Photogr., **132** (29.3.85): 358–359, 367, 3 fig.; London. (BJPb)

Watkins, R. S., Hoyle, F., Wickramasinghe, N. C., Watkins, J., Rabilizirov, R. & Spetner, L. M. (1985c): *Archaeopteryx* – further evidence – Brit. J. Photogr., **132** (26.4.85); 468–470, 3 fig.; London. (BJPc)

Wellnhofer, P. (1974): Das fünfte Skelettexemplar von *Archaeopteryx.* – Palaeontogroaphica, A **147**: 169–216, 13 Abb., Taf. 20–23; Stuttgart.

Author's address: Prof. Dr. Siegfried Rietschel, Direktor der Landessammlungen für Naturkunde (Museum am Friedrichsplatz), Erbprinzenstr. 13, Postfach 40 45, D-7500 Karlsruhe 1, Federal Republic of Germany

Contents

Preface 7

Acknowledgements 8

Introduction
OSTROM, JOHN H.: Introduction to *Archaeopteryx* 9
CHARIG, ALAN J.: Analysis of the Several Problems Associated with *Archaeopteryx* 21

The Habitat of *Archaeopteryx*: The Solnhofen Environs
VIOHL, GÜNTER: Geology of the Solnhofen Lithographic Limestones and the Habitat
of *Archaeopteryx* 31
DE BUISONJÉ, PAUL: Climatological Conditions During Deposition of the Solnhofen Limestones . . 45

The Biology of *Archaeopteryx*
REGAL, PHILIP J.: Common Sense and Reconstructions of the Biology of Fossils:
Archaeopteryx and Feathers 67
FEDUCCIA, ALAN: On Why the Dinosaur Lacked Feathers 75
THULBORN, RICHARD A. & HAMLEY, TIM L.: A New Palaeoecological Role for *Archaeopteryx* 81
YALDEN, DERIK W.: Forelimb Function in *Archaeopteryx* 91
PETERSON, ALLEN: The Locomotor Adaptations of *Archaeopteryx*: Glider or Cursor? 99

Anatomical Issues of *Archaeopteryx*
HOWGATE, MICHAEL E.: Problems of the Osteology of *Archaeopteryx*: Is the Eichstätt Specimen
a Distinct Genus? 105
WELLNHOFER, PETER: Remarks on the Digit and Pubis Problems of *Archaeopteryx* 113
WALKER, A.: The Braincase of *Archaeopteryx* 123
BÜHLER, P.: On the Morphology of the Skull of *Archaeopteryx* 135
HINCHLIFFE, J. R.: „One, two, three" or „Two, three, four": An Embryologist's View of the
Homologies of the Digits and Carpus of Modern Birds 141

The Phylogenetic Significance of *Archaeopteryx*
HECHT, MAX H.: The Biological Significance of *Archaeopteryx* 149
OSTROM, JOHN H.: The Meaning of *Archaeopteryx* 161
MARTIN, LARRY D.: The Relationship of *Archaeopteryx* to Other Birds 177
GAUTHIER, JACQUES & PADIAN, KEVIN: Phylogenetic, Functional and Aerodynamic Analysis of the
Origin of Birds 185
BOCK, WALTER J.: The Arboreal Theory for the Origin of Birds 199
MOLNAR, RALPH E.: Alternatives to *Archaeopteryx*: A Survey of Proposed Early or Ancestral Birds . 209
RAATH, MICHAEL A.: The Theropod *Syntarsus* and Its Bearing on the Origin of Birds 219
TAQUET, PHILIPPE: Two New Jurassic Specimens of Coelurosaurs (Dinosauria) 229

Archaeopteryx *and the Evolution of Flight*
PETERS, D. STEPHAN & GUTMANN, WOLFGANG FR.: Constructional and Functional Preconditions
for the Transition to Powered Flight in Vertebrates 239
PETERS, D. STEPHAN: Functional and Constructive Limitations in Early Evolution of Birds 243

RIETSCHEL, SIEGFRIED: Feathers and Wings of *Archaeopteryx*, and the Question of Her Flight Ability ... 249
STEPHAN, BURKHARD: Remarks on Reconstruction of *Archaeopteryx* Wing 261
BALDA, RUSSEL P., CAPLE, GERALD & WILLIS, WILLIAM R.: Comparison of the Gliding to Flapping Sequence with the Flapping to Gliding Sequence 267
RAYNER, JEREMY M. V.: Mechanical and Ecological Constraints on Flight Evolution 279
RAYNER, JEREMY M. V.: Cursorial Gliding in Proto-Birds: An Expanded Version of a Discussion Contribution 289
NORBERG, ULLA M.: Evolution of Flight in Birds: Aerodynamic, Mechanical, and Ecological Aspects ... 293
NORBERG, R. ÅKE: Function of Vane Asymmetry and Shaft Curvature in Bird Flight Feathers; Inferences on Flight Ability of *Archaeopteryx* 303
TARSITANO, SAMUEL F.: The Morphological and Aerodynamic Constraints on the Origin of Avian Flight 319
SCHALLER, DIETRICH: Wing Evolution 333

Historical Illuminations

VIOHL, GÜNTER: CARL F. and ERNST O. HÄBERLEIN: The Sellers of the London and Berlin Specimens of *Archaeopteryx* 349
WELLNHOFER, PETER: The Story of ALBERT OPPEL'S *Archaeopteryx* Drawing 353
OSTROM, JOHN H.: The Yale *Archaeopteryx*: The One that Flew the Coop 359
BUFFETAUT, ERIC: The Strangest Interpretation of *Archaeopteryx* 369

Addendum

RIETSCHEL, SIEGFRIED: False Forgery 371

List of Participants

BALDA, RUSSELL P., Flagstaff, Arizona U.S.A.
BOCK, WALTER, New York, U.S.A.
BÜHLER, PAUL, Stuttgart, Federal Republic of Germany
BUFFETAUT, ERIC, Paris, France
BUISONJÉ, PAUL H. DE, Amsterdam, Netherlands
CAPLE, GERALD, Flagstaff, Arizona, U.S.A.
CHARIG, ALAN J., London, Great Britain
CLARK, JAMES, Chicago, U.S.A.
COLBERT, EDWIN H., Flagstaff, Arizona, U.S.A.
CORNEA, AUREL, Eichstätt, Federal Republic of Germany
CURRIE, PHILIP, Drumheller, Alberta, Canada
DODSON, PETER, Philadelphia, U.S.A.
FEDUCCIA, ALAN, Chapel Hill, North Carolina, U.S.A.
FREY, EBERHARD, Tübingen, Federal Republic of Germany
GATESY, STEPHEN, Watchung, New Jersey U.S.A.
GAUTHIER, JACQUES, San Francisco, U.S.A.
GOTH, KURT, Frankfurt, Federal Republic of Germany
GRIFFITHS, PETER, Aberystwyth, Dyfed, Great Britain
GROISS, JOSEF TH., Erlangen, Federal Republic of Germany
GUTMANN, WOLFGANG F., Frankfurt, Federal Republic of Germany
HAUBITZ, BERND, Hildesheim, Federal Republic of Germany
HECHT, MAX K., Flushing, New York, U.S.A.
HESSE, ANGELIKA, Frankfurt, Federal Republic of Germany
HINCHLIFFE, J. R., Aberystwyth, Great Britain
HOCH, ELLA, Kobenhavn, Danmark
HOMBERGER, DOMINIQUE, Baton Rouge, Louisiana, U.S.A.
HOPSON, JAMES A., Chicago, U.S.A.
HOTTINGER, LUKAS, Basel, Switzerland
HOWGATE, MIKE, London, Great Britain
JURCSAK, TIBOR, Orandea, Roumania
KLEPSCH, PETER, Spalt, Federal Republic of Germany
KREBS, BERNARD, Berlin, Federal Republic of Germany
LIENAU, HANS-WERNER, Hamburg, Federal Republic of Germany
MÄUSER, MATTHIAS, Würzburg, Federal Republic of Germany
MARTIN, LARRY D., Lawrence, Kansas, U.S.A.
MÜLLER, GERD, Wien, Austria
NORBERG, R. ÅKE, Göteborg, Sweden
NORBERG, ULLA, Göteborg, Sweden
OELOFSEN, B. E., Stellenbosch, South Africa
OSTROM, JOHN H., New Haven, Connecticut, U.S.A.
PADIAN, KEVIN, Berkeley, California, U.S.A.
PARKES, KENNETH C., Pittsburgh, Pennsylvania, U.S.A.
PETERS, D. STEPHAN, Frankfurt, Federal Republic of Germany
PETERSON, ALLEN R. D., Owego, New York, U.S.A.
RAATH, MICHAEL A., Johannesburg, South Africa
RAYNER, JEREMY M. V., Bristol, Great Britain
REGAL, PHILIPP J., Minneapolis, Minnesota, U.S.A.
REICHHOLF, JOSEF, München, Federal Republic of Germany
RICQLÈS, A. DE, Paris, France

Riess, Jürgen, Tübingen, Federal Republic of Germany
Rietschel, Siegfried, Karlsruhe, Federal Republic of Germany
Rolshoven, Marianne, Eichstätt, Federal Republic of Germany
Senn, David G., Basel, Switzerland
Schaller, Dietrich, München, Federal Republic of Germany
Taquet, Philippe, Paris, France
Tarsitano, Samuel F., Flushing, New York, U.S.A.
Thulborn, Richard A., St. Lucia, Queensland, Australia
Viohl, Günter, Eichstätt, Federal Republic of Germany
Vos, John de, Haarlem, Netherlands
Walker, Alick D., Newcastle upon Tyne, Great Britain
Walker, Cyril A., London, Great Britain
Wellnhofer, Peter, München, Federal Republic of Germany
Westphal, Frank, Tübingen, Federal Republic of Germany
Wild, Rupert, Stuttgart, Federal Republic of Germany
Willis, William R., Flagstaff, Arizona, U.S.A.
Yalden, Derik W., Manchester, Great Britain